Henry Austin Wilson

Navigation New Modelled

A Treatise of Geometrical, Trigonometrical, Arithmetical, Instrumental and

Practical Navigation

Henry Austin Wilson

Navigation New Modelled
A Treatise of Geometrical, Trigonometrical, Arithmetical, Instrumental and Practical
Navigation

ISBN/EAN: 9783741180309

Manufactured in Europe, USA, Canada, Australia, Japa

Cover: Foto ©berggeist007 / pixelio.de

Manufactured and distributed by brebook publishing software
(www.brebook.com)

Henry Austin Wilson

Navigation New Modelled

Navigation New Modelled:

OR, A

TREATISE

OF

Geometrical, Trigonometrical, Arithmetical,
Instrumental, and *Practical*

NAVIGATION;

TEACHING

How to keep a *Reckoning*, both in *Latitude* and
Longitude, without TABLES or INSTRUMENTS, by a NEW
METHOD never yet Published: Illustrated with several
Examples of keeping a JOURNAL, and correcting it by an
Observation, with a New Way of finding the *Variation*, and
Time of *High-Water* at any known Port.

TOGETHER WITH

All necessary TABLES, calculated to the New STILE, and
the Projection of the Sphere *Orthographic* and *Stereographic*. Also
Current Sailing, with other Pleasant Questions, and how to correct
the *Longitude* by a Solar Observation.

The NINTH EDITION, with the Addition of
Spherical Trigonometry, and *Astronomy.*

By HENRY WILSON,

Revised and corrected by *William Mountains,* Teacher of the
Mathematics, and F. R. S.

LONDON:
Printed for J. MOUNT, and T. PAGE on *Tower-Hill,*
M,DCC,LXIX.

To the Right Honourable

P A T T E E,

LORD VISCOUNT

TORRINGTON.

MY LORD,

I SHOULD *have been at a Lofs to approach your* Lordfhip *with this DEDICATION, if I had not firft had the Honour immediately to Serve your Honourable Father Admiral* Byng, *in the Quality of* Mathematician, *aboard his Majefty's Ship the* Barfleur, *in the* Mediterranean; *I at the fame Time having the Honour to attend and affift your Lordfhip in your Mathematical Studies*
Aboard

The DEDICATION.

Aboard of the said Ship, and his Lordship after his Return to England, *being pleased to Accept and Approve of a Dedication of this Book of Navigation : I humbly crave Leave to do myself the additional Honour, to make an humble Offering of the sixth Impression to the Patronage of your Lordship, and to continue it under the same Noble Family and Title, where it has been so much encouraged, and beg Leave to be allowed the Honour to subscribe myself,*

My Lord,

Your Lordship's most humble,

And most obedient Servant,

HENRY WILSON.

THE

EPISTLE

TO THE

READER.

THE general Approbation that this Book has met with, both in the Navy and amongst Merchant Ships, hath encouraged the Book-seller, not only to print a large Number, but also to be at the Charge of the Addition of *Spherical Trigonometry, Astronomey,* &c. and engraving all the Cuts upon Copper Plates, with such other Improvements, as might be thought necessary to make it a *Compleat Treatise of Navigation.* But if there are yet some in the World, that not being capable of Improvements themselves, envy what is made by

A 3   others:

others : Let ſuch conſider, that there is Room enongh for every induſtrious Navigator, to employ his Talent that Way: Nor is it poſſible to conclude, that Navigation is arrived at its *ne plus ultra*, ſo long as the Longitude remains to be ſuch a puzling Subject to our beſt Mathematicians, eſpecially now, in a Day, when the Parliament hath been pleaſed to offer ſuch a liberal and plentiful Gratuity to him that ſhall firſt diſcover it : For the ſupplying of which Want of an exact Method to find the Longitude, I have, toward the latter End of this Treatiſe, inſerted a very uſeful Method for finding the Longitude by the Sun, any Day at Noon, when the Sun can be ſeen. But to wave this Subject, we find that Mathematical Studies did hitherto, and do yet admit of Improvement : For in former Times, when Mathematical Sciences were in their Non-age, and *Minerva's* Fruit did not grow in ſo many *Engliſh* Gardens as it doth now, it was thought (and indeed was) an extraordinary Attainment in Trigonometrical Operations, when the Antients found out the Way of applying Right Lines to Arches of Circles, which we call Sines, Tangents, and Secants, and could thereby perform their Trigonometrical Calculations, although with the vaſt Trouble of multiplying by 5, 6, 7, or 8 Figures, which took up a large Quantity both of Time and Room ; but this Manner of Operation was much facilitated by my Lord *Napier's* admirable Invention of Logarithms, with the Improvement of Mr. *Henry Briggs*; the Uſe of which is now ſo common, that he would be thought but a ſlender Mathematician, that knows no other Way of Operation than by Multiplication and Diviſion, the

Trouble

Trouble whereof is now thought intolerable: To
this we may add the Invention of *Gunter's* Scale,
both long and ſliding, upon which, with a Turn
of a Pair of Compaſſes on the one, or only removing the other, are performed thoſe Propoſitions in *Trigonometry* which formerly required
Abundance of Figures in the Operation; but all
this not being thought ſufficiently expeditious at
Sea, there have been alſo invented Tables ready
calculated, called Traverſe Tables, or Tables of
Difference of Latitude and Departure, wherein by
Inſpection only you may anſwer any Queſtion in
any of the ſix Caſes of *Plane Sailing*, and alſo find
the Difference of Longitude, as is taught at large
in this Treatiſe : But of all the Improvements that
have been made, I never heard of any that ever attained to a Method for ſolving all the Caſes in
Trigonometry, and *Practical Navigation, viz.* the
keeping of a Reckoning both in Latitude and
Longitude, with ſufficient Exactneſs, without any
Tables, Books or Inſtruments, or any other Help,
but only Pen and Ink, or a Piece of Chalk; and
this is what I have preſented to the World in this
Treatiſe, and which I take for granted was never
taught or printed before, excepting that Proportion by which I have wrought the ſix Caſes of
Plane Sailing by the *New Method*, which we find
hinted at by *Snellius*, in his *Cyclometria*, quoted
by Mr. *Collins*, in his Book entitled, *The Plane
Scale New Plained, &c.* in the Title Page of which
Book he enumerates among the reſt of the Con ·
tents, [*Arithmetical Navigation*, *or Navigation
performed by the Pen, if Tables were wanting, &c.*]
at the firſt Sight of which I thought my Deſign
had been anticipated, till looking further into the

A 4   **Book**

Book where he is upon that Subject, (*viz.* in Page 118 of the firſt Book) I find he touches upon no more but only that Caſe, where three Sides of a right-angled Triangle are given (or found) to find an Angle; but in (in Page 120) ſpeaking of that Caſe of one Side, and two Angles given, to find the other Side, he ingeniouſly owns, that it cannot be done without a Table of Sines, and that is what I have laid down in this Treatiſe, *viz.* The Solution of all the other five Caſes of *Plane Sailing*, or any other Operation that depends upon a right angled plane Triangle, with as much (if not more) Eaſe and Exactneſs as that is performed there, which encouraged me to proceed to the compleating of that New Method, which I deſigned ſhould only have been the Subject of this Treatiſe, when I firſt thought of appearing in Public. But ſuppoſing that thoſe who had learned the whole Art of Navigation, both *Geometrical*, *Trigonometrical* and *Inſtrumental*, as it is commonly taught in Schools, and practiſed at Sea, ſhould be diſcouraged from buying the Book, when they found nothing in it that they had formerly learnt; therefore for the Sake of ſuch, and to make the Book univerſally Uſeful, I reſolve to make it a compleat Treatiſe of Navigation, as performed,

Firſt, Geometrically, by Projection, with Chords and equal Parts.

Secondly, By Trigonometrical Calculation, by Logarithms, Sines and Tangents.

Thirdly, By Inſpection, in the Traverſe Table.

Fourthly,

Fourthly, Arithmetically, by given Numbers.

Fiftbly, Inſtrumentally, by Scale and Compaſſes.

Sixthly, By a New Method, with only Pen and Ink.

Laſtly, Practical Navigation in ſeveral Parts; as,

- *Firſt*, How to find the Variation of the Compaſs by an eaſy Method, without Trigonometrical Calculation, Azimuth, or Amplitude, whereby they that can neither write or read may do it as well as the beſt Mathematician.

Secondly, How to divide the Log-line, and Reaſons given for the Length aſſigned for one Knot thereof.

- *Thirdly*, The beſt Method for taking and working an Obſervation, contracted into four Caſes or Varieties, which anſwers all Places in the World, and at all Times of the Year.

Fourthly, How to reckon the Tides, by a Method ſo eaſy, that you may thereby find the Time of High-Water, at any known Port only by a Sight of the Moon, at any Time of the Day or Night, illuſtrated by a Diagram, or Figure, in which, indeed to have been exact, the light Part of the Moon's Body ſhould have been determined by Ellipſis's, and not by Arches of Circles, as is evident from the Analemma, if we compare the Poſition of the Eye, with reſpect to the Moon, with the Poſition of the Eye, with reſpect to the
Ortho-

Orthographic Projection; but I have contented
myfelf with thefe circular Divifions, becaufe they
are eafier performed, and ferve as well to illuftrate
the Matter for which they are projected.

And here by the Way, we may take Notice of
one Obfervation with Refpect to the Tides, though
for fome Reafons omitted in its proper Place, *viz.*
Whether the Moon in Perigeon, by Reafon of her
nearnefs to the Earth, may not have a greater attractive Influence upon the Water; as alfo whether a greater Congrefs of Planets about the Sun
or Moon, at the New or Full Moon, may not
conduce fomething to the augmenting the Tides,
by contributing their (although but fmall) Attraction, when other more fubftantial Reafons
there mentioned do concur; and this I am rather induced to believe, becaufe the Truth hereof has been confirmed to us by an Inftance within the Reach of our own Experience; for at the
New Moon in *September* 1709, there was fuch a
Confluence of natural Caufes of the Tide : For
firft, the Moon was then with the Sun. Secondly, It was the Lunation next after the Autumnal
Equinox. Thirdly, the Moon was at or near her
Perigeon. Fourthly, The three Planets *Mars*,
Venus and *Mercury*, were all within lefs than
thirty Degrees of the Sun and Moon; yea,
fome within fifteen Degrees, and *Jupiter* was
within thirty-five Degrees of the Sun and Moon;
and this Lunation was attended with fuch exuber-
ant Tides, that in the Low Ground in *Yorkfhire*,
over againft *Stockton*, where the Tide never comes
but in the very high Spring-Tides, it flowed two
or three feet Water, and doubtlefs at other Places

it was proportionable, which ſeems to confirm the Truth of this more than probable Conjecture, and render it worthy of the Obſervation of the Ingenious.

Fifthly, How to make your own Charts for any Voyage ſeveral Ways.

Laſtly, The Application of the Whole to the actual Keeping of a Reckoning, both in Latitude and Longitude, ſailing Large, or upon a Wind ; and how to make all proper Allowances for Lee-Way, Variation, *&c.* and how to correct your Reckoning by an Obſervation, with Reaſons and Demonſtrations of the different Ways of doing it. One and the ſame Days Work in the Journal, as well as one and the ſame Queſtion elſewhere in the Book, being performed both by the common Method, and alſo by the New Method, that the Reader may ſee the Truth and Certainty thereof.

With Directions ſhewing how to project and anſwer the various Caſes and Queſtions relating to Currents, with other uſeful Queſtions, and a Method for finding and correcting the Longitude without the Help of the Dead Reckoning.

And that the Book may not be imperfect, or inferior to any of its Kind and Volume, I have added the neceſſary Problems in *Spherical Geometry,* with the twenty-eight Caſes of *Spherical Triangles,* and the Demonſtration of the Concluſions, by which ſeveral of the Oblique are performed, deduced from the Catholic Propoſitions made uſe

of

of in the ſixteen Caſes of Right-angled Triangles:
I have alſo given Directions for the Solution of
Quadrantal Triangles, and the Application of the
Whole to ſo much Practical Aſtronomy, as is uſe-
ful in Navigation; all which is what was not in
the former Impreſſions.

And although here are Ways propoſed for anſwering all neceſſary Caſes in Navigation without
Logarithms, &c. yet to make the Book univerſally uſeful, and as much as can be to anſwer the
End of all Buyers, as well the Learner at School,
as the accompliſhed Mathematician in his Trigonometrical Operations, I have added a correct
Table of Logarithms, Sines, and Tangents, and
ſome other uſeful Tables, which concludes the
Whole. Thus deſiring a favourable Conſtruction
to be put upon what Faults may eſcape either
the Pen or the Preſs; I queſtion not the noble
hearted Sailor's Acceptance; and that it may be
a Benefit to the Buyer, a Help to the Learner, a
fit Companion for the Mariner, and a Credit to
the Author, is the Deſire of,

Yours to ſerve you,

HENRY WILSON.

THE
CONTENTS.

CHAP. I.

Navigation Geometrical.

CHAP. II.

Navigation Trigonometrical.

Sect.

CHAP. III.

Navigation by Inspection in the Traverse Table.

CHAP. IV.

Navigation Arithmetical.

CHAP. V.

Navigation Instrumental.

CHAP.

CHAP. VI.

A New Method for solving any Question in Plane Trigonometry, and keeping a Reckoning both in Longitude and Latitude, without Book, Table, or Instrument, &c. only with Pen and Ink.

CHAP. VII.

Practical Navigation, or the Application of the foregoing Rule, to the actual keeping of a Reckoning, &c.

A

CHAP. VIII.

Of the Projection of the Sphere.

TABLES

In the Second PART.

NAVIGATION

New Modell'd.

CHAP. I.

N A V I G A T I O N *Geometrical.*

SECT. I.

Geometrical Problems.

PROB. I. *From a Point given in the Middle of a Line,*
 * *to erect a Perpendicular.*

WITH your Compasses opened at any Extent, and one Foot in the given Point A, make the Marks *b* and *c*; then with any greater Extent, and one *Fig. 1.* Foot in *b*, make an Arch, as at *d*, and with the same Extent, and one Foot in *c*, cross the said Arch at *d*; then from the crossing of the said Arches to the Point A, draw a Line, it is the Perpendicular required.

PROB. II. *From a Point given above a Line, to let fall*
 a Perpendicular.

With your Compasses opened at any Extent (more than the nearest Distance from the given Point to the given Line) and one Foot in the given Point A, cross the given Line in *b* and *c*; then with any Distance more than half the Distance between *b* and *c*, and

B   one

one Foot in *b* draw the Arch *d*, and with
Fig. 2.   the fame Extent, and one Foot in *c*, crofs the
said Arch at *d*, then lay a Scale from the Point
A to the crofsing of the Arches at *d*, fo draw the Line
AE, which is the perpendicular required.

PROB. III. *How to raife a Perpendicular upon the End*
of a given Line.

With your Compafses at any Extent, and one Foot
in the given Point A, make a Mark at any
Fig. 3.   convenient Diftance above the Line, as at *b*;
then keeping one Foot in the Mark *b*, with
the fame Extent crofs the given Line at *c*, and turning
your Compafses about, make the Arch *d*; then lay a
Scale from the Crofsing at *c* to the Mark at *b*, and
make a Mark where it crofses the Arch *d*; fo a Line
drawn from that Interfeftion to the Point A is the
Perpendicular required.

PROB. IV. *How to let fall a Perpendicular upon the*
End of a given Line, from any given Point over the End
of the faid Line.

Draw a Line from the given Point A to interfeft the
given Line at any convenient Diftance, as at *b*,
Fig. 4.   then divide the Line A *b* into two equal Parts
at *c*; and upon *c*, with the Extent *c* A or *c b*,
crofs the given Line in *d*; then a Line drawn from the
given Point A to the Interfeftion at *d*, is the Perpendicular required.

PROB. V. *To draw a Line parallel to another Line at*
any given Diftance.

Take in your Compafses the given Diftance, and with
one Foot in the given Line, towards each End
Fig. 5.   of it, draw the two little Arches *a* and *b*; a
Line drawn from the Extremity of thefe
Arches, is the Parallel required,

                                              PROB.

PROB. VI. *To bring any three Points (not situate in a right Line) into the Circumference of a Circle.*

Suppose the three Points be A, B, C; first take more than half the Distance A B, and with one Foot in A, sweep an Arch, then with the same Distance, and one Foot in B, draw another Arch to cross the aforesaid Arch in the Points *d* and *e*; then with more than Half the Distance BC, and one Foot in B, draw the Arch *g h*, and with the same Distance, and one Foot in C, draw an Arch to cross the aforesaid Arch   *Fig. 6.* in the Points *g* and *h*; then draw the Lines *a d e*, and *a g h*, and where these Lines cross, as at *a*, is the Center of the Circle required.

SECT. II.

The Construction of Sines, &c.

EVERY great Circle is (supposed to be) divided into 360 equal Parts, called Degrees, whereof the Half, or Semicircle, contains 180 Degrees, and the Quarter or Quadrant is 90 Degrees, and upon one Quadrant is projected the Sines, Tangents, &c.

The Radius is the Semidiameter of a Circle, upon which the Projection is made; as A K, or A 90. or A S, is the Radius of the projected Diagram, and is commonly supposed to contain 10000 equal Parts.

A Sine is a Perpendicular let fall from the given Degree to the Base or Semidiameter of the Circle; as the Line *g* 30, is the Sine of 30 Degrees, and the Line *b* 80, is the Sine of 80 Degrees, &c. the Sine of 90 Degrees is equal to the Radius.

A Tangent or Touch-line, is a Perpendicular erected upon the End of the Semidiameter, just so as to touch the Periphery or the Circle: Thus   *Fig. 7.* the Line K *r* is a Tangent Line.   The Tan-

gent

gent of any Degree is the Diſtance from the Beginning
or Foot of the Tangent Line, to that part of it where
the Line, drawn from the Center over the given Degree, cuts the Tangent Line: Thus the Part of the
Tangent Line *k p* is the Tangent of 50 Degrees. The
Tangent of 45 is equal to the Radius.

A Secant is a Line drawn from the Center through
the given Degree 'till it interſect the Tangent Line:
Thus the Line A *p* is the Secant of 50 Degrees. The
Secants begin at the Radius, and proceed to Infinite.
The Secant of o Degrees being equal to the Ra-
dius.

A Chord is the neareſt diſtance in a ſtrait Line between any quantity of Degrees; or from o Degrees to
the Degree whoſe Chord is required: Thus the Line
S. 90 is a Line of Chords, and the Diſtance S. upon
that Line is the Chord of 30 Degrees, &c. the Chord
of 60 Degrees is equal to the Radius; and hence it is,
that the Chord of 60, commonly called the Sweep of 60,
is generally taken in the Compaſſes to draw any Great
Circle, or Arch of a Circle, whoſe quantity in Degrees
is to be meaſured.

The Sine-Complement of any Arch, is the Sine of
the Complement of that Arch to 90: Thus the Sine-
complement of 30 Degrees is the Sine of 60, and the
Sine-complement of 70 is the Sine of 20, &c. And
in the Diagram the Line W 40 (equal to the Diſtance
A *f* upon the Baſe) is the Sine-complement of 40 Degrees, and ſo in others.

A verſed Sine is a Segment of the Baſe, contained
between the Sine of the Degree and the End of the
Baſe where the Tangent Line begins; thus the Segment of the Baſe *e* K is the verſed Sine of 50, &c

From this Projeċtion is reduced the following *Axioms*
for the Solution of all the Caſes in Plane Trigonometry: and which (if well underſtood) are the Ground of
the whole Art of Navigation, ſo far as it depends upon
Triangles, and is ſolved by a Canon; which I ſhall
firſt

firſt ſhew in the following Caſes, and then proceed to
ſhew how it may be done without.

A X I O M I.

In all right-angled Plane Triangles, if one Side be
made Radius, the other Sides will be Sines, Sine-com-
plements, Tangents or Secants, as is evident from the
Diagram ; for,

Suppoſe in the Diagram the Triangle A *p* K ; here
the Side A K the Radius, and the Angle at A being 50
Degrees, the Side *p* K is the Tangent of 50, and the
Hypotenuſe A *p* is the Secant of 50 ; and what Proportion the *Radius* hath to the Side A K, the ſame Proportion hath the Tangent of 50 to the Side K *p*, and
the ſame Proportion hath the Secant to the Hypotenuſe
A *p*.

Again, if you will make the Hypotenuſe Radius, and
ſuppoſe the Angle at A. be 40 Degrees ; then in the
Diagram it is repreſented by the Triangle A *f* 40, and
then the Side *f* 40 is the Sine of the Angle at A, and
the Side A *f* equal to W 40 is the Sine-complement of
the Angle at A, and then what Proportion the Hypotenuſe hath to the Radius, the ſame Proportion will
the Side 40 *f* have to the Sine of 40, and the ſame Proportion will the Side A *f* have to the Sine-complement
of 40, &c. and from this Proportion proceeds the ſecond Axiom.

A X I O M II.

In all Plane Triangles the Sides are proportional to
the Sines of their oppoſite Angles, and the contrary ; as
in the Diagram in the Triangle above-mentioned, A *f*
40, it is demonſtrated, As the Radius or Sine of 90, is
to the Hypotenuſe, or Side oppoſite ; ſo is the Sine of
40, to the Side oppoſite to the Angle at A, &c.

This Proportion commonly called Oppoſite Sides and
Oppoſite Angles, holds true alſo in Oblique Plane Triangles,

angles; only obferve, that where you have an Obtufe
Angle, *viz.* more than 90 Degrees, the Sine of it is
found by fubtracting the Obtufe Angle from 180 Degrees, the Sine of the Remainder is the Sine of the
Obtufe Angle required.

AXIOM III.

In all Triangles, as the Sum of the Legs containing
any Angle is to their Difference, fo is the Tangent of
half the Sum of the other two Angles, to the Tangent
of half their Difference; and therefore,

When there are given two Sides, and an Angle included, to find the other Angles, the Proportion is,

As the Sum of the Sides is to the Difference of the
Sides, fo is the Tangent of half the Sum of the unknown Angles, to the Tangent of half their Difference;
which half Difference added to the half Sum, is the
greater Angle, and fubtracted leaves the leffer.

AXIOM IV.

In all Triangles, as the Bafe or greater Side, to the
Sum of the other two Sides; fo is the Difference of the
Sides, to the Difference of the Segments of the Bafe,
which Difference fubtracted from the whole Bafe, the
Perpendicular falls in the Middle of the Remainder;
and fo the Oblique Triangle is reduced to the Right-
angled ones, and may be wrought after the fame Manner.

By thefe Axioms are all the following Cafes of Plane
Triangles folved; in which obferve, In Right-angled
Triangles, the two Sides including the Right-angle are
called Legs or Sides, or fometimes Bafe and Perpendicular; and the flope Line is called the Hypotenufe, &c.

S E C T. III.

Plain Trigonometry Geometrical.

Plane Trigonometry Right-angled.

C A S E. I. *The Hypotenuse, and one acute Angle being given (consequently both) to find either Leg.*

Note, IN all Plane Triangles, the three Angles make up 180 Degrees, therefore in all Right-angled Triangles, because the Right-angle is always 90, the Sum of the other two Angles is also 90, and therefore subtract the acute Angle given from 90, the Remainder is the other acute Angle

     Given Hypotenuse AC 550
     The Angle at —— A 35d 00m.   *Fig.* 8.
     Required the Leg AB.
     And the Leg BC.

1. In this, and all other Cases, draw the Base AB at Pleasure.

2. With the Chord of 60, and one Foot in A, sweep the Arch *d e.*

3. With the Chord of the given Angle 35, and one Foot in *d,* with the other Foot cross the Arch at *e.*

4. From A through *e,* draw the Line AC, upon which set off the Hypotenuse 550 from A to C.

5. From C let fall the Perpendicular upon the Line AB to cut it in B, and 'tis done.

The Leg AB measured on the same equal Parts from which the Hypotenuse was taken, will be found to be 450.5, viz. 450 and an half, and the Leg BC is 315.5.

CASE II. *Given the Angles and a Leg to find the Hypotenuse.*

     Given the Angle A —— 33d. 45m.
       the Leg AB —— 459   *Fig.* 9.
     Required the Hypotenuse AC.

1. Draw

1. Draw AB 449.

2. With the Sweep of 60, and one Foot in A, draw the Arch *d e*.

3. With the Chord of the given Angle, and one Foot in *d*, set off the given Angle 33d. 45m. from *d* to *e*.

4. From B erect a Perpendicular BC.

5. Through A and *e* draw the Hypotenuse AC, till it cross BC, and 'tis done.

The Hypotenuse AC measured on the Scale of equal Parts will be found to be 540.

CASE III. *Given the Angles and Leg, to find another Leg.*

Given the Angle A ——— 33d. 45m.
the Leg AB ——— 449d.

Required the Leg BC.

This is laid down as Case II. the same Things being given, and the required Leg BC being measured, will be found to be 300.

CASE IV. *Given the Hypotenuse and the Leg to find an Angle.*

Given the Hypotenuse AC 540

Fig. 10.   the Leg AB ——— 449

Required the Angle at A.

1. Draw AB at Pleasure.

2. Upon AB set off the given Leg 449, from A to B.

3. From B erect a Perpendicular.

4. With the Hypotenuse 540, and one Foot in A cross the Perpendicular BC in C.

5. From the said crossing to A draw the Line CA.

6. With the Sweep of 60, and one Foot in A, draw the Arch *d e*, and 'tis done.

The Arch *d e* measured on the Chords gives 33d. 45m. the Angle required.

CASE V.   *Given the Hypotenuse and a Leg, to find other Leg.*

Given the Hypotenuse AC 540
the Leg AB ——— 449.

Required the Leg BC   The

The laying down of this is the same as the former, because the same Things are given; and the Leg BC measured on the same equal Parts from whence the given Parts are laid down, will be found to be 300.

CASE VI. *The Legs given, to find the Angle.*

Given the Leg AB —— 449
    the Leg BC —— 300   *Fig.* 11.
Required the Angles.

1. Draw AB 449.
2. Upon B erect the Perpendicular BC, and thereupon set off 300 from B to C.
3. Draw the Hypotenuse AC.
4. With the sweep of 60, and one Foot in A, draw the Arch *d e*, and 'tis done.

The Arch *d e* measured on the Chords gives 33d. 45m. the Angle required.

CASE VII. *The Legs given, to find the Hypotenuse.*

Given the Leg AB —— 449
    the Leg BC —— 300
Required the Hypotenuse AC.

This is laid down as *Case* VI. and the Hypotenuse AC measured on the same Scale of equal Parts is 540.

Note; *These seven Cases may be reduced to four, for making the first Case as above to be the first Case; the second and third having both the same Things given may be called the second, and for the same Reason, the fourth and fifth may be comprised in one, and called the third; and the sixth and seventh having also the same Things given, may then be called the fourth Case, and at this Rate the seven Cases are comprehended in four, because four Times laying down includes all the Varieties of Right-angled Plane Triangles.*

SECT.

SECT. IV.

Of Oblique Plane Triangles.

CASE I. *Two Angles and a Side oppofite to one of them given, to find the Side oppofite to the other.*

Given the Angle at A —— 30d. om.
Fig. 12.   the Angld at B —— 45d. om.
      the Side BC —— 290
Required the Side AC.

1. Draw AB at Pleafure.

2. With the Chord of 60, and one Foot in B (the Angle next the given Side) draw the Arch *d e.*

3. Upon that Arch fet off the Chord of the given Angle 45, from *d* to *e.*

4. From B through *e* draw the Line BC, and upon it fet off the given Side 290 from B to C.

5. The Angles A and B being given, the Angle C is alfo known, being the Supplement of the other two Angles to 180, and will be found to be 105d. and therefore with 60 off the Chords, and one Foot in C, fweep the Arch F *g*, upon which fet off 105d. from F to *g*, and through *g* draw the Line CA to cut AB in A, and 'tis done.

The Side AC meafured on the fame equal Parts is 410, the Side required.

Note; When you have any Chord above 90 to fet off, you muft do it at twice, becaufe the Chords go but to 90, &c.

CASE II. *Two Sides and an Angle oppofite to one of them given, to find the Angle oppofite to the other.*

Given the Side AB —— 560
Fig. 13.   the Side AC —— 410
      the Angle at B —— 45d. om.
Required the Angle at C.

1. Draw AB 560.

2. Make the Angle B 45, and draw the Line BC.
                      3. With

3. With the Side AC 410, and one Foot in A, cross the Line BC in C, and draw the Line AC, and 'tis done.

The Angle at C measured on the Sweep *d e*, gives 105d. the Angle at C required.

Note; This Case is ambiguous, and will admit of two Answers; so that it is necessary that it be known whether the Angle at C be acute (that is, less than 90) or obtuse (that is, more than 90) for if it be obtuse, the Angle marked C is the Angle required ; but if acute, the Angle at O is it ; for the Side AC 410 taken in the Compasses, and one Foot in A will cross the Line BC (continued) both in C and O, *&c.*

CASE III. *Given as in Case II. to find the third Side.*

The laying down is the same as in *Case* II. because the same Things are given ; and the Side BC measured on the equal Parts is 290, if the Angle be obtuse ; or 503, if it be acute.

CASE IV. *Two Sides and an Angle between them given, to find either of the other Angles.*

Given the Side AC ———— 410   *Fig.* 14.
      the Side AB ———— 560
      the Angle at A ———— 30d. 0m.
   Required the Angle B.
1. Draw AB 560 from A to B.
2. Make the Angle A 30d. 0m.
3. Draw AC 410 from A to C.
4. From C to B draw the Side CB, and 'tis done.
The Angle B measured on the Chords is 45d. 0m.

CASE V. *Given two Sides and the Angle between them, to find the third Side.*

Given the Side AC ———— 410
      the Side AB ———— 560
      the Angle A ———— 30d.
   Required the Side BC.
Lay down as in *Case* IV. and the Side BC measured is 290.

CASE

CASE VI. *Three Sides given to find an Angle.*

Given the Side AB —— 560
Fig. 15.   the Side AC —— 410
             the Side BC —— 290
Required the Angle at A.

1. Draw AB 560 from A to B.

2. Take the Side AC 410 in your Compasses, and with one Foot in A, draw an Arch at C.

3. With the Side BC 290 in your Compasses, and one Foot in B, cross the Arch at C.

4. From the crossing of the said Arches to A and B, draw CA and CB, and 'tis done.

The Angle at A measured on the Chords is 30d. 00m.

SECT. V.

Plane Sailing Geometrical.

IN laying down all the Cases of Plane Sailing, Traverse, Mercator, &c. observe to make the Top of the Book or Slate North, and the Bottom South; the Right-hand East, and the Left-hand West, &c.

Observe that in the Application of Right-angled Triangles, to Questions in Plane Sailing, the Course is always the Angle at the Base: The Distance is the Hypotenuse: The Difference of Latitude is the North and South Line; and the Departure is the East and West Line of the Triangle.

Thus in the four Triangles, *Fig.* 16, N° 1, represents a Course in the N. E. Quarter; N° 2, *Fig.* 16. a Course in the S. E. Quarter, N° 3, a Course in the S. W. Quarter; N° 4, a Course in the N. W. Quarter: And in each of these Triangles the Point A represents the Place sailed from, and the Point C the Place sailed to: And the Angle at A is the Course, the Angle at C its Complement, the Hypotenuse AC, the Distance, the Leg AB the Difference of Latitude, and the Leg BC the Departure, &c.

CASE

C A S E I. *Courfe and Diftance given, to find the Difference of Latitude and Departure.*

A Ship fails S. W. by S. 540 Miles, I demand the
Difference of Latitude and Departure.

1. In this and all other Cafes of Plane Sailing, draw
the North and South Line A B.

2. The Courfe being three Points from the
Meridian, *viz.* 33d. 45m. make the Angle at   *Fig.* 17.
A 33d. 45m. from *d* to *e*.

3. Thro' *e* draw the Hypotenufe (or Diftance) AC
540 from A to C.

4. From C let fall a Perpendicular CB to cut AB in
B, and 'tis done.

The Difference of Latitude AB meafured is 449,
and the Departure BC is 300, &c.

C A S E II. *Courfe and Difference of Latitude given,
find the Diftance and Departure.*

A Ship fails S. W. by S. till her Difference of Latitude be 449, I demand her Diftance and Departure.

1. Make AB the Difference of Latitude 449,   *Fig.* 18.
from A to B.

2. Upon B erect the Perpendicular BC at Pleafure.

3. Set off the Courfe 33d. 45m. and draw the Hypotenufe AC to cut BC in C, and 'tis done.

The Diftance A C meafured will be found to be 540,
and the Departure BC meafured is 300 Miles.

C A S E. III. *Courfe and Departure given, to find the Diftance and Difference of Latitude.*

A Ship fails N. E. by N. till her Departure be 300
Miles, her Diftance and Difference of Latitude is required.

1. Draw AB at Pleafure.

2. At the Diftance 300 the Departure given,
Fig. 19. draw the parallel Line *d e*.

3. With the Angle of the Courfe 33d.
45m. draw the Hypotenufe AD, to cut the Parallel in C.

4. From

· 4. From the croffing at C, let fall the Perpendicular CB, to cut AB in B, and 'tis done.

Another Way to lay it down.

1. Draw A B at Pleafure.

2. Upon the North End thereof (becaufe the Ship fails to the Northward) erect a Perpendicular BC, upon which fet off the Departure 300.

3. Make the Angle at C 56d. 15m. the Complement of the Courfe, and draw the Hypotenufe CA, to cut AB in A, and 'tis done.

The Diftance AC is 540.

● The Difference of Latitude AB is 449.

CASE IV. *Diftance and Difference of Latitude given, to find the Courfe and Departure.*

A Ship fails in the S. E. Quarter 540 Miles, and then finds her Difference of Latitude is 449, her Courfe and Departure is required.

1. Make the Difference of Latitude AB 449, from A to B, and upon B erect the Perpendicular BC.

Fig. 20.   2. Take the Diftance in your Compaffes, and with one Foot in A crofs the Perpendicular at BC in C.

3. From the croffing C to A draw AC.

4. With the Chord of 60 make the Arch *d e*, and 'tis done.

The Arch *d e* meafured on the Rhumbs, is three Points, or on the Chords is 33d. 45m.

The Departure BC is 300.

CASE V. *Diftance and Departure given, to find the Courfe and Difference of Latitude.*

A Ship fails between the N. and W. 540 Miles, her Departure is 300; I demand her Courfe, and Difference of Latitude.

1. Draw AB at Pleafure.

2. A[

2. At B erect the Perpendicular BC, and set off the Departure 300 from B to C.   *Fig.* 21.

3. With the Distance 540 in your Compasses, and one Foot in C, cross the Line AB in A.

4. From C to the Crossing at A draw AC.

5. With the Chord of 60, and one Foot in A, make the Arch *d e*, and 'tis done.

AB the Difference of Latitude is 449, and the Course *d e* is three Points, or 33d. 45m.

CASE VI. *Difference of Latitude and Departure given, to find the Course and Distance.*

A Ship sails in the S. W. Quarter, till her Difference of Latitude be 449, and her Departure 300; her Course and Distance is required.

1. Draw the Difference of Latitude AB 449, from A to B.

2. Upon B erect the Perpendicular BC, upon which set off the Departure 300, from B to C.   *Fig.* 22.

3. From A to C draw the Hypotenuse AC.

4. With the Chord of 60, make the Arch *d e*, and 'tis done.

The Distance AC is 540, and the Angle at A, the Course three Points, or 33d. 45m.

SECT VI.

Traverse Sailing Geometrical.

TRaverse Sailing is of Use, when a Ship having set sail from one Port intending for another, whose Course and Distance from the Port sailed from, is given or known; but by Reason of contrary Winds, or other Accidents is forced to sail upon several Courses, which are required to be brought into one Course, to know thereby (after so many various Turnings and Windings) the true Course and Distance made good from the Place sailed from, and the true Point or Place where the Ship is, (that so the Wind coming fair) it may be known

how

how to shape a Course for the Place intended; and this, which is the chief Subject of Traverse Sailing, may be performed Geometrically two Ways, both which I shall briefly lay down.

The first is performed by drawing new Meridians through the Extremity of every Course, parallel to the first Meridian, or North and South Line, that you make at the first, and so set off every Course with a Sweep of 60, as if it were a Question of Plane Sailing. You may also let fall Perpendiculars to every new Meridian from the Point that the Ship sailed to upon that Course, and so you have the Course, Distance, Difference of Latitude and Departure, to every Course; and this Method is very useful where the Courses tend generally all one Way, without intersecting one another. But if your Courses frequently cross one another, 'tis best to lay them down without new Meridians, *viz.* to set off one Course by another; for which Mr. *Atkinson* in his Epitome laid down four Rules; but they being so burthensome to the Memory of young Learners, I shall not so much as name them; but shall lay down one Rule which is universally useful in all Cases, and easily remembered, and 'tis this;

Observe how many Points are between the Point next to be laid down, and the Point opposite to the Course last laid down, for that is the Point for laying down; therefore with the Chord of 60, and one Foot in the Point the Ship is last come to, describe an Arch; upon which set off the Points found by the abovesaid Rule, and through that draw the Line for the next Course, &c.

I shall explain both Ways by Example; and first,

How to lay down a Traverse by new Meridians.

A Ship bound for a Port distant 120 Miles N. E. ¼ E. sails S. S. E. 30 Miles, then N. E. by N. 40, then E. by N. 25, then N. N. E. 44; I demand the Course and Distance made good, and also the Course and Distance to the Port bound for.

First,

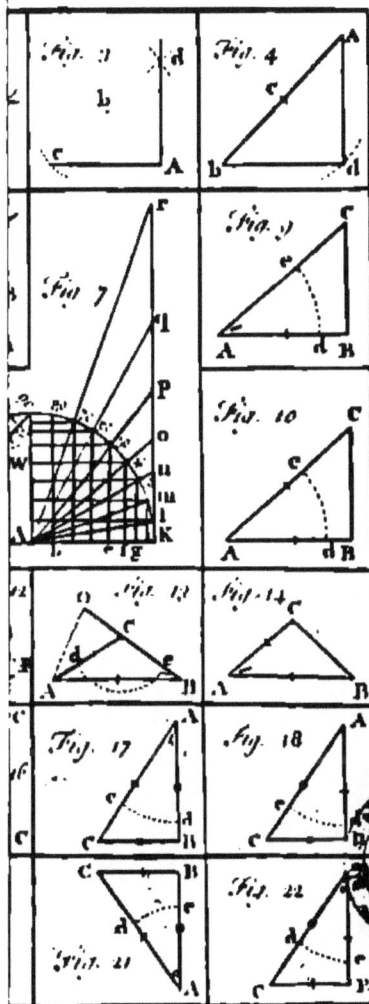

First, Draw the Line K H at pleasure, for a Meridian, or North or South Line; and therein assume a Point, as at A, for the Port sailed from; then with 60 from the Chords, and one Foot in A, draw the Arch L *m*, upon which set off two Points, (because the Course is S.S.E.) from L to *m*, and draw the Line A *m*, upon which set off the Distance 30 from A to B, then is your Ship at B; then let fall the Perpendicular BK, then is AK, 27.7. the Difference of Latitude, and BK, 11.5 the Departure for the first Course———Then for the second Course, with the Distance KB draw the Parallel BN, and thereby, with the Chord of 60, as before, set off the second Course and Distance N E. by N. 40, from B to C, and let fall the Perpendicular CL, then is your Ship at C; the Difference of Latitude upon that Course is BL 33.3, and the Departure CL 22.2 :——— Then proceed in the same Manner for the third Course; with the Parallel CO, set off E. by N. 25, from C to D, and draw the Line D P, (from which set off the last Course, N.N.E.) 44. then is your Ship at E. Now seeing the Ship came from A, and is now at E, the Line A E measured on the same equal Parts, upon which all the other Distances were taken, will be found to be 91 Miles, and the Arch R Q measured on the Rhumbs is five Points, *viz.* N.E. by E. so that the Ship is now 91 Miles, N.E. by E. from the Port sailed from.

Now to find her Course and Distance to the Port bound for, set off four Points and a Half upon the Arch RQ, from R to S, and from A thro' S. draw the Line ASF; upon which set off 120 (the Distance from the Port sailed from to the Port bound for) from A to F, then is F the Port bound for; now the Port bound for being at F, and the Ship being but at E, therefore the Line EF measured on the same equal Parts that the rest were taken from, will be found *Fig.* 23. to be 31, and the Arch TV measured on the Chords is 35d. 12m. or N.E. by N. somewhat Easterly, *&c.*

How

How to lay down a Traverse without new Meridians.

First, Draw a North and South Line, as in the former, as the Line RM; in which assume a Point, as at
A, for the Port fail'd from; then from A set off the first
Course and Distance, *viz.* N.N.W. 68, from A to B;
then for the second Course, with the Chord of 60, and
one Foot in B, draw the Arch TW, upon which, for
setting off the next Course S.S.W. 70, observe the Rule
at the Beginning of Traverse Sailing Geometrical, *viz.*
Take the Number of Points between the Point opposite
to the last Course fail'd, and the Point you are next to
fail. The Reason of this Rule is this: If from A to B
your Course be N.N.W then back from B to A must
needs be S.S.E (the opposite Point) and then if you
were to fail S. by E. it must be one Point to the Southward of that S.S.E. Line; if South it is two Points,
and consequently my next Course being S.S.W. I set
off four Points from T to W, and through W draw the
Line BC, which is as a S.S.W. Line, upon which set off
70 Miles from B to C, and then is your Ship at C:
Then for the third Course, if from B to C be S.S.W.
then from C to B is N.N.E. but my next Course being
E. half N. the Points between N.N.E. and E. half N.
are five Points and a half; therefore, with the Chord of
60, and one Foot in C, draw the Arch *x y,* upon which
set off five Points and an half from *x* to *y,* and thro' *y*
draw the Line CD, upon which set off 90 Miles from
C to D, then is your Ship at D; after the same Manner
lay down all the rest as DE W.N.W. half N. 70; then
EF South 25; then FG, E. half S. 45; then lastly,
GH South 30, which is the last Course: Then your
Ship being at H, and the Port failed from at A, the
Line AH 28 Miles is the Distance made good, and the
Angle at A is four Points, *viz.* S.E. but the Port intended for being S.W. 55. I set it off from A to K, but
the Ship being at H the Line HK 62 Miles is the Di-
stance from the Ship to the Port bound for, and the
                                                    Course

Courſe is found by meaſuring the Angle at H 71d. 48m. or W.S.W. more than one Quarter Weſterly.

The Queſtion.   A Ship at A bound for a Port at K, which bears S.W. from A diſtant 55 Leagues
but meeting with contrary Winds, ſhe ſails N. *Fig.* 24.
N.W. 68 Leagues, then S.S.W. 70 Leagues
then E. half N. 90 Leagues, then W.N.W. half N. 70 Leagues, then S. 25 Leagues, then E. half S. 45
Leagues, then South 30 Leagues; I demand the Courſe and Diſtance made good, and Courſe and Diſtance to the Port bound for.

How to tranſcribe a Traverſe, or any other right-lined Figure from the Slate to a Book, or from one Book to another.

In the firſt Line RM make two or three Marks, as at R, A and M; then in the Book into which
you would lay down, draw alſo the ſame Line, *Fig.* 25.
and take the Diſtance RA, and ſet from R to
A in the new Draught; likewiſe ſet the Diſtance AM from A to M; then for laying down the Courſes; as firſt, from A to B, with the Diſtance AB in the old Draught, and one Foot in A in the new Draught, make an Arch at B; then with the Diſtance MB in the old Draught, and one Foot in M in the new Draught, croſs the former Arch at B; then from A, and the croſſing of theſe Arches, draw the Line AB; then for the Line BC, with the Diſtance RC, and one Foot in R, draw a ſmall Arch, and with the Diſtance BC, and one Foot in B, (or with the Diſtance MC, and one Foot in M) draw an Arch to croſs the former in C, then from B to that croſſing draw the Line BC, and ſo in the Reſt of the Courſes.   And note, you need not regard what two Points you take in the whole Draught for the fixed Point of your Compaſſes, provided you take the ſame Points in the new Draught, only obſerve to take two Points, ſo as that the Arches may fairly croſs one another; or elſe you cannot ſo well find the Point of Interſection; as ſuppoſing you would find the Point B, if you take the Diſtance AB, and ſet from A to B,

and

and make an Arch; and then if you take the Distance RB, and set from R to B and cross the former
Arch, these two Arches will run so along one another,
that you cannot exactly discern the Place of Intersection;
to remedy which, take two Marks that bear more square
from B, as A and M, and then the Arches will cross
more directly, and the Operation will be more exact.

S E C T. VII.

Mercator's *Sailing Geometrical.*

MErcator's Sailing is laid down by a right-angled
Triangle, as Plane-Sailing, only the Triangle
hath two Perpendiculars, the shorter representing the
Departure; as in Plane-Sailing, and the longer is the
Difference of Longitude.

Thus in the Triangle ADE, A represents the Place
sailed from; the Angle at A is the Course,
Fig. 26. and the Side AB is the proper Difference of
Latitude, and BC the Departure, as in Plane-
Sailing.   The whole Base AD is the Meridional Difference of Latitude; and the Perpendicular DE is the
Difference of Longitude; and seeing the Angle at A is
common to both Triangles, ABC and ADE, therefore
the Base AB is in Proportion to the Perpendicular BC,
as the whole Base AD is to the Perpendicular DE; and
hence comes the Proportion; as the proper Difference
of Latitude, to the Meridional Difference of the Latitude; so is the Departure, to the Difference of Longitude.   *Euclid. lib.* 6. *Prop.* 6.

The Meridian Line, with the Scale of equal Parts
next to it, upon the Gunter's Scale, are of use for laying down Questions in Mercator's Sailing; they are the
two lowest Lines upon the Scale, the Meridian Line
marked [Merid.] and the equal Parts marked [Eq. P.]
the Graduations of the Meridian Line increases, as the
Degrees of Latitude grow bigger near the Poles; the
Distance

Diſtance and proper Difference of Latitude may be taken off the equal Parts, and the Meridional Difference of Latitude off the Meridian Line, if the Queſtion be ſo large as to permit it. Nevertheleſs, in ſmall Queſtions, and ſhort Diſtances, where the Diſtance between the two Latitudes upon the Meridional Line is too little to make a handſome Figure, you may find the Meridional Difference of Latitude, by the Table of Meridional Parts, and ſo take both it and the proper Difference of Latitude, with the Departure, Diſtance, and Difference of Longitude, off any equal Parts that you think will be proportional to the Dimenſions that you would have your Queſtion to contain; only take Care, that from whatſoever equal Parts you take the given Sides, you muſt meaſure the required Sides, (when found) upon the ſame equal Parts. I ſhall inſtance both Ways, in the following Examples.

CASE I. *One Latitude, Courſe and Diſtance given, to find the other Latitude, and Difference of Longitude.*

A Ship in Latitude 50d. 0m. North, ſails N. W. by N. 987 Miles, I demand the Latitude come to, and Difference of Longitude.

1. Draw ABD at pleaſure.

2. At an Angle of 33d. 45m. draw AC continued, upon which from the equal Parts ſet off the Diſtance 987 from A to C.

3. Let fall the Perpendicular CB.

4. Meaſure AB the Difference of Latitude 820 min. or 13d. 40m. which added to Latitude 50, gives the Latitude come to, 63d. 40m.

5. Extend the Compaſſes upon the Meridian Line from 50 to 63d. 40m. and ſet that Extent upon the Line AD, from B to D; then is AD *Fig.* 27. 1519, the Meridional Difference of Latitude.

6. From D erect the Perpendicular DE, to cut A, C, E, and 'tis done.

Then is AB 820 min. or 13d. 40m. the Difference of Latitude, which added to the Latitude ſailed from

C 3 (becauſe

(becaufe the Ship fails to the Northward) produces 63d. 40m. the Latitude come to.

DE 1017 is the Difference of Longitude required.

CASE II.   *Both Latitude and Courfe given, to find the Diftance and Difference of Longitude.*

A Ship in Lat. 56d. 25m. North, fails N.N.E. into Latitude 68d. 30m. I demand as above.

1. Extend the Compaffes on the Meridian Line from Lat. 56d. 25m. to Lat. 68d. 30m. and fet *Fig.* 28. that Extent on the Line AD, from A to D.

2. Take 725 (the proper Difference of Latitude, in Minutes) from the equal Parts under the Meridian Line, and fet upon the Line AD, from A to B.

3. At an Angle of 22d. 30m. (the Courfe given) draw the Hypotenufe ACE.

4. At B and D erect the Perpendiculars BC and DE, and 'tis done.

Then is AC 785 the Diftance, and DE 660 is the Difference of Longitude required.

Note, Upon the equal Parts under the Meridian Line, every Degree being 60 Minutes or Miles, every 10 Degrees is 600 Miles; and fo for 660, fet one Foot in 10, the other extended one whole Divifion beyond the Cypher towards the Right-hand is 660, and two Divifions is 720, &c.

CASE III.   *One Latitude, Courfe, and Difference of Longitude given, to find the other Latitude and Diftance.*

A Ship in Lat. 50 fails N.N.W. till her Difference of Longitude be 7 Deg. or 420 Minutes; I demand as above.

1. Draw ABD at pleafure.

2. At an Angle of 22d. 30m. draw the *Fig.* 29.   Hypotenufe AC continued.

3. At the Diftance of 420 (the Difference of Longitude) draw a Line parallel to AD.

4. From the Point where that Parallel interfects the Hypotenufe, as at E, let fall the Perpendicular DE.

5. With the Extent AD, and one Foot in the given Latitude 50 upon the Meridian Line, extend the other upward,

becaufe

becaufe the Ship fails to the Northward, and the other Foot will light upon 59d. 40m. the Latitude come to.

6. With the proper Difference of Latitude 580 (off the equal Parts) and one Foot in A, find the Point B in the Line ABD.

7. At B erect the Perpendicular BC, and 'tis done.

Then is AB 580 the proper Difference of Latitude, which reduced into Degrees and Minutes is 9d. 40m. which added to the Latitude failed from, *viz.* 50d. 00m. the Sum 59d. 40m. is the Latitude come to, and AC 628 is the Diftance required.

CASE IV.   *Both Latitudes and Diftance given, to find the Courfe and Difference of Longitude.*

A Ship in Latitude 50d. North, fails 3505 Miles, and is then by Obfervation in Lat. 13d. 12m. I demand as above.

This Queftion being larger than the Book can conveniently contain, if taken off the Meridian Line, I fhall take it off a Scale of leffer equal Parts, and find the Meridional Difference of Latitude by the Table.

1. Draw ABD, upon which fet off the proper Difference of Latitude 2208 from A to B, and the Meridional Difference of Latitude, 2676 *Fig.* 30. from A to D.

2. At B and D erect the Perpendiculars BC and DE.

3. With the Diftance 3505, and one Foot in A, with the other crofs the Perpendicular BC, in C.

4. Draw AC continued to E, and 'tis done.

The Angle BAC meafured on the Chords is 50d. 57m. and the Perpendicular DE 3298, is the Difference of Longitude required.

CASE V.   *Both Latitudes and Departure given, to find Courfe, Diftance, and Difference of Longitude.*

A Ship in Latitude 55d. North, fails in the North Eaft Quarter into Latitude 56d. 10m. North, her Departure 50 Miles; I demand as above.

This and all the following Queftions in *Mercator's Sailing Geometrical* contain too fhort Diftances to be taken

off

off the Meridian Line, I shall therefore project them from larger equal Parts.

1. Draw ABD upon which set off the proper Difference of Latitude 70 from A to B, and the Meridional Difference of Latitude 124 from A to D.

Fig. 31.

2. From the Points B and D erect the Perpendiculars BC and DE.

3. Upon BC set off the Departure 50 from B to C.

4. Thro' the Point C draw the Line AC continued, till it cut the Perpendicular DE in E, and 'tis done.

Then the Angle at A measured on the Chords, is 35d. 32m. for the Course, which is N. E. by N. something Easterly, and the Hypotenuse AC 86 is the Distance, and the Perpendicular D E 89 is the Difference of Longitude required.

CASE VI. *Both Latitudes and Difference of Longitude given, to find the rest.*

Suppose two Places, one in Latitude 56d. 15m. and the other in Latitude 58d. 35m. their Difference of Longitude 2d. 30m. I demand the Course from the Southermost to the Northermost, and also their Distance and Departure.

1. Upon the Line ABD set off the proper Diff. Lat. and Merid. Diff. Lat. to B and D, and there erect Perpendiculars, as in Case the Fifth.

Fig. 32.

2. Upon DE set off the Difference of Longitude 150 from D to E.

3. From A to E draw the Hypotenuse ACE to cut BC in C. Then is BC 80 the Departure; the Angle at A 29. 53m. is the Course, and AC 161 is the Distance.

Note. The Figure placed between B and D (as here is 260) is the Meridional Difference of Latitude represented by the whole Line AD.

CASE VII. *One Latitude, Course and Departure given, to find the other Latitude, Distance and Diff. Longitude.*

A Ship in Latitude 58d. 0m. North, sails S.S.W. ¼ W. till her Departure be 60 Miles or Minutes; I demand as above.

1. Draw

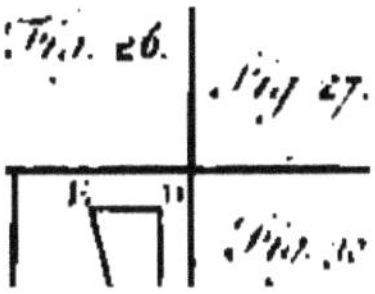

Fig. 26.

Fig. 27.

Fig. 28.

1. Draw ABD at pleasure.
2. At the Angle of 28d. 7m. the given Course draw ACE continued.   *Fig.* 33.
3. At 60 Distance from ABD, draw the Parallel *e g*, and where it cuts ACE, as in C, let fall the Perpendicular CB.
4. Then is AB the proper Difference of Latitude, by which you may find the Latitude, Course, and also the Meridional Difference of Latitude, which (found as directed in *Mercator's Sailing Trigonometrical,* Case I.) is 206, to be set from A to D.
5. Raise the Perpendicular DE, to cut ACE in E, and 'tis done.

The Latitude come to is 56d. 8m. the Difference of Longitude DE 110, &c.

CASE VIII.   *One Latitude, Distance, and Departure given, to find the other Latitude, Course, and Difference of Longitude.*

A Ship in Lat. 58d. om. North, sails between the South and West 127 Miles, her Departure 60; I demand the rest.

See the Figure in Case the Seventh.
1. Draw ABD at pleasure.
2. At 60 (the Departure) Distance from ABD, draw the Parallel *e g*.
3. With the Distance 127, and one Foot in A cross the said Parallel in C.
4. Let fall the Perpendicular CB to cut ABD in B; then is AB the proper Difference of Latitude, by which find the Meridional Difference of Latitude, as in Case VII. which set from A to D, and raise the Perpendicular DE.
5. From A through the crossing at C, draw ACE, to cut DE in E, and 'tis done.

The Latitude come to is 56d. 8m. the Angle A the Course 28d. 7m.

The Difference of Longitude DE is 110.

SECT. VIII.

Parallel Sailing Geometrical.

THERE are only two ways moſt uſeful and intelligible for laying down *Parallel Sailing*; the one
called Bell-faſhion; the other Plane Triangle.

In the firſt of theſe it is laid down by a Figure ſome-
what like a Bell, from whence I ſuppoſe it
Fig. 34. take its Name. In this Figure the two Sides
AD and AE are equal (*viz.* equal to the Radius or Sine of 90; or if you think the Projection will
be too ſmall, you may make them twice the Sine of 90;
(provided you take all the other double alſo) and then
the Line BC repreſents their true Diſtance in the
Parallel; and DE repreſents their Diſtance in the Equinoctial, or Difference of Longitude.

The other Way is by Plane Triangle, in which two of
the Angles are equal to the Complement of
Fig. 35. the Latitude in which their Diſtance is required; and therefore if their Difference of Longitude, or Diſtance in the Equinoctial, be given or required, the Triangle is right-angled, as in the Figure;
and there the Angle at A is equal to the Complement
of Latitude, the Perpendicular BC repreſents the Diſtance, and the Hypotenuſe AC the Difference of Longitude.

But when the Diſtance in one Parallel is given, and
the Diſtance in another Parallel is required, it makes an
oblique Triangle, *viz.* with two acute Angles, (becauſe
the Complement of all Latitudes, except Latitude od.
om. *viz.* under the Equinoctial, is leſs than 90 Degrees)
as in the Figure, wherein the Angle at A re-
Fig. 36. preſents the Complement of one Latitude,
and its oppoſite Side BC their Diſtance in that
Parallel, and the Angle at B is equal to the Complement of the other Latitude, and its oppoſite Side AC
repreſents their Diſtance in that Parallel, ſuppoſing each
Ship keeps always under the ſame Meridian; and this
by Oblique Triangles is uſeful only in the two laſt Caſes.

CASE

CASE I.   *Two Ships or Places in one Parallel, their Latitude and Distance given, to find their Difference of Longitude.*

Two Ships in Lat. 50, distant 76 Miles; I demand as above.

1. With the Sine of 90, (or Chord of 60) and one Foot in A, sweep the Arch DE.

2. With the Sine 40 (the Complement of La-   *Fig.* 37. titude) and one Foot in A, sweep the Arch BC.

3. Take the Distance 76 from any equal Parts, and set from B to C, and draw the Line AE through the Point C, to cut the Arch DE in E, and draw AD through B.

4. Draw the Lines BC and DE, then is BC 76 the Distance, and DE 118 the Difference of Longitude.

The same Question by Plane Triangle.

1. Draw the Line AB at pleasure.   *Fig.* 38.

2. At an Angle of 40 Degrees the Complement of Latitude, draw the Hypotenuse AC continued.

3. Draw the Parallel *e g* distant from AB, 76 the Distance.

4. Where that parallel intersects the Hypotenuse AC, as in the Point C, let fall the Perpendicular CB to cut AB in B, and 'tis done.

Then is BC the Distance 76, and AC 118 the Difference of Longitude.

CASE II.   *Latitude and Difference of Longitude given, to find the Distance.*

Two Ships in Latitude 50, their Difference of Longitude 118: I demand their Distance.

See the Figures in Case I.

Bell-Fashion.

1. With the Sine of 90 draw the Arch DE; and with the Sine of 40, the Complement of Latitude, draw the Arch BC, as in Case the first.

2. Set the Difference of Longitude 118, from D to E.

3. Draw the Lines AD and AE, which will cut the Arch BC in B and C.   4. Draw

4. Draw alfo the Lines BC and DE, and 'tis done. The Line BC 76 is the Diftance required.

By Plane Triangle.

1. Draw AB at pleafure.

2. With 40 Degrees, the Complement of Latitude, make the Angle at A, and draw the Hypotenufe AC, upon which fet off the Difference of Longitude 118 from A to C.

3. From C let fall the Perpendicular CB to cut AB in B, and 'tis done.

Then is BC 76 the Diftance required.

CASE III. *Diftance and Difference of Longitude given, to find the Parallel or Latitude.*

A Ship fails due W. 2700 Miles, her Difference of Longitude 4200 min. I demand what Latitude the Ship fails in.

1. With the Sine of 90, and one Foot in A, draw the Arch DE.

Fig. 39.   2. Set the Difference of Longitude 4200 from D to E, and draw the Line DE.

3. Draw the Lines ABD and ACE.

4. Biffect the Line DE in *m*, and draw A *m*.

5. Draw the Lines *i k* and *g b* parallel to A *m*, and diftance from A *m* equal to half the given Diftance 1350.

6. Where thefe Parallels crofs the Lines AD and AE, with the fixed Point of your Compaffes in A draw the Arch BC, and then draw the Line BC.

Then place one Foot in A, and the other in B or C, that Extent applied to the natural Sides will reach to 40, the Complement of Latitude, fo that 50 is the Latitude required.

By Plane Triangle.

1. Draw AB continued.

Fig. 40.   2. Any where as at B, erect the Perpendicular BC, and fet off the Diftance 2700 from B to C.

3. With

3. With the Difference of Longitude 4200, and one Foot in C crofs the Line AB in A, and draw the Line AC.

4. With the Chord of 60, and one Foot in A draw the Arch *d e.*

The Angle at A meafured on the Arch *d e,* is 40 Degrees, the Complement of Latitude required.

CASE IV. *Two Ships failing directly North or South, their Diftance in one Parallel given, to find their Diftance in another Parallel.*

Suppofe two Ships in Latitude 50 North, diftant 200 Miles, they both fail directly North into Latitude 73 North; their Diftance in that Parallel is required.

1. With the Sine Complement of the Latitude in which their Diftance is given (*viz.* Sine Complement of 50, which is Sine of 40) and one Foot in   *Fig.* 41. A, defcribe the Arch DE, and upon that fet off the Diftance in that Parallel 200 from D to E.

2. With the Sine of 90 draw the Arch FG.

3. From A through D and E draw the Line AF and AG.

4. With the Sine-Complement of the other Latitude, *viz.* 73 (which is the Sine of 17) and one Foot in A, draw the Arch BC to cut AF in B, and AG in C, and draw BC.

The Line BC meafured on the fame equal Parts from which you took the firft Diftance DE, will be found to be 91, the Diftance of the Ships in Latitude 73; and if their Difference of Longitude were required, it may be found by meafuring the Line FG.

By Plane Triangle.

Here is given the Angles A and B, the Complement of each Latitude, and the Side BC their Diftance in Latitude 50, to find their Side AC, their   *Fig.* 42. Diftance in Latitude 73: So that here is two Angles and a Side oppofite to one of them given, to find
the

the Side oppofite to the other, by Cafe I. of *Oblique Plane Triangles*, to which I refer you.

CASE V. *Two Ships in one Parallel with their Diftance in that Parallel given, failing both directly North or South, and then their Diftance in the Parallel come to is given, to find the Parallel arrived at.*

Two Ships in Latitude 50, diftant 200 Miles, fail North, till they are but 91 Miles Diftant; I demand the Latitude come to.

1. With Sine-Complement of 50, and one Foot in A, draw the Arch DE, and fet off 200 from D to E, being the given Diftance in that Parallel. *Fig.* 43.

2. Through D from the Point A, draw the Line AF, and from A through E draw AG, having firft drawn the Arch FG with the Sine 90.

3. Biffect FG in *m*, and draw the Parallels *g b* and *i k*, diftant from A *m* equal to half the Diftance in the Latitude come to, *viz.* half 91, which is 45d. ½

4. Obferve where thefe Parallels cut the Line AF and AG, as in B and C, therefore draw the Arch BC, and the Line BC, and 'tis done.

Then AB, or AC, taken in your Compaffes, and applied to the Sines, will reach to the Sine of 17, the Complement of the Latitude required; fo that the Latitude come to is 73.

By Plane Triangle. See the Triangle in Cafe IV. in which there is given the Side BC 200, their Diftance in Latitude 50, and the Angle oppofite, *viz.* at A the Complement of that Latitude: There is alfo given the Side AC, their Diftance in the Latitude come to, to find the Angle at B the Complement of the Latitude come to; fo that here are two Sides and an Angle oppofite to one of them given, to find the Angle oppofite to the other by Cafe II. *Oblique Plane Trigonometry Geometrical.*

SECT.

SECT IX.

Middle Latitude Sailing Geometrical.

THERE are innumerable Ways of projecting *Middle Latitude Sailing.* I shall not trouble the Reader with them all, becaufe many of them are altogether ufelefs and impracticable at Sea; I shall therefore only infert thofe Methods that may ferve as a Demonftration and Proof of the Proportions laid down in the Trigonometrical Part of *Middle Latitude Sailing,* and omit the reft; being unwilling to put the Reader to the Charge of any Thing but what may be ufeful and profitable.

The firft Proportion mentioned in *Middle Latitude Sailing Trigonometrical,* is, As the Difference of Latitude is to the Difference of Longitude; fo is Sine-Complement of Middle Latitude, to the Tangent of the Courfe. Or (which is the fame) As the Difference of Latitude is to the Sine Complement of Middle Latitude, fo is the Difference of Longitude, to the Tangent of the Courfe.

The firft Queftion there, is in Cafe I. *One Latitude, Courfe, and Diftance given, to find the other Latitude, Departure and Difference of Longitude.*

A Ship in Latitude 50d. om. North, fails N. E. by N. 987 Miles: I demand the Latitude come to, with Departure and Difference of Longitude.

Note; I shall inftance, in this Queftion, Examples of all the different Ways that I shall infert for projecting *Middle Latitude Sailing,* and for demonftrating the Proportions generally ufed therein, fuppofing it too tedious to infert all the different Projections and Demonftrations in every Cafe thereof, and firft for the Proportion above-mentioned.

Geometrical Conftructions.

Draw the North and South Line AB, and with the given Courfe and Diftance lay down the Triangle ABC,

as in *Plane Sailing Geometrical,* Case I. and hav-
Fig. 44.   ing thus found the Difference of Latitude, and
consequently the Latitude come to, and the Middle Latitude; set off the Sine Complement of the Middle Latitude from A to *b,* and set off the Radius or Sine
of 90, from A to *e,* and erect the Perpendiculars *b i* and
e g parallel to BC, and continue BC at pleasure: Then
observe where the Hypotenuse AC cuts the Perpendicular *e g* which is erected from the Sine of 90) as is *k*; then
take in your Compasses the Extent *e k,* and place from
b to *i,* and from A through *i* draw the Line AD to cut
the Perpendicular BD in D; then is BD 1093 the Difference of Longitude required, and BC 548 is the Departure, and AB 821 is the Difference of Latitude, the
Hypotenuse AC 987 the Distance, and the Angle BAC
33d. 45m. is the Course from the North Eastwards, or
N.E. by N.

Now that this is a demonstrative Proof of the Analogy or Proportion before mentioned, is evident; for
as AB the Difference of Latitude, is to A *b* the Sine
Complement of Middle Latitude; so is BD the Difference of Longitude, to *b i,* which by Construction is always equal to *e k,* the Tangent of the Course: And that
it is, As AB to A *b,* so BD to *b i,* See *Euclid. Lib.* 6.
Prop. 4. Therefore as the Difference of Latitude is to
the Sine Complement of Middle Latitude, so is the Difference of Longitude to the Tangent of the Course.

Another Proportion is, as the Sine Complement of
the Middle Latitude is to Radius; so is the Departure,
to the Difference of Longitude, which may easily be
projected and demonstrated in one Right-angled Triangle, to which if we affix another Right-angled Triangle; which is common in *Plane Sailing,* (viz. as Radius,
to the Distance; so is Sine of the Course, to the Departure) making the Perpendicular, which is the Departure common to both, you will thereby have constituted an Oblique Triangle, whose Proportion between
its opposite Sides and Angles plainly proves the third
Proportion,

Proportion, *viz.* As Sine Complement of Middle Latitude, is to the Sine of the Courfe ; fo is the Diftance to the Difference of Longitude : Or (which is the fame) As Sine Complement of Middle Latitude is to the Diftance ; fo is the Sine of the Courfe, to the Difference of Longitude. I fhall inftance in the firft Cafe beforementioned, where one Latitude, Courfe and Diftance is given, to find the other Latitude, Departure and Difference of Longitude.

A Ship in Latitude 50d. 0m. fails N. E. by N. 987 Miles ; I demand as above.

With the Courfe 33d. 45m. and Diftance 987, lay down the Triangle ABC, as in *Plane Sailing,* Cafe I. and having found the Difference of *Fig.* 45. Latitude, and confequently Middle Latitude, which in this Queftion is 56d. 51m. therefore draw the Line CD to make an Angle of 56d. 51m. with the Line CB, and draw the faid Line CD, till it cut the Line AB continued, then is CD 1003 the Difference of Longitude required. Becaufe if the Angle BCD be the Middle Latitude, the Angle BDC muft needs be the Complement of Middle Latitude, becaufe DBC is a Right-Angle : And the Side BC being the Departure, is common to both Triangles ; therefore in the Triangle ABC, it is, as Radius is to the Diftance AC ; fo is the Sine of the Courfe BAC, to the Departure BC ; and then it muft needs be alfo, As Sine of BDC, the Complement of Middle Latitude, to its oppofite Side BC the Departure, fo is Radius, to the Difference of Longitude DC.

Hereby is alfo projected by Confequence a Demonftration of that Proportion ; As Sine Complement of Middle Latitude, is to the Diftance ; fo Sine of the Courfe, to the Difference or Longitude : Or as it is expreffed in the Beginning of *Middle Latitude Sailing Trigonometrical,* as Sine Complement of Middle Latitude is to the Sine of the Courfe ; fo is the Diftance, to the

D   Difference

Difference of Longitude. For by these two Right-angled Triangles, with one Perpendicular common to both, is constituted the Oblique Triangle ACD, in which the Angle DAC is the Course, and its opposite Side DC is the Difference of Longitude, and the Angle ADC is the Complement of Middle Latitude, and its opposite Side AC is the Distance; and hence it must needs follow by the known Proportion of Sides to their opposite Angles, and the contrary, that as Sine of ADC, is to AC; so is Sine of DAC, to DC: Therefore, as Sine Complement of Middle Latitude, is to the Distance; so is the Sine of the Course, to the Difference of Longitude.

These Methods of projecting Middle Latitude being the most demonstrative Proofs of the Analogies or Proportions generally made Use of in the Trigonometrical or Arithmetical Calculation thereof, and therefore the most useful, I shall add one or two more, although there are almost infinite Methods for doing it; as one Way that is commonly used by Semicircle, a Way very much in Practice, of which I shall instance in the foregoing Case. A Ship sails from Latitude 50d. om. North, and sails N. W. by N. 987 Miles; I demand the Latitude come to, with Departure and Difference of Longitude.

With the Chord of 60, and one Foot in H, draw the Semicircle AME, and at one End thereof, as at *Fig.* 46. A, make an Angle of 33d. 45m. the Course given, and lay down the Triangle ABC, with the Course and Distance, as in *Plane Sailing*; and having found the Middle Latitude, which in this Question is 56d. 51m. Set off the Chord thereof both Ways, from M to I and K, and draw I K to cut H M in L; then set the Distance L M from A to D, and from B to E, and continue the Line M H till it cut the Line AC as in N; then from E erect the Perpendicular E G, and from D through N draw D G to cut E G in G; then is E G the Difference of Longitude, B C the Departure, A C the Distance, A B the Difference of Latitude, and the Angle CAB the Course.

Another

Another Projection of the same Question.

With the given Course and Distance lay down the Triangle ABC, as in *Plane Sailing, Case* I. and
Fig. 47. having found the Middle Latitude, as before,
set off the Sine of the Complement of Middle
Latitude upon the Line AB, from A to D, and sweep
the Arch DE, then with the Departure BC in your Compasses, and one Foot in D, extend the other along the
Arch DE to E; and draw the Line DE equal to BC,
and through E draw AE continued: and then with the
Sine of 90, and one Foot in A, draw the Arch GH, till
it cut the Line AE continued in H, and draw the Line
GH, which measured on the same equal Parts from
which the rest were projected, gives the Longitude required.

These four Methods are sufficient to entertain the
Reader with Variety; and I suppose it altogether needless to instance in any more Cases; it being very easy,
from what is given in any other Case, whether both Latitudes and Course, or both Latitude and Difference of
Longitude, &c. to make the same Projection, as a little
Practice will make evident.

CHAP.

C H A P. II.

Navigation Trigonometrical.

S E C T. I.

PLANE TRIGONOMETRY.

IN the Solution of all Plane Triangles, whether right angled or oblique, there are always three Things given to find a fourth, suppofing the Radius in right-angled Triangles to be always given or known; which three given Terms are always the firft in the Proportion, and the required Term is always the fourth or laft; and they are always to be placed in fuch an Order, that the Proportion may be, as the firft Term is to the fecond; fo the third to the fourth. And before the admirable Invention of Logarithms, the Method was, as in other Cafes in the Rule of Three to multiply the two laft Numbers by each other, and divide the Product by the firft Number, and the Quotient was the Anfwer to the Queftion; but by the Help of Logarithms the Work is much facilitated: For having fet down the Numbers in their proper Order, and their Logarithms againft them, as you fee in the following Examples in *Plane Sailing*, you have no more to do, but add the two laft Logarithms of the three that are given, and from their Sum fubtract the firft, the Remainder is the Logarithm of the fourth Number required. And again, obferve in any three given Terms, where the Radius is the firft Term, you need but add the two laft toge-ther, and from the Sum abate one towards the Left-hand, and the Remainder is the Logarithm of the fourth Number required. But if Radius be the fecond or third Term, place one on the Left-hand of the faid third or fecond Term, not Radius, and from that Sum fubtract the firft Term, the Remainder is the Logarithm

of

of the fourth Term required. But if none of the Terms
be Radius, as it frequently happens in Oblique Plane
Triangles, but that every Term be the Logarithm,
Sine or Tangent of some intermediate Degree or Number,
then instead of the Logarithm of the first Term,
whether Number, Sine, or Tangent, &c. take its Complement
Arithmetical, which is done thus: Begin at
the Left-hand, and subtract each Number from 9, and
set down the Remainder, till you come at the Figure
next the Right-hand, which subtract from 10, and set
down the Remainder as before. This Number thus
found is called the Complement Arithmetical of the
Logarithm from whence it was taken, and is to be set
down against the first Term, instead of the Logarithm
itself, and the Logarithm of the second and third Terms
in their proper Order under it; which being done, add
up all the three Sums together, and from their Sum
subtract the Radius or 10 towards the Left-hand, the
Remainder is the Logarithm of the fourth Term sought.

Plane Trigonometry right-angled.

CASE I. *Hypotenuse and one Acute Angle given (consequently
both) to find a Leg. See Plane Trigonometry
Geometrical.* Case I.

Given $\begin{cases} \text{Hypotenuse} \longrightarrow 550 \\ \text{The Angle at A} - 35\text{d. om} \end{cases}$ $\begin{cases} \text{Required AB} \\ \text{and BC.} \end{cases}$

1. Making the Hypotenuse the Radius, the Leg BC
is the Sine of the Angle at A, and the Leg AB is the
Sine Complement of the said Angle by the Explanation
of Axiom the first, therefore the Proportion is, as Radius
is to the Hypotenuse; so is the Sine Complement of
the Angle at A, to the Leg AB.

Again, As the Radius to the Hypotenuse; so is the
Sine of the Angle at A, to Leg BC.

2. Making the Base Radius, the Perpendicular BC is
the Tangent of the Angle at the Base, and the Hypotenuse
the Secant of the said Angle. By Explanation
of Axiom the first, therefore the Proportion is, As the
Secant

Secant of the Angle at A, is to the Hypotenuse, fo is the Radius, to the Leg AB.

Again, As the Secant of the Angle at A, to the Hypotenuse; fo is the Tangent of the faid Angle at A, to the Leg BC.

Note, Making the Perpendicular Radius, is the fame as making the Bafe Radius, the Hypotenuse ftill continuing to be a Secant.

In this Part I have purpofely omitted the Operations by Logarithms, they being inferted at large in *Plane* and *Mercator's Sailing*, which is the Application of *Plane Trigonometry* to *Navigation*; but the Things given and required are the fame in both. For whereas in *Cafe* I. of *Plane Trigonometry Right-angled*, there is given the Angles and Hypotenuse to find a Leg, fo in *Plane Sailing*, *Cafe* I. there is given the Courfe (which is the Angle at the Bafe) and confequently its Complement (the other acute Angle) and the Diftance (which is the Hypotenuse) to find either Leg or both (which are Difference of Latitude and Departure) and the Rules for Calculation are the fame.

CASE II. *Given the Angle at* A, *and a Leg* AB, *to find the Hypotenuse.* See *Plane Trigonometry Geometrical,* Cafe II.

1. Making the Hypotenuse Radius, the given Leg will be the Sine Complement of the Angle at A by the Explanation of Axiom the firft; therefore, As the Sine Complement of the Angle at A is to the Bafe; fo is Radius, to the Hypotenuse,

2. Making the Bafe Radius, the Hypotenuse is Secant of the Angle at A; therefore, as Radius is to the Bafe; fo is the Secant of the Angle at A, to the Hypotenuse.

CASE III. *Given the Angles and a Leg, to find the other Leg. Given the Leg* AB, *as before, required the Leg* BC.

Note, Each Cafe here always refers to the fame Cafe in *Plane Trigonometry Geometrical,* both in Right-angled and Oblique.

1. Making

1. Making the Hypotenuse Radius, the given Leg is Sine Complement of the Angle at A, and the required Leg BC is the Sine of the same Angle; therefore, as Sine Complement of the Angle at A, is to the Leg AB; so is the Sine of the Angle at A, to Leg BC.

2. Making the Base Radius, the Leg BC is the Tangent of the Angle at A; therefore, as Radius is to the Leg BA; so is the Tangent of the Angle at A, to the Leg BC.

Note, Both these Cases, the second and third, are contained in *Case* II. of *Plane Sailing,* where the Angles and a Leg (*viz.* Course and Difference of Latitude) are given, to find the Distance (which is the Hypotenuse) by Case Second, and the other Leg (which is the Departure) by Case the third. They are also both comprehended under the third Case of *Plane Sailing,* there being also the Angles and a Leg given, *viz.* Course and Departure, to find Distance and Difference of Latitude.

CASE IV. *Given the Hypotenuse and a Leg, to find an Angle, Given Hypotenuse* AC *and Leg* AB, *required* BAC

1. Making the Hypotenuse Radius, the Leg AB will be the Sine Complement of the required Angle, therefore, as the Hypotenuse, is to Radius; so is the Leg AB, to Sine Complement of the Angle at A.

2. Making the Base Radius, the Hypotenuse is the Secant of the Angle at A, therefore as the Base AB, is to Radius; so the Hypotenuse AC is to the Secant of the Angle at A.

CASE V. *Given the Hypotenuse and a Leg, to find the other Leg. Given* AC *and* AB, *required* BC.

1. Making the Hypotenuse Radius, the required Leg BC is the Sine of the Angle at A; therefore having found the Angle at A by Case the fourth, the Proportion is, as Radius is to Hypotenuse; so is the Sine of the Angle at A, to the Leg BC required.

2. Making

2. Making the Base Radius, the required Leg BC is the Tangent of the Angle at A, therefore having found the Angle at A, as before, the Proportion is, as Radius, to the Base, so is the Tangent of the Angle at A, to Leg BC required.

These two Cases are jointly comprehended in the fourth and fifth Cases of *Plane Sailing*, in each of which there is given the Hypotenuse and a Leg, to find the other Leg and the Angles.

Note, It is usual, in most Cases in *Plane Sailing*, to work after the Method here delivered, making the Hypotenuse Radius, because that Way of stating brings it under that known Proportion of opposite Sides to opposite Angles.

CASE VI. *The Legs given to find the Angles. Given BA and BC, required BAC.*

Note, When a Side is required, any Side may be made Radius; but to find an Angle, a given Side must be made Radius; and here being obliged to find an Angle before we can find the Hypotenuse, we shall make the Leg AB Radius, and then BC is the Tangent of the Angle at A. Therefore as AB, is to Radius; so is BC, to the Tangent of the Angle at A required.

CASE VII. *Given the Legs, to find the Hypotenuse, Given AB and BC, to find AC.*

First, Find the Angle at A as before, and then,

1. Making the Hypotenuse Radius, the Base AB is the Sine Complement, and the Perpendicular BC is the Sine of the Angle at A; therefore, as Sine Complement of the Angle at A, is to the Base AB; so is the Radius, to the Hypotenuse AC required. Or, as Sine of the Angle at A, is to the Leg BC; so is Radius to the Hypotenuse required.

2. Making Base Radius, and having found the Angle at A, it is, as Radius, is to the Base; so is the Secant of the Angle at A, to the Hypotenuse.

These

Thefe two Cafes are both contained under the fixth Cafe of *Plane Sailing*; and if you carefully obferve the Rules laid down in the Beginning of *Plane Sailing Geometrical*, about the Application of Plane Triangles to Queftions in Navigation, the Connection between them will appear fo plain, and the Directions given in one will be fo plain and pertinent in the other, that there needs no more to be faid to inftruct the Reader as to the Solution of the Right-angled Plane Triangles.

SECT. II.

Of Oblique Plane Triangles.

I Shall be a little more particular in this Section of Oblique Triangles, becaufe we have them not anfwered numerically elfewhere in this Book, except in fome Queftions at the latter End.

CASE I. *Two Angles and a Side oppofite to one of them given, to find the Side oppofite to the other.* Fig. 12.

Given $\begin{cases} \text{The Angle at A} - 30\text{d. om.} \\ \text{The Angle at B} - 45\text{d. om.} \\ \text{The Side BC} - 290 \end{cases}$ Required the Side AC.

By *Axiom the Second.*

As Sine of A ——— 30d. om. *Co. Ar.* 0.30103
To Side oppofite BC ——— 290 ——— 2.46239
So Sine of B ——— 45 00 ——— 9.84948

To Side oppofite AC required 410 ——— 2.61290

Note, Here being no Radius in this Proportion, I have taken the Complement Arithmetical of the firft Logarithm, and added all the three Sums together, omitting one to the left Hand, according to the Directions in the Beginning hereof.

CASE

C'A S E II. *Two Sides and one Angle oppofite to one of them given, to find the Side oppofite to the other.* Fig. 13.

Given $\left\{\begin{array}{l}\text{The Side AB — 560}\\ \text{The Side AC — 410}\\ \text{The Angle B 45d. om.}\end{array}\right\}$ Required the Angle at C.

By Axiom the Second.

			Co. Ar.
As Side AC	410	—	7 38722
To Sine of the Angle oppofite 45	0	—	9.84948
So Side AB	560	—	2.74818

To Sine of the Angle oppofite 74d. 58m. — 9.98488

Note, This Anfwer is ambiguous, and therefore it is neceffary firft to know whether the Angle fought be a cute or obtufe, which if acute, it is 74d. 58m. but if obtufe, it is found by fubtracting 74d. 58m. from 180, the Remainder 105d. 2m. is the obtufe Angle required.

C A S E III. *Given two Sides, and an Angle oppofite to one of them, to find the third Side.* Fig. 13.

Given $\left\{\begin{array}{l}\text{The Side AB — 560}\\ \text{The Side AC — 410}\\ \text{The Angle at B 45d. om.}\end{array}\right\}$ Required the Side BC.

Find the Angle at C, as in Cafe II. which, if fuppofed acute, is 74d. 58m. then the two Angles B and C being given, the third is found by Subtraction to be 60d. 2m. Then by Axiom the fecond.

	d.	m.	Co. Ar.
As the Sine of B	45	0	0.15052
To Side oppofite AC	410		2.61278
So Sine of A	60	2	9.93767
To Side oppofite	502		2.70097

But if the Angle at C had been fuppofed obtufe, *viz.* 105d. 2m. the Angle at A would have been 29d. 58m. and then the Operation is,

As

$$\text{d. m.}$$

As Sine of B ——— —— 45   0   *Co. Ar.* 0.15052
To Side oppofite AC ——— 410 ——— 2.61278
So Sine of A ——— —— 29 58 ——— 9.69853

To Side oppofite BC——— 290 ——— 2.46183

C A S E IV.   *Given Two Sides and a contained Angle, to*
  find either of the other Angles. Fig. 14.

Given { The Side AC — 410 } Required the Angles
      { The Side AB — 560 }   B and C.
      { The Angle A 30d.0m. }

By Axiom the Third.

As Sum of the given Sides — 970 *Co. Ar.* 7.01323
To their Difference ——— 150 ——— 2.17609
So Tangent of half Sum of un-}
 known Angles ——— —— } 75d. 0m. — 10.57194

To Tang. of half their Difference 29 59 — 9.76126
 The Sum is the greater Angle C —— 104d. 59m.
 The Difference is the leffer Angle B —— 45   1

C A S E V.   *Two Sides and a contained Angle given, to*
  find the third Side. Fig. 15.

Given { The Side AC — 410 }
      { The Side AB — 560 } Required the Side BC.
      { The Angle A 30d. 0m. }

Find the Angle B by *Cafe* IV. *viz.* 45d. 1m. Then,

$$\text{d. m.}$$

As the Sine of B ——— 45   1   *Co. Ar.* 0.15039
To the Side oppofite AC — 410 ——— 2.61278
So Sine of A ——— —— 30   0 ——— 9.69897

To Side oppofite BC——— 290 ——— 2.46214

C A S E

CASE VI. *Three Sides given, to find an Angle.* Fig. 15.

Given $\begin{cases} \text{The Side AC } 410 \\ \text{The Side CB } 290 \\ \text{The Side AB } 560 \end{cases}$ Required the Angle at A.

Let fall a Perpendicular from C the greateſt Angle, upon AB the greateſt Side, by Axiom the fourth.

		Co. Ar.
As the Baſe AB	—— —— ——	560 — 7.25182
To the Sum of the Sides AC 410, and CB 290 —— ——	$\Big\}$	—700 — 2.84509
So the Difference of the ſaid Sides	——	120 — 2.07918
To the Difference of the Segments of the Baſe AD —— ——	$\Big\}$	—150 — 2.17609

The half Difference 75, added to half the Baſe 280, the Sum 355 is the Baſe AE.

The half Difference 75, ſubtracted from half the Baſe 280, the Difference 205 is the Baſe EB.

Then in the Right-angled Triangle AEC is given AE 355, AC 410, to find the Angle A.

		Co. Ar.
As AC	—— —— ——	410 — 2.61278
To Radius	—— —— ——	90d. om. 10.00000
So AE	—— —— ——	355 — 2.55021

To Sine Comp. of the Angle at A 30   1 — 9.93744

After the ſame Manner you may find the Angle E.

SECT III.

Plane Sailing Trigonometrical.

ALthough the Method here propoſed be new and ſhort without the Trouble of Logarithms, Sines, Tangents and Secants; yet I ſhall by the Way ſet down

an

an Example or two in every Cafe, both in *Plane Sailing* and *Mercator*, according to the old and common Method, partly for the Help of thofe, who either through Negligence or Want of Opportunity to practife, have forgot at Sea what they have learned at School; but chiefly, that in all the Examples wrought by the new Way; I may refer to thofe wrought by the old Way, that fo the Reader, feeing the exact Harmony and Concord which is between them, may (to his great Satisfaction) be convinced of the Verity and Infallibility of this new Method, equal to the beft for Truth and Certainty, and fuperior to all for Expedition and Readinefs.

Plane Sailing.

CASE I. *Courfe and Diftance given, to find Difference of Latitude, and Departure.*

A Ship fails N. W. by N. 123 Miles; I demand Difference of Latitude and Departure.

Firft, For the Difference of Latitude.

	d.	m.	
As Radius ———————	90	00	10.00000
To the Diftance failed ————		123	2.08990
So Sine Comp. of the Courfe — 56	15		9.91985
To the Difference of Latitude —	102½		2.00975

For the Departure.

	d.	m.	
As Radius ———————	90	00	10.00000
To Diftance failed ———— ——		123	2.08990
So Sine of the Courfe ———— 35	45		9.74474
To Departure —————		68	1.83464

Example 2. A Ship fails South 25 Degrees Eafterly 96 Miles, I demand the Difference of Latitude and Departure.

For the Difference of Latitude.

	d.	m.		
As Radius —————————	90	00	—	10.00000
To Diſtance ſailed ———— ———		96	—	1.98227
So Sine Comp. of Courſe——	65	00	—	9.95728
To Diff. of Latitude —— ——		87	—	1.93955

For the Departure.

	d.	m.		
As Radius ——————— ——	90	00	—	10.00000
To Diſtance ſailed ——— ———		96	—	1.98227
So Sine of the Courſe ————	25	00	—	9.62595
To Departure ——— ———		40½	—	1.60822

CASE II. *Courſe and Difference of Latitude given, to find Diſtance and Departure.*

A Ship ſails S. W. by S. till her Difference of Latitude be 174 Miles, I demand her Diſtance and Departure.

For the Diſtance.

	d.	m.		
As Sine Comp. of Courſe——	56	15	—	9.91985
To Difference of Latitude ——		174	—	2.24055
So Radius ——— —————	90	00	—	10.00000
To the Diſtance ——— ——		209	—	2.32070

For the Departure.

	d.	m.		
As Sine Comp. of Courſe——	56	15	—	0.08015
To Difference of Latitude——		174	—	2.24055
So Sine of the Courſe ———	33	45	—	9.74474
To Departure ——— ———		116	—	2.06544

Example 2. A Ship ſails North 38 Degrees Weſterly, till ſhe raiſe the Pole 2 Degrees: I demand how far ſhe hath ſailed, and how much ſhe is departed from her firſt Meridian.

The 2 Degrees of Latitude reduced into Miles, is 120 Miles. Then for the Diſtance.

A

```
                              d.   m.
As Sine Comp. of Courfe —  52   00   —   9.89653
To Difference of Latitude ——  120   —   2.07918
So Radius —— ——— ——— 90   00   —   10.00000
                                         ————————
To the Diftance —— ——  152   —   2.18265
```

For the Departure.

```
                              d.   m.         Co. Ar.
As Sine Comp. of Courfe —  52   00   —   0.10347
To Difference of Latitude ——  120   —   2.07918
So Sine of the Courfe ——  38   00   —   9.78934
                                         ————————
To Departure —— ——  94   —   1.97199
```

CASE III. *Courfe and Departure given, to find Diftance and Difference of Latitude.*

A Ship fails S.E. by S. till her Departure be 103 Miles; I demand her Diftance and Difference of Latitude.

For the Diftance.

```
                              d.   m.
As Sine of the Courfe ——  33   45   —   9 74474
To Departure —— ——  103   —   2.01284
So Radius —— ——  90   00   —   10.00000
                                         ————————
To the Diftance —— ——  185   —   2 26810
```

For the Difference of Latitude.

```
                              d.   m.         Co. Ar.
As Sine of the Courfe ——  33   45   —   0.25526
To Departure —— ——  103   —   2.01284
So Sine Comp. of Courfe ——  56   15   —   9.91984
                                         ————————
To Difference of Latitude ——  154   —   2.18794
```

Example 2. A Ship fails North 19 Deg. 41 Min. Eafterly, which is NNE.¼ Northerly, till her Departure be 72 Miles; her Diftance and Difference of Latitude is required.

For

For the Distance.

	d.	m.		
As Sine of the Course	19	41	—	9.52739
To Departure		72	—	1.85733
So Radius	90	00	—	10.00000
To the Distance		214	—	2.32994

For the Difference of Latitude.

	d.	m.		Co. Ar.
As Sine of the Course	19	41	—	0.47260
To Departure		72	—	1.85733
So Sine Comp. of Course	70	19	—	9.97385
To Difference of Latitude		201	—	2.30378

CASE V. *Distance and Difference of Latitude given, to find Course and Departure.*

A Ship sails between the North and East 110 Miles, and then finds by Observation that she hath raised the Pole one Degree; I demand the Course and Departure.

The Difference of Latitude reduced to Miles or Minutes is 60.

Then for the Course.

As the Distance		110	—	2.04139
To Radius	90d.	00m.	—	10.00000
So Difference of Latitude		60	—	1.77815
To Sine Comp. of Course	33	03	—	9.73676

So that the Course is N.E. by E. nearest, or N. E. by E. 42 Min. Easterly.

For Departure.

	d.	m.		
As Radius	90	00	—	10.00000
To the Distance		110	—	2.04139
So Sine of Course	56	57	—	9.92334
To Departure		92	—	1.96473

Example

Example the Second.

A Ship fails between the North and East, till her Difference of Latitude be 103 Miles, and then is diftant from the Place fhe fail'd 117 Miles, her Courfe and Departure is required.

For the Courfe.

As the Diftance————————— 117 — 2.06818
To Radius ——————— 90d. 00m. — 10.00000
So Difference of Latitude———— 103 — 2.01284
———————
To Sine Comp. Courfe ——— 61 41 — 9.94466

For the Departure.

        d. m.
As Radius—————————— 90 00 — 10.00000
To the Diftance ——————— 117 — 2.06818
So Sine of the Courfe ——— 28 19 — 9.67609
———————
To Departure ——————— 55¼ — 1.74427

The Courfe is N.N.E. fomewhat more than half a Point Eafterly, and the Departure 55¼ Miles.

CASE V. *Diftance and Departure given, to find Courfe, and Difference of Latitude.*

A Ship fails between the South and Weft 124 Miles, her Departure 95 Miles: I demand the Courfe and Difference of Latitude.

For the Courfe.

As the Diftance————————— 124 — 2.09342
To Radius ——————— 90d. 00m. — 10.00000
So Departure ——————— 95 — 1.97772
———————
To Sine of the Courfe ——— 50 00 — 9.88430

E   *For*

For the Difference of Latitude.

		d.	m.	
As Radius ——	——	90	00	10.00000
To the Distance ——	——	124		2.09342
So Sine Comp. of Course —	40	00		9.80806
To Difference of the Latitude ——		80		1.90148

The Course is S.W. half Point Westerly, or 4½ Points nearest, the Difference of Latitude 80 Miles.

Example the Second.

A Ship sails North Easterly 100 Miles; till her Departure be 38 Miles, her Course and Difference of Latitude is required.

For the Course.

As the Distance ——	——	100	——	2.00000
To Radius ——	90d. 00m.		——	10.00000
So the Departure ——	——	38	——	1.57978
To Sine of the Course ——	22	20	——	9.57978

For the Difference of Latitude.

As Radius ——	——	90d. 00m.	——	10.00000
To the Distance ——	——	100	——	2.00000
So Sine Comp. of Course —	67	40	——	9.96613
To Difference of Latitude ——	92½		—	1.96613

The Course N.N.E. *fere*; and the Difference of Latitude 92½.

CASE VI. *Difference of Latitude and Departure given, to find the Course and Distance.*

A Ship sails between the North and West, till her Difference of Latitude be 220 Min. and Departure 108 Min. or Miles; I demand the Course and Distance.

For

For the Course.

As Difference of Latitude —— 220 —— 2.34242
To Radius ——— 90d. 00m. —— 10.00000
So Departure ——— 108 —— 2.03342
_____
To Tangent of the Course 26 9 —— 9.69100

For the Distance.

As Sine of the Course — 26d. 9m. —— 9.64416
To Departure ——— 108 —— 2.03342
So Radius ——— 90 00 —— 10.00000
_____
To the Distance ——— 245 —— 2.38926 .

The Course N.N.W. somewhat more than ¼ Westerly, the Distance 245 Miles.

Example the Second.

A Ship sails between the South and West, until her
Difference of Latitude be 309 Min. and Departure 206
Min. I demand the Course and Distance.

For the Course:

As Diff. of Latitude ——— 309 —— 2.48995
To Radius ——— 90d. 00m. —— 10.00000
To Departure ——— 206 —— 2.31386
_____
To Tangent of the Course 33 41 —— 9.82391

For the Distance.

As Sine of the Course — 33d. 41m. —— 9.74398
To the Departure ——— 206 —— 2.31386
So Radius ——— 90 00 —— 10.00000

To the Distance ——— 371 —— 2.56988

E 2   The

The Courfe S.W. by S. *fere,* the Diftance 371 Miles.

Thus have I fet down two Examples in every Cafe of Plane Sailing, according to the old and common Way of working by Logarithms, Sines, and Tangents, having obferved in the three firft Cafes where the Courfe is given, to fet one Queftion with the Courfe given in Points of the Compafs as N.W. by N. &c. and another with the Courfe given in Degrees, as South 25 Degrees Eafterly, &c. intending in the Sequel hereof, to proceed to fhew the Reader how to anfwer all the fame Queftions, (where the Courfe is either given or required) in Points, Half-points, or Quarter-points of the Compafs, without any Canon, Scale or Compaffes, only by two Tables, each of which (or both if need require) might be contain'd in one Page hereof; which Table you have at the Beginning of the fecond Part of this Book; by which Tables only you may work all the Cafes in Plane-Sailing, and keep a Journal by that way, without a Traverfe Table, or the voluminous Tables of Logarithms, Sines, Tangents, &c. which fill up the far greater Part of fome of our Epitomies of Navigation now in Print: You may alfo by the fame, and a Table of Meridional Parts, work all the Cafes in *Mercator*; and keep a Reckoning throughout all the World; and not only fo, but that the Work may juftly deferve the Title of a New Method, I fhall add fome Rules teaching how to work any Queftion in Plane Trigonometry, or Navigation, only by the Pen or a Piece of Chalk, without any Geometrical Projection by Scale and Compaffes, or Arithmetical Calculation by Sines and Tangents; fo that if at Sea you have loft all Books and Inftruments, or if a Shore you be in Company where you have no Books nor Inftruments prefent, you fhall work any Queftion as eafily as a Queftion in Common Arithmetic, a Thing never yet known, nor publifh'd by any: But I fhall now proceed to propofe an Example, or two, in the feveral Cafes of,

S E C T.

S E C T. IV.

Mercator's *Sailing Trigonometrical.*

CASE I. *One Latitude, Course and Distance given, to find the other Latitude, and Difference of Longitude.*

IN this, and the other Cases, we shall have Occasion to make Use of the Meridional Difference of Latitude; for the finding of which, observe, if two Places (whose Meridional Difference of Latitude you seek) be both in North Latitude, or both in South Latitude, subtract the Meridional Parts of the lesser Latitude, from the Meridional Parts of the greater Latitude, the Remainder is the Meridional Difference of Latitude; but if one be in North Latitude, and the other in South, add the Meridional Parts of both Latitudes together, the Sum is the Meridional Difference of Latitude; the following Examples will make it plain.

A Ship in Latitude of 50d. 00m. North, sails N.W. by N. 987 Miles; I demand the Latitude come to, as also the Difference of Longitude.

As Radius ———————— 90d. 00m. ——— 10.00000
To the Distance ——— ·— 986 ——— 2.99432
So Sine Comp. of Course — 56   15 ——— 9.91984

To proper Diff. of Lat. ——— 820 ——— 9.91416

The Difference of Latitude 820 min. divided by 60, to bring it into Degrees, is 13d. 40m. which added to 50, Deg. (because the Ship sails towards the Pole, and increases her Latitude) the Sum 63d. 40m. is the Latitude come to. Then to find the Meridional Difference of Latitude between Lat. 63d. 40m. and Lat. 50d.

　　　　　　　　Meridional

Meridional Parts for Lat. 63d. 40m. ——— 4994
Meridional Parts for Lat. 50 00 ——— 3475

Subt. one from t'other, refts Merid. Diff. Lat. — 1519

Then for the Difference of Longitude.

	d. m.	
As Radius —— ——	90 00	— 10.00000
To Merid. Diff. of Latitude —	1519	— 3.18155
So Tangent of the Courfe——	33 45	— 9.82489
To Difference of Longitude —	1015	— 3.00644

The Latitude come to is 63d. 40m. and Diff. Longitude 1015m. which divided by 60, is 16d. 55m. Weft Longitude, becaufe the Courfe is Wefterly.

CASE II. *Both Latitudes and Courfe given, to find the Diftance and Difference of Longitude.*

A Ship fails from a Port in Latitude 56d. 25m. N. and Longitude 14d. 12m. W. and fails away N.N.W. feveral Days, and then finds herfelf by Obfervation to be in Latitude 68d. 30m. I demand how far fhe hath failed, and what is her Difference of Longitude.

	d. m.	
Meridional Parts for ——	68 30	——— 5712
Meridional Parts for ——	56 25	——— 4119
Meridional Difference of Lat.	——	1593
Proper Difference of Latitude	12 05, or ——	725

Then for the Diftance.

	d. m.	
As Sine Comp. of Courfe —	22 30	— 9.99561
To proper Diff. of Latitude —	725	— 2.86033
So Radius —— ——	93 00	— 10.00000
To the Diftance —— ——	785	— 2.89472

For

For the Difference of Longitude.

	d.	m.	
As Radius —— ——	90	00	— 10.00000 ·
To Meridian Diff. Latitude ——1593		—	3.20221
So Tangent of the Course — 22	30	—	9.61722

To Diff. of Longitude —— 660 — 2.81943

The Distance is 785 Miles, or 262 Leagues almoft, and the Difference of Longitude is 660 Min. or 11 Degrees Eaft, which added to 14d. 12m. the Longitude failed from, the Sum 25d. 12m. is the Longitude the Ship is in, which being found, find in a Mercator's Chart the Longitude fo found, and the Latitude by Obfervation 68d. 30m. and that is the Point in the Chart that your Ship is in.

CASE III. *One Latitude, Courfe and Difference of Longitude given, to find the other Latitude and Diftance.*

A Ship in Latitude 50, fails N.N.W. till her Difference of Longitude be 7 Degrees, or 420 Min. I demand the other Latitude and Diftance.

For the other Latitude.

	d.	m.	
As Tangent of the Courfe —— 22	30	—	9.61722
To Difference of Longitude ——	420	—	2.62324
So Radius ——	90	00	— 10.00000

To Meridian Diff. Latitude —— 1014 — 3.00602

Merid. Parts anfwering to Latitude 50d. 00m. — 3475
To which add Meridional Diff. Lat. —— 1014

The Sum ·—— —— — 4489

Which found in the Table of Meridional Parts, anfwers to 59d. 40m. the Latitude come to.

Then

Then for the Distance.

	d.	m.	
As Sine Comp. of Course ——	67	30	— 9.96561
To proper Diff. of Latitude ——	580		— 2.76343
So Radius—————————	90	00	— 10.00000

To the Distance —— ——	628	— 2.79782

The Latitude come into is 59d. 40m. the Distance 628 Miles.

CASE IV. *Both Latitudes and Distance given, to find the Course and Difference of Longitude.*

A Ship in Latitude 50d. 00m. N. sails 3505 Miles, and then is found by Observation to be in Latitude 13d. 12m. North, I demand her Course, and Difference of Longitude.

The proper Difference of Latitude is found by subtracting the Lesser Latitude from the greater, the Remainder is the Difference of Latitude in Degrees and Minutes, which multiplied by 60, gives the Difference of Latitude in Miles.

Example.

Greater Latitude—50d. 00m. ⎫	36	48
Lesser Latitude —13 12 ⎬	60	
Remains ———— 36 48 ⎭	2208 proper Diff. Lat.	

	d.	m.	
Meridional Parts for Latitude ——	50	00	—— 3475
Meridional Parts for Latitude ——	13	12	——— 799

Meridional Difference of Lat.————————	2676
Proper Difference of Latitude———————	2208

For

For the Course.

As the Distance —— —— 3505 —— 3.54469
To Radius ——————— 90d. 00m. —— 10.00000
So proper Diff. of Lat.——— 2208 —— 3.34400

To Sine Comp. of Course — 39 03 —— 9.79931

For the Difference of Longitude.

d. m.
As Radius —— —— —— 90 00 —— 10.00000
To the Meridional Diff. of Lat. 2676 —— 3.42749
So Tangent of the Course — 50 57 —— 10.09086

To Diff. of Longitude —— 3298 —— 3.51835

In the four foregoing Cases I have purposely omitted the Canons for finding the Departure, it being sufficiently taught in the Cases of Plane-Sailing; and the Canons for finding it are exactly the same in both, according to what is given. Nevertheless in all the following Cases (except the 6th) the Departure is given, and therefore I shall make it one necessary Term in each of the ensuing Cases. And when we come to the practical part, *viz.* the keeping of a Journal, I shall lay down some easy and compendious Ways for reducing Departure to Difference of Longitude, both by Mercator and Middle Latitude, without any Canon.

CASE V. *Both Latitudes and Departure given, to find Course, Distance, and Difference of Longitude.*

A Ship departs from Latitude 55d. 00m. N. and sails between the North and East into Latitude 56d. 10m. N. her Departure 50 Min. or Miles; I demand the Course steered, the Distance sail'd, and the Difference of Longitude.

For

For the Courfe:

As proper Diff. of Latitude		70	1.84509
To Radius	90	00	10.00000
So Departure		50	1.69897

To Tang. of the Courfe ——— 35 32 — 9.85388

For the Diftance.

	d.	m.	
As Sine of the Courfe	35	32	9.76430
To Departure		50	1.69890
So Radius	90	00	10.00000

To the Diftance ——— 86 — 1.93467

For the Difference of Longitude.

As Radius	90	00	10.00000
To Meridional Diff. of Lat.		124	2.09342
So Tangent of the Courfe	35	32	9.85380

To Difference of Longitude —— 89 *fere* — 1.94722

Or, which is the fame.

As proper Difference of Latitude	70 Co. Ar.		8.15499
To Meridional Difference of Lat.	124		2.09342
So Departure	50		1.69897

To Difference of Longitude —— 89 *fere* — 1.94738

CASE VI. *Both Latitudes and Difference of Longitude given, to find Courfe, Diftance and Departure.*

Suppofe two Places, one in Latitude 56d. 15m. North, the other in Latitude 58d. 35m. North, their Difference of Longitude 2d. 30m. I demand the Courfe and Diftance from the Southermoft to the Northermoft, and alfo the Departure;

Proper Difference of Latitude ——— 140
Meridian Difference of Latitude —— 260

For

For the Course.

As Merid. Diff. Lat. ——— 260 —— 2.41664
To Radius ——— ——— 90 om. —— 10.00000
So Diff. Long. ——— 150 —— 2.17609

To Tang. Course ——— 29 53 —— 9.75945

For the Distance.

As Sine Comp. of Course —— 60 07 —— 9.93804
To proper Diff. Lat. ——— 104m. —— 2.14612
So Radius ——— ——— 90 00 —— 10.00000

To the Distance ——— ——— 161 —— 2.20808

For the Departure.

As Sine Comp. of Course——60 07 Co. Ar. 0.06196
To proper Diff. Lat. ——— 140 —— 2.14612
So Sine of Course ——— 29 53 —— 9.69743

To Departure ——— ——— 10 —— 1.90551

Or, as Radius to proper Difference of Latitude, so Tangent of the Course to Departure.

CASE VII. *One Latitude, Course and Departure given, to find the other Latitude, Distance and Difference of Longitude.*

A Ship in Latitude 58d. om: N. sails S.S.W. half W. till her Departure be 60 Min. I demand the Latitude come to, Distance sail'd, and Difference of Longitude.

☞. If your Difference of Longitude be East, add it to, or if West, subtract it from the Longitude sailed from, and the Sum or Remainder is the Longitude come into, if you begin to reckon your Longitude from *Pico Teneriff* Eastwards to 360 Degrees, (which is seldom done in *England*) but if you begin your Longitude at the Place sail'd from, (according to the *Mariner's Compass Rectified*) you only reckon your Longitude West, according to the Denomination of your Departure, whether East or West.

For

For the Difference of Latitude.

	d.	m.	Co. Ar.
As Sine of the Course ——— 28		07	0.32673
To the Departure ——— —		60	1.77815
So Sine Comp. of Course ——— 61		53	9.94546
To Difference of Latitude ———	112		2.05034

Difference of Latitude 112 Miles, or Minutes; or, 1d. 52m. which subtracted from the Latitude sailed from 58 Deg. (because the Course is Southerly) the Remainder 56d. 8m. is the Latitude come to.

Then for the Distance.

	d.	m.	Co. Ar.
As Sine of the Course —— 28		07	9.67326
To Departure ——— ——		60	1.77815
So Radius ——— ——— 90		00	10.00000
To the Distance —— ———	127		2.10489

For the Difference of Longitude.

	d.	m.	Co. Ar.
As Radius ——— ——— 90		00	10.00000
To Meridional Diff. of Lat. ———206			2.31386
So Tangent of the Course——28		07	9.72780
To Difference of Longitude ——	110		2.04166

Or, Difference of Longitude may be thus found.

		Co. Ar.
As proper Difference of Lat. ——	112	7.94965
To Meridional Diff. Lat ——	206	2.31386
So Departure ——— ——	60	1.77815
To Difference of Longitude ———	110	2.04166

CASE

CASE VIII. *One Latitude, Distance and Departure given, to find the other Latitude, Course and Difference of Longitude.*

A Ship in Lat. 58d. om. N. sails between the South and West 127 Miles, and then finds her Departure to be 1 Degree, or 60 Minutes; I demand as above.

For the Course.

As the Distance ———————	127	——	2.10380
To Radius ————— ———	90 00	——	10 00000
So Departure ——— ———	60	——	1 77815
To Sine of the Course ——	28 11	——	9.67435

For the Difference of Latitude.

	d. m.		Co. Ar.
As Sine of the Course ——	28 11	——	0.32579
To Departure ——— ———	60	——	1.77815
So Sine Comp. of the Course	61 49	——	9.94519
To Difference of Latitude ——	112	——	2.04913

For the Difference of Longitude.

	d. m.		
As Radius ——— ———	90 00	——	10.00000
To Meridional Diff. of Lat. ——	206	——	2.31386
So Tangent of the Course	28 11	——	9 72901
To Difference of Longitude ——	110	——	2.04288

The Difference of Latitude is 112 Minutes; or 1 Degree 52 Minutes, which subtracted from the Latitude sailed from 58 Deg. the Remainder 56 Deg. 8 Min. is the Latitude come to; and the Difference of Longitude is 110 Minutes, or 1d. 50m. which is to be subtracted from the Longitude sailed from, if you are decreasing your Longitude, &c.

How to work Mercator's Sailing, without a Table of Meridional Parts.

The Table of Meridional Parts may be supplied by the Table of Artificial Tangents; for every half Degree of the Tangents above 45 answers to one Degree of Latitude, (the Characteristic or Index of the Tangent being rejected) and therefore to find the Meridional Leagues for any Latitude, and half the given Latitede to 45, the Sum is a Degree, whose Tangent multiplied by 10, and the Product divided by 376, the Quotient is the Meridional Leagues belonging to that Latitude, which multiplied by 3, gives (nearly) the Meridional Parts found in the Table.

Note, If you have a Table that hath 6 Figures, or more, in the Tangents, besides the Index, you may use the first six Figures, and need not multiply them by 10, as before.

But because this Way is not of Use, except when both Latitudes are given or found, to find the Difference of Longitude, 'tis best to find the Tangents answering to both Latitudes, and subtract the less from the greater, and divide the Remainder by 376, which gives the Meridional Difference of Latitude in Leagues, which if you please you may reduce to Miles, (by multiplying by 3) without further trouble.

Example.

In the foregoing Question in the eighth Case of Mercator's Sailing Trigonometrical, the two Latitudes are found to be 58d. 0m. and 56d. 8m.

Half of 58d. 0m. is 29d. 0m.

To which add —   45   0
_____

The Sum is ——   74   0 whose Tangent is 542503

Half of 56d. 8m. is   28   4
To which add ——   45   0
_____

The Sum is ——   73   4 whose Tangent is 516471

Remainder

Remainder —— 26032
376)26032(69 Merid. Diff. Lat. in Leagues —— 69
   3472   Which multiply by   —— 3

   ·· 88   Gives Merid. Diff. Lat. in Miles — 207
Nearly agreeing with that found by the Table.

And although this be not exactly the same with the Table, yet the Difference is scarce discernable, and therefore it may be of great Use where Tables of Meridional Parts are wanting.

SECT. V.

Parallel Sailing Trigonometrical.

SOME rank the Cases of *Parallel Sailing* amongst the rest of the Cases in *Mercator's Sailing*, and indeed not without good Reason, seeing it is grounded upon the same Projection, and is really a Branch of the same. Nevertheless, seeing both *Data* and *Quæsita* relate only to the Distance or Difference of Longitude, &c. of Ships or Places in the same Parallel or Latitude, I chuse rather to speak of it in so many distinct Cases under that Head, and shall (as before) set down an Example in every Case, and work it in the common Way, (*viz.* by Logarithms, &c. here, and shall hereafter proceed to work it also, as well as *Plane* and *Mercator's*, without any Canon, only by common Arithmetic.

CASE I.   *Two Ships (or Places) in one Parallel, their Latitude and Distance given, to find their Difference of Longitude.*

Suppose two Ships in Latitude 50d. 00m: distant 76 Miles; I demand their Difference of Longitude.

A3

d. m.

As Sine Comp. of Latitude —— 40 00 —— 9.80807
To Radius ——— ———— 90 00 —— 10.00000
So Diftance —— —— 76 —— 1.88081
_____

To Difference of Longitude —— 118 —— 2.07274
    The Difference of Longitude 118m. or 1d. 58m.

CASE II. *Two Places in one Parallel, their Latitude
and Difference of Longitude given, to find their Diftance.*

This is only the former Cannon inverted, as in the following Example:
The Latitude of two Ships (or Places) and their Difference of Longitude given, to find their Diftance.

Example. Suppofe two Places in the Latitude of 50d.
0m. their Difference of Longitude 118m. I demand their
Diftance.

d. m.

As Radius —— —— 90 00 —— 10.00000
To Sine Comp. Lat. —— —— 40 00 —— 9.80806
So Difference of Longitude —— 118 —— 2.07188
_____

To the Diftance —— 76 —— 1.87994

CASE III. *Two Places in one Latitude, their Diftance
and Difference of Longitude given, to find the Latitude
they are in.*

Example. A Ship fails due Weft 2700 Miles, and
then finds her Difference of Longitude to be 4200 min.
I demand what Latitude the Ship fails in.

As Diff. Longitude —— 4200 —— 3.62325
To Radius —— —— 90d. 00m. —— 10.00000
So is the Diftance —— 2700 —— 3.43136
_____

To Sine Comp: Lat. —— 50 00 —— 9.80811

CASE

CASE IV. *Two Ships failing directly North or South, their Distance in one Parallel given, to find their Distance in another Parallel.*

Example. Suppose two Ships in Latitude 50d. 00m. North, distant 200 Miles, sail both directly North into Latitude 73d. 00m. I demand their Distance in that Parallel.

d. m.

As Sine Comp. of Lat. failed from 40 00 *Co. Ar.* 0.19193
To their Distance in that Parallel   200 —— 2.30103
So Sine Comp. of Lat. sail'd to — 17 00 —— 9.46594

To their Distance in that Parallel — 91 —— 1.95890

CASE V. *Two Ships in one Parallel, with their Distance in that Parallel given, Sailing both directly North or South, with their Distance in the Parallel sail'd to given, to find the Latitude come to.*

Example. Two Ships in Latitude 50d. 00m. distant 200 Miles, sail both directly North, till their Distance is but 91 Miles; I demand the Latitude come to.

This is but the last Canon inverted, *viz:*

Co. Ar.

As their Distance in Lat. sail'd from — 200 — 7.69897
To Sine Comp. of Lat. sail'd from 40d. 00m. 9.80807
So Distance in Lat. come to ——— 91 — 1.95904

To Sine Comp. of Lat. come to — 17 00 — 9.46608

In these five Cases of *Parallel Sailing*, I have set down but one Example in each Case, it being of no great Use in the Practice of Navigation; and yet it was not fit it should be omitted, as being a good Help towards a right Notion and Apprehension of the Terrestrial Globe; for in *Plane Sailing* there is no such Thing as Difference of Longitude, being distinct from

F   Depar-

Departure, but here the Difference between the one and the other is plainly evident ; the Difference of Longitude being sometimes twice or thrice as much as the Departure, it bearing always the same Proportion to the Departure, as Radius does to Sine Complement of the Latitude or Parallel sailed in.

S E C T. VI.

Middle Latitude Sailing Trigonometrical.

ALtho' *Middle Latitude Sailing* is not exactly true, yet becaufe of its (general) Conformity to the Globe, I fhall not altogether omit it ; for notwithftanding it comes fhort of *Mercator's* (or rather *Wright's* Sailing, in which the Degrees of Longitude bear exactly the fame Proportion to the Degrees of Latitude in any Parallel, as it does upon the Globe) yet it exceeds *Plane Sailing* (which fuppofeth the Earth to be a plane Superficies, and the Degrees of Longitude and Latitude to be every where equal;) and therefore feeing it hath been fometimes taught and practifed, as one of the Kinds of Sailing (or keeping a reckoning) now in Ufe ; I' fhall proceed to an Example in each Cafe thereof. And the rather I take Notice of it, becaufe when I come to go over this in the new Way of Working without any Canon, I fhall fhew how it may be done, not only with more Eafe and Expedition, but alfo much more nearly (if not exactly) agreeing with *Mercator*, and confequently with the Globe itfelf, than what has been commonly taught and practifed in that Kind of Sailing ; but fhall refer that to its proper Place in the fecond Part hereof, and fhall here proceed (as in *Plane* and *Mercator*, &c.) to work the feveral Cafes of *Middle Latitud Sailing* the common Way, and the Analogies or Proportions made Ufe of herein, befides thofe taught in *Plane Sailing* (which are applicable and ufeful here, as well as in *Mercator's Sailing*) are commonly thefe two, *viz.*

As

As Difference of Latitude,
To Difference of Longitude;
So Sine Complement of Middle Latitude,
To Tangent of the Course from the Meridian.

The other Proportion is,

As Sine Complement of Middle Latitude,
To Radius;
So is the Departure,
To Difference of Longitude.

And these Proportions may be inverted accordin to
what is given, and what is required, by *Euclid, Lib. 5.
Prop.* 14. & 16 & *Corol.* always observing to make the
required Term the last of the Four in the Proportion;
As Suppose both Latitudes and Course be given, and
Difference of Longitude required, then the Proportion
will be,

As Sine Complement of Middle Latitude,
To Tangent of the Course;
So Difference of Latitude,
To Difference of Longitude.

And so in others, as in the ensuing Operations will
more plainly appear.
There is also another Proportion of Use for finding
the Course, Distance, or Difference of Longitude, when
any two of them, together with both Latitudes, are
given or found; and the Proportion is,

As Sine Complement of Middle Latitude,
Is to the Sine of the Course from the Meridian;
So is the Distance failed,
To the Difference of Longitude.

And so by inverting the Canon, any three of them
being given, the fourth may be found.

CASE I. *One Latitude, Courſe, and Diſtance given, to find the other Latitude, Departure and Difference of Longitude.*

A Ship in Latitude 50d. 00m. North, ſails N.W. by N. 987 Miles; I demand the Latitude come to, with the Departure and Difference of Longitude.

See the ſame Queſtion in *Mercator's Sailing, Caſe* the firſt.

For the Difference of Latitude.

	d.	m.	
As Radius	90	00	— 10.00000
To Diſtance		987	— 2.99432
So Sine Comp. of Courſe — 56	15		9.91984
To Difference of Latitude —	821		— 2.91416

The Latitude come to is 63 Deg. 41 Min.

For the Departure.

	d.	m.	
As Radius	90	00	— 10.00000
To the Diſtance		987	— 2.99432
So Sine of Courſe	33	45	— 9.74473
To Departure —	548		— 2.73905

For the Difference of Longitude.

Becauſe this Canon for finding the Difference of Longitude is grounded upon the Middle Latitude, obſerve that it is always found by adding the two Latitudes together, and half the Sum is the Middle Latitude required:

Thus in this Example.

	d.	m.
The Latitude ſailed from	50	00
The Latitude ſailed to is	63	41
The Sum is	113	41
The Half of which is the Middle Latitude, *viz.*— 56	51

The

The exact Half is 56d. 50m. ½; but you need not be careful to a half Minute; but where the Sum is an odd Minute, take always the bigger Half, for the Middle Latitude is a Latitude rather too little to work with, and consequently the Sine Complement of it is always a Sine rather too great, and makes the Longitude come out lefs than really it fhould be: And although there is no Way to find a Mean Latitude to work with as Middle Latitude (without too much Trouble) that will bring out the required Longitude the fame exactly with *Mercator*; yet the Middle Latitude found as before, being always too little, the taking the greater Half, where an odd Minute happens, will be fo far from augmenting the Error, that it rather leffens it, and makes the Longitude fo found more nearly (though not exactly conformable to the Longitude found by *Mercator's Sailing*; which for Inftance in this firft Queftion, we fhall work by each of the three Canons before-mentioned.

	d.	m.	Co. Ar.
As Sine Comp. of Mid. Lat. —	33	09	— 0.26215
To Tang. of the Courfe —	33	45	— 9.82489
So Diff. of Latitude —		821	— 2.91434
To Diff. of Longitude —		1003	— 3.00138

Again,

	d.	m:	
As Sine Comp. of Mid. Lat. —	33	09	— 9.73785
To Radius —	90	00	— 10.00000
So Departure —		— 548	— 2.73878
To Difference of Longitude —		1003	— 3.00096

Again,

	d.	m.	Co. Ar.
As Sine Comp: of Mid. Lat. —	33	09	— 0.26215
To Sine of Courfe —	33	45	— 9.74473
So is the Diftance —		— 987	— 2.99432
To Difference of Longitude —		1003	— 3.00120

C A S E

CASE II.   *Both Latitudes and Course given, to find Distance, Departure, and Difference of Longitude.*

A Ship takes her Departure from an Island, in Latitude 56 deg. 25 min. North, and fails N.N.W. for several Days, and then finds herself by Observation to be in Latitude 68d. 30m. North; I demand the Distance sailed, with her Departure and Difference of Longitude.

See the same Question in *Mercator's* Sailing. Case the Second. The Difference of Latitude is 12d. 5m. or 725 Min. or Miles.

Then for the Distance.

	d.	m.	
As Sine Comp. of Course —	67	30	—— 9.96561
To Difference of Latitude —	725		—— 2.86033
So is Radius ——	90	00	—— 10.00000
To the Distance ——		785	—— 2.89472

Then for the Departure.

	d.	m.	
As Sine Comp. of Course —	67	30	—— 0.03439
To Difference of Longitude —	725		—— 2.86033
So Sine of the Course ——	22	30	—— 9.58284
To Departure ——		300	—— 2.47756

The two Latitudes { 56d. 25m.
                   { 68   30

Sum —	124	55	
Half Sum —	62	28	the Middle Latitude

	d.	m.	
As Sine Comp. of Mid. Lat.	27	32	—— 9.66489
To Radius ——	90	00	—— 10.00000
So Departure ——		300	—— 2.47712
To Difference of Longitude —		649	—— 2.81223

CASE

CASE III. *Both Latitudes and Distance given, to find the Course, Departure and Difference of Longitude.*

A Ship in Latitude of 55d. 00m. North, sails between the North and the East 86 Miles, and is then come into Latitude 56d. 10m. North; her Course, Departure, and Difference of Longitude is required.

See the same Question, (only inverted, what is required there being given here) in *Mercator's Sailing, Case* the Fifth.

For the Course.

As Distance		86 —	1.93449
To Radius	— 90d. 00m. —		10.00000
So Difference of Latitude	—	70 —	1.84509
To Sine Comp. of Course —	54	29 —	9.91060

For the Departure.

	d.	m.	
As Radius	90	00 —	10.00000
To the Distance	—	86 —	1.93449
So Sine of the Course —	35	31 —	9.76413
To Departure		50 —	1.69862

For the Difference of Longitude.

	d.	m.	Co. Ar.
As Sine Comp. of Mid. Lat.—	34	25 —	0.24779
To Tang. of the Course —	35	31 —	9.85086
So Difference of Latitude —		70 —	1.84509
To Diff. of Longitude —		88 —	1.94374

	d.	m.	
As Sine Comp. of Mid. Lat.	34	25 —	9.75220
To Radius	90d. 00m,		10.00000
So Departure		50 —	1.69897
To Difference of Longitude —		88 —	1.94677

F 4   CASE

CASE IV.   *Both Latitudes and Departure given, to find Course, Distance, and Difference of Longitude.*

A Ship in Latitude 55d. oom. North, sails between the North and East into Latitude 56d. 10m, North, her Departure 50 Minutes; the Course, Distance, and Difference of Longitude are required.

See the same Question in *Mercator's Sailing*, Case the fifth.

For the Course, by Case the Sixth of Plane Sailing.

As Difference of Latitude —— 70 — 1.84509
To Radius ———— 90d. 00m. — 10.00000
So Departure ———— 50 — 1.69897
                                    ——————
To Tangent of the Course — 35 32 — 9.85388

For the Distance.

                                    d. m.
As Sine of the Course ———— 35 32 — 9.76430
To Departure———— 50 — 1.69897
So Radius ———— 90 00 — 10.00000
                                    ——————
To the Distance———— 86 — 1.93467

For the Difference of Longitude.

                                    d. m.
As Sine Comp. of Mid. Lat. — 34 25 — 9.75220
To Radius —— 90 00 — 10.00000
So Departure———— 50 — 1.69897
                                    ——————
To the Difference of Longitude — 88 — 1.94677

CASE V.   *Both Latitudes and Difference of Longitude given, to find Course, Distance and Departure.*

I demand the Course, Distance and Meridian Distance between two Places, one in Latitude 56d. 15m. North, and the other in Latitude 58d. 35m. North, their Difference of Longitude 2d. 30m.

See *Mercator*, Case the Sixth.

For

Then for the Course.

As Difference of Latitude ——— 140 *Co. Ar.* 7.85387
To Difference of Longitude ——— 150 ——— 2.17609
So Sine Comp. of Mid. Lat.—57d. 25m. — 9.73120
_____
To Tangent of Course ——— 32   25   —   9.76116

For the Distance.

                            d.   m.
As Sine Comp. of Course ——— 60   01 —   9.93760
To Difference of Latitude ———   140 —   2.14612
So Radius ——— ——— ——— 90   00 — 10.00000
_____
To the Distance ——— ——— 162 —   2.20852

For the Departure.

                        . . d.   m.   *Co. Ar.*
As Sine of the Course ——— 60   01 —   0.06239
To Difference of Latitude ———   140 —   2.14612
So Sine of the Course ——— 29 ' 59 —   9.69875
_____
To Departure ——— ——— ——— 81 —   1.90726

These Five are the most necessary and useful Cases in *Middle Latitude*, and indeed all that are of Use in keeping a Reckoning; yet there are several other Cases, which I shall Insert for Variety; but because they are not so necessary in Practice, I shall only set down the *Data* and *Quæsita*, with the Canons or Proportions for finding what is required, and shall leave the Operations for the Reader's Practice.

CASE VI.   *One Latitude, Course and Departure given, to find the other Latitude, Distance and Difference of Longitude.*

A Ship in Latitude 55d. 00m North, sails North 35d. 32m. Easterly till her Departure be 50; I demand as above.

For

For the Difference of Latitude.

As Sine Courfe is to Departure; fo Sine Comp. of Courfe; to Diff. of Latitude.

The Difference of Latitude thus found, the Latitude come to is eafily found, as in *Mercator*, Cafe the firft; and both Latitudes being found, the Middle Latitude is found, as in *Middle Latitude Sailing*, Cafe the firft. Then,

For the Diftance.

As Sine of the Courfe is to Departure; fo is Radius to the Diftance.

For the Difference of Longitude,

As Sine Comp. of Mid. Lat. to Radius; fo is Departure to Diff. of Longitude.

CASE VII   *One Latitude, Diftance and Departure given, to find the other Latitude, Courfe and Difference of Longitude.*

For the Courfe.

As Diftance is to Radius; fo is Departure to the Sine of the Courfe.

For the Difference of Latitude.

As Radius is to the Diftance; fo is Sine Comp. of the Courfe, to the Difference of Latitude.

For the Difference of Longitude.

As Sine Comp. of Middle Latitude is to Radius; fo is Departure to Difference of Longitude.

CASE VIII.   *Courfe, Departure and Difference of Latitude given, to find both Latitudes and Diftance.*

For the Diftance:

As Sine of the Courfe is to Departure; fo is Radius, to the Diftance.

                                                    For

For the Difference of Latitude.

As Sine of the Course is to Departure; so is Sine Comp. of the Course, to Difference of Latitude.

For the Middle Latitude.

As Difference of Longitude is to Departure, so is Radius to Sine Comp. of Middle Latitude.

The Middle Latitude and Difference of Latitude thus found, both Latitudes are easily found; for adding half the Difference of Latitude to the Middle Latitude, the Sum is the greater Latitude, and subtracting the said Half from the Middle Latitude, the Remainder is the lesser Latitude.

CASE IX. *Middle Latitude, Course and Distance given, to find both Latitude, Departure and Difference of Longitude.*

For Difference of Latitude, and consequently both Latitudes.

As Radius is to the Distance; so is Sine Comp. of Course to Diff. of Latitude.

Then find both Latitudes, as in Case the Eighth.

Then for Departure, by Case the first of Plane Sailing.

As Radius is to the Distance; so is Sine of the Course, to Departure.

For the Difference of Longitude.

As Sine Com. of Middle Latitude, to the Sine of the Course; so is the Distance to the Difference of Longitude.

CASE X *Course, Distance, and Difference of Longitude given, to find both Latitudes and Departure.*

For the Middle Latitude.

As Difference of Longitude is to the Distance; so is Sine of the Course, to Sine Comp. of Middle Latitude.

For

For the Difference of Latitude.

As Radius is to the Distance; so Sine Comp. of Course to Difference of Latitude.

Then find both Latitudes as in Case the Eighth.

For the Departure.

As Radius is to the Distance; so Sine of the Course, to Departure.

CASE XI. *Course, Difference of Latitude, and Difference of Longitude given, to find both Latitudes, Distance and Departure.*

For the Middle Latitude.

As Difference of Longitude is to Difference of Latitude; so is Tangent of the Course, to Sine Comp. of Middle Latitude.

By Difference of Latitude and Middle Latitude find both Latitudes as in Case the Eighth.

For the Distance.

As Sine Comp. of the Course is to the Difference of Latitude; so is Radius to the Distance.

For the Departure.

As Sine Comp. of the Course, is to the Difference of Latitude, so Sine of the Course, to the Departure.

CASE XII. *Middle Latitude, Course and Difference of Longitude given, to find both Latitudes, Distance and Departure.*

As Tangent of the Course is to Sine Comp. of Middle Latitude; so Difference of Longitude, to Difference of Latitude.

Then find Distance and Departure, as in Case the Eleventh.

CASE

CASE XIII. *One Latitude, Departure, and Difference of Longitude given, to find the other Latitude, Course and Distance.*

As Difference of Longitude is to Departure; so is Radius, to Sine Comp. of Middle Latitude.

Subtract the given Latitude and Middle Latitude, the less from the greater, the Remainder doubled is the Difference of Latitude.

Then for the Course.

As Difference of Latitude is to Radius; so is Departure, to the Tangent of the Course.

Then for the Distance.

As Sine of the Course is to Departure; so is Radius to the Distance.

CHAP. III.

Navigation by Inspection.

S E C T. I.

How to work all the Cases in Plane Sailing, and to work a Traverse; and also to keep a Reckoning both in Longitude and Latitude, only by Inspection in the Table of Difference of Latitude and Departure.

IN all Cases of Plane Sailing, there are always four Terms, *viz.* Course, Distance, Difference of Latitude and Departure, two of which are always given, to find the other two; and in the Traverse Table, the Course is to be found in the Degrees, Points, or
Quarter

Quarter Points, at the Top or Bottom of each Leaf. The Diftance is found in the firft and laft Column of each Page, and the Difference of Latitude and Departure are found in the reft of the Columns in the Body of each Leaf Side; and which two of the Terms foever you have given, they being found in their proper Places, the two required Terms may alfo be found in the corresponding Columns: I fhall inftance an Example in all the fix Cafes of Plane Sailing.

CASE I. *Courfe and Diftance given, to find Difference of Latitude, and Departure.*

A Ship fails S. 25 Deg. E. 96 Miles; I demand her Difference of Latitude and Departure.

In this Table, the Left-hand Page in the Columns of Diftance proceeds from 1 to 50, and the Right-hand Page from 50 to 100, and the given Diftance here being above 50, *viz.* 96, look along the Top of the Right-hand Pages till you find 25 Deg. and right under it, and againft 96, the Diftance you find 87 Min. or Miles for the Difference of Latitude, and 40.6 Min. or Miles for the Departure, as the Title at the Top directs.

But suppofe your Courfe hath been South 65 Degrees Eaft, you find 65 Degrees at the Bottom of the fame Column where you find 25 at the Top, and againft 96 the Diftance, you find Difference of Latitude, and Departure the fame, *viz.* 87 Miles, and 40. 6 Miles only with this Alteration, that now 87 Miles is the Departure, and 40.6 Miles is the Difference of Latitude; whereas before it was contrary: But to avoid Miftakes herein, obferve where you find the Degree of the Courfe whether at the Top or Bottom of the Table, there find alfo the Title of the Columns of Difference of Latitude and Departure: Thus, in the firft Example of 25 Degrees, you find the Degree at the Top, and therefore the Title alfo at the Top fhews that the firft of the two Columns is Difference of Latitude, and the fecond the Departure: but in the fecond Example of 65 Degrees, you find the Degree at the Bottom, and therefore the Title being

alfo

alfo found at the Bottom, fhews the firft Column is the Departure, and the fecond the Difference of Latitude, &c.

But fuppofe my Diftance be above 100 Miles, as in the firft Example of *Plane Sailing* in the Book, which is, a Ship fails N.W. by N. 123 Miles; I demand her Difference of Latitude and Departure.

Here, becaufe the Table reaches but to 100, you muft take it out at twice; thus the Courfe being N. W. by N. 3 Points, I look for 3 Points, which I find at the Top of the Leaf, and under it againft 100, I find,

	Diff. Lat.	Dep.
	83 . 1	55 . 6
Then againft 23 I find under 3 Points	19 . 1	12 . 8
	102 . 2	68 . 4

Thefe Sums added feverally, give 102 . 2 for the Difference of Latitude and 68 . 4 for the Departure.

But fuppofe your Diftance was fo great a Number, that it would require to be taken out at a great many Times, as in the Queftion in Cafe I. of *Mercator's Sailing Trigonometical.* A Ship fails 987 Miles N.W. by N. &c. you may alfo take this out at twice; for having found 3 Points, you may take out firft for 980, and then for 7 thus: look under 3 Points, till you come down againft the Diftance 98, and againft it you have 81 . 5, for Difference of Latitude, and 54 . 4 for the Departure; but becaufe the 98 is advanced a Figure or Place towards the Left-hand, and called 980, and therefore advance the other alfo a Place to the Left-hand, then will the Diff. of Latitude,

Diff. Lat.	Dep.
815 ‖ 544	81 . 5, become 815, and the Departure 54 . 4, will be 544; then finding
5 : 8 ‖ 3 . 9	alfo for the 7, becaufe the whole Diftance is 987, the Work will ftand as
820 . 8 ‖ 547 : 9	in the Margin, and the Difference of Latitude is 820 : 8, and Departure,

547 . 9.

Another

· Another Method very proper when your given Diſtance exceeds what is found in the Table, is to take, ½ or ⅓, &c. of your given Diſtance, and find the Diff. of Lat. and Departure belonging to that, which being doubled, if you take half, or tripled, if you take One-third, or multiplied by 4, if you took One-fourth, &c. of the given Diſtance; the Numbers ſo found ſhall be the Difference of Latitude and Departure required.

Example.

A Ship ſails S.S.W. 120; I demand Diff. of Latitude and Departure.

Here, becauſe the Table goes not ſo far as 120, I take the Half of it, *viz.* 60, and againſt it under two Points (becauſe S.S.W. is two Points from the Meridian) I find 55.4, for Difference of Latitude, and 23.0, for the Departure; which two Numbers being doubled, (becauſe I take but Half the Diſtance) gives 110.8, for the Difference of Latitude, and 46.0 for the Departure, &c.

CASE II.   *Courſe and Difference of Latitude given, to find Diſtance and Departure.*

A Ship ſails North 38 Deg. Weſterly, till her Difference of Latitude be 120 Miles; I demand her Diſtance and Departure.

· Becauſe 120 is not to be found in the Table, I take half of it, *viz.* 60, and finding 38 Deg. at the Top, I look under it till I find 60 [under Lat.] for Difference of Latitude; and right againſt it, in the Column of Diſtance, I find 76, which doubled is 152, the Diſtance required, and againſt it [under Dep.] I find 46.8, which doubled is 93.6, the Departure required.

CASE

CASE III.   *Course and Departure given, to find the Distance and   Difference of Latitude*

A Ship sails S.W. by S. till her Departure be 40 Miles; I demand Distance and Difference of Latitude.

Find the Points at the Top, and under it, in the Column of Departure, find 40, against which you find 72 for the Distance, and 59.9, for the Diff. of Latitude.

CASE IV.   *Distance and Difference of Latitude given, to find Course and Departure.*

A Ship sails 71 Miles in the S.W. Quarter, her Difference of Latitude 59 Miles; I demand Course and Departure.

Look for 71 in the Column of Distance, and run over the several Columns of your Table, till you find 59 in one of the other Columns against it, which being found, observe where you find the Title [Lat.] whether at the Top or Bottom of the Table; for there you find also the Course, as in this Example. I run over the Table, till against 71, I find 59.0, and the Title [Lat.] being at the Top, I find also the Course at the Top, *viz.* 3 Points, or S.W. by S. and Departure 39.4.

CASE V.   *Distance and Departure given, to find Course, and Difference of Latitude.*

A Ship sails between the South and East 100 Miles, her Departure 53 Miles; I demand the Course and Difference of Latitude.

Look for 100 in the Column of Distance, and against it run over your Table till you find 53 for the Departure, and over it you find 32 Degrees for the Course, and 84.8 for the Difference of Latitude required.

G   CASE

CASE VI.   *Difference of Latitude and Departure given, to find the Course and Distance.*

A Ship sails between the North and East, till her Difference of Latitude be 56.0, and her Departure 35.0; I demand the Course and Distance.

Look through the Columns of Diff. of Lat. and Departure, till you find 56.0, against 35.0, and over that you find 32 Degrees for the Course, and 66 Miles for the Distance, &c.

Note 1st.   You cannot always find exactly the Numbers given, but find the nearest you can, which will generally be of sufficient Exactness for common Use.

Note 2d.   If the Departure be greater than the Difference of Latitude, the Course is more than four Points, but if the Departure be less, then the Course is less than four Points; and the greater the Difference is between these two, the more in Proportion will the Course differ from four Points; by duly observing this, a little Practice will make you find the Course and Distance with Readiness.

Note 3d.   If either Difference of Latitude, or Departure, or both, be greater than the Extent of your Tables, then Half or Quarter, &c. both of them, and find out as above, and the Course will be the same, but the Distance will be the Half, Quarter, &c. accordingly.

SECT. II.

How to work a Traverse by the Tables of Difference of Latitude and Departure.

THIS is the principal Thing for which these Tables are of Use, and this Way of working a Traverse is equal to the best for Exactness, and much more expeditious, and therefore I have on Purpose omitted

Traverse

Traverse Sailing in the Trigonometrical Part hereof, as being too tedious for Practice at Sea.

I shall instance in Question the first of *Traverse Sailing Geometrical*, and shall solve the same here by the Table.

A Ship S.S.E. 30 Miles, then N. E. by N. 40 Miles, then E. by N. 25 Miles, then N.N.E. 44; I demand the Course, Distance, Difference of Latitude and Departure, from the Place sailed from.

First, Make a little Table with six Columns, as you see in this Page.

The first is the Course.
The second the Distance.
The third the Northing.
The fourth the Southing.
The fifth the Easting.
The sixth the Westing.

Course.	Dist	North.	South.	East.	West.
S. S. E.	30	———	27 . 7	11 . 5	———
NE. by N.	40	33 . 3	———	22 . 2	———
E. by N.	25	4 . 9	———	24 . 5	———
N. N. E.	44	40 . 6	———	16 . 8	———
		78 . 8	27 : 7	75 . 0	
		27 . 7			
		51 : 1			

Then by *Case* I. find the Difference of Latitude and Departure in every Course, and set them in their proper Columns; as where the Course is Northerly, set the Difference of Latitude under Northing, or in the North

G 2   Column;

Column, and when the Course is Southerly, set the Difference of Latitude in the South Column.

Again, when the Course is Easterly, set the Departure in the East Column, and when the Course is Westerly, set the Departure in the West Column; then adding up each Column by itself, subtract the North and South Columns, the lesser from the greater, the Remainder is the Northing or Southing made good. Also subtract the East and West Columns, the lesser from the bigger, the Remainder is the Easting or Westing made good; then have you Difference of Latitude and Departure given, to find the Course and Distance, by *Case* VI.

In this Example, the first Course is E.S.E. 30 Miles; or two Points 30 Miles; for which, by *Case* I. I find the Difference of Latitude 27.7. Now the Course being between South and East, I place my Difference of Latitude in the South Column, and my Departure 11.5, in the East Column, leaving the North and West Columns blank.

Then for the second Course, N.E. by N. or 3 Points, 40 Miles, here my Difference of Latitude 33,3, is to be placed in the North Column, and the Departure 22.2, in the East Column, because the Course is between the North and East.

Then the third Course being E. by N. or 7 Points, 25 Miles, I place my Difference of Latitude, 4.9, in the North Column, and Departure, 24.5, in the East Column.

And so for the fourth Course, N.N.E. or two Points, 44 Miles, I place my Difference of Latitude, 40.6, in the North Columns, and my Departure, 16.8, in the East Column, and then adding up each Column, the Sum of the Northing Column is 78.8, and the Sum of the Southing Column is 27.7, which subtracted from the Northing 78.8, the Remainder 51.1, is the Difference of Latitude made good, which is Northing, because the Northing was the greater Number.

Again, the Sum of the Easting Column is 75.0, which (because there is no Westing to subtract from it) is the Easting made good. Thus you have the Northing, 51.1,

and

and the Easting, 75.0, given, to find Course and Distance, by *Case* VI. and although you cannot find in the Table the exact Numbers of 51.1, and 75.0, together, yet find the nearest you can, which is 75.4, and 50.9, under which, at the Bottom, you find 56 Degrees for the Course, which is N.E. by E. nearest, and the Distance 91 Miles.

But if you have a Place proposed that you are bound for, whose Course and Distance from the Place sailed from is given, find thereby the Difference of Latitude and Departure to the Place bound for, from which subtract the Difference of Latitude and Departure, made good, (if it be the lesser Number, or the other from it, if that be the greater Number) the Remainder is the Difference of Latitude and Departure from the Ship to the Place bound for, by which you may find the Course and Distance by *Case* VI.

Example. In the first Question in *Traverse Sailing Geometrical,* there is a Place proposed to be sail'd to, distant 120 Miles N.E. half E. the Difference of Latitude for that Course and Distance is found by *Case* I. to be 76.2, and the Departure, 92.8. Now the Difference of Latitude made good by the Ship being but 51.1, if subtracted from the whole Difference of Latitude to the Place bound for, 76.2, it is evident that the Remainder 25.1, must be the Difference of Latitude from the Ship to the Place bound for. Also the Departure made good by the Ship, 75.0 subtracted from the whole Departure, 92.8, the Remainder 17.8, is the Departure from the Ship to the Port bound for. And thus you have the Difference of Latitude, 25.1, and the Departure, 17.8, to find the Course and Distance by *Case* VI. and the nearest to these two Numbers that can be found is 25.4, and 17.8, over which you have 35 Deg. for the Course, and against it 31 Miles for the Distance, *viz.* N.E. by N. 1d. 15m. Easterly, 31 Miles is the Course and Distance to the Place bound for, *&c.*

But suppose the Difference of Latitude made good by the Ship was more than the Difference of Latitude

to

to the Port bound for, you muſt ſubtract the leſſer from
the greater, and the Remainder is the Difference of Latitude from the Ship to the Port, but of a contrary Denomination, which ſeems to be ſo plain, that it needs
no Example; for if a Man intends to travel to a Place
that bears due North from him 12 Miles; if a Man tra-
vels 15 Miles due North, his Courſe from thence to the
Place bound for is South 3 Miles; even ſo if a Ship intending for the Port aforeſaid, whoſe Difference of Latitude
is 76.2 North, and the Departure 92.8 Eaſt, ſhould ſail
between the North and Eaſt, till her Difference of Latitude be 100 North, and Departure 75.0 Eaſt, 'tis plain
ſhe is to the Northward of her Port; becauſe ſhe has got
more Difference of Latitude Northerly; and ſhe is alſo to
the Weſtward of her Port, becauſe ſhe hath not ſo much
Departure Eaſterly; and therefore ſubtracting the leſſer
Difference of Latitude from the greater, the Remainder
23.8, is the Difference of Latitude Southerly from the
Ship to the Port bound for; and the Departure 57.0,
ſubtracted from the whole Departure, 92.8, the Remainder, 17.8, is the Departure Eaſterly from the Ship
to the Port bound for; and in this Caſe the Courſe and
Diſtance would be found by *Caſe* VI. to be S.E. by S.
¼ S. 29.5 Miles.

S E C T. III.

*How to find the Difference of Longitude by the Table of
Difference of Latitude and Departure.*

HAVING one Latitude, Courſe and Diſtance
given, or both Latitude and Courſe, &c. you may
find the Reſt according to *Plane Sailing*, by the foregoing
Rules, which being done, find the Middle Latitude,
(always obſerving this Caution, to take a Latitude rather too big than too little, where the Middle Latitude
cannot exactly be had without a Fraction, then the Rule
is:

Find

Find the Complement of Middle Latitude amongst the Degrees at the Top or Bottom of the Table, and (under it if you find it at the Top, or above it, if you find it at the Bottom) in the Column of Departure, find your Departure, and right against it, in the Column of Distance, is the Difference of Longitude.

And although this Method is not so practicable in great Numbers, yet it is very useful at Sea, for working a Day's Run, or the like, which seldom exceeds 50 or 60 Leagues; and therefore I shall recommend it to the Learner, as the most expeditious Method, and of sufficient Exactness for short Distances; here being Methods sufficient laid down in this Treatise for more exactly correcting the Reckoning once or twice in a Week, of which more when we come to that Part.

However, should the Departure be too great for the Tables, the Difference of Longitude may be very nearly found by this small additional Trouble of Halfing, Quartering, &c. the same, and proceed as before, whence you will have in the distant Column, the Half, Quarter, &c. of the Difference of Longitude.

Note, If you think it too much Trouble to subtract the Middle Latitude from 90, to find its Complement; you may find the Middle Latitude amongst the Degrees, as before; and find the Departure in the Column of Difference of Latitude, and against it, in the Column of Distance, you have the Difference of Longitude as before.

I shall give an Example or two, and refer the Reader for further Practice of this Method to the Examples laid down for keeping a Reckoning in the latter Part of this Book, where I shall illustrate this Method by Example, and shew its general Agreement with the Longitude, as found by *Mercator's Sailing.*

Example 1. A Ship in Latitude 55.0 North, sails in the North East Quarter, into Latitude 56.10, her Departure 50; I demand Course, Distance, and Difference of Longitude.

See the Queſtion in *Caſe* IV. of Middle Latitude Sailing Trigonometrical.

Difference of Latitude 70, Departure 50, the Courſe will be found by *Caſe* VI. of Plane Sailing to be 36 Degrees neareſt; omitting the Minutes, and the Diſtance 86.

The Middle Latitude is 55.35, but becauſe we have only whole Degrees, we muſt uſe either 55 Degrees, or 56; and becauſe, as I ſaid before, it is beſt to take a Degree too great rather than too little, we ſhall call it 56, whoſe Complement 34 found in the Degrees, look under it for the Departure 50, the neareſt Number to which is 49.8, againſt which in the Column of Diſtance, I find 89 for the Difference of Longitude required.

Example 2. In *Caſe* V. of Middle Latitude Sailing Trigonometrical.

I demand the Courſe, Diſtance and Difference of Longitude between two Places, one in Latitude 56.15 North, the other in Latitude 58.35 North, their Meridian Diſtance 81.

Difference of Latitude 140, Departure 81, the Courſe is found by *Caſe* VI. of Plane Sailing to be an Angle of 30 Degrees from the Meridian, and the Diſtance 162, as you may ſee by the Traverſe Table.

Then for Difference of Longitude, the Middle Latitude is 57.25, but we muſt call it in this Caſe 58, which found in the Degrees, there is not ſuch a Number as 81 in the Column of Latitude; therefore find its half 40.5, or the neareſt to it, which is 40.3, and againſt it, in the Column of Diſtance, you find 76, which doubled is 152, the Difference of Longitude required.

Note, Although in particular Queſtions the Longitude found this Way differs ſometimes a Minute or two from that found by Calculation; becauſe we are obliged to uſe whole Degrees in the Table for Middle Latitude; yet the Error being ſometimes a Minute or two too much, and ſometimes as much too little, the Difference is not diſcernable in a long Reckoning.

CHAP.

CHAP. IV.

Arithmetical Navigation.

SECT. I.

Plane Sailing Arithmetical.

HAVING thus finished all the Kinds and Cases of Navigation in the foregoing Book, according to the Method commonly taught, and practised both at Sea and a-shore, *viz.* by a Canon of Logarithmetical Sines, Tangents, and Secants, (together with the Help of the Logarithms and natural Numbers) my next Work is (according to my Promise) to shew, how all the aforesaid Kinds of Navigation may be compleatly performed without any Canon, only by the Help of some given Numbers, (which shall be inserted in their proper Places) which are of Use where the Course is either given or required: And first in Order, I shall begin with *Plane Sailing*, in which the three first Cases are wrought only by the Table of given Numbers, as also are those Questions in the three last Cases where the Course is required: But the Solution of all Questions in *Plane Sailing*, when two Sides are given to find a third (as Distance and Difference of Latitude given to find Departure, &c.) are grounded upon that known Proposition in *Euclid, Lib.* 1. *Prob.* 47. *viz.* That the Square of the Hypotenuse (called there the Base, because it is the longest Side) of right-angled Triangle, is equal to the Squares of both Legs added together. Hence then (the Hypotenuse representing the Distance, and the two Legs the Difference of Latitude and Departure (if you square each Leg, that is, multiply each Leg by itself) these two Sums added together shall be

equal

equal to the Square of the Hypotenuse; and consequently, if from the Square of the Hypotenuse you subtract the Square of one Leg, the Remainder is the Square of the other Leg, as may be proved by those three known Numbers, 3, 4, and 5, which three Numbers make a Right-angled Triangle; the two Legs being 3 and 4, and the Hypotenuse 5. Now if you square the two Legs, 3 Times 3 is 9, and 4 Times 4 is 16, which two Squares of 16 and 9 added together, the Sum is 25, the Square of 5 the Hypotenuse. Again, if from 25, the Square of the Hypotenuse, you subtract 16, the Square of one Leg, there remains 9, the Square of the other Leg, &c. And when the Square of any Side is thus found, the Square Root thereof is the Side required.

Also with respect to the given Numbers, if Course and Distance be given, multiply the Distance by the given Numbers, as the Title of the Table directs for Latitude and Departure, (whose Use follows in several Examples) the Product, abating as many Figures to the Right hand as you multiply by, is the Difference of Latitude or Departure required: Now if the Hypotenuse multiplied by the given Numbers, it produces a Leg, only two Figures are to be cut off; the said Leg being also given, and two Figures (or Cyphers) added to it, and divided by the said given Number, the Quotient must needs be the Hypotenuse; and if so, then if the said Leg, with two or more Figures (or Cyphers) added to it, be divided by the Hypotenuse, the Quotient must needs be the given Number, as is apparently evident to almost every Body that have but learned the first Rudiments of Arithmetick, and understand (what every School Boy is taught) how Multiplication and Division will undo and prove each other; and hence any two Parts of a Question of *Plane Sailing* being given, the Rest may be found by the Square Root, and one Table of given Numbers; all the Cases except the sixth, being wrought immediately by the given Numbers; and even in the sixth Case, where the two Legs are gi-

ven,

ven, the Hypotenuse or Distance may be found by the Square Root, and then the Course as in the other Cases; as I shall explain by the following Example.

A TABLE of Given Numbers.

Points	*Latit.*	*Depar.*	*Points*
0 ¼	99.88	04.91	7 ¾
0 ½	99.52	09.79	7 ½
0 ¾	98.92	14.66	7 ¼
1	98.08	19.51	7
1 ¼	97.01	24.30	6 ¾
1 ½	95.70	29 02	6 ½
1 ¾	94.16	33.68	6 ¼
2	92.39	38.27	6
2 ¼	90.40	42.76	5 ¾
2 ½	88.20	47.13	5 ½
2 ¾	85 78	51.40	5 ¼
3	83.15	55.56	5
3 ¼	80.32	59.57	4 ¾
3 ½	77.30	63.44	4 ½
3 ¾	74.10	67.15	4 ¼
4	70.71	70.71	4
Points	*Depar.*	*Latit.*	*Points*

The Use of this Table is for working the six Cases of *Plane Sailing*, in which observe, that in the first and last Columns you have the Rumbs or Points of the Compass, with the half Points, and quarter Points, numbered from the Meridian, or North and South Line, either East or West, according as they stand in Order; thus S. S. W. is two Points, because South is upon the Meridian, then S. by W. is one Point, and S. S. W. two Points, *&c.* So likewise N. N. E. and N. N. W. is two Points from the North. Again, suppose a Ship sails S. W. ½ S. that is, three Points and an half, and N.N.E. ¼ E. is two Points and a Quarter; which being a Thing so commonly known I shall not need to add any more Examples.

The Number of Points thus found, observe on which Side of the Table they are found, . for if you find the Course at the Left-hand, then find the Denomination of Latitude and Departure at the Top: But if the Course be on the Right-hand find the Denomination of Latitude and De-

ture

parture at the Bottom; for the Left-hand Column contains all Points under four Points, and the Right-hand Column contains all the Rest of the Points to eight. Now if the Course be found in the first Columns, (viz. less than four Points) then the second Column contains the given Numbers for Difference of Latitude, and the third Column is the given Numbers to find Departure. But if the Course be in the last Column, (viz. above the four Points) then the third Column is the given Numbers to find Difference of Latitude, and the second is for the Departure, &c. And the Use of the Table being thus known for the first Case, the Rest is easily found by Consequence, as we have hinted before, and shall make it evident and intelligible to the meanest Capacity, by the following Examples.

Note; If you have a given Number, with a Cypher on the Left-hand, (as in 7¼ and 7¾ Points) although a Cypher adds Nothing to the Product (being on the Left-hand) yet it causeth a Figure more to be cut off from the Decimal Fraction; and having cut off as many Figures towards the Right-hand as you multiply by, the Rest is the Latitude or Departure sought, and those cut off are a Numerator of a Decimal Fraction to a Denominator, consisting of as many Cyphers as you cut off Figures, with a 1 at the Left-hand; so if you cut off 97, it is $\frac{97}{100}$, or if you cut off 342, it is $\frac{342}{1000}$, and so of the Rest.

Note; When you have found the given Number in the Table, you need not always use all the four Figures, especially, if the third or fourth be a small Digit, as 1 or 2, as suppose your given Number be 8315, here the third Figure being a 1, you may cut off the two last Figures, and use only 83 to work with. Again, if the third Figure be one of the highest Digits, as 8 or 9, you may cut off the two last, only add one to the second Figure, and so work as before; as suppose your Number be 2899, you may add 1 to the second Figure, viz. 8, and then cut off the two last, and then your Num-

ber

ber to work with will be 29 ; the Reafon of all which is very evident to any that underftand the Nature of Decimals, and will by a little Practice become eafy to the meaneft Capacity.

Plane Sailing.

CASE I.   *Courfe and Diftance given, to find the Difference of Latitude and Departure.*

The R U L E.

Multiply the Diftance by the given Numbers for the Courfe, and from the Product cut off as many Figures as you multiply by, and the Reft is Difference of Latitude and Departure required.

Queftion.   A Ship fails N.W. by N. 123 Miles; I demand her Difference of Latitude and Departure?

The given Number for Difference of Latitude being 8315, we fhall only ufe 83, and cut off the other two Figures, (towards the Right-hand) and multiplying the Diftance 123, by the given Number 83, and from the Product cut off two Figures, (becaufe you multiply by two Figures) you have Difference of Latitude ; but for Departure the given Number is 5556, the third Figure being a 5, we fhall for more Exactnefs ufe three Figures, *viz.* 555, and work as before : The Operation for Difference of Latitude and Departure will ftand thus :

For the Diff. of Latitude.	For Departure:
123	123
83	555
———	———
369	615
984	615
———	615
102|09	———
	68|265

The

The Difference of Latitude is 102 $\frac{99}{100}$, the Departure is 68 $\frac{34}{100}$; but you need not regard the Fractions, only if a Fraction be considerably above Half of one Mile, or Minute, you may add 1 to the whole Number; as suppose your Fraction be 78, *viz.* $\frac{78}{100}$, you may add 1 to the whole Number, and reject the Fraction, *&c.*

Another Example in Case I.

A Ship fails N.N W. $\frac{1}{2}$ W. 165 Miles; I demand the Difference of Latitude and Departure.

N.N.W. $\frac{1}{2}$ W. is 2 $\frac{1}{4}$ Points.

For Diff. of Lat.	*For Depart.*
165	165
88	47
———	———
1320	1155
1320	660
———	———
145|20 Diff. Lat. 145.	77|55

Dep. 77, or rather 78, because the Fraction is $\frac{55}{100}$.

CASE II. *Course and Difference of Latitude given, to find Distance and Departure.*

The R U L E.

For the Distance, add as many Cyphers to the Difference of Latitude as the Number of Figures you think to make use of in the given Number, and then divide it by the given Number, and the Quotient is the Distance, which being found, find the Departure as in *Case* I.

Question. A Ship fails S.W. by S. till her Difference of Latitude be 174; I demand the Distance and Departure.

The

The Courfe is 3 Points, the given Number 8315 and 6556.

For the Diftance.	For Departure.
83)17400(209 *Dift.*	209
· · 800	555
———	———
53	1045
	1045
	1045
	———
	115|995

The Diftance 209, and Departure 115, or rather 116, becaufe the Fraction is fo great, *viz.* above $\frac{9}{10}$.

CASE III. *Courfe and Departure given, to find Diftance, and Difference of Latitude.*

The R U L E.

Add fo many Cyphers to the Departure as you think to ufe Figures in the given Number; and then divide by the given Number anfwering to Departure, and the Quotient is the Diftance, which being found, find the Difference of Latitude, as in *Cafe* I.

Queftion. A Ship fails N. by E. ¼ E. till her Departure be 72 ; her Diftance and Difference of Latitude is required.

See Example the Second, in *Cafe* III. of *Plane Sailing.*

For the Distance.	For Diff. Lat.
337)72000(213	214
·460	94
1230	
———	856
219	1926
	———
	201|16

The Distance is 214, being so large a Fraction, and the Difference of Latitude is 201.

CASE IV. *Distance and Difference of Latitude given, to find the Course and Departure.*

The RULE.

Square the Distance (that is, multiply by itself) and likewise the Difference of Latitude, and subtract the lesser Number from the greater, the Square Root of the Remainder is the Departure required; and then for the Course, add two or three Cyphers to the Difference of Latitude, and divide that Sum by the Distance, the Quotient is the given Number for the Course, which being found in the Table, you have against it the Course required.

Note; If you cannot find exactly the same Numbers (which seldom happens) find the nearest to it, which is sufficient, and find the Course to the nearest Quarter Point.

Question. A Ship sails between the North and East 110 Miles; and then finds by Observation, that she hath raised the Pole 1 Degree, or 60 Min. or Miles; I demand the Course and Departure.

For

For the Course.   For the Departure.

```
110)60000(545      110      60    Sq. of Dist. 12100
     .50            110      60    Sq. Diff. Lat. 3600
     600           ____     ____
    ____           1100     3600   Diff. of Sq. 8500
     .50           110              . .
                  ____             8500(92
                  12100            81
                                  _____
                                  182)400
                                      364
                                  _____
                                      36
```

The Quotient 545 fought in the Table, the neareft given Number is 555, and againft that for the Courfe, you have 5 Points, which is N. E. by E. becaufe the Courfe is between North and Eaft, and the Departure is 92, for you need not regard the Fraction in the Square Root, it being but a fmall Part of a Mile or Minute.

CASE V.   *Diftance and Departure given, to find Courfe and Difference of Latitude.*

The RULE.

Subtract the Square of the Departure from the Square of the Diftance, the Square-Root of the Remainder is the Difference of Latitude; and then find the Courfe as in *Cafe* IV. only mind to find the Quotient in the Departure Column.

Queftion. A Ship fails in the S. W. Quarter 124 Miles, her Departure 95; I demand the Courfe and Difference of Latitude.

H   *For*

For the Course.   *For the Diff. of Lat.*

```
124)95c00(766    124        95    Sq. of Dist.—15376
    ·82          124        95    Sq. of Depart. 9025
    ·760         ——         ——    Diff.of Squares 6351
    ——                                    · ·
     16          496        475
                 248        855        6351(79
                 124        ——          49
                 ——         9025        ——
                 15376               149)1451
                                         1341
                                         ——
                                         110
```

The Course is S.W. ¼ W. nearest; for the greatest
given Number to the Quotient 766 is 773, which is the
Number for four ¼ Points, or S.W. ¼ W. and the Difference of Latitude is 79, or rather 80, because the
Remainder of the Square Root is so great: And here
observe, for a general Rule, that because the last Remainder in the Extraction is a Numerator to twice the
Root for a Denominator; therefore if the last Remainder be more than the Root, add one to the Root, and
that will be the Root exact enough in this Case; but if
the Remainder be less than the Root found, the Root
is the nearest whole Number that answers the Question,
and you need not regard the Fraction at all: Thus in
the Example, *Case* IV. the Root is 92, and the Remainder is but 36, and therefore I keep 92, for the Departure, and reject the Fraction. But in the Example
Case V. the Root is 79, and the Remainder is more than
the Root, *viz.* 110; and therefore I add one to 79, and
the Sum 80 is the Difference of Latitude required, and
so in others.

                                        CASE

C A S E VI. *Difference of Latitude and Departure given,*
to find the Course and Distance.

The R U L E.

Square the Difference of Latitude, and also the Departure, and add the two Squares together; the Square
Root of the Sum is the Distance required; and then find
the Course, as in *Case* IV. or *Case* V.

Question. A Ship sails in the N.W. Quarter, till her
Difference of Latitude be 220, and her Departure 108;
I demand the Course and Distance.

For the Distance.

220	108	Square of Diff. Lat.	48400
220	108	Square of Departure	11664
44	864	Sum of the Squares	60064
44	1080		
		60064(245	
48400	11664	4	
		44)200	
For the Course		176	
245)220000(867		485)2464	
2400		2425	
1950			
		39	
395			

The Course is 2 Points and a Quarter, or N.N.W.
¼ W. the Distance 245 Miles.

Although the Operation in this Case seems the most
tedious of all the rest, because we have assumed greater
Numbers for the *Data* of the Question, yet even in the
greatest Numbers, this Way is much more expeditious

H 2   than

than working by a Canon, as a little Practice will make manifest.

SECT. II.

Mercator's Sailing Arithmetical.

WHAT I have wrote of *Plane Sailing* is also of use in *Mercator's Sailing*, so far as Course, Distance, Difference of Latitude, and Departure, are Terms either given or required; and the same Table of given Numbers is of equal Use in both; and as for the Difference of Longitude, it bears the same Proportion to the Departure, that the Meridional Difference of Latitude bears to the true Difference of Latitude, by *Euclid. Lib.* 6. *Prop.* 4. and indeed by the Help of a Table of Meridional Parts, the whole Practice of *Plane Sailing*, as performed by given Numbers, may be reduced to *Mercator*, by *Euclid. Lib.* 6. *Prop.* 4. For both being projected by a Plane Triangle, and the Course, whether given or required, being the same in both, there must needs be the same Proportion between the enlarged Distance, and the true Distance, (they being supposed to be the Hypotenuse of 2 Equiangular Triangles) or between the Difference of Longitude and Departure, (the Perpendiculars of the said 2 Equiangular Triangles) as there is between the Meridional Difference of Latitude, and the proper Difference of Latitude, (being the Bases of the said two Triangles) and upon these Propositions the whole Fabrick of *Mercator's Sailing* is grounded; the exact Truth of which is proved not only by the Demonstration of *Euclid*, in the Proposition above-named; but also by comparing the Questions thereby performed with those in the first Part, performed by a Canon, in which I shall, for the Reader's Satisfaction, refer to the said Question, as in the following Examples.

CASE

C A S E I. *One Latitude, Course and Distance given, to find the other Latitude, Departure and Difference of Longitude.*

The R U L E.

Find Difference of Latitude and Departure, as in *Case* I. of *Plane Sailing Arithmetical*; and then for Difference of Longitude, the Proportion is, As proper Difference of Latitude, is to Meridional Difference of Latitude; so is Departure, to Difference of Longitude.

Question. A Ship in Lat. 50d. 00m. North, sails N.W: by N. 987 Miles; I demand the Latitude come to; together with her Departure, and Difference of Longitude.

For Difference of Latitude: *For Departure.*

```
      987                        987
       83                        555
   ─────                    ─────
     2961                       4935
     7896                       4935
   ─────                       4935
   819|2 L                  ─────
                             547|785
```

Diff. of Lat. 819 or rather 820, Dep. 548, Merid. Diff. of Lat. 1519, then for Diff. of Longitude; As 820 to 1519, so 548 to Diff. of Longitude.

```
     1519            820)832412(1015
      548               .124
   ─────              ─────
    12152               4212
     6076             ─────
     7595               112
   ─────
    832412
```

H 3   CASE

CASE II. *Both Latitude and Course given, to find the Distance, Departure, and Difference of Longitude.*

The RULE.

Find the Distance and Departure, as in Case the Second of *Plane Sailing Arithmetical*, and then find the Difference of Longitude, as in the first Case of *Mercator's Sailing.*

Question. A Ship sails from a Port in Latitude 56d. 25m. North, and sails away N.N.E. several Days, and then finds herself by Observation in Latitude 68d. 30m. North; I demand how far she hath sailed, and what is her Difference of Longitude?

For the Distance.	*For the Departure.*
924)725000(784	785
·7820	383
4280	———
———	2355
584	6280
	2355
	———
	300│655

As 725 to 1593, so 300 to Diff. Longitude.

300	725)477900(659
———	4290
477900	·6650
	———
	125

The Distance is 784 $\frac{464}{924}$, or rather 785, The Departure 300, and the Difference of Longitude 659 $\frac{425}{725}$, or 660.

CASE

CASE III: *One Latitude, Courſe, and Difference of Longitude given, to find the other Latitude, Diſtance and Departure.*

The RULE.

Add ſo many Cyphers to the Difference of Longitude as you intend to uſe Figures in the given Number to divide by, and work as in Caſe III. of *Plane Sailing Arithmetical*; only obſerve, that whereas the Quotient there is the true Diſtance, becauſe you work by Departure, here the Quotient is the enlarged Diſtance or the whole Hypotenuſe contained between the Angle of the Courſe, and the Top of the longeſt Perpendicular repreſenting the Difference of Longitude; and then the enlarged Diſtance being found, from the Square of the enlarged Diſtance, ſubtract the Square of the Difference of Longitude, the Square Root of the Remainder is the Meridional Difference of Latitude, which being found, the one Latitude given, the other Latitude is eaſily found: Then for Departure, it is only the former Proportion inverted, *viz.* As Meridional Difference of Latitude to proper Difference of Latitude, ſo Difference of Longitude to Departure.

Queſtion. A Ship in Lat. 55d. 00m. North, ſails N.N.W. half W. till her Difference of Longitude be 1d. or 60m. I demand the Diſtance, Departure, and Latitude come to.

For the enlarged Diſtance.

$$47)6000(127\tfrac{31}{47}$$

130
360
———
31

Square of the enlarged Diſtance	16215
Square of Diff. of Longitude	3600
Difference of the Squares ——	12685

12685

$$12685(112$$
$$1$$

$$21)026$$
$$21$$

$$222)585$$
$$444$$

$$141$$

Merid. Parts of Lat. fail'd from 54d. oom. — 3968
Merid. Diff. Lat. add ———— ———— 113

Merid. Parts of Lat. come to—56   04 ——— 4081

As Merid. Diff. Lat. 113 to proper Diff. Lat. 64, fo
is the enlarged Dift. 128 to the true Dift. 72¼.

$$
\begin{array}{ll}
128 & 113)8192(72 \\
64 & .282 \\
\hline
512 & .56 \\
768 & \\
\hline
8192 &
\end{array}
$$

For the Departure.

As Merid. Diff. Lat. 113, is to proper Diff. Lat. 64
fo is Diff. of Longitude 60, to Departure.

$$
\begin{array}{ll}
64 & 113)3840(33 \text{ Dep. or rather } 34. \\
60 & .450 \\
\hline
3840 & 111
\end{array}
$$

CASE

CASE IV. *Both Latitudes and Distance given, to find the Course, Departure and Difference of Longitude.*

The R U L E.

Find the Course and Departure, as in *Case* VI. of *Plane Sailing Arithmetical*, and then find Difference of Longitude, as in *Case* I. of *Mercator's Sailing Arithmetical*.

Question. A Ship in Lat. 55d. 00m. North, sails between the North and East 86 Miles, and then is by Observation in Lat. 56d. 10m. N. I demand as above.

See the same Question in *Mercator's Sailing, Case* V.

For the Departure.

86		Sq. of Dist. ———— 7396	2496(49
86	70	Sq. of Diff. of Lat. 4900	16
———	70		———
516	———	Diff. of Sq. 2496	89)896
688	4900		801
———			———
7396			95

For the Course.   *For the Difference of Longitude.*

86)70000(813   As 70 to 124, so 50 to Diff. Long:
  120   50   70)6200(88⅝
  340   ———   .600
 ——   6200   ———
  82   40

The Departure 50, the Course N.E. ¾ N. the Diff. of Longitude 88 ⁴⁄₇ or rather 89.

CASE

CASE V. *Both Latitudes and Departure given, to find the Course, Distance, and Difference of Longitude.*

The RULE.

Find the Course and Distance, as in *Case* VI. of *Plane Sailing Arithmetical*; and Difference of Longitude, as in the first Case of *Mercator's Sailing Arithmetical*.

Question. A Ship in Lat. 55d. om. North sails between the North and East into Lat. 56d. 10m. North, her Departure 50 Miles; I demand the Course, Distance and Difference of Longitude.

For the Distance.

70	50	Sq. of Diff. Lat.	4900	7400(86
70	50	Square of Dep.	2500	64
4900	2500	Sum of Squares	7400	166)1000
				996
				004

For the Course.

86)70000(813
120
340
——
82

For Diff. of Long.

As 70 to 124, so 50 to Diff. of Long.

50
——
6200

70)6200(88⅞
60
——
40

. The Course is N.E. ¼ N. the Distance 86; the Difference of Longitude 88 ⅞ or rather 89.

CASE

CASE VI. *Both Latitudes and Difference of Longitude given, to find Course, Distance, and Departure.*

The RULE.

Square the Difference of Longitude, and also Meridional Difference of Latitude, the Square Root of the Sum of these two Squares is the enlarged Distance: Then, as Merid. Diff. of Latitude, is to proper Diff. of Latitude; so is the enlarged Distance, to the true Distance: And then find the Course as in *Plane Sailing Arithmetical, Case* IV.

Question. Suppose two Places, one in Lat. 56d 15m. North, the other in Lat. 58d. 35m. their Diff. of Long. 2d. 30m. or 150 Min. I demand the Course and Distance from the Southermost to the Northermost.

For the Distance.

$$
\begin{array}{cc}
150 & 260 \\
150 & 260 \\
\hline
75 & 156 \\
15 & 52 \\
\hline
22500 & 67600 \\
\end{array}
$$

Sq. of Mer. Diff. of Lat. 67600
Sq. of Diff. of Long. — 22500

Sum of the Squares —— 90100

$$
\begin{array}{l}
90100(300 \\
9 \\
\hline
60)001 \\
000 \\
\hline
600)100 \\
\end{array}
$$

As

As 260 to 140, so 300 to the Diſtance.

```
        300        260)420000(161    For the Courſe.
                         160         161)140000(869
      42000            . . 400            1120
                      ————                1540
                         140          ————
                                          091
```

The Diſtance 161.   The Courſe N.N.W. ¾ W.

CASE VII. *One Latitude, Courſe, and Departure given, to find the other Latitude, Diſtance, and Difference of Longitude.*

The RULE.

Find the Diſtance by *Caſe* III. of *Plane Sailing Arithmetical*, and the Difference of Latitude as in *Caſe* V. of the ſame, and Difference of Longitude as in all the other Caſes hereof.

Queſtion. A Ship in Lat. 58d. 00m. North, ſails S.S.W. half W. till her Departure be 60 Miles, I demand as above.

```
   For the Diſtance.          For Diff. of Lat.

  47)6000(127            127        60
     130                 127        60
     360                ————      ————
    ————                 889       3600
      31                 254
                         127
                        ————
                        16129
```

Square of Diftance —— 16129   12529(111

Square of Departure — 3600   1

Diff. of the Squares — 12529   21)025

     21

     221)429

     221

     208

For the Difference of Longitude.

As proper Diff. of Lat. 112, to Merid. Diff. Lat. 206, fo Departure 60 to Diff. of Long.

     206   112)12360(110

     60   .116

     12360   ..40

The Difference of Latitude 112 Min. or 1d. 52m. which fubtracted from the Latitude failed from, becaufe the Courfe is Southerly, the Remainder is the Latitude come to, 56d. 8m. the Diftance 127; the Diff. Longitude 110.

CASE VIII. *One Latitude, Diftance, and Departure given, to find the other Latitude, Courfe, and Difference of Longitude.*

The   R U L E.

Find the Courfe and Difference of Latitude as in *Cafe* V. of *Plane Sailing Arithmetical*; then find the Difference of Longitude as before.

Queſtion.

Queftion. A Ship in Lat. 58d. 00m. North, fails between the South and Weft 127 Miles, her Departure
60, I demand the reſt.

For the Difference of Latitude.

```
                                              . . .
  127        Sq. of Dift. — 16129        12529(111
  127    60  Sq. of Dep. — 3600          1
         60
 ───         ───        Diff.of Squares 12529      21(025
 889        ───                               21
 254    3600
 127                                          ───
 ───         .                            221)429
16129                                         ───
                                              208
```

For the Courfe   *For Difference of Longitude.*

```
127)60000)472     · As 112 to 206, fo 60 to Diff. Long.
    ·920                  60      112)12360(110
    · 310                 ───         ·116
    ───                  12360         ───
      56                             ·· 40
```

The Courfe S. S. W. ¼ W. the Lat. come to 56d.
8m. and Diff. of Long. 110.

S E C T. III.

Parallel Sailing Arithmetical.

BEFORE I proceed to the Solution of the feveral
Cafes of *Parallel Sailing*, I fhall infert a Table of
given Numbers, which is equally ufeful in *Parallel* and
Middle Latitude Sailing ; for by thofe Numbers the Departure in *Middle Latitude Sailing*, (which is the Diftance
in a Parallel, or Eaft and Weft Courfe) is reduced to
Difference of Longitude without any Logarithms, Sines,
or Tangents, only by one Operation in Divifion, the Ufe
of which Table in *Parallel Sailing* is thus :

When you have the Diftance in the Parallel given,
add fo many Cyphers to the Diftance as you intend to
ufe Figures in the given Number, and then divide the
one by the other, and the Quotient is the Difference of
Longitude : But if Difference of Longitude, and the
Latitude is given, multiply the Difference of Longitude
by the given Number belonging to that Latitude, and
from that Product cut off as many Figures as you multiply by, and the reft is the Diftance ; and confequently
if Diftance and Difference of Longitude be given,
and Latitude required, add two, three or four Cyphers to the Diftance, and divide that Sum by the Difference of Longitude, the Quotient found in the Table
is the given Number, anfwering to the Latitude required, *&c.*

A TABLE

A TABLE of given Numbers for Parallel and Middle Latitude; the given N° for Lat. ood. oom. being 10000.

Lat.	N°	Diff.	Lat.	N°	Diff.	Lat.	N°	Diff.
1	9998	. . 2	31	8572	. : 88	61	4848	. 152
2	9994	. . 4	32	8480	. . 91	62	4695	. 153
3	9986	. . 8	33	8387	. . 93	63	4540	. 155
4	9976	. . 10	34	8290	. . 96	64	4384	. 156
5	9962	. . 14	35	8191	. . 99	65	4226	. 158
6	9945	. . 17	36	8090	. 101	66	4067	. 159
7	9925	. . 20	37	7986	. 104	67	3907	. 160
8	9902	. . 23	38	7880	. 106	68	3746	. 161
9	9877	. . 26	39	7771	. 109	69	3584	. 162
10	9848	. . 29	40	7660	. 111	70	3420	. 163
11	9816	. . 32	41	7547	. 113	71	3256	. 164
12	9781	. . 35	42	7431	. 116	72	3090	. 165
13	9744	. . 37	43	7313	. 118	73	2924	. 166
14	9703	. . 41	44	7193	. 120	74	2756	. 167
15	9659	. . 44	45	7071	. 122	75	2588	. 168
16	9613	. . 47	46	6946	. 124	76	2419	. 169
17	9563	. . 50	47	6820	. 126	77	2249	. 170
18	9511	. . 53	48	6591	. 129	78	2079	. 170
19	9455	. . 56	49	6560	. 131	79	1908	. 171
20	9397	. . 59	50	6428	. 133	80	1736	. 172
21	9336	. . 61	51	6293	. 135	81	1564	. 172
22	9272	. . 65	52	6157	. 137	82	1392	. 172
23	9205	. . 67	53	6018	. 139	83	1219	. 173
24	9135	. . 70	54	5878	. 141	84	1045	. 174
25	9063	. . 72	55	5735	. 143	85	0871	. 174
26	8988	. . 75	56	5592	. 144	86	0697	. 174
27	8910	. . 78	57	5446	. 146	87	0523	. 174
28	8829	. . 81	58	5299	. 147	88	0349	. 174
29	8746	. . 83	59	5150	. 149	89	0175	. 174
30	8660	. . 86	60	5000	. 150	90	0000	. 175

The

The Use of this Table of given Numbers for *Parallel* and *Middle Latitude Sailing*, is partly laid down in the Page preceeding the Table, and shall be further exemplified in the several Cases of *Parallel* and *Middle Latitude Sailing*; and the same Cautions are to be observed in the Use of this, that were to be observed in the first Table of given Numbers, *viz.* That for *Plane Sailing Arithmetical*; for if the third Figure of the given Number be a Cypher, or 1 or 2, you need use but two Figures (unless your Question be proposed in very great Numbers;) or if the third Figure be 8 or 9, you may add 1 to the second Figure, and so proceed in your Work: But in this Table, as also in that of *Plane Sailing*, if your Question be propounded in Numbers consisting of three or four Figures, it is best to use at least as many Figures of the given Number, as there is in the *Data*, or given Terms of your Question, as we shall make evident by the succeeding Examples.

I have set down the given Numbers, only to whole Degrees of Latitude; the Numbers to the intermediate Minutes being easily found by proportioning the Differences set down in the third, sixth, and ninth Columns of the Table, the Use of which is this: Suppose in a Parallel Course, the Latitude of the Parallel (or in a *Middle Latitude* Question, the *Middle Latitude*) be 62 Degrees, and I desire to find in the Table the given Number to that Latitude; I look in the Table under [*Lat.*] and find 62, and against it, under [*Num.*] you find 4695, the given Number required: But if you would find the given Number for Lat. 59d. 30m. I look for the Number for Lat. 60, and find 5000, and it being a Latitude between 59 and 60, I look against 60, under [*Diff.*] for the Difference (for the Difference between the given Numbers for both Latitudes is always against the greater Latitude) and the Difference is 150, therefore I take Half of 150, because 30 Minutes is half a Degree, and add it to the given Number for 60 Degrees, *viz.* 5000, the Sum 5075 is the given Number

I   ber

ber for the Lat. 59d. 30m: which was required to be known.

Note; The given Number for Lat. od. om. is 10000, and from thence, as the Latitude encreafes, the Numbers decreafes, fo that the Number for Lat. 90 is 0000 ; hence the greater the Latitude is, the leffer is the given Number; therefore obferve always when you would find the Number to any Latitude confifting of Degrees and Minutes, take the given Number to the bigger Latitude, and to that add the proportional Part of the Difference found againft the bigger Latitude, the Sum is the given Number required. One Example will make it plain.

Suppofe I defire to know the given Number for Latitude 45d. 40m. the Number for 46 (the bigger Latitude) is 6946, and the Difference 124 ; one third of which, *viz.* 41 neareft, added to the Number 6946, the Sum 6987 is the Number required

But if you want the given Number to any odd Minutes, which are not an even Part of a Degree, as half, or a third, *&c.* you may find it by Proportion if you would be very exact, and the Proportion is, As 60 Min. is to the whole Difference; fo is the Minutes, which the given Latitude wants of the bigger Latitude, to the Number to be added to the given Number of the bigger Latitude.

I defire to know the given Number for Latitude 43d. 24m.

The Latitude 43d. 24m. wants 36 Min of 44 Deg. therefore, as 60 to the Difference 120, fo 36 to 72 ; which added to 7193, the given Number for the bigger Latitude, the Sum 7265 is the Number for Lat. 43d. 24m. required.

Note; If you ufe all the four Figures of the given Number, the Difference is as is expreffed in the Table; but if you cut off one Figure or two, *&c.* from the given Number to the Right-hand, you muft alfo cut off as many from the Difference, and the Remainder

is

is the Difference; only if the laft Figure of the Difference be 8 or 9, you may add one to the third Figure,
as in the other Cafes before mentioned; the Reafon is
evident to them that underftand Decimals.

Example. I defire to know the given Number for Latitude 69d. 45m. I cut off one Figure to the Right-hand
of each collateral Number, *viz.* Latitude 69, and Latitude 70, and then the Numbers are 358 and 342:
And feeing I cut off a Figure from the Number, I
muft alfo cut one off from the Difference, and then the
Difference is 16, which ufed according to the foregoing
Rules, the given Numbers for Latitude 69d. 45m. will
be found to be 346, and fo in others.

CASE I. *Two Ships or Places in one Parallel, their Latitude and Diftance given, to find their Difference of
Longitude.*

The RULE.

Add fo many Cyphers to the Diftance as you think
to ufe Figures in the given Number and divide by the
given Number for the Latitude of the Parallel, the
Quotient is the Difference of Longitude.

Queftion. Two Ships in Latitude 50d. 0m. diftant
76 Miles; I demand their Difference of Longitude.

The given Number for Latitude 50 is 6428, but if
I cut off one Figure to the Right-hand, it will be 642;
but becaufe the laft Figure, *viz.* that to the Right-
hand, which I cut off, is 8, I add one to the third, and
then it will be 643; and becaufe I ufe three Figures of
the given Number, I add three Cyphers to the Diftance,
and then divide one by the other, according to the a-
forefaid Rule, and the Work will ftand thus:

I 2   643)76000(118

643)76000(118 The Diff. of Long. 118Min. or 1d. 58m.
     1170
      5270
      ────
       126

Or if I cut off two Figures, the reſt to work with will
be 64, and then I add but two Cyphers to the Diſtance
(becauſe I uſe but two Figures of the given Number)
and the Work will ſtand thus:

64)7600(118   The ſame with the fore-
 . 120   going Operation.
   560
   ───
    48

C A S E II. *Two Places in one Parallel, the Latitude,
and Difference of Longitude given, to find their Diſtance.*

The R U L E.

This is only the former Caſe inverted; for having
found the given Number for the Latitude, multiply
the Difference of Longitude by it, and from the Product cut off as many Figures as you multiply by, and
the reſt is the Diſtance required; and thoſe cut off are
a Decimal Fraction to be underſtood as in the Rules
laid down at the Beginning of *Plane Sailing Arithmetical.*

Queſtion. Two Places in Latitude 50, their Difference
of Longitude 118; I demand their Diſtance.

        643
        118
        ────
       5144 The Diſtance 75.874, or rather 76.
       643
      643
      ─────
      75|874

                              C A S E

CASE III. *Two Places, their Distance and Difference of Longitude given, to find the Latitude they are in.*

The RULE.

Add three or four Cyphers to the Distance, and divide that Sum, by the Difference of Longitude, the Quotient is the given Number for the Latitude required.

Question. A Ship sails due West 2700 Miles, and then finds her Difference of Longitude to be 4200 Min. I demand what Latitude the Ship is in?

$$4200)2700000(642$$
$$180$$
$$12000$$
$$\overline{}$$
$$3600$$

By adding three Cyphers to the Distance, and working as before directed, you find the Quotient 642; or if you add four Cyphers, it will bring out four Figures in the Quotient, *viz.* 6428, which found in the Table, you find it against Latitude 50, which is the Latitude of the Parallel required.

CASE IV. *Two Ships sailing directly North or South, their Distance in one Parallel given, to find their Distance in the other Parallel.*

The RULE.

Find the given Numbers for each Latitude, and say by the Rule of Three: As the given Number for the first Latitude, is to the given Number for the second Latitude; so is the Distance in the first Latitude, to the Distance in the second Latitude.

I 3

Question

Queſtion. Suppoſe two Ships in Lat. 50, diſtant 200 Miles, ſails both directly North into Lat. 73d. 0m. I demand their Diſtance in that Parallel.

Here we ſhall uſe but the two firſt Figures of the given Numbers, becauſe the third Figure in each Number is but ſmall, *viz.* 2 : And then by the foregoing Rule.

As 64 to 29, ſo 200 to the Diſtance in Lat. 73.

$$\begin{array}{ll} \begin{array}{r} 29 \\ 2|00 \\ \hline 58c0 \end{array} & \begin{array}{l} 64)5800(90 \\ \hline \quad \cdot\cdot40 \end{array} \end{array}$$

Their Diſtance in Latitude 73, is 90 $\frac{40}{64}$, or (if you will avoid Fractions) 91 Miles.

CASE V. *Two Ships in one Parallel, with their Diſtance in that Parallel given, and ſailing both directly North or South, with their Diſtance in the Parallel ſailed, given, to find the Latitude come to.*

The RULE.

This is only the former Proportion in (*Caſe* IV.) inverted, *viz.* As the Diſtance in one Latitude, is to the Diſtance in the other Latitude ; ſo is the given Number for one Latitude, to the given Number for the other Latitude.

Queſtion. Two Ships in Lat. 50, diſtant 200 Miles, ſail both directly North, till they are but 91 Miles diſtant ; I demand the Latitude come to?

As 200 to 91, ſo 64 to the given Number for the Latitude come to.

```
   91
   64                200)5824(29
 ------                   182
   364                 -------
   546                   24
 ------
  5824
```

The Quotient 29 is the two firft Figures of the given Number for the Lat. come to, *viz.* Lat. 73d. om.

SECT. IV.

Middle Latitude Sailing Arithmetical.

IN working the feveral Cafes and Queftions of *Middle Latitude Sailing*, we fhall have occafion to make Ufe of both the Tables of given Numbers ; for in all Cafes or Varieties, where the Courfe, Diftance, Difference of Latitude, and Departure, are Terms either given or required, the Operation is performed by the firft Table, as in *Plane Sailing* ; but when Difference of Longitude is a Term in the Queftion, (whether given or required) the fecond Table is of Ufe, as will appear by the following Examples.

CASE I. *One Latitude, Courfe, and Diftance given, to find the other Latitude, Departure, and Difference of Longitude.*

The RULE.

By Courfe and Diftance, find Difference of Latitude and Departure, (and confequently the other Latitude) as in *Cafe* I. of *Plane Sailing Arithmetical* ; and then for the Difference of Longitude, it is found by the Second

I 4   Table

Table thus : Find the given Number for the Middle Latitude, of which you may use two or three, or four Figures, according to the Rules laid down at the Beginning of *Plane Sailing*, and also at the Beginning of *Parallel Sailing Arithmetical*; then to the Departure add so many Cyphers to the Right hand, as you intend to use Figures in the given Number; that Sum divided by the given Number, the Quotient is the Longitude required.

Question. A Ship in Lat. 50d. 00m. North, sails N.W. by N. 987 Miles: I demand the Latitude come to, with Departure, and Difference of Longitude.

For the Difference of Latitude and Departure by the first *Case of Plane Sailing Arithmetical.*

For Diff. of Lat.	*For the Departure.*
987	987
83	556
———	———
2961	5922
7896	4935
———	4935
8 19|2 1	———
	548|772

Although the three first Figures of the given Number (to find your Departure) for that Course be 555, yet the last Figure being 6, I cut it off and add 1 to the third Figure, and call it 556, as in the Operation above, and so the Difference of Latitude is 819, or rather 820, and the Departure 549, allowing one for the Fraction.

Then for the Difference of Longitude.

The Diff. of Latitude being 820 Min. or 13d. 40m. the Latitude come to is 63d. 40m. and consequently the Middle Latitude is 56d. 50m. the given Number for
that

that Latitude (found by the Rules laid down in the Beginning of *Parallel Sailing*) is 5470; but the laft Figure being a Cypher, I cut it off, and work with the reft, as in the Rule above, and the Operation ftands thus:

547)549000(1003   The Difference of Longitude
..2000   is 1003 Min. or 16d. 43m.
————   Weft.
    359

CASE II. *Both Latitudes and Courfe given, to find the Departure, and the Difference of Longitude.*

The R U L E.

Find the Diftance and Departure, as in Cafe the fecond of *Plane Sailing Arithmetical*, and then find the Difference of Longitude, as in the firft Cafe of *Middle Latitude Sailing Arithmetical.*

Note; In *Middle Latitude Sailing Arithmetical*, this Way of finding the Longitude anfwers in all the Cafes thereof, and therefore I fhall not need to mention it in any of the reft.

Queftion. A Ship takes her Departure from a Port in Latitude 56d. 25m. North, and fails N.N.E. till fhe finds herfelf to be by Obfervation in Latitude 68d. 30m. North; I demand the Diftance failed, with her Departure, and Difference of Longitude,

The Difference of Latitude is 12d. 05m. or 725 Min.

For the Diftance.	*For the Departure.*
924)725000(784	785
·7820	383
4280	————
————	2355
584	6280
	2355
	————
	300|655

Tho

The Middle Latitude 62d. 28m.
The given Number for 62d. 28m. is 4622.

Then for the Difference of Longitude.

461)301000(652
    2440
    ˙1350
    ——
    428

The Diff. 785, the Departure 301; the Diff. of
Longitude 653 Min. or 10d. 53m.

CASE III. *Both Latitudes and Distance given, to find
the Course, Departure and Difference of Longitude.*

The RULE.

Find the Course and Departure, as in *Case* IV. of
Plane Sailing Arithmetical, and the Difference of Longitude, as in other Cases hereof.

Question. A Ship in Lat. 55d. 00m. sails between
the North and East 86 Miles, and then finds herself by
Observation in Lat. 56d. 10m. her Course, Departure,
and Difference of Longitude is required.

For the Departure by the Square Root.

86	70	Square of the Distance	7396
86	70	Square of the Departure	4900
——	4900	Diff. of the Squares ——	2496
516			
688			
——			
7396			

2496

2496(49
16
──────
89)896
801

The Departure 50 neareft, becaufe the Fraction is more than the Root by the general Rule in *Cafe* V. of *Plane Sailing Arithmetical.*

──────
95

For the Difference of Long.

The Midd. Lat. is 55d. 35m.
The given Number 5652
565)50000(88
4800
──────
280

For the Courfe.

86)70000(813
120
340
──────
82

The Departure 50; the Courfe N.E. ¾ N. and the Difference of Longitude 88 ½ *fere.*

CASE IV. *Both Latitudes and Departure given, to find the Courfe, Diftance and Difference of Longitude.*

The RULE.

Find the Courfe and Diftance by *Cafe* VI. of *Plane Sailing Arithmetical*; and Longitude as in the other Cafes.

Queftion. A Ship in Lat. 55d. 0m. North, fails between the North and Eaft into Lat. 56d. 10m. North, her Departure 50 Min. or Miles, the Courfe, Diftance and Difference of Longitude are required.

The Difference of Latitude is 70 Miles.

For the Diftance by the Square Root.

70	50	Square of Diff. Lat. 4900	7400(86
70	50	Square of Dep. — 2500	64
────	────	────────	────
4900	2500	Sum of Squares — 7400	166)1000
			996
			────
			004
			Fo'

For the Course.   *For the Diff. of Long.*

86)70000(813   The Mid. Lat. 55d. 35m.
120   The given Numb. 5652
340
——   565)50000(88
82   4800
                                      ——
                                      280

The Diftance 86 ; the Courfe N.E. ¼ N. the Diff. of Longitude 88 ¼ Min. or 1d. 28m. Eaft.

CASE V. *Both Latitudes and Difference of Longitude given, to find Courfe, Diftance, and Departure.*

The R U L E.

In all Cafes where the Difference of Longitude is given, and Departure required, the Operation is but the Inverfe of that for finding Difference of Longitude when Departure is given ; for if the Departure, with Cyphers added to the Right-hand, and divided by the given Number for the Middle Latitude, produce the Difference of Longitude; then the Difference of Longitude, multiplied by the given Number for the Middle Latitude, muft needs produce the Departure, with Figures to be cut off; becaufe any Quotient multiplied by the Divifor produceth the Dividend; therefore, to find the Departure in this Cafe, multiply the Difference of Longitude by the given Number for the Middle Latitude, and from the Product cut off as many Figures to the Right-hand as you multiply by, and the reft is the Departure : thus having the Difference of Latitude and Departure, find the Courfe and Diftance, as in *Cafe* VI. of *Plane Sailing Arithmetical.*

Question. I demand the Course, Distance and Departure between two Places, one in Lat. 56d. 15m. North, the other in Lat. 58d. 35m. North, their Difference of Longitude 150 Min. or 2d. 30m.

For the Departure.

The Middle Latitude being 57d. 25m. and the given Number 5384.

$$
\begin{array}{r}
5384 \\
150 \\
\hline
269200 \\
5384 \\
\hline
80|7600
\end{array}
$$

For the Course.		*For the Distance by the Square Root*	
161)14000(86)	140	81 Sq. of Diff. Lat.	19600
·1120	140	81 Square of Dep.	6561
·1540·			
———	5600	81 Sum of Squares	26161
··91	140 648	· · ·	
	———	26161(161	
	19600 6561	1	
		26(161	
		156	
		———	
		321(561	
		321	
		———	
		240	

The Distance is 161 ; the Departure 80 $\frac{74}{112}$, or (to avoid Fractions) 81 ; the Course 2 ¼ Points to the Eastward or Westward of the North or South, according to the Denomination of the Longitude, whether East or West.

These

These being the most useful Cases, I shall (as in *Middle Latitude Sailing Trigonometrical*) set down only the *Data* and *Quæsita* of the rest, with Rules for working them, and leave the Operation for the Learner's Practice.

CASE VI. *One Latitude, Course, and Departure given, to find the other Latitude, Distance, and Difference of Longitude.*

The R U L E.

Find the Distance and Difference of Latitude, (and consequently the other Latitude) by *Case* III. of *Plane Sailing Arithmetical,* and Difference of Longitude as in the other Cases.

CASE VII. *One Latitude, Distance and Departure given, to find the rest.*

The R U L E.

Find the Course and Difference of Latitude by *Case* V. of *Plane Sailing Arithmetical,* and Difference of Longitude as before.

CASE VIII. *Course, Departure and Difference of Longitude given, to find both Latitudes and Distance.*

The R U L E.

Find Distance and Difference of Latitude by *Case* III. of *Plane Sailing Arithmetical*; then divide the Departure (with Cyphers added on the Right-hand) by the Difference of Longitude, the Quotient is the given Number for the Middle Latitude, consisting of as many Figures as you added Cyphers; and the Middle Latitude, and Difference of Latitude, thus given, or found, the other Latitude is easily found; but because Departure and Difference of Longitude are given, to find the Middle Latitude, (and consequently both Latitudes) is a

Variety

Variety that we have not had an Example in, I shall instance in one Question for the Reader's Practice.

Question. A Ship sails N.N.E. till her Departure be 301, and Difference of Longitude 651; I demand both Latitudes and Distance?

For the Distance.	*For Diff. Lat. by the Square Root.*	
383)301000(785	785	301
·329	785	301
·2260		
	3925	301
345	6280	9030
	5495	
		90601
	616225	

Square of the Distance —— —— 616225
Square of the Departure —— —— 90601

Difference of the Squares —— —— 525624

For the Middle Latitude 525624(724
                                            49

651)301000(462	142)356
4060	284
1540	
	1444)7224
238	5776
	1448

The Number 462 is the three first Figures of the given Number for the Middle Latitude, which answer to Latitude 62d. 28m. then half the Difference of Latitude added to, and subtracted from the Middle Latitude, gives the two Latitudes 56d. 25m. and 68d. 30m. and the Distance is 785.

Note;

Note, Although the Difference of Latitude comes out but 724, yet the Fraction is so great that if the Number had been but one Unit more, the Root had been 725.

CASE IX. *Middle Latitude, Course and Distance given, to find both Latitudes, Departure and Difference of Longitude.*

The RULE.

By Course and Distance find Difference of Latitude and Departure, by *Case* I. of *Plane Sailing Arithmetical,* and then find Difference of Longitude, as in other Cases.

CASE X. *Course, Distance, and Difference of Longitude given, to find both Latitudes and Departure.*

The RULE.

Find Difference of Latitude and Departure by *Case* I. of *Plane Sailing Arithmetical,* and then find Middle Latitude, and consequently both Latitudes, by *Case* VIII. hereof.

CASE XI. *Course, Difference of Latitude, and Difference of Longitude given, to find both Latitudes, Distance and Departure.*

The RULE.

Find the Distance and Departure by *Case* II. of *Plane Sailing Arithmetical,* and then find Middle Latitude, and thereby both Latitudes by *Case* VIII. hereof.

CASE

CASE XII. *Middle Latitude, Course, and Difference of Longitude given, to find both Latitudes, Distance and Departure.*

The R U L E.

For the Departure, multiply the Difference of Longitude by the given Number for the Middle Latitude, and from the Product cut off as many Figures to the Right-hand as you multiply by, and the Rest is the Departure; and thus having the Course and Departure, you may find the Distance and Difference of Latitude, and, (by the Help of Middle Latitude) both Latitudes by *Case* III. of *Plane Sailing Arithmetical.*

C A S E XIII. *One Latitude, Departure and Difference of Longitude given, to find the other Latitude, Course, and Distance.*

The R U L E.

Find the Middle Latitude by *Case* VIII. hereof, and then having one Latitude, and Difference of Latitude, the other Latitude is easily found; and thus having both Latitudes and the Departure, find Course and Distance by *Case* VI. of *Plane Sailing Arithmetical.*

And thus have I gone through the thirteen Cases of *Middle Latitude Sailing Arithmetical,* wherein the first five Cases, in which there are Questions proposed and wrought, I have set down the very same Questions here that are in *Middle Latitude Sailing Trigonometrical,* that the Reader may see the exact Harmony between the *Trigonometrical* and *Arithmetical Operations,* and consequently may be convinced of the Truth and Certainty of this Way of working, so far as *Middle Latitude Sailing* is to be depended upon; and although *Middle Latitude Sailing* deviates somewhat from *Mercator,* especially at a great Distance from the Equinoctial, or where the Question is proposed in great Numbers, yet it may be of great Use, in keeping a Reckoning; because, in a single Day's Work (which seldom exceeds 30 or 40

Leagues)

Leagues) the Difference is ſcarce diſcernable; and always obſerve, that if through Haſte you cannot take Time to reckon the Middle Latitude exactly to a Minute, take rather a Latitude too great than too little for Middle Latitude, and that will leſſen the Error, as I hinted in *Middle Latitude Sailing Trigonometrical, Caſe* I.

C H A P. V.

Inſtrumental Navigation.

S E C T. V.

Plane Sailing Inſtrumental.

MY Intent in this Section of *Inſtrumental Naviga-tion*, is not to write a large Treatiſe of all the Inſtruments that are uſeful in Navigation, for that would take up too large a Part in this ſmall Treatiſe; but I ſhall here ſhew, how all the Propoſitions in *Plane Sailing, Mercator's, &c.* (which are in other Parts of this Treatiſe ſolved *Geometrically, Trigonometrically, Arith-metically, &c.* may be ſolved *Inſtrumentally, viz.* by Scale and Compaſſes; and here I ſhall confine myſelf only to the Uſe of *Gunter's Scale*, it being an Inſtrument ſo common and expeditious for Inſtrumental Operation, that few that are inclined to improve themſelves in the Knowledge of Navigation, go to Sea without it. In order to the right Underſtanding of which, it is neceſ-ſary, firſt, to give a Deſcription of that Part of the Scale which ſerves for our preſent Purpoſe, and then proceed to the Uſe of it.

The

The firft Line next the Top, commonly marked *S.R.*
is the Sine Rumbs, and is numbered 1, 2, 3, 4, 5,
6, 7, and ends at the Brafs Nail with 8, which is Radius, becaufe 8 Rumbs or Points of the Compafs, is a
Quarter of the whole Circle, and therefore equal to 90
Degrees.

The next Line marked *T. R.* is the Tangent Rumb,
and this Line is numbered 1, 2, 3, and then at the
Brafs Nail 4, the Tangent of 4 Points, or 45 Degrees,
being equal to Radius, and then proceed back towards
the Left-hand with 5, 6, 7, the Tangent of 8 Points,
or 90 Degrees being infinite:

Thefe Sine Rumbs and Tangent Rumbs, are only
ufeful in Navigation. The Sine Rumb, or Tangent
Rumb, being only the Sine or Tangent of the Degrees
and Minutes anfwering to thofe Rumbs: Thus, becaufe
1 Point of the Compafs contains 11d. 15m. therefore
the Sine of 1 Point is equal to the Sine of 11d. 15m. &c.
Alfo three Points are 33d. 45m. therefore the Sine of
3 Points, in the Sine Rumbs is exactly againft the
Sine of 33d. 45m. in the Sines, and the Tangent of 3
Points is right againft the Tangent of 33d. 45m. &c.

The next Line being the third in order, marked
(*Numb.*) is the Line of Numbers, which is numbered
from the Left-hand with 1, 2, 3, 4, 5, 6, 7, 8, 9;
1, 2, 3, 4, 5, 6, 7, 8, 9, 10. The Divifions of 1, 2, &c.
being the greateft, and growing lefs towards 9, &c.
Thefe Divifions being divided each into 10 Parts, and
every one of thefe again fubdivided (where Room
will permit) into ten Parts, and are thus read: If
you call 1 at the Left-hand 1, then the next 2 is
2, and fo on 3, 4, 5, &c. to 9, and the 1 on the Middle is 10, and the next 2, 3, 4, &c. are 20, 30, 40,
&c. and the 10 at the Right-hand is 100, and thus,
when 1 in the Middle is 10, and the 2 is 20, &c.
then every one of the 10 lefler Divifions between 1 and
2 is 1: Thus 1 being 10, the next fmaller Divifion is
11, and the next 12, where there is commonly a Brafs

Nail,

Nail, and fo on, to 13, 14, 15, &c. till you come at
the 2 for 20, and 3 for 30, and fo in the reft.

But if you make 1 at the Left-hand 10, then 1 in
the Middle is 100, and 10 at the Right-hand is 1000,
and then every of the leffer Divifions, which before
were reckoned 1, are now reckoned 10, &c. Thus the 1
at the Middle being 100, the next fmall Divifion (which
before was called 11) is now 110, and the next where
the Brafs Nails is 120, &c. and then every one of the
fmalleft Subdivifions, is called 1.   Thus fuppofe I
would find 125 on the Scale, I call it 1 in the Middle (or
at the Left-hand) 100; and counting 2 of the fmall
Divifions further (which is juft at the Brafs Nail) and
then 5 of the fmalleft Subdivifions, there I hold my
Compaffes, it being the Number required; but it muft
be obferved that all Scales are not graduated alike, as
to the Subdivifions; fome of them (for the Sake of
Cheapnefs, or faving Trouble) having one Subdivifion
put for two, and therefore the Value of thefe muft be
eftimated accordingly; however a little Practice and
Attention, will make the whole very eafy.

The next Line is the Sines marked at the Right-hand
[*Sines*] and numbered from the Left-hand, 1, 2, 3, 4, 5,
6, 7, 8, 9, 10, 20, 30, 40, 50, 60, 70, 80, 90, the longer
Subdivifions between 10 and 20, and between 20 and
30, &c. being each 1 Degree; but the Degrees under
20 are fubdivided into 6 Parts, each Part being 10 Minutes; and where the Degrees are fubdivided into 4 Parts,
each Part is 15 Min. &c.   Thus fuppofe I feek the
Sine of 23d. 30m. firft I find 20, and then count 3 of
the bigger Subdivifions which are 23; and then 2 of
the fmalleft Subdivifions for 30 Min. (becaufe the Degree is divided into 4 Parts) and that is the Point reprefenting the Sine of 23d. 30m. where there is commonly
a Brafs Nail, as at 12 on the Line of Numbers.

The next Line is Verfed Sines marked [*V. Sines*] of
which there is feldom Ufe made in the common Practice
of Navigation, and thefe are reckoned from the Right-
hand towards the Left.

The

The next is Tangents, marked at the Right-hand [*Tang.*] they proceed from the Left-hand to the Right, from 1 to 45, and then back to the Left-hand towards 90, the Subdivisions are underſtood in all Reſpects, as thoſe of the Line of Sines, and therefore need be no further explained.

By this Scale, and a Pair of Compaſſes, may be performed all the Problems in *Trigonometry* and *Navigation*, in all their Parts.   As they are ſolved elſewhere in this Treatiſe by Geometrical Conſtruction, and by Arithmetical Calculation, ſo here I ſhall ſhew how they may be ſolved by Inſtrumental Operation; and here, whatever the Queſtion be, the Proportion for anſwering it is the ſame as in *Trigonometry*, by Logarithms, Sines and Tangents; only whereas in Natural Numbers, the Operation is performed by Multiplication and Diviſion (where three Numbers are given to find a fourth) and in Logarithms it is done by Addition and Subtraction; ſo here it is performed by taking in your Compaſſes the Diſtance upon the Scale, between the firſt and ſecond given Terms, and the ſame Extent will reach from the third given Term to the fourth Term required; but becauſe, if the ſecond Term have the ſame Proportion to the firſt, that the fourth hath to the third, then the third will have the ſame Proportion to the firſt that the fourth hath to the ſecond, by *Euclid*, *Lib.* 5. *Prop.* 16. and therefore when the firſt and third Terms are both of one Denomination, *viz.* both Sines, or both Tangents, or both Numbers, as it happens in moſt Queſtions in Navigation, and where the ſecond and fourth are alſo of one Denomination, I would rather adviſe to extend from the firſt to the third, and by that Means you keep along one Line, whether Sines, Tangents, or Numbers; and then the Proportion and Anſwers will come out the ſame, as by Trigonometrical Calculation.

And becauſe the Proportions in this Manner of working are the ſame as in the Logarithmetical Operations, I ſhall not be at the Pains of running over all the parti-

cular

cular Cafes of *Plane Sailing* and *Mercator*, &c. as in the *Geometrical* and *Trigonometrical* Parts hereof; but fhall only inftance in fome few Queftions, and thereby fhew the Learner how all the reft may be deduced from the Proportions elfewhere laid down in the *Trigonometrical* Part.

Example in Plane Sailing.

CASE I. *Courfe, and Diftance given, to find the Diffe-rence of Latitude and Departure.*

A Ship fails N. W. by N. 123 Miles; I demand her Difference of Latitude and Departure.

Firft, For the Difference of Latitude.

You may obferve in the *Trigonometrical* Part, that the Proportion is, As Radius is to the Diftance; fo is Sine Complement of the Courfe, to Difference of Latitude; here the Courfe being N. W. by N. that is, three Points therefore from the Meridian, its Complement is five Points; therefore extend on the Sine Rumbs from Radius (the Brafs Nail at the End) to five Points, the fame Extent will reach on the Line of Numbers from 123 to 102, the Difference of Latitude required.

Then for the Departure.

The Proportion in *Trigonometry* is, As Radius is to the Diftance; fo is Sine of the Courfe, to Departure; therefore extend from Radius to 3, on the Sine Rumbs. the fame Extent will reach from the Diftance 123 (on the Line of Numbers to 68, the Departure required.

Note; The fame may be performed on the Line of Sines; for if you extend from Radius to Sine of 33d. 45m. (the Courfe reckon'd in Degrees and Minutes, allowing 11d, 15m. to a Point) the fame Extent will reach on the Line of Numbers from the Diftance 123 to the Departure 68, as before; and for Difference of Latitude, extend from Radius, to Sine Complement of the Course, 56d. 15m. the same Extent will reach from the Distance 123, to the Difference of Latitude 102, and so in others.

Thus in *Plane Sailing, Case* II. where the Course and Difference of Latitude is given, to find the Distance and Departure, you see in the *Trigonometrical* Part, that the Proportion for finding the Distance is, as Sine Complement of the Course, is to the Difference of Latitude; so is Radius, to the Distance; therefore extend from Sine Complement of the Course to Radius, the same Extent will reach from the Difference of Latitude to the Distance.

Example.   A Ship sails S.W. by S. till her Difference of Latitude be 174 Miles; I demand her Distance and Departure.

First, For the Distance.

Extend from 5 on the Sine Rumbs (or from 56d. 15m. on the Sines) the Complement of the Couse, to Radius; The same Extent will reach from the Difference of Latitude 174, to the Distance 209; then for the Departure the Proportion is, As Sine Complement of the Course, is to Difference of Latitude; so is Sine of the Course, to the Departure: Therefore extend from Sine Complement of the Course 5 on the Sine Rumbs, (or 56d. 15m. on the Sines) to the Sine of the Course 3 on the Rumbs, or (33d. 45m. on the Sines) the same Extent will reach on the Line of Numbers from the Difference of Latitude 174, to the Departure 116.

In the like Manner are all the Cases in *Plane Sailing* and *Mercator* performed on the Scale, if you do but know the Proportions, (which you have at large in the *Trigonometrical* Part) and therefore I need not enlarge any further upon it: Only this is to be observ'd, that you

K 4   must

muſt mind always to extend your Compaſſes the ſame Way (whether backwards or forwards) at the laſt Extent that you do at the firſt; that is, if your Extent from the firſt Number to the ſecond be from the Left-hand towards the Right, then your fixed Point being placed at the third Number the moveable Point muſt be extended towards the Right-hand alſo, and the contrary, unleſs in caſe you have to extend from a Tangent above 45, becauſe they increaſe from the Right-hand towards the Left (as hath been ſaid before) of which you have ſometimes Occaſion to make Uſe, I ſhall therefore give you an Example.

A Ship ſails between the South and Weſt till her Difference of Latitude be 80, and her Departure 95; I demand the Courſe and Diſtance.

The Proportion being, As Difference of Latitude is to Radius; ſo is Departure, to Tangent of the Courſe; therefore extend from Difference of Latitude 80, to the Departure 95, (which is towards the Right-hand, becauſe increaſing, *viz.* 95 is more than 80;) the ſame Extent will reach from Radius to the Tangent of the Courſe 50; and although Radius is at the End of the Line, ſo that you cannot extend any further towards the Right-hand, yet in your extending towards the Left-hand, you muſt count the Degree that the Compaſs Point falls upon, on thoſe Degrees that are increaſing above 45 towards the Left-hand; thus in this Caſe, you find the moveable Point of the Compaſs fall upon a Stroke marked 40, 50, *viz.* it is either the Tangent of 40 or of 50; but the Extent from 80 to 95 being towards the Right-hand increaſing, and your fixed Point being placed in the ſecond Extent at Radius, *viz.* Tangent of 45, you muſt account the ſecond Extent alſo increaſing, and the Tangent that is above 45, *viz.* 50, is the Anſwer to your Queſtion; and this (with the Help of *Plane Sailing Trigonometrical* for finding the Proportions, till you have them by heart) may be ſufficient for *Plane Sailing Inſtrumental*

SECT.

S E C T. II.

Mercator's Sailing Instrumental.

IN *Mercator's Sailing*, as well as in *Plane Sailing*, the Proportions are the same as in the *Trigonometrical Part*, and may be learned in *Mercator's Sailing Trigonometrical*; but because my Design is to shew how all the Cases hereof may be solved by Scale and Compasses, without Tables; I shall first shew how the Meridional Difference of Latitude may be found by the Scale and Compasses, both Latitudes being first given or found.

There is upon the Scale two other Lines, besides what we have yet spoken of, which are next the lower Edge of the Scale; the first of them next the Tangent Line, is the Meridian Line, marked [*Merid.*] and the next and lowest is a Line of equal Parts, marked |*Eq. P.*] these are of Use for graduating a *Mercator's* Chart, (of which more elsewhere in this Book) and also for finding the Meridional Difference of Latitude. The Equal Parts are Degrees of the Equinoctial, and every Division or Degree contains 60 Miles, or Minutes; and hence every 10 Degrees marked 10, 20, 30, &c. contains 600 Miles; but the Degrees of Latitude growing larger towards the Poles, in a *Mercator's* Chart, for holding the same Proportion to the Degrees of Longitude that they do upon the Globe, every Degree nearer the Pole, contains so many more Meridional Miles, according as the Degrees of Latitude upon the Meridional Line grow larger, as the Latitude increases; hence, to find the Meridional Difference of Latitude, when the two Latitudes are given or found, extend the Compasses between the two given Latitudes on the Meridional Line; which done, apply the same Extent to the equal Parts, and accounting every 10 Degrees 600 Miles, every Degree 60, and each half Degree 30 Miles, &c. you have the Meridional Difference of Latitude between

the

the two Places, and fo may find what is required, and anfwer your Queftion *Inftrumentally*, as well as *Trigona-metrically*, if you be careful in the Operation.

Example in Mercator's Sailing, Cafe I.

A Ship in Latitude 5cd. oom. N. fails. N.W. by N. 987 Miles; I demand the Latitude come to, and Dif-ference of Longitude.

According to the former Directions, extend from Ra-dius to Sine Complement of the Courfe, *viz.* 5 on the Sine Rumbs, (or 36d. 15m. on the Sines) the fame Ex-tent will reach from the Diftance 987, to Difference of Latitude 820 Miles or Minutes, or 13d. 40m. which added to the Latitude failed from, finds the Latitude come to 63d. 40m. then to find the Meridional Diffe-rence of Latitude, extend the Compaffes upon the Me-ridian Line from 50d. om. to 63d. 40m. that Extent applied to the Equal Parts, accounting (as before di-rected) every 10 Degrees for 600, and each Degree for 60 Miles or Minutes, will be found to be 1519, the Meridional Difference of Latitude. Then for Diffe-rence of Longitude, proceed according to the Propor-tion in *Mercator's Sailing Trigonometrical,* Cafe I. *viz.* Ex-tend from Radius to the Tangent of the Courfe, *viz.* 3 on the Tangent Rumbs, or 33d. 45m. on the Tangents, the fame Extent will reach from the Meridional Diffe-rence of Latitude 1519 to the Difference of Longitude 1015, and fo in others.

In *Cafe* II. where both Latitudes and Courfe is given, the Diftance is found, as in *Cafe* II. of *Plane Sailing In-ftrumental,* and the Difference of Longitude as in *Cafe* I. hereof, which is fo eafy, that it needs no Example.

CASE III. *One Latitude, Courfe, and Difference of Lon-gitude given, to find the other Latitude and Diftance.*

A Ship in Latitude 50 North, fails N.N.W. till her Difference of Longitude be 7 Degrees, or 420 Minutes, I demand as above.

                                        Extend

Extend from the Tangent of the Course, two Points to Radius, the same Extent will reach from 420 the Difference of Longitude, to 1014 the Meridional Difference of Latitude.

Take the Meridional Difference of Latitude 1014 from the Scale of Equal Parts, then with that Extent, and one Foot in Lat. 50, on the Meridian Line, the other will reach to 59d. 40m. the Latitude come to. Then for the Distance extend from Sine Complement of the Course (6 on the Sine Rumbs, or 67.30 on the Sines) to Radius, the same Extent will reach from the proper Difference of Latitude 580 to 628, the Distance required.

SECT. III.

Parallel and Middle Latitude Sailing Instrumental.

THE Proportion for solving all the rest of the Cases both in *Mercator* and *Parallel Sailing*, and also in Middle Latitude, may be so easily deduced from the Proportions laid down in the *Trigonometrical* Part, that I need say no more about them; only in the Instrumental Operation, where a Tangent is required, the Point of the Compass will sometimes fall beyond the End of the Line; as suppose the Proportion was at Tangent 21d. 30m. to Tangent 37d. 20m. so Tangent 42d. 40m. to a Tangent required. And here, if you extend from the first Term 21d. 30m. to Tangent 37d. 20m. the same Extent will reach from 42d. 40m. to beyond the End of the Scale; but to remedy this, extend from Tangent 21d. 30m. to Tangent 37d. 20m. then with that Extent, and one Foot in Tangent 45, extend the other back towards the Left-hand, and where-ever it lights, keep it fixed and close the other to the third Term, *viz.* Tangent 42d. 40m. then with this Extent, and one Foot in Tangent 45, the other will fall upon Tangent 60d. 45m. the Tangent required.

As

As for *Traverse Sailing*, it being compounded of feveral Queſtions in *Plane Sailing*, I ſhall not need to ſpeak of it in this Place, ſuppoſing that by what is already laid down in this Secton of *Inſtrumental Navigation*, the Learner will be able to make Application, as Occaſion requires, without any further Inſtructions.

C H A P. VI.

Navigation New Modell'd;

O R,

The W H O L E A R T Performed

B Y A

N E W M E T H O D.

S E C T. I.

Rules and Grounds of this Method.

IN order to the right Underſtanding of this new Method of *Trigonometry*, I ſhall proceed according to the uſual Manner, and ſhall, for the Help of Memory, laid down ſome fundamental Rules or Axioms, upon which the whole Operation depends, and by which all the Caſes in *Plane Trigonometry*, both Right and Oblique, may be ſolved, without any Book, Table, or Inſtrument whatſoever. But before I come to the Axioms, I ſhall premiſe, that whenever a Side and an Angle is given, to find another Side, (which is the firſt and moſt uſeful Caſe in Navigation) there muſt firſt be a
Number

Number found, which I call the Natural Radius, not only becaufe it is the Original, from whence the Solutions are deduced, but alfo becaufe being found, it produces the fame Anfwer in Natural Numbers, that the Radius, or Sine of 90, produces in a Sinical Proportion; and this Natural Radius is thus found.

METHOD the First.

Take the Angle whofe oppofite Side is either given or fought, and divide four Times the Square of its Complement to 90 Degrees, by 300 added to three Times the faid Complement, and then the Quotient added to the faid Angle, is the Natural Radius required; and this Rule is univerfally true in all Angles from 0 to 90.

METHOD the Second.

But becaufe in Angles under 45, the Complements are above 45, and their Squares amount to greater Numbers than the Squares of the Complements of the Angles above 45; therefore to render the Work as eafy, and the Contrivance as ufeful as poffible, I fhall fhew another Way to find the Natural Radius for all Angles under 45, and the Rule is,

Divide three Times the Square of the Angle (whofe oppofite Side is given or fought) by 1000, the Quotient added to 57.3, that is 57 $\frac{3}{10}$) the Sum is the Natural Radius required.

This being premifed, the Rules are thefe;

RULE the First, In Right-angled Triangles.

An Angle and a Side given, to find another Side.

The Natural Radius bears always the fame Proportion to the Hypotenufe, that the Angle (by which the Natural Radius was found) bears to its oppofite Side.
Therefore

Therefore if the Angles and Hypotenuse be given, it is, As Natural Radius is to Hypotenuse; so is the Angle, to its opposite Side. But if the Angles and a Leg be given, then it is; As the Angle is to its opposite Side; so is Natural Radius to the Hypotenuse.

RULE *the Second.* In Right-angled Triangles.

Two Sides given, to find a Third.

The Hypotenuse is equal in Power to the two Legs; That is, the Square of the Hypotenuse is equal to the Square of both Legs added together; of which see more in *Plane Sailing Arithmetical.*

RULE *the Third.* In Right-angled Triangles.

The Hypotenuse and a Leg given, to find the other Leg.

Multiply the Sum of the Hypotenuse, and given Leg by their Difference: The Square Root of the Product is the other Leg required.

RULE *the Fourth.* In Right-angled Triangles.

Three Sides given, to find an Angle.

Add half the longer Leg to the Hypotenuse; Then, As that Sum is to 86; so the shorter Leg, is to its opposite Angle.

RULE *the Fifth.* In Oblique Triangles.

Three Sides given, to find where the Perpendicular must fall.

Multiply the Sum of the two shortest Sides by their Difference, divide the Product by the third Side, which is the greatest, and upon which the Perpendicular is to fall: The Quotient added to the greatest Side, or subtracted from it, shall be double the greater or lesser Segment, on each Side of Perpendicular.

Another

Another Way.

Add the Squares of the biggest and least Sides together, and from their Sum subtract the Square of the middlemost; Half the Remainder, divide by the biggest or longest Side, the Quotient is the lesser Segment, which subtracted from the whole Base, leaves the greater Segment.

See another Way in *Axiom* IV. *of Plane Trigonometry.*

SECT. II.

Plane Sailing by a New Method.

I Shall now proceed to some Examples in Right-angled Triangles; applied to *Plane Sailing*; and here Note, That to avoid Fractions, I shall propose the given Angle always in whole Degrees, that being sufficiently exact in all Uses in Navigation; nay, a far more exact Way than reckoning by Points, Half Points, or Quarter Points; one Degree being a much smaller Part of a great Circle than a Quarter Point of the Compass. And I shall make Use of Degrees rather than Quarter Points, not only for its Exactness, but also because it is a Method much in Use aboard of the Men of War, to reckon the Course in Degrees, and not in Points and Quarter Points.

CASE I. *Course and Distance given, to find the Difference of Latitude and Departure.*

Note; I shall in every Case hereof propose the same Questions that are inserted in the second Example of each Case in *Plane Sailing Trigonometrical,* except in *Case* V: and VI. where I have made Use of the first Example.

A Ship sails South 25 Degrees, Easterly 96 Miles; I demand as above.

The

The Operation at large.

Note; Where both Legs are required, chuse always to find the lesser first, because the Natural Radius is more easily found, and then find the longer Leg by *Rule the Third.*

For the Departure.		*The Rule* I.
The Angle ———— 25		As Natural Radius 59.2
Multiplied by itself — 25		to the Distance 96: So the
————		lesser Angle 25, to its oppo-
125		site Side the Departure.
50		

Square of the Angle — 625
Multiplied by ———— 3   96   592)24000(40½
                     25   .320
The Product ———— 1875
1875 divid. by 1000 is 1.875   480   320
To which add ——— 57 3   192   Nearest which 40½
                         is the Departure
Sum is Natural Rad. 59.175   2400   required.
Or rather briefer — 59.2
    Which is exact enough.

Then for the Difference of Latitude by Rule III.

Hypotenuse or Distance 96.0
The Dep. in Decimals 40.5   7575(87   The Diff.
                     64   of Latit.
Their Sum ——— 136.5
Their Difference — 55.5   167)1175
                   1169
          6825   (6)
          6825
          6825

The Product ———— 7575.75

    The Departure is 40 ½ (or in Decimals 40.5) The Difference of Latitude 87.

CASE

CASE II. *Courfe and Difference of Latitude given, to find the reft.*

A Ship fails North 38 Degrees Weft, her Difference of Latitude 120; her Diftance and Departure is required.

Here the Side oppofite to the bigger Angle is given, therefore we muft make Ufe of *Method* I. to find Natural Radius, becaufe *Method* II. ferves only to Degrees under 45.

```
   38        38     414)5776(13¦¦¦     Becaufe the Frac-
   38         3         1636          tion is fo great, I
 ─────      ─────       ─────         fhall call the Quo-
   304       114         394          tient 14, which ad-
   114       300                      ded the biggeft An-
 ─────      ─────                     gle 52, to the Sum
  1444       414                      66 is the Natural
     4                                Radius.
 ─────
  5776
```

Then by Rule I.

As 52 the greater Angle, is to 120 its oppofite Side; fo is Natural Radius 66, to the Diftance required.

Then for the Departure, by Rule III.

```
   120 | Diftance — 152            . .
    66 | Diff. of Lat. 120      8704(93 Departure
 ─────                            81
   720 | Sum ——— 272
   720 | Difference — 32        183(604
 ─────                             549
  7920                           ─────
                          544      (55)
                          816.
 52)7920(152  ─────
   272      Product — 8704
   120
 ─────
    16
```

The Diftance 152. The Departure 93.

L **CASE**

CASE III: *Courſe and Departure given, to find Diſtance and Difference of Latitude.*

A Ship ſails North 19 Degrees Eaſterly, her Departure 72 Miles; I demand as above.

Here the ſhorter Leg is given; therefore I ſhall find the Natural Radius by Method II.

Three Times the Square of the given Angle is 1083, this divided by 1000, which is done by cutting off three Figures to the Right-hand, the Quotient is 1.083; which becauſe the ſecond Figure in the Fraction is above 5, I add one to the firſt Figure, which is a Cypher, and then call the Quotient 1.1, which added to 57 3, the Sum 58.4, is the Natural Radius required.

$$
\begin{array}{r}
19 \\
19 \\
\hline
171 \\
19 \\
\hline
361 \\
3 \\
\hline
1.083
\end{array}
$$

Then for the Diſtance by Rule I.

As the Angle 19, is to its oppoſite Side 72; So is Natural Radius 58.4, to the Diſtance.

$$
\begin{array}{r}
58.4 \\
72 \\
\hline
1168 \\
4088 \\
\hline
4204.8
\end{array}
\qquad
\begin{array}{l}
19)4204(221 \\
\cdot 40 \\
\cdot 24 \\
\hline
5
\end{array}
$$

The Diſtance required.

Then find the Difference of Latitude, by Rule III.

Diſtance ———— 221
Departure ———— 72

Sum —— · — 293
Difference —— · — 149

$$
\begin{array}{r}
2637 \\
1172 \\
293 \\
\hline
43657
\end{array}
$$

43657(208
4

40)03657
408)3264

(393)

The Root 208, but the Fraction being ſo large, I rather call it 209, the Difference of Lat. required.

CASE

CASE IV.   *Distance and Difference of Latitude given,*
to find the Course and Departure.

A Ship fails between the North and East 117 Miles,
her Difference of Latitude 103 Miles; I demand her
Course and Departure.

For the Departure, by Rule III.

```
Distance  ———  ——  117|          · ·
Diff. of Lat.  ———  103|        3080(55
                                  25
Sum  ———  ———  220         ——————
Difference  ———  14        105(580
                                 525
                 880          ——————
                 220            (55)
              ——————
Product  ———  ——  3080| The Departure 55½:
```

Then for the Course, by Rule IV.

Hypotenufe or Distance 117
Half the longeſt Leg——51.5
                        ——————
Sum ——— ——— — 168.5

As 168.5 is to 86 : So is ſhorteſt Leg, or Departure,
55.5, to its oppoſite Angle, the Courſe.

```
55.5    168.5(4773.0(28     The Courſe 28 .¹¹³De-
 86          1403.0         grees, which is almoſt ¹⁄₃,
————         ——————         viz. 28 Degrees, 19 Mi-
333.0        · · 550        nutes.
4440
————
4773.0
```

CASE V. *Distance and Departure given, to find the rest.*

A Ship sails in the South West Quarter, 124 Miles, her Departure 95 Miles; I demand her Course and Difference of Latitude.

For the Difference of Latitude, by Rule III.

Distance ———	124	. .	
Departure ——	95	6351(79	
Sum ———	219	49	
Difference ——	29	149|1451	
	1971	1341	
	438	110	
Product———	6351		

The Square Root of 6351 is 79 or 80 (because the Remainder 110 is more than the Root) the Difference of Latitude required.

Then for the Course, by Rule IV.

Hypotenuse or Distance ———	124 ——	124	
Longest Leg, which here is Dep.	95 its half	47.5	
	Sum ———	171.5	

As 171.5, is to 86; So is the shortest Leg (which here is the Difference of Latitude) 80, to its opposite Angle, the Complement of the Course.

86	171.5(6880.0(40 Degrees
80	.0200
6880	20.0

The Complement of the Course is 40 Degrees, and consequently the Course is 50 Degrees, from the South Westerly, and the Difference of Latitude is 80 Miles *fere,*

CASE

CASE VI. *Difference of Latitude and Departure given,*
to find the Course and Distance.

A Ship sails in the North West Quarter, till her Difference of Latitude be 220 Miles, and her Departure 108 ; I demand the rest.

For the Distance, by Rule II.

| 220 | 108 | Square of Diff. Lat. —— 48400 |
| 220 | 108 | Square of Departure —— 11664 |

4400	864	
440	1080	Sum of the Squares — 60064(245
48400	11664	4

$$44)200$$
$$176$$
$$485)2464$$
$$2425$$
$$(39)$$

Then for the Course by Rule IV.

As 355 is to 86 ; So is 108 to the Angle of the Course.

108	355)9288(26
86	2188
648	..58
864	
9288	

The Distance is 245, and the Course 26 $\frac{14}{15}$ Degrees, or 26d. 9m. or N.N.W. somewhat more than $\frac{1}{4}$ West.

SECT.

S E C T. III.

Oblique Triangles, by a New Method.

How to solve all the Cases in Oblique Plane Triangles, by this new Method, without any Canon, Book, Instrument, &c.

IN the Solution of *Oblique Triangles*, by this New Method, it is necessary that they be first divided into two right-angled Triangles, by a Perpendicular let fall, in which observe:

Let it fall from the End of a given Side, and opposite to the given Angle.

By this Means, the Perpendicular will sometimes fall within, and sometimes without; when it falls within, it falls upon some intermediate Part of the Base, or longest Side, but when it falls without, it falls upon one of the shortest Sides continued; in either Case, there is two right-angled Triangles produced, and then the Angles or Sides sought, are found as if they were Parts of a right-angled Triangle.

C A S E I.

Given $\left\{\begin{array}{l}\text{The Angle at A} \quad 30\ 0 \\ \text{The Angle at B} \quad 35\ 0 \\ \text{The Side BC} \qquad 290\end{array}\right\}$ Required the Side AC.

Here the Perpendicular falls from the End of the given Side B C, and opposite to the given

Fig. 48. Angle at B; and then in the Triangle B D C is given the Angle at B, and the Side BC, to find the Side CD, which being found, there is given in the Triangle ADC, the Angle at A, and the Side D C, to find AC, the Side required.

I shall not trouble the Reader with the Operation for finding Natural Radius, it having been often enough repeated in the former Part; but being found by *Method* I. the Natural Radius for the Angle 45 is 6362; therefore,

therefore, As 63.62, to Side CB, 290; fo 45, the Angle at B, to Side oppofite CD.

```
    45      63.62)13050.00(205   The Side CD 205.
   290           ..326.00

  4050              ..790
   900

 13050
```

Now in the Triangle ADC, is given the Angle at A 30d. om. and the Side DC 205, to find AC.

The Natural Radius found by *Method* II. for the Angle 30 is 60. Therefore,

As the Angle at A 30, is to the Side oppofite DC 205; fo is 60 to the Hypotenufe AC required.

```
    205     30)12300(410    The Side AC required.
     60         .30
                .00
  12300         ..
```

CASE II.

Given $\left\{\begin{array}{l}\text{The Side AB } 560 \\ \text{The Side AC } 410 \\ \text{The Angle B } 45.0\end{array}\right\}$ Required the Angle at C.

Here the Perpendicular falls without, upon the Side BC continued, and in the Triangle BDA there is given the Angle at B 45, and the Hypote-   *Fig.* 49. nufe AB 560, to find AD, which being found you have given in the Triangle CAD, the Hypotenufe CA, and the Leg AD to find the Leg CD, and the Angle DAC; and by Subtraction, the Angle DCA, and the Angle BCA, in their proper Cafe of right-angled Triadgles.

The

The Operation.

The Natural Radius for the given Angle 45, is 63.62, as found by *Method* I : Therefore, as 63.62, to the Hypotenuse 560; so is the Angle at B 45, to the Side opposite AD.

```
   560        63.62)25200.00(396 the Side AD:
    45            6114.0
  ─────            388.20
  2800            ──────
  2240            . .6.48
  ─────
 25200
```

Then you have AD 396, and AC 410, to find CD by *Rule* III. thus :

```
Side AC ── ── 410      . . .
Side AD ── ── 396      11284(106,theLegCD.
             ───           1
Sum ── ── 806          ──────
Difference ── ── 14    20)012
             ───           00
            3224       ──────
             806       206)1284
             ───           1236
Product ── 11284       ──────
                          (48)
```

Then for the Angle CAD, by *Rule* IV.

```
Hypotenuse AC ── 410    As 608 is to 86; so is
Half the Leg AD ── 198  106 to the Angle CAD.
                ───     106    608)9116(14:⅜
Sum ── ── 608           86        3036
                       ─────     ─────
                        636       604
                        848
                       ─────
                        9116
```

The Angle CAD is 14⅘⅗, or rather 15 Deg. which subtracted from 90, leaves ACD 75 Deg. and that subtracted from 180, leaves 105 Deg. the Angle ACB req.

CASE III: *Given as in Case* II. *to find the third Side* BC.

In the whole Triangle BDA, you have given the Angle ABD 45, and the Hypotenuse BA 560; also by Consequence, the Angle BAD, which is also 45, to find the whole Side BD; but in this Case the acute Angle being equal, *viz.* 45 Degrees, the Leg BD is equal to AD 396; then having found CD 106, by the second Operation in *Case* II. subtract it from the whole Side BD 396, the Remainder 290, is the Side BC required.

C A S E   IV.

Given $\begin{cases} \text{The Side AC}\text{——}410 \\ \text{The Side AB}\text{——}560 \\ \text{The Angle at A}\text{—}30.0 \end{cases}$ Required the Ang. at B.

In the Triangle AEC, is given the Angle at A 30, and the Hypotenuse AC 410, to find CE, which by the first Case hereof is found to be *Fig.* 50. 205, and therefore I need not repeat the Operation. Then in the same Triangle ACE, there is given the Sides AC 410, and CE 205, to find the Side AE, by *Rule* III.

AC	—— ——	410	. : .	
CE	—— ——	205	126075(355, the Leg AE	
			9	
Sum	—— ——	615		
Difference	——	205	65)360	
			325	
		3075		
		12300	705)3575	
			3525	
Product	—— =	126075		
			50	

The

The Leg AE 355 subtracted from the whole Side AB 560, rest EB 205; then in the Triangle BEC, you have given BE 205, and EC 205, to find CB by *Rule* II. and the Angle B by *Rule* IV. but in this Case, BC and EB being equal, the Angle at B is proved 45 Degrees, without Calculation.

CASE V.

Given as in Case IV. to find the third Side BC.

Although this is the fifth *Case* in the *Trigonometrical* Operation, yet the Side BC is necessarily found in *Case* IV. before the Angle at B can be found; and therefore, although the Operation in *Case* IV. be somewhat tedious, yet both the Fourth and Fifth *Cases* are included in it.

CASE VI:

Given $\begin{cases} AB \text{ ——— } 560 \\ AC \text{ ——— } 410 \\ BC \text{ ——— } 290 \end{cases}$ Required the Angle at A.

Find AE by the Rule laid down in *Axiom* IV. of *Plane Triangles.* As the Base AB 560, is to the Sum of the other two Sides, 700; So is the Difference of the said Sides 120, to the Difference of the Segments of the Base AD 150, as by the Operation below.

Fig. 51.

120	56.0)14000(150	To the half Diff. 75 add the half Base 280, the Sum 355 is the greater Base AE, but subtracted the Difference is the lesser Base EB 205.
700	280	
84000	..000	

Then in the Triangle AEC, there is given AC 410, and AE 355, to find CE by *Rule* III. and the Angle at A by *Rule* IV.

The

```
The Side AC ——— 410  |        . : .
The Side AE ——— 355  |      42075(205
                     |    405)02075
Their Sum   ——— 765  |        2025
Their Difference— 55 |
                     |        ————
                     |          50
            3825     |    The Square Root of
            3825     |  42075, viz. 205, is the
            ————     |  Side CE required.
Product  ——— 42075   |
```

Then, by Rule IV. find the Angle at A.

Hypotenuse ————— 410
Half the longeſt Leg — 177.5

Their Sum ——— ——— 587.5

As 587.5 is to 86 ; ſo is CE 205, to the Angle oppoſite at A 30.

```
205        587.5)17630.0(30  The Angle at A.
 86          . . . 50          required.

————        ————
1230          50
1640

————
17630
```

Although this Method be not altogether ſo expeditious for Oblique Triangles, as the Calculation by Logarithms, becauſe you are obliged to divide every Oblique Triangle into two Right-angled ones, which ſometimes requires two Operations ; yet I thought fit to inſert it to make the Method compleat, it being of great
Uſe when Tables are wanting, and of ſufficient Exactneſs
for moſt Uſes in Navigation ; but the Right-angled Caſes, are performed hereby. I ſhall recommend to the
Reader, being very uſeful, ſufficiently exact, and as expeditious as any Method commonly in Uſe.

SECT.

SECT. IV.

How to find the Difference of Longitude, and keep a Reckoning both in Latitude and Longitude, by this New Method of Trigonometry, (as applied to Navigation) without the Help of any Tables or Instruments whatsoever, according to Middle Latitude, which is of sufficient Exactness for the working so short a Distance as a Day's Run, and consequently of great Use in Navigation.

YOU may remember, that in *Middle Latitude Sailing Trigonometrical*, there is a Proportion for finding the Difference of Longitude, which is, as Sine Complement of Middle Latitude, is to the Departure; So is Radius, to the Difference of Longitude. And therefore, in *Middle Latitude Sailing Geometrical*, one Way which I have proposed for projecting *Middle Latitude Sailing*, is, by constituting a Right-angled Plane Triangle, whose Angle at the Base is equal to the Complement of Middle Latitude, and the Perpendicular is equal to the Departure: And then by that known Proportion of opposite Sides, and opposite Angles, it will necessarily follow that the Hypotenuse must needs represent the Difference of Longitude; which being granted, there is no more to do for finding the Difference of Longitude, but only the Solution of the said Right-angled Triangle: Of the several Varieties of which you have had sufficient Instances in the six Cases of *Plane Sailing* before going, where any two Parts being given, the other two are easily found. Nevertheless, that nothing may be wanting for the Reader's Instruction, I shall instance in one Question for Example's Sake, which I shall first work by this new Method, and then by the Method proposed in *Middle Latitude Sailing Trigonometrical*; and lastly, shall work the same by *Mercator*, to

let

let the Reader fee the Sufficiency and Exactnefs of this
new and ufeful Invention.

Queftion. A Ship in Latitude 38d. oom. North, fails
South 25 Degrees Eafterly 96 Miles; I demand the Latitude come to, and alfo her Departure and Difference
of Longitude

The Courfe and Diftance is the fame as in the Example 2d of *Cafe* I. in *Plane Sailing*, as performed by this
New Method, and therefore I fhall refer you to the
Operation there, for finding the Difference of Latitude
and Departure. The Difference of Latitude being
there found to be 87, and the Departure 40½ or 40,5,
and therefore (the Courfe being Southerly) the Latitude
come to is 56d. 33m. and confequently the Middle Latitude, found by the Direction and Caution laid down
in *Cafe* I. of *Middle Latitude Sailing Trigonometrical* is
57d. 17m. and the Complement of Middle
Latitude is 32d. 43m. From hence by the *Fig.* 52:
foregoing Directions, is conftituted the Triangle ABC, wherein the Angle at A, is equal to the
Complement of Middle Latitude 32d. 43m. and the
Side oppofite BC, is equal to the Departure 40.5, both
which are given to find the Hypotenufe AC, equal to
the Difference of Longitude required; and here the
Side oppofite to the leffer Angle being given, I fhall find
Natural Radius by *Method* II. And here obferve, that
although in Queftions of *Plane Sailing*, you need not
regard Minutes in the Angle of the Courfe, becaufe
whole Degrees are exact enough to keep Account of a
Ship's Way; yet in this Cafe you muft not omit the
odd Minutes in the Angle; and therefore reduce the
odd Minutes to Tenths of a Degree, accounting 6 Minutes for one Tenth of a Degree, and 12 for two
Tenths, &c. And then 42 Minutes is 7 Tenths; and
this Angle being 32d. 43m. I fhall call it 32.7, *viz.* 32
and 7 Tenths, it being but one Minute more, which
cannot caufe any great Error in the Operation.

For

For finding Natural Radius, by Method II.

The Angle ———— 32.7		Then becauſe the Hypo-
Mult. by itſelf ——— 32.7		tenuſe is required, the Pro-
		portion is,
2289		As the Angle 32.7, is to
654		its oppoſite Side 40.5; So
981		is Natural Radius 60.5, to
		the Hypotenuſe.

Sq. of the Angle — 1069.29

Mult. by ———— 3   60.5   327)2450.2(74¹¹²⁄³²⁷

The Product — 3207.87   40.5   1612

                                          3025   304
Divided by 1000, the Quo-   24200
tient is 3.207, but it is   —————
exact enough to call it   2450.25
3.2, which added to 57.3   The Product of the Mul-
the Sum is 60.5, the Na-   tiplication is but 2450, the
tural Radius required.   other two Figures being but
a Decimal Fraction, are to be cut off; but becauſe the
Diviſor 32.7 is a Decimal, I add one Figure to the Dividend, as you ſee in the Operation above.

The Difference of Longitude is 74¹¹²⁄³²⁷, which without exactly regarding the Fraction, may be ſet down
75: And this Operation may be performed with great
Eaſe and Readineſs, with a little Practice, although I
have here ſet it down in Words at large, to make it
more intelligible.

The ſame Queſtion anſwered by Middle Latitude
Sailing Trigonometrical.

FIND Difference of Latitude and Departure, and
conſequently the Latitude come to, with Middle
Latitude and Complement of Middle Latitude, as in
Caſe I. of *Middle Latitude Sailing Trigonometrical,* which
will be found to be as above expreſſed.

                                                          Then

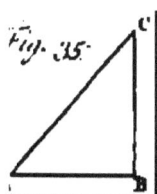

Fig. 35.

Fig. 36.

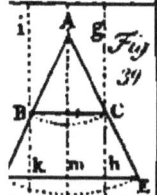

Fig. 39

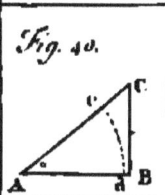

Fig. 40.

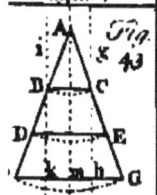

Fig. 43

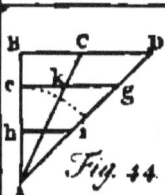

Fig. 44.

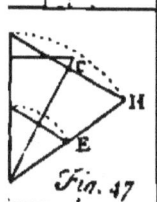

Fig. 47

Fig. 48.

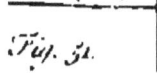

Fig. 51.

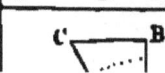

Then for the Difference of Longitude.

	d.	m.	
As Sine Comp. Mid. Lat. ——	32	43	—— 9.73278
Is to the Departure ——		40.5	—— 1.60745
So is the Radius —— ——	90	00	— 10.00000

To the Difference of Longitude — 74.9 —— 1.87467

The foregoing Question answered by Mercator's Sailing.

FIND Difference of Latitude, and consequently the Latitude come to, as in *Mercator's Sailing Trigonometrical*, and you will find the Latitude come to is 56d. 33m and the Departure 40½ or 40.5.

Latitude sailed from—58d. oom.	⎫	Merid. Parts	⎧ 4294
Latitude come to——56 33	⎭		⎩ 4133

Meridional Difference of Latitude —— — 161

Then,   *Co. Ar.*

As proper Difference of Latitude 87 ——	8 06048
To Merid. Diff. of Latitude ——161 ——	2.20682
So is the Departure —— 40.5 ——	1.60745

To the Difference of Longitude 74.9 —— 1.87475

And thus you see the exact Agreement of this with the true Operation, as performed by *Mercator's Sailing*; it not differing from it so much as one Quarter of a Minute in Longitude, in so great a Distance as 96 Miles, and in a Latitude so near the Pole as 58 Degrees, where there is much more Danger of contracting an Error than in lesser Latitudes, or Voyages nearer the Equator.

CHAP.

C H A P. VII.

Practical Navigation.

O R,

The Application of the foregoing Rules to the actual keeping of a Reckoning, according to the several Kinds of Navigation.

THERE are four Things very neceſſary to be known by all that take upon them the Charge of conducting a Ship from one Part of the World to another, which may properly be called the Practical Part of Navigation.

The firſt is a right Underſtanding of the Compaſs, with the Variation thereof, in order to the true Knowledge of the Courſe made good.

The ſecond is the Log-line and Half-minute Glaſs, that the Knots on the Log-line be a due Length, and that the Glaſs be a juſt Half-minute, that thereby you may, as near as poſſible find the true Diſtance.

The third is the right Manner of taking and working an Obſervation by the Sun by Day, or by the Stars by Night, thereby to find the true Latitude, and to correct the Dead Reckoning, if there has been an Error contracted either in the Courſe or Diſtance.

And, fourthly, having by theſe Means and Helps finiſhed your Reckoning, and being come near the deſired Port, it is alſo neceſſary, that there be a right Underſtanding of the Tides, which Way the Ebbs and Floods ſet, and what Moon makes full Sea upon any Coaſt, that ſo it may be known how long to ride at

Anchor,

Anchor, or to lie by to wait the Tide, if you know you are too foon; or what fail to make to fave your Tide, if you fear being too late: And thefe four Things I fhall handle in this Part, and that in fuch a Manner, as may be intelligible to the meaneft Capacity, and moft ufeful as well as moft eafy to be put in Practice at Sea.

SECT. I.

Of the Variation of the Compass.

THE Way commonly taught for finding the Variation of the Compafs, is by the Sun's Azimuth, or Amplitude; but thefe Ways not being attainable by any but thofe who have learnt fomething of Aftronomy, and being alfo treated of in other Books, I fhall not trouble the Learner with them, but proceed to a far eafier Way, which is this:

When you are at Sea, and defire to know the Variation of the Compafs, take your Quadrant about 8, 9, or 10 a Clock, when you fuppofe the Sun is near about half up from the Horizon to the Meridian, and take an Obfervation of the Sun's Altitude, as you would do at Noon to find the Latitude of the Place, which being done, lay by your Quadrant (letting the Vanes remain unremoved) and by your Azimuth Compafs (if you have one on board) fet the Sun, and mind what Point of the Compafs the Sun is upon at that Obfervation. This done, wait till the Afternoon, that the Sun grows almoft as low as he was when you obferved in the Forenoon, and then with your Vanes fix'd, as in the firft Obfervation, obferve till the Sun be fo low, as that the Vanes fo fixed will juft take the Sun's Altitude without altering them; which done, obferve immediately (as before) upon what Point of the Compafs the Sun is, at that Obfervation: Then the Space between that Point

M   which

which the Sun was upon at the first Obfervation, and that Point upon which the Sun was at the laft Obfervation, divided into two equal Parts, the Middle is the true South Point of the Compafs, and the Diftance between that and the South Point of the Card, is the Variation required.

Example.

Suppofe at the Forenoon Obfervation, I find the Sun by the Compafs to be South Eaft (it matters not what his Altitude be, fo you mind what it be, or elfe let the Vanes ftand unremoved, till the Afternoon) and fuppofe in the Afternoon I find, when the Sun hath the fame Altitude, that he bears Weft South Weft; now the Diftance between South Eaft and Weft South Weft is ten Points, the Half of that is five Points, which reckoned from South Eaft towards the Weft South Weft, it falls upon the South and by Weft; therefore I conclude that the South and by Weft Point of the Compafs points to the true South Point, and the Diftance between the South and by Weft Point (which is the true South) and the South Point of the Compafs (which we may call falfe South) or (magnetical South) is the Variation of the Compafs; and becaufe the magnetical South is Eaftwards from the true South; therefore the magnetical North is Weftward from the true North. Here I conclude, that the Variation is one Point Wefterly, &c.

In this Cafe there is only this Caution to be obferved, *viz.* that this Obfervation be not made when the Ship is running very faft Northwards or Southwards, which may make fome fmall Error, though fcarce difcernable; for if the Ship ftood ftill, the Sun would have exactly the fame Altitude at 8, 9, or 10, in the Forenoon, that it would have at 2, 3, 4, in the Afternoon; but if the Ship fails very faft to the Southward in North Latitude, or to the Northward in South Latitude, fhe raifeth the Sun a little, and by Confequence the Sun will be fomewhat higher at 4 in the Afternoon than at 8

in

in the Morning, and may cause some Error, but it is little, and a Thing that seldom happens: And if it do happen that your Course be North or South, and the Wind so fair, you may defer your Observation till another Day, (it being not necessary to set the Variation every Day) and thereby the Error may be avoided, and yet the Variation exactly found as often as it is necessary.

Note; I shall shew how to find the Variation by the Sun's Azimuth and Amplitude hereafter in the Astronomical Part.

SECT. II.

How to divide the Log-line, and try a Half-Minute-Glass.

THEY that take upon them to be Master, Mate, or Pilot of a Ship, and would be very exact and accurate in the keeping a Reckoning or Account of the Ship's Way, ought to take care before they go from the Shore, to be furnished with all Things necessary for that Purpose. For the best of Scholars, the greatest Artists, the most profound Mathematicians, or the most experienced Navigators may be deceived, and carried into the grossest Errors, by a Defect in their Instruments, or Means for keeping a Reckoning, as well as the most ignorant may be by a Defect in their Knowledge, and so far as they design that their Account shall not depend upon their Dead Reckoning, they ought chiefly to be careful in these two Things.

First, the Half-Minute Glass, that it be of a due Length; for if it be longer than it should be, it makes the Ship by Estimation to run so much more than indeed she does, and by that Means perhaps, in a Month or 6 Weeks Sailing, you will expect to arrive at your Port, or make such or such Land, when in fact, you are 30, 60, or 80 Leagues, or more or less short of it,

according

according as the Error of your Glaſs is more or leſs.
And altho' it is hard to know a true Half-Minute Glaſs,
yet there are theſe two Ways to prove them, and know
whether they are right or wrong.

The firſt Way is by an Expedient mentioned by Mr.
Henry Philips, in his *Advancement of the Art of Navigation*, and alſo quoted by *Seller*, in his *Practical Navigation*, and 'tis this: Take a Bullet of any competent
Weight, it matters not what, and make faſt to it a Piece
of fine Thread, or Silk, of the juſt Length of 39 ½ Inches; let there be a Nooſe on the End of the Thread, and
let the very End of the Nooſe be juſt 39 ½ Inches from
the Center of the Bullet (as I ſaid before) then hang it
up by the Nooſe upon a ſmall Pin, where it may hang
at Liberty, and ſwing freely, and ſo give it Way, and
each Swing ſhall be a true Second of Time; that is, each
Time that it paſſes by the Perpendicular let fall from
the Pin on which it hangs, ſhall be a Second; and every
Time of its Return to the Place where it firſt began its
Motion, is two Seconds of Time, and a Glaſs that runs
till the ſaid Bullet hath made 30 Swings ſhall be a true
Half-Minute-Glaſs.

A ſecond Way (if it may properly be ſo called) is, by
the Experience of thoſe that have had Occaſion to uſe
a Glaſs in long Voyages; and having a Line rightly
divided, by a Glaſs of ſo true a Length, that their Dead-
Reckoning when carefully kept, hath agreed with the
Truth of their Obſervations; and that their making of
Land, &c. hath fallen out according to Expectation, by
the Dead-Reckoning; I ſay ſuch a Glaſs, or another of
the ſame Length, ought to be preferred before any other,
as a true Half-Minute Glaſs.

A ſecond Thing neceſſary, in order to the Keeping of
a true Reckoning, is to take Care that the Log-line be
rightly divided; for although the Glaſs be true, yet if the
Log-line be divided into Knots too long or too ſhort, it
muſt needs make an Error in the Reckoning, according
to the Proportion of the Error in Diviſions of the Line,

if you work by a true Half-Minute-Glass. Indeed, if the Divisions of the Log-line be too short, and the Glass also too short, (or if both be too long, which is the same) then the one Error helps to compensate the other ; but if the Faults in the Line and in the Glass be contrary, that is, the one too long, and the other too short, the Fault is intolerable.

As for the Length of each Knot on the Log-line, or how it should be divided, there are different Opinions amongst different Authors and Navigators. Indeed it is an undeniable Truth, and apparent to every Man's Reason, that one Knot upon the Log-line should be the 120th Part of a Mile; because half a Minute is the 120th Part of an Hour; (for as the Whole is to the Whole, so is a Part to the Part, &c.) but the Difficulty arises from the different Opinions, as to how many Feet, Yards, &c. there are in one Degree of a great Circle upon the Earth. Mr. *Oughtred*, in his Circles of Proportion, will have $66\frac{1}{4}$ Miles to answer one Degree upon the Earth, each Mile containing 5280 Feet. Hence there is by his Account, 349800 Feet in one Degree of a great Circle upon the Earth, and 5830 Feet in one Minute, or 60th Part of such a Degree, and consequently the 120th Part of a Minute, or Length of one Knot upon the Log-line, must be $48\frac{7}{12}$ Feet.

But Mr. *Norwood*, in his *Seaman's Practice*, P. 43. (relating an Experiment of his for finding the Quantity of one Degree of a great Circle upon the Earth) saith, that one Degree contains 367200 of our *English* Feet to a Degree, which Account, without any Allowance would give 51 Feet to one Knot of the Log-line, although, for Reasons there mentioned, he allows 1 Foot out of the 51, and so would have 1 Knot of the Log-line to be just 50 Feet: But how far that Experiment of his is to be depended upon, (considering the Unevenness of the Ground, and Crookedness of the Ways and other Inconveniencies, which he could only give Allowance for according to his Judgment) and also how far that one Foot in

5 I may compensate the Way that the Log. makes after the Ship, I shall not take upon me to determine. In the mean Time I shall, with Submission to better Judgments, rather adhere to the Way of dividing the Log-line, that is commonly received, and used by most Mariners, I mean that of 42 Feet, or 7 Fathom to one Knot. Now it will presently be objected, that according to that Division for Half a Minute, multiplying that by 120, for 1 Mile, and that Product by 60, for 1 Degree, there is by Consequence but 302400 Feet in a Degree, which seems to contradict, and intolerably to vary from the Opinions of the ingenious Mr *Norwood*, and other experienced Men in these Matters.

I answer, it doth not contradict them, at least so much as at first Appearance it seems to do; for one grand Reason why I agree to these shorter Divisions is, to give Allowance for the Way that the Log makes after the Ship; for altho' there is so much stray Line allow'd, as may be supposed to veer the Log moderately well out of the Eddy of the Ship's Wake, yet it cannot be supposed, but that still the Log must have some Way after the Ship, if but by the Weight of the Line, which altho' but light, yet the Water being but a soft fluid Substance, the Log must needs have a motion after the Ship, and especially sailing large, and in a fresh Gale, it cannot be but that the Wind will have so much Effect upon the Log, and so much of the superficial Part of the Water, as to shove it along after the Ship; and that in my Opinion, much more than one Foot in 51.

Indeed another Consideration, which may be accounted a second Reason why I do adhere to that Way of dividing the Line, is, because if there is an error, it is on the safer Side; for although the Truth is best, if it could be attain'd, yet if an Error must be, 'tis better that the Reckoning be a-head of the Ship, than that the Ship should be a-head of the Reckoning; and better to look out for Land before we come at it, than to be a-shore before we expect it.

But

But my third Reason is, the Confirmation of this my Opinion, by the daily Practice and Experience of many, if not most Mariners, who use this Way, and find the Success to answer their Expectations, at least much nearer than a much larger Division would do. 'Tis true, if the Generality of Glasses be so much too short, as to countervail those two short Divisions (if they are too short) it were to be wish'd that the Errors in one were rectified, and then the Faults in the other might be a-mended; but till then I shall recommend that Way of following 42 Feet, or 7 Fathom, to one Knot of the Log-line. Indeed if it be, as it is reported by some, that to make Amends for the shortness of the Knots, the Glasses are commonly made but 27 Seconds; and if so, then if the Glasses were regulated and increased from 27 Seconds to 30, the Knots, by the same Proportion, should also be increased from 42 Feet to 46 ¼ Feet to one Knot, which seems more agreeable to Reason, and to Mr. *Norwood*'s Observation.

Note; When you divide the Log-line you must allow 12, 15, or 18 Fathom of Stray-Line, according to the Bigness of your Ship, accounted from the Log, before you begin to set out the Knots, and then put in a red or white Rag, and from thence begin to divide the Line into Knots. The Reason of the Stray-Line is to veer the Log pretty well out of the Ship's Wake, least the Eddy should suck the Log after the Ship, and deceive you in your Reckoning.

SECT. III.

How to make a Plane Chart.

THE Log-line being thus divided, and the Half-Minute-Glass examined and regulated, the next Thing is to make a Chart for the Voyage intended; and of Charts there are several Sorts.

The first is commonly called a Plane Chart, in which the Degrees of Longitude and Latitude are every where

M 4   equal

equal without any Respect to the Globularness of the Earth, but rather supposing the Earth and Sea to be a p ane Superficies; and hence it is, that this Projection is false, except in Places under or very near the Equinoctial. However it being much in Use, I shall lay down the Projection of it as followeth.

If you would make a Plane Chart for all the Earth and Sea, it is best to do it upon two Sheets of Paper; one Half upon each Sheet. Through the Middle of each Sheet, cross-ways, from the Right-hand to the Left, draw the spotted Line AB, which divided into 180 equal Parts, by any equal Parts of the Scale, as large as your Paper will contain, which you may mark at every 10 Degrees with Figures, 10, 20, 30, &c. to 180, beginning at any Place where you intend to reckon your Longitude from, which suppose let it be in *London*, let your Chart begin at the West or Left-hand, and so reckon Eastwards, as in the following Example, where the two Charts contain all the 360 Degrees of Longitude in Compass, and 90 Degrees of Latitude each Way from the Equinoctial Then from the Middle of the Chart, as a Center, draw the 32 Points of the Compass, as you see done.   Then for inserting any known Place in the Chart, find by the Table of Latitudes and Longitudes of Places what Latitude and Longitude your Place hath, and place it in that Latitude and Longitude in the Chart; as *Fig* 53. for Example, suppose I would insert the *Lizard*, and the West-end of *Cyprus* in the *Straits*, and would find their Bearing and Distance; according to the Plane Chart, I find the Latitude of the *Lizard* in the Table is 49d. 55m. North, and Longitude 5d. 14m. W. I reckon upon the Line AB, which is the Equator, till I come at 5 Degrees, *viz.* 5 of the small Divisions, and as near as I can compute, somewhat less than ¼ of another small Division for the 14 Minutes, and setting one Foot of the Compasses in that Mark, I extend the other to the next North and South Line, and running them up into Latitude 49d. 55m. I make the Mark ☉ to represent the

Lizard,

Lizard; and by the same Method I find the
Fig. 53.   West-end of *Cyprus* at the Mark △ then the
· Distance between these two Marks taken in
your Compass, and applied to the Equator AB, accounting every Degree 60 Miles, and every 10 Degrees 600
Miles, gives the Distance between the two Places, according to the Plane Chart; and then for the Course,
observe what Line, or Point of the Compass, a Line supposed to be drawn between these two Places would be
parallel to, or nearest thereunto, which in this Case you
see is an E.S.E. and W.N.W. Line, the nearest Point
representing the Course required.

In like manner you may insert any other Place whose
Latitude and Longitude is given; as for Instance, I have
inserted the following Places, which because the Draught
is too small to contain the Names at large. I have represented them by the following Letters annexed to them.

⊙ The *Lizard.*
△ The Island of *Cyprus*, in the *Straits.*
a *Majorca*, an Island in the *Straits.*
b *Barbadoes.*
c *Jamaica.*
h Cape *Henry* in *Virginia.*
o *Bengal*, in the *East-Indies*,
d Cape *Bona Esperanza.*
e The *Naze* of *Norway.*

These are sufficient to let the Learner see how to set
down the Places in a Plane Chart.

But suppose you are to sail from any one of these
Places, or any other Place, to some other Port, it is best
to make a Chart for the particular Voyage, to contain
only so much of the Earth and Sea as is contained between the two Places, or little more, and then you may
make your Degrees of Latitude and Longitude larger,
as in the following Example.

A Ship sets sail from *Flamborough-Head*, in Lat. 54d.
8m. North, and Longitude od. 10m. E. intending for
the

the *Naze* of *Norway*, in Lat. 57d. 50m. and Long. 7d. 22m. E; I defire a Chart made of that Voyage.

This Chart is made to contain from 54 to 58° of North Latitude, and nearly 9 Deg. of Weft Longitude; and then by the former Rules, the Point *Fig.* 54. A reprefents *Flamborough-Head*, and the Point ☉ upon the oppofite Corner, is the *Naze of Norway*: Then at *Flamborough-Head* the Place failed from, as a Center, I defcribe a Quarter of a Compafs, which in this Cafe is always fufficient, and thus is the Chart ready for the Voyage.

As for fetting off the Courfes and Diftance, or Difference of Latitude and Departure, upon a Chart, commonly called pricking a Chart, I fhall refer it till I come to give fome Examples of keeping a Journal by the Log and Compafs.

How to make a Mercator's Chart.

A *Mercator's* Chart appears fomewhat like a Plane Chart, only with this Difference, that whereas, in a Plane Chart, the Degrees of Latitude and Longitude are every where equal, it is not fo in a *Mercator's* Chart, for in it the Degrees of Latitude bear the fame Proportion to the Degrees of Longitude, that they do upon the Globe; and the Invention of this Chart is moft properly owing to our worthy and ingenious Countryman Mr. *Edward Wright,* as may be feen in his Corrections of the Errors of *Navigation:* However, (I know not well for what Reafon) unjuftly afcribed to *Mercator.* Now, although upon the Globe the Degrees of Latitude are every where equal, and the Degrees of Longitude grow lefs nearer the Poles; yet in this Chart it is not fo, for the Meridians are Parallel, and every where equal, as in a Plane Chart, but the Degrees of Latitude grow bigger near the Poles; fo that in a *Mercator's* or *Wright's* Chart, there is always the fame Proportion between a Degree of Latitude, and a Degree of Longitude, in any Parallel, as there is upon the Globe itfelf, though the Diftances are extravagantly diftended, efpecially near the Poles.

For

For the Projection of this Chart, there are two Lines upon *Gunter's* Scale, commonly placed next the Bottom, on that Side upon which the Logarithmic Numbers, Sines, and Tangents are; the lowest of the two is a Scale of equal Parts, and the next to it is called the Meridian Line, by the Help of which, if you would draw a large Chart of all the World between 85 Deg. of North Latitude, and 85 Degrees of South Latitude, you must prepare large Paper; or paste Sheets of Paper together, till your Sheet contain about four feet each Way, through the Middle of which draw the Equinoctial Line, as you see in the Plane Chart before inserted, and graduate it with 10, 20, 30, &c. by those Divisions upon the Line of equal Parts now mentioned, which done, when you see how far the 180 Degrees of Longitude reach (if you make the Chart in two Parts, or to the 360 Degrees, suppose you make it all in one Draught, as your Sheet of Paper four Feet square will be large enough for that Purpose) there draw Lines at Right-angles, with the Equinoctial Line, as you see in the foregoing Charts, the Lines CAE, and DBF, *Fig.* 53, then these Lines graduated from A the Equinoctial, both Ways; by the Graduations upon the Meridian Line upon the Scale, shall set off every Parallel of Latitude according to *Mercator*; and when the Lines on both Sides are so graduated upwards and downwards, from the Equinoctial, the Lines drawn from every Degree on one Side to the same Degree on the other Side, shall represent the Parallels of Latitude required.

But because these Graduations would be too small to make a Chart by, for any particular Voyage, it is better to make the Chart larger, and in that Case the Want of a Meridional Line so large, may be supplied by a Table of Meridional Parts; for having drawn the Equinoctial, or any other Parallel of Latitude, which is the same, for one Out-line of your Chart; set off 60 of any equal Parts for every Degree of Longitude, both at the Top and Bottom of your Chart; this done, find (by the

Rules

Rules laid down in *Mercator's Sailing Trigonometrical)* the Meridional Difference of Latitude, or the Meridional Parts contained between that Degree and the next, thele taken from the fame Scale of equal Parts, and fet from the laft Degree marked, fhall find the next Degree, and fo in the reft.

The Rumbs, of Points of the Compafs, are exactly the fame as in a Plane Chart.

I fhall explain what hath been faid by an Example of a *Mercator's* Chart, from *Flamborough-Head* to the *Naze* of *Norway* before-mentioned.

	Lat.	Long.
Flamborough-Head	58d. 8m. N,	od. 10m. W:
Naze of Norway	57 50 N.	7 22 E.

Now it is beft to make the Chart to a whole Degree, and therefore in this Example I fhall, as before, make it from Lat. 54 to 58, and to contain 9 Degrees of Longitude.

Merid. Parts for Lat. 58d. 0m. ———— 4294
Merid. Parts for Lat. 54d. 8m. ——— 3871

Merid. Diff. of Lat. in the whole Chart — 416
Diff. of Longitude in the whole Chart —— 540

Having drawn the Line *d a*, fet off the whole Meridional Difference of Latitude 416, of any convenient Scale of equal Parts, from *d* to *a*, and draw *a b*, and *d c*, perpendicular to *d a*, and fet off the whole Difference of Longitude 9 Degrees, or 540 Min. from *d* to *c*, and *a* to *b*, and draw *c b*; thus you have the whole Subftance of your Chart.

Then fet 60 of the fame Parts from *d* to o, and from o to 1, and from 1 to 2, &c. both upon the Line *d c*, and upon the Line *a b*, and draw the Line o———o 1 — 1, &c. thus is your Longitude graduated.

Then for the Latitude, find the Meridional Difference of Latitude, between Lat. 54 and 55, which is 103, therefore fet 103 of the fame equal Parts from *a* to 55,

and

and from *b* to 55, and draw the Line 55—55; then find the Meridional Difference of Latitude between Lat. 55, and Lat. 56, which is 106, which set from 55 to 56, on both Sides, and draw the Line 56—56, and so find also the Line 57—57. Then upon the Center A, which reprefents the Place failed from, draw a Quarter of a Compafs, the Rhumb upon that Compafs fhews the true Courfe from *Flamborough-Head*, at the Point A, to the *Naze* of *Norway*, at the Point ☉, which you fee differs much from the Courfe found by the Plane Chart:

The Diftance between any two Places in a *Mercator's*. Chart is thus found; find the Difference of Latitude between the two Places, which here   *Fig. 55.* is 3d. 42m. or 222 Minutes, which fet off upon the Lines *a b*; take the 3 whole Degrees from the 3, and the 42 Min. in the fubdividing Degree, accounting every Part 10 Min. becaufe the Degree is divided into 6 Parts; then keeping your Compaffes at that Extent, lay a Ruler fo, as it may juft cut the two Places, whofe Diftance is required; then fet one Foot of your Compaffes at the Ruler's Edge, fo as that the other turned about may juft touch fome Eaft and Weft Line, then keeping that Foot faft that ftood againft the Ruler, open the other to the croffing of the Ruler, and the faid Eaft and Weft Line, that Extent meafured on the Parallel *a b*, allowing 60 Miles to every Degree, gives the true Diftance required.

Thus a Scale laid from *a* to ☉, defcribes the pricked Line *a* ☉, then with 3d. 42m. in your Compaffes, and one Foot in *b*, the other turned about will juft touch the Line 57——57; then obferve where the Line 57—— 57 cuts the prick'd Line *a* ☉, as in *k*, the Extent *b k*, meafured on the Line *a b*, gives the true Diftance.

How to make a true Plane Chart.

BUT a Third Sort of Charts I fhall now defcribe, which is as true as *Mercator's*, and yet as plain, eafy,

eafy, and expeditious for Practice, as the Plane Chart, but it cannot be made but by a particular Voyage, and generally when you intend to come back the same Way you go, and it is thus made.

Having the Latitude and Longitude of the two Places between which you are to fail, find the true Courfe and Diftance, by Cafe the Sixth of *Mercator's Sailing*, which found, fet off the Courfe and Diftance between the Place failed from, and the Place bound for, as you are taught in *Traverfe Sailing Geometrical*; fo have you two Points reprefenting the two Places, and if your Ship fail upon feveral Courfes and Diftances, in Form of a Traverfe, as you are taught in *Traverfe Sailing Geometrical,* you may every Day fet off the Courfe and Diftance from the Ship to the Place bound for, according to *Mercator,* and yet it is done as eafily as in *Plane Sailing.* I fhall inftance in the fore-mentioned Voyage.

Lat.   Long.

A Ship fails from *Flamborough-Head* 54d. 8m. 0d. 10m E. Intending for the *Naze* of *Norway* 57 50   7 22 E.

For the Courfe, according to Mercator.

As Merid. Diff. Lat. ——— 401 ——— 2.603144
Is to Radius ——— ——— 90d. 00m. — 10.000000
So is Diff of Long. - — —— 432 ——— 2.635484

To the Tangent of the Courfe—47 08 —— 10.032340

Then for the Diftance.

As Sine Comp. of Courfe —— 42d. 52m. — 9.832697
To proper Diff. of Lat. · ·—— 222 —— 2.346353
So is Radius ——— ·——— 90 00 —— 10.000000

To the Diftance ——— ——— 326$\frac{1}{8}$ — 2.513656

Here A reprefents *Flamborough-Head*, B the *Naze* of *Norway*, the Line AB the true Diftance, 326 *Fig. 56.* Miles, the Angle CAB 47d. 08m. the Courfe from the Meridian, and in this Chart the Degrees of Latitude are equal Divifions, and eafily reprefented,

sented, but the Degrees of Longitude, if they are inserted here, would be curved Lines, and hard to project; nor is it needful, seeing upon this Chart you need not regard the Longitude as you go along, but only the Course, and Distance upon each Course, according to the Rules laid down in the second Question of *Traverse Sailing Geometrical.*

Having thus shewed how to make the several Charts, I shall shew the Use of them, when I come to give some Examples of keeping a Journal, in which I shall instance in the same Voyage, for which these Charts are projected, and shall shew the Learner how to prick off every Day's Work upon each Chart; of which hereafter.

SECT. VI.

How to take an Observation.

THE most useful and easy Way to take an Observation by the Sun, is with a Quadrant, commonly called *Davis*'s Quadrant *, consisting of two Arches, in some a 30 Arch, and 60 Arch; but more commonly of late, an Arch of 25 Degrees, and another of 65; but the Way of using them is all one and the same; for by these Quadrants you do not so readily find the Sun's Altitude, but the Complement of the Sun's Altitude, commonly and (and properly) the Sun's Zenith Distance, being the Sun's Distance from the Zenith, or Point right over your Head, which is easily found by one of these Quadrants; for having fitted your Vanes so, as that when you have the Horizon right through the Sight Vane and Horizon Vane, the Shadow

may

---

* When this Book was first published, *Davis*'s Quadrant was the best Instrument then known by Mariners, for this Purpose; but *Hadley*'s is now so much in Use, so superior in Practice, and so well understood, that nothing more need here be said of it.

may at the same Time fall directly from the Top of the Shadow Vane to the Top of the Slit (which is the Middle) of the Horizon Vane, still removing the Sight Vane downwards, as you observe the Sun to rise, till you find that the Sun is upon the Meridian, and then you have done your Observation for that Day. This done, observe what Number or Figure you have upon the 60 (or 65) Arch, just under the upper Edge of the Shadow Vane, (which should always be placed upon an even 10 Degrees, to save Trouble in the Addition) and also to observe what Degree and Minute you have upon the 30 (or 25) Arch, just under the Middle of the Sight Vane, even with the Line or Stroke that goes from the Hole along the Middle of the Vane; and these two Numbers added together, the Sum is always the Zenith Distance, which is to be used as in the following Directions for working an Observation.

How to Work an Observation.

THERE are but four Cases, or Varieties in working an Observation, in whatsoever Part of the World you be, or whether the Sun's Declination be North or South.

The first is, when the Sun is between the Horizon and the Equinoctial, and then the Rule is, subtract the Declination from the Zenith Distance, the Remainder is the Latitude of the Place.

Demonstration.

The Latitude of any Place is the nearest Distance between the Equinoctial and the Zenith of that Place. The Sun's Declination is the Sun's nearest Distance from the Equinoctial: And the Sun's Distance from the Zenith of the Place, is the Zenith Distance or Complement of the Sun's Altitude, Now, if from ☉ Z the Zenith Distance, you subtract ☉ E, the Sun's Declination, there remains EZ, the Latitude of the Place of Observation.

Fig. 57.

Example.

Example.

Suppose the Sun's Zenith Distance be Z ☉ 76, and the Sun's Declination South E ☉ 16, subtract ☉ E the Declination 16, from ☉ Z the Zenith Distance 76, there rests EZ 60, the Latitude required.

The second Variation is, when the Sun is between the Equinoctial and the Zenith, and then the Rule is, Add the Zenith Distance to the Sun's Declination, the Sum is the Latitude of the Place : For if to ☉ Z, the Zenith Distance, you add ☉ E the Sun's Declination, the Sum is EZ, the Distance between the Zenith and the Equinoctial, which is the Latitude required.

Example.

Suppose the Sun's Zenith Distance ☉ Z 35, the Sun's Declination North E ☉ 10, if to ☉ Z 35 the Sun's Zenith Distance, you add E ☉ 10 the   *Fig. 58.* Sun's Declination, the Sum must needs be EZ 45, the Latitude required.

The third Variety is when the Sun is between the Zenith and the elevated Pole, then the Rule is, subtract the Zenith Distance from the Declination, the Remainder is the Latitude of the Place.

Example.

Suppose the Sun's Declination be E ☉ 20, and the Zenith Distance Z ☉ 10 : Subtract the Zenith Distance Z ☉ 10, from the Sun's Declinati-   *Fig. 59.* on E ☉ 20, there rests EZ 10, the Latitude required.

The fourth Variety is when the Sun is between the elevated Pole and the Horizon, and then the Rule is, subtract the Sun's Complement of Declination from the Zenith Distance, the Remainder is the Complement of Latitude.

N

Example.

Suppofe the Sun's Declination be 22, (its Complement is 68) the Zenith Diftance is 85, then
from Zenith Diftance ⊙ Z 85, fubtract the   *Fig.* 60.
Complement of Declination ⊙ P 68, the Remainder Z P 17, is the Complement of Latitude,
which fubtracted from 90, leaves 73, the Latitude required.

But becaufe 'tis feldom that any fail fo far North or
South, as that they can conveniently take a backward
Obfervation by the Sun, under the elevated Pole, in
this Cafe it may be done by a forward Obfervation, and
work with the Sun's Altitude or Heighth above the
Horizon, and then the Rule is, add the Sun's Altitude
to the Complement of Declination, the Sum is the
Latitude; thus if in the laft Example, you add the
Sun's Altitude, O ⊙ 5, to the Complement of Decli-
nation ⊙ P 68, the Sum O P 73, is the Height of the
Pole above the Horizon, which is the Latitude of the
Place.

This fame Operation will hold in taking an Obfervation by a Star, when under the elevated Pole, becaufe
both here, and in all other Cafes in obferving by the
Stars, we are obliged to take the Obfervation forward,
becaufe a Star cafts no light fufficient for a backward
Obfervation, and then, inftead of the Zenith Diftance,
work with the Altitude.

In forward Obfervations, whether by Sun or Stars,
there are alfo Varieties, as in a backward Obfervation,
which I fhall only fpeak to, the Demonftrations being
pretty evident from the foregoing Figures.

The firft Cafe is, when the Star is between the Equinoctial and the Horizon, and then the Rule is, Add the
Altitude of the Star to its Declination, the Sum is the
Complement of Latitude, which fubtracted from 90,
leaves the Latitude required.

Secondly,

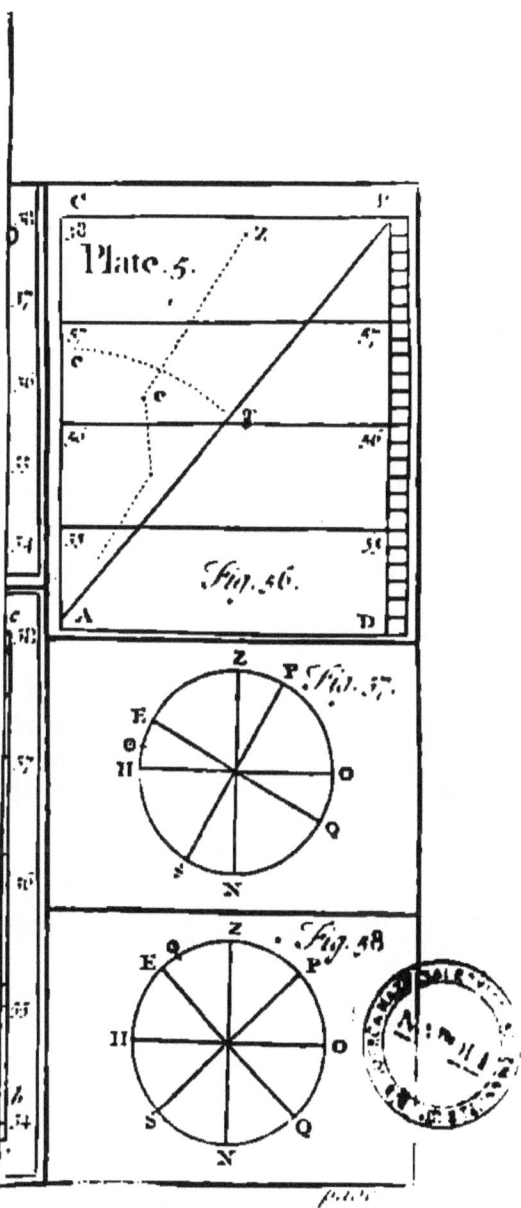

Plate 5.

Fig. 56.

Fig. 57.

Fig. 58.

Secondly, When the Star is between the Equinoctial and the Zenith, subtract the Star's Declination from its Altitude, the Remainder is the Complement of Latitude.

Thirdly, When the Star is between the Zenith and the Elevated Pole, then subtract the Complement of Declination from the Altitude, the Remainder is the Latitude.

The fourth Variety is spoke to in the fourth Variety before going; and so much shall serve for working an Observation; these Rules if well observed, being sufficient for working Observations in all Latitudes, whether North or South, and at all Times of the Year, a little Practice will therefore make them evident.

It is possible to take an Observation by the Moon; but there are so many Things to be accounted for, as Paralax, Refraction, &c. and the Moon seldom to be seen but when some known Star may also be seen, and I having in this Book inserted a Table of the principal fixed Stars, their Declinations, &c. I shall refer the Reader to it, the Sun and Stars being sufficient in all Cases for taking Observations in common Practice.

SECT. V.

How to reckon the Tides.

Of the General Motion of the Tides, and how to know the Time of High-Water at any known Port, only by a Sight of the Moon at any Time of the Day or Night.

THOSE who desire to give a good Account of the Tides, or of the Time of High-Water or Low Water, at any Port or Harbour proposed, it is necessary, in order thereunto, that they should have a right Understanding of the original Cause of the Motion of the Tides, or Ebbing and Flowing of the Sea; a Thing which hath been often in Dispute among

the

the Learned, both Mathematicians and Philofophers, whofe different Sentiments have rendered the Thing as dubious as when they firft began with it, fome afcribing the Fluctuation of the Sea to the fwift Motion of the Earth, according to the *Copernican* Syftem, and the Water being a Fluid Body, and not prefently acquiring fo fwift an Agitation as the Earth itfelf, it muft confequently be higher Water upon one Part of the Globe than upon another, and this they illuftrate thus: Suppofe a Boat under Sail, with frefh Way, and a fmall Quantity of Water in the Boat, it feems very plain, that the fwift Motion of the Boat would make the Water incline rather towards the Stern of the Boat, but if the Boat, when failing with that Speed, fhould be coming afhore, or fome other Accident, fhould meet with a fudden Interruption, and at once ftand ftill, the Water ftill retaining in fome Meafure its former Motion will prefently run to the fore Part of the Boat, and by this they would fome Way or other Demonftrate, that the Motion of the Tides depends upon the Motion of the Earth: But to ufe no other Argument for the Confutation of this Opinion, the Abfurdity hereof will appear in this; that if the Motion of the Earth was the original Caufe of the Motion of the Tide, then the Tide muft neceffarily follow the Motion of the Earth (or to our Appearance the Motion of the Sun) and confequently it muft always be High-Water, at one Place, at one and the fame Time of the Day; but the contrary is fo evident to all, that there needs no more to be faid to difprove it.

Others fay that the Flowing of the Tide is occafioned by a great Confluence of Water proceeding from the Mael-Stream, called by fome the Navel of the Sea, being (as is reported) an Eddy or Whirl-pool, under the Weft-Coaft of *Norway*, or *Finmark*, from whence (it is faid during the fix Hours Flood, the Water iffues violently out, and occafions the Rifing or Flowing of the Water in all the adjacent Parts, and finks with the

the fame Violence during the Ebb; fo that it is faid,
that during the Flood, the heavieft Metal will not fink;
and during the Ebb, the lighteft Subftance, or the beft
of Ships will not fwim; but what Reafon they will
give why this ftrange Ebbing and Flowing (if it is fo)
fhould be regulated and governed by the Motion of the
Moon, I do not underftand, unlefs this Iffue of Water
be fupplied with fome Communication that it hath with
the *Eaftern* Seas, or by fome Paffage under the Main
Continent of *Norway*, but this being an Uncertainty, we
fhall wave it, as well as the Opinion of a third Sort,
who affirm, that God, who created all Things, gave
Life to all his Creatures; and that the Ebbing and
Flowing of the Sea is no more but the Breathing of
the Earth, which feems to me a very odd Fancy, and
not worth inferting, had it not been in this Collection
of the various Opinions of the Learned upon this
Subject. I fhall among the reft, deliver my own Opinion in the Matter; and although I do not think my
felf able, infallibly to give a definitive Sentence in the
Cafe, yet I fhall endeavour to prove it to be confiftent
with the Obfervations and Experience of our Mariners,
and fhall anfwer what Objections can eafily be made
againft it.

It is evident to all that own the Rotundity of the
Earth, (a Thing generally out of Controverfy among
the Learned) that there is a Principle of Gravitation
towards the Center of the Earth, and that this attractive Influence is diffufed to all beings whatfoever within the Orb thereof; and hence it is, that we who in-
habit the Earth find no fuch Thing as an upper Side
and an under Side of the Earth, but in all Parts of the
Superficies thereof we find a like natural Tendency
towards the Center, as it is evident by the Experience
of thofe who have failed about the World, and yet in
their fo far different (if not diametrically oppofite)
Places that they have failed to, have found themfelves,
and every particular Thing to have the fame preffing

Inclination towards the Center of the Earth, which feemed to them to be downwards, as well as it doth to us. Now if we grant the Earth this ftrong Principle of Gravitation, Inclination, or Attraction towards its Center, which Reafon and common Experience proves, we have Reafon from thence to believe, that the other Bodies, as the Sun and Moon, have the fame Principles of Gravitation towards their Centers, (which may be proved by fome Reafons, which, to infert here, would be too great a Digreffion from the prefent Subject) and which being granted, I fuppofe the Ebbing and Flowing of the Sea to be occafioned by the Attraction of the Sun and Moon (efpecially the Moon being a fecondary Planet, which moves far nearer the Earth, and refpects it as her Center) the Strength of which Attraction, although it cannot have any influence upon the folid Parts of the Earth, yet the Water being a fluid Subftance, is more eafily affected with this attracting Power, and by Virtue thereof, (while the Earth continues round) the Water is gently fucked and drawn into an oval Form, by reafon of its inclining Tendency towards thofe attractive Bodies, thereby caufing High-Water where the Ends of the Oval is, and confequently Low-Water at the Middle of the Oval, as may be demomftrated thus ; fuppofe two Hoops made of any flexible Subftance, as Wood, fine Steel, &c. of equal Dimenfions, and laid directly one upon the other; and then if from oppofite Points the uppermoft was extended into an Oval Form, it is evident, that as the extended Part or End of the Oval is drawn without the round Hoop, the two Sides, or Middle of it, will be contracted and pulled within it; and ftill, as the traverfe Diameter of the Oval was extended and augmented, the Conjugate or fhorteft Diameter, will be contracted and diminifhed; which Plainly demonftrates how the Water, when by an Attractive Power it is drawn above its mean Elevation, at the End of the Oval, it muft needs be depreffed below its ufual

Pofition

Pofition at the two Places a Quarter diftant from thefe two oppofite Points, which Elevation and Depreffion is the Occafion of High and Low-Water, which happen at (near) fix Hours Diftance, between High and Low-water, and confirms the Truth hereof, which is alfo evident by the following Reafons.

First, Becaufe the Motion of the Tides generally follows, and is govern'd by the Motion of the Moon; fo that the Moon being upon one and the fame Point of the Compafs, makes High-Water at any particular Place, at one and the fame Time, (unlefs accelerated, or retarded by Winds, Land-floods, or the like) by which it feems very probable, that the Moon is the principal Agent in this regular Motion of the Sea; and if fo, then Nothing is more likely than that the Influence of her attractive Power, drawing the flexible Subftance of Water into an eliptical or oval Form, as before afferted, is the Caufe thereof.

But Secondly, It is obferved by all, that the Spring-Tides, (viz. at the New and Full Moon) are greater than the Neap-Tides (which are at the firft and laft Quarters) by which it feems evident, that although as I faid in the laft Paragraph, that the Moon is the principal moving Caufe of the Tides, yet the Sun having alfo the like Attractive Power, hath an Influence upon the Waters alfo, although not fo great as that of the Moon, (becaufe the Moon is much nearer the Earth) and is a fecondary Planet, refpecting the Earth as her Center, and therefore, when the Sun and Moon are in Conjunction (as the New Moon) or in Oppofition, (as at the Full Moon) then the Tides are greateft; becaufe the attractive Influence of the Sun is added to that of the Moon, and both raife the Water at one and the fame Place, making the Spring-Tides. But at the Quarters, viz. when the Sun and Moon are about 90 Degrees, or a Quarter of the Zodiac diftant, then the attractive influence of the Sun, rather impairs that of the Moon; the one raifing the Water where

the

the other depreſſes it; ſo that although, for Reaſons
before given, the Influence of the Moon's Attraction
upon the Water is greater than that of the Sun, and
therefore the Tides follow the Motion of the Moon,
and not of the Sun, yet their contrary Influence leſſens
the greater Influence; *viz.* that of the Moon; which is
the Cauſe of the Neap-Tides; and this alſo may be
illuſtrated thus: Suppoſe a Hoop of Steel, or any flexible Metal, be faſtened to any Place, if a Man with a
Rope fixed to one Side thereof ſhould pull with all his
Strength at the ſaid Rope, it would make the Hoop
decline into an eliptical or oval Form; but if another
Man, though of an inferior Strength, ſhould fix a Rope
in the ſame Place where the firſt was fixed, his Strength
added to the firſt would make the Ring or Hoop yet
more eliptical, and would ſtretch out the transverſe Diameter thereof yet longer, in Compariſon of the Conjugate. But if the firſt Rope remain fixed as before,
and the other managed by inferior Strength, be re-
moved a Quarter of the Hoop's Diſtance from the firſt,
then although the Elipſis will in ſome Meaſure retain
its own oval Form, inclining to the ſtronger Attraction,
yet the Power of the contrary Attraction will depreſs
the other, and cauſe the Hoop to retain a Form at leaſt
nearer to a true Circle, than as if the Attraction were
all in one Place; and this plainly illuſtrates the different Attractions, cauſing different Tides, *viz.* when
the attractive Powers are united their Influence is greater,
cauſing the Water to be more eliptical, thereby occa-
ſioning Spring Tides, at the New and full Moon; but
when the attractive Powers are ſeparated, as at the firſt
and laſt Quarter of the Moon, the Influence of the
greater is not ſo apparent, which is at the Neap-Tides.

But my third Reaſon is, becauſe we ſee that the
Spring Tides at the Equinoxes, *viz.* in *March* and *Sep-
tember*, are commonly higher Tides than the Spring
Tides at the Solſtices, *viz.* in *June* and *December*; the
Sun, and alſo the New and full Moon, moving in or

near

near the Equinoctial, in a right Aspect to the Earth; whereas their Influence is somewhat impeded at the Solstices, by their more oblique Position to the Earth. All which seems plainly to prove that as the Tides are regulated and governed by the Motion of the Moon, so her attractive Power, together with that of the Sun when joined with it, is the original Cause thereof.

Now against what hath been said, there seems two grand Objections to arise; the first is, that if the Tides be governed by the Moon, and if her Attraction be the Cause thereof, her apparent diurnal Motion being from East to West, it would follow that the Flood-Tides should in all Places set Westwards; but daily Experience proves the contrary; for it sets in some Places South, as upon the Coast of *England*, and in some Places North, as upon the Coast of *Holland*; yea, and in some Places East, as in the *British* Channel, which makes against my former Assertion.

A second Objection is, that if the Tides be caused by the Moon's Attraction, it should be High-Water at all Places, when the Moon is upon the Meridian of the Place, or that a North or South Moon should make full Sea in every Place; but the contrary is evident by the Tide-Tables, and by the Experience of all Sailors.

In answer to the first, *viz.* that the Flood-Tides should set Westward in all Places, (the Answer to which also partly implies an Answer to the second, *viz.* that it should always be High-Water at any Place, when the Moon is upon the Meridian of the Place) I grant that if the Tides be occasioned as before asserted, then this regular Motion of the Sea, from East to West, must necessarily follow, if this terraqueous Globe was equally environed with Water, and that the aforesaid Motion of the Sea was not interrupted by the Land; but it is evident, especially where the Sea joins immediately upon the West of any main Continent, that the Flood cannot set Westwards there, because that would be right from the Shore; as for Example, suppose the whole Land of *Europe* and *Africa* were

were all one Main Continent; and suppose the North
Cape of *Finmark*, in about Lat. 71 North, and the
South Cape of *Africa*, in about Lat. 34 South, were
both under one Meridian, and that this Main Continent
were terminated on the West with a strait Coast, lying
under the same Meridian which the two extreme Points,
the North and South Capes lie under, now it is plain
that the Tide cannot come from the Eastward upon
this West Coast; but if the Tide be caused by this
Westerly Motion of the Sea, according to the Motion
of the Moon, it must set Westwards, about the North
and South Parts of the said Continent, and so proceed
from the North Cape Southward, and from the South
Cape Northward, along this West Coast; and the
Truth of this seems evident also from our common Experience; for we find, that the Flood proceeds from the
Northward, along the Coast of *Norway*, and there finding
a Passage between *Scotland* and *Norway*, marches along
the East Coast of *England*, and hence it is High-Water
sooner in *Scotland* than in *England*, and sooner on the
Coast of the North Part of *England* than on the South
Part; thus it is High-Water at *Aberdeen* in *Scotland*,
45 Min. after Moon's Southing, but at *Tinmouth-Bar*,
not till three Hours after, and at the *Spurn* 5 Hours
and 15 Min. and at *Cromer* 6 Hours, and at *Yarmouth
Pier* 9 Hours, and at *Harwich* 10 Hours 30 Min. after
the Moon's Southing; the Tide at the same Time being rolling along the West Coast of *Scotland*, and from
thence to the West Coast of *Ireland*, that Passage between the North-East Part of *Ireland*, and the South-
West Part of *Scotland*, being so narrow, that the Tide
finds little Passage; and therefore the East Coast of *Ireland* is supplied by a Tide which sets Southward along
the West Coast; and hence the Tide flows Eastward,
along the South Coast, and Northward along the East
Coast thereof, as is evident by the Time of High-Water observed in the Tide-Table; for upon the Full
and Change Days, it is High-Water at *Seyn-Head* at
10 Hours 20 Minutes, from thence in 6 Hours it

passes

paſſes along the Weſt Coaſt, and butts in upon the
South Coaſt, ſo that at 4 Hours 30 Minutes it is High-
Water at *Kingſale, Cork,* and *Waterford,* and from thence
in 45 Minutes more, it paſſes over to *Milford, Lundy, &c.*
and at the ſame Time it ſlides Northwards, along the
Eaſt Coaſt of *Ireland;* ſo that in 3 Hours 45 Min. it
flows from *Waterford* to *Dublin,* and in 45 Min. more
to the *Iſle* of *Man, &c.* Thus the general Motion of
the Tide to the Southward, having, as by a Branch
proceeded from thence, filled up all the Vacancy between *England* and *Ireland,* it proceeds ſtill to the Southward, and upon the Full and Change Days, it is High-
Water at the Land's End of *England* at 7 Hours 30
Min. and its unwearied Motion to the Southward
it finds a Paſſage up the *Britiſh* Channel, and 45 Minutes after High-Water at the Land's End it is High-
Water at *Portland,* and in 3 Hours 45 Min. more it is
got up to the Iſle of *Wight, Southampton,* and *Portſmouth,*
being High-Water at the ſame Time on the other Side
of the Channel, upon the Coaſt of *France,* about *Guern-
ſey* and *Jerſey,* and ſo proceeds ſlowly along the Coaſt
of *France* and *Holland* ſtill to the Northward, being ſup-
plied by a Tide up the Channel; and this (to follow it
no further) may ſufficiently prove, that the General
and Original Motion of the Tide Weſtwards occaſions
its Motions to the Southward along the Coaſt of *Norway,* and conſequently all its compounded Motions and
Branches into the *Iriſh* Sea, the Chanuel, *&c.* And it
is evident, that although the Tide, if not interrupted,
ſhould be at the Height where the Moon is upon the
Meridian; yet it finding ſo many whirling Motions to
and fro between the Lands, the Moon at the ſame Time,
keeping her ſtrait and uninterrupted Courſe to the Weſtward, all the Time that thoſe irregular Vacuities are
filling up, it is plain that the Moon muſt needs be paſt
far from the Meridian before it can be High-Water at
ſome of thoſe Places, and yet the Tide occaſioned as
before aſſerted; which I think will ſufficiently anſwer
the two Objections before mentioned, and prove this
Hypotheſis

Hypothefis to be very confiftent with Reafon itfelf, and the Experience of all Obfervers.

Now if any Body will ftill infift further upon what I hinted upon before, *viz.* That the Tide is occafioned by a Confluence of Water arifing at the Mael-ftream, or Navel of the Sea, and proceeding from thence, *&c.* and for Confirmation thereof will alledge that they have feen and obferved this regular Rifing and Sinking of the Water (before fpoken of) at that Place; I anfwer, I can eafily approve of this Opinion, without denying mine own hitherto afferted; for they are very eafily reconcilable; fince it is poffible, and very probable, that this Ebbing and Flowing at the Mael-ftream, may be occafioned by fome fubterraneous Cavern, whereby this Place hath fome Communication with the Eaftern Seas, and is fupplied from thence; which being granted, it will follow, that while the Tide (whofe general Motion is Weftwards) is interrupted by the Continent of *Norway,* &c. and thereby is forced to find a Way about the North Cape; yet by the Way, finding the Paffage under Ground, and the Tide, by Virtue of the Moon's Attraction, inclining that Way, there may be fuppofed to be fo much Water conveyed that Way as that obfcure Paffage can contain, though far fhort of fo much as to occafion that Flowing and Ebbing that is obferved; and this being granted, it is evident, that this Rifing and Falling at the Mael-ftream, muft needs keep Time with the Motion of the Moon, and of the reft of the Tide; becaufe it proceeds from the fame Original, and is a fmall Branch thereof; and this we may fee illuftrated by our common Rivers, whofe Natural Motion, when interrupted by Banks, or other Impediments, caufeth the main Body of the Water to find a Paffage fome other Way; yet if at the fame Time any fmall Holes be found in the faid Banks, there will always fo much Water pafs thereby as the faid Holes can contain; though the main Body of Water is forced another Way: Hence, from what has been faid, it is plain, that this

fmall

small Flood and Ebb obferved at the Mael-ftream, muft needs have the fame regular Motion with the reft of the Tide, which I fuppofe has at firft occafioned, and fince feemed to confirm that Opinion that the Tide proceeds only from thence.

To what has been faid I fhall only add, that this which I have here inferted feems yet to be further confirmed by our Obfervations Abroad; for it is obferved, that there is little or no Tide at the *Straits* of *Gibraltar*, nor upon the Coaft of *Guiney*; nor can it be expected, according to this Hypothefis; for if the Tide comes from the Northward, it may be fuppofed, that by the Time it hath paffed fo far along, and having fo many Vacancies to fill up, as the *German* Ocean, the *Irifh* Sea, the *Britifh* Channel, the Bay of *Bifcay*, &c. its Power muft be very much impaired, if not totally exhaufted, before it comes fo far as the Coaft of *Africa* before mentioned.

Now it may be queftioned, what is the Reafon that there is little or no Tide in the *Baltick* Sea, feeing there are ftrong Tides almoft on every Side of it, *viz.* upon the Coaft of *Norway* on the North of it, on the Coaft of *Holland* on the South, and on the Coaft of *England* on the Weft of it, &c.

I anfwer, it is hard to determine abfolutely what is the Caufe thereof; but I fhall lay down fome Conjectures, which may conclude very much that Way; for if it was High-Water at the *Naze* and Coaft of *Norway*, at the fame Time that is High-Water upon the Coaft of *Jutland* Southward from the *Baltick* Sea, it muft needs force a Tide into the *Baltick* Sea, as well as High-Water (upon any Coaft) forces a Tide up the Inland Rivers there; but this cannot reafonably be expected, if we confider, that the Tide along the Coaft of *France*, *Holland*, and *Jutland*, proceeds from the *Britifh* Channel, and comes from thence Northwards, along the Coaft of *Holland*, &c. but the Tide upon the Coaft of *Norway* lies to the Southward, as hath been largely proved, the two Tides both terminating at the

Mouth

Mouth of the *Baltick* Sea; hence it is very probable, that it may be High-Water upon the Coaft of *Jutland,* when it may perhaps be Low-Water, or fome intermediate Tide at the *Naze,* or South Coaft of *Norway*; and if fo, it cannot force a Tide into the *Baltick* Sea, but rather a fucking Current, or Inclination of the Sea from that Place on one Side of the Mouth of the *Baltick* Sea, where it is High-Water, to that Coaft on the other Side where it is Low-Water, and hence it will neceffarily follow, that this Current muft fet fometimes one Way, and fometimes another. And this may be affigned as a Reafon why many have found themfelves deceived in their Reckonings, when intending to make the *Naze,* or other Lands thereabouts; and when they have imputed their Miftake to a Current fetting towards the Point of the Compafs, towards which they have found themfelves unexpectedly carried, thinking thereby to regulate their future Reckonings, they have found themfelves at another Time under a quite contrary Error, and hence have concluded that there was no Current at all, but fome other Thing hath been the Caufe of their Error; whereas, if the Tides upon the Coaft of *Norway* and *Jutland,* viz. on each Side of the Mouth of the *Baltick* Sea, were carefully obferved and determined, and the Current allowed to run or fet from the Higheft Water to the Loweft, (upon which of thefe Coafts foever it were) and to be at a Stand only when the Water upon both the faid Coafts were of equal Height (whether Rifing or Falling) I queftion not but that the Motion of this Current might thereby be limited and determined, as well as the Motion of the Tides elfewhere, and due Allowance might be given for the Current there, as well as for the Ebbs and Floods in other Places, to the great Satisfaction and Advantage of thofe that Ufe the *Eaft-Country* Trade.

This Motion of the Tide thus granted, I fhall next fhew how to find the Time of the Moon's Southing, and (with a little Application) the Time of High Water by a Sight of the Moon, at any Time of the Day or Night.

How

How to know the Time of the Moon's coming to the South,
only by a Sight of the Moon, at any Time of the Day or
Night

IT is commonly known, that the New Moon being
in Conjunction with the Sun, Souths at Noon, and
the Full Moon being opposite to the Sun,
comes to the South at Midnight; and at the   *Fig.* 61.
Quarters, when she is just half full, *viz.* to
the Line N. 6. S. which crosses the Figure directly in
the Middle, then she is South at 6 o'Clock, and if the
light Half be on the West Side N.W.S. and she half
full, she Souths at 6 in the Evening, as at the first
Quarter; but if the light Half be on the East-side,
N.E.S. she Souths at 6 in the Morning, which granted, the Moon's coming to South upon any other Phasis,
or at any intermediate Age, may be easily gathered from
the Figure, observing this General Rule; when you see
the Moon at any Time of the Day or Night, observe
how much of the Moon is light, whether on the East
or West-side, and compare it as near as you can with
the Figure, considering to which Line in the Figure
you suppose the light Part will reach, and observe what
Number is upon that Line, for that is the Hour at
which the Moon will come to the Meridian that Day or
Night; as for Example;

Suppose I observe the Moon, and find her somewhat
more than half light on the East-side, so that compa-
ring the Moon with the Figure, I suppose the light
Part to appear like the light Part of the Figure; then
I observe to what Number the light Part reaches, and
I observe it reaches to the Number 5; and hence I
conclude, that the Moon comes to the South about 5
o'Clock; and because the light Part is on the East-side,
I conclude it is at 5 in the Morning she will be upon the
Meridian:

Again, suppose I see her in the Evening, and the
light Part on the West-side, as near as I can compute,

to

to be like the darker Part of the Figure, towards the Right-hand, then obferving how far the light Part extends, I fee it comes to the Figure 5, and becaufe the light Part is on the Weft-fide, I conclude fhe will be South at Five in the Evening, &c.

And thus you obferve the Moon two or three Days after the Change to appear in the Weft in the Evening, with a very little Light on the Weft-fide, which by Computation may be fuppofed to come as far over the Moon's Body as the Figure 3, I conclude fhe has been upon the South that Afternoon about Three o'Clock, and the next Night you will obferve the Light to encreafe, and come towards the Figure 4, I conclude fhe was South that Afternoon between 3 and 4, or near 4, and then perhaps 5 or 6 Days after, I obferve the light Part encreafed beyond half the Body of the Moon, as far as the Arch N. 3. S. I conclude from hence, that the Moon is South about Eight that Night; and although this Method is not fufficient to find the Moon's Southing exactly to a Minute, yet it is of fufficient Exactnefs for Reckoning the Tides, where a Quarter of an Hour, or Half an Hour, make no Material Error, it being generally impoffible to predict the Tides to the abfolute Exactnefs of a Minute, although you had the Moon's Southing exactly, becaufe Winds or Land Floods, &c. may alter the Tides, and few that have the Charge of a Ship will truft to the firft or laft Scruple of the Tide for going into an Harbour, or coming out, but will endeavour if poffible, to have the beft of the Tide, and to be ready for it againft it comes, whether Ebb or Flood.

The Moon's Southing being thus found, the next Thing is to fhew how thereby to find the Time of High-Water at any known Port; and for an Help thereunto, I have inferted a Tide-Table, in an Alphabetical Order, in which you need but find the Name of the Port, at which you would know the Time of High-Water, and againft it you have a Number of Hours and Minutes, which added to the Time of the Moon's South-

ing

ing, gives the Time of High-Water at that Place that Day; as for Example.

Suppose I was lying before *Tinmouth* Bar, waiting for half Flood to go in, I happen to see the Moon in the Morning, and I observe the East-side of the Moon to be light, like the light Part of the Figure before spoken of, *viz.* It is by Computation so much above half the Body of the Moon that the whole light Part reacheth to the Figure 5; and because the Light is on the East-side, I conclude she is South at 5 in the Morning, then I look in the Tide-Table, in the Letter *T*, and find *Tinmouth*, and against it I find 3 Hours o Minute, which added to 5, the Time of the Moon's Southing, the Sum, which is 8 Hours o Min. is the Time of High-Water at *Tinmouth-Bar*; so that I find I may go in about 5, 6, or 7 o'Clock, with the Flood Tides according to the Draught of Water that my Ship requires.

Now it is not necessary, that you should always have this Book or Figure about with you, for you may, with a very little Practice, get the Nature and Reason of it imprinted in your Memory, always remembring that the New Moon being with the Sun, souths at Noon, and the Full Moon being opposite to the Sun, souths at Midnight. The Moon in the first Quarter is South at 6 in the Evening, and at the last Quarter is South at 6 in the Morning. All the intermediate Times of her Age may be easily computed according to the Figure, exact enough for finding the Time of High-Water at any Port mentioned in the Tide-Table.

If any will object, that what hath been said serves only for finding the Time of High-Water, at Places mentioned in the Tide-Table, but no where else; I answer, that by this Way of finding the Moon's Southing, and consequently the Time of High-Water at any Port mentioned in the Table, together with a right Apprehension of the General Motion of the Tides, as you have it sufficiently described and illustrated, in the Beginning of this Section, you may be able to give a very

O

good

good Account of the Tide both at thofe Places inferted
in the Table, and of thofe that are not, provided you
know but upon what Coaft the Places are, and how
fituate from fome known Place expreffed in the Table,
and whether they are upon the Sea Coaft, or up fome
River, and the like; as for Inftance, fuppofe coming
from the Weft of *England*, up the Channel, intending
for *Boulogne*, and not having a Tide-Table that hath
the Port of *Boulogne* expreffed in it, I am at a Lofs to
know how the Tide falls, (fuppofing it at New Moon.)
Now fuppofe in my Tide-Table, I find *Diepe* and *Dunkirk* (Ports on each Side of *Boulogne*) and upon Exami-
nation I find it is High-Water at *Diepe*, on the Full and
Change Days, at 9 Hours 45 Min. Again, I find it
is High-Water at *Dunkirk* the fame Day at 12 o'Clock ;
now I conclude that *Boulogne* lying betwixt thefe two
Ports hath alfo High-Water between thefe two Times,
and not long after High-Water at *Diepe*, as, if you look
in the Table, you will find it High-Water at 10 Hours
30 Min. at the Full and Change.

In like Manner, if you obferve the General Motion
of the Tide to the Southward, along the Coaft of *England*, and to the *Eaftward*, up the Channel, &c. you
may, by knowing the Time of High-Water at any
Port, very eafily compute the Flowing and Ebbing of
the Tide at any adjacent Port, and with Allowance for
deep Bays, or inland Rivers, you may very nearly determine the Time of High-Water at any defired Port.

But for Variety, and the univerfal Satisfaction of all
Navigators, I fhall infert another (though common)
Method of finding the Moon's Age and Southing, by
the Epact, in order to which you muft firft find the
Golden Number, which is thus done.

Add 1 to the Year of our Lord, and divide that
Sum by 19, the Remainder is the Golden Number

2. For the Epact, fubtract 1 from the Golden Number,
multiply the Remainder by 11, divide the Product by
30, (neglecting Fractions) the Remainder will be the
Epact for the Year propofed.

3. For

3. For the Moon's age and the Epact, the Day of the Month, and the Number for the Month (as expressed below) together, the Sum if under 30, is the Moon's Age. But if the Sum exceeds 30, take 30 from the Sum as often as may be, and the last remainder will be the Age of the Moon.

4. For the Moon's Southing, multiply the Moon's Age by 4, and divide the Product by 5, the Quotient is the Hour, and for every one that remains add 12 Min. gives the Time of the Moon's Southing.

The Number for the Months are,

0 2 1 2 3 4 5 6 and 8

8 10 and 10 these are the Numbers right.

That is, *Jan.* 0, *Feb.* 2, *March* 1, *April* 2, *May* 3, *June* 4, *July* 5, *August* 6, *Sept.* 8, *Oct.* 8, *Novem.* 10. *Decem.* 10.

Note, The Epact is here supposed to change the first of *January*.

Exemple.

I desire to know the Time of High-Water at *Berwick* the 16th Day of *October*, 1770.

First for the Golden Number.

The Year ——— 1770   19)1771(93
Add —— ——— 1   61
Sum —— ——— 1771   4

The Quotient 93 is of no Use in this Case; the Remainder 4 is the Golden Number required.

To find the Epact.

The Golden Number ——— 4
Multiply by —— ——— 1
                      3
Multiply by —— ——— 11
Which divide by ——— 30)33(10
                         3

The Remainder 3 is the Epact required.

Note, The Rule for finding the Epact according to New Style, will not hold good after the Year 1800 begins.

For

For the Moon's Age.

The Epact ——— —— ——	3
The Number for *October* ——	8
The Day of the Month ——	16

Sum .. —— ——— —— 27 which being under
30 is the Moon's Age required, the Moon changing
on the 19th of *September*, as may be seen by the Table
of the Moon's changing immediately following the
Tide Table.

For the Moon's Southing.

Multiply the Moon's Age 27 by 4, the Product 108,
divide by 5, the Quotient is 21, and 3 remains, *viz.*
21 Hours and 36 Min. past Noon, to which add 1 Hour
30 Min. found against *Berwick* in the Tide-Table, the
Sum 23 Hours 6 Min. is the Time of High-Water at
Berwick, *October* 16, at Eleven o'Clock in the Morning
nearly:

But because the Spring Tides do not shift so much
as the Neap Tides, you may yet be more exact in using
the Table annexed.

Moon's Age.		Time. H. M	
1	16	0	42
2	17	1	21
3	18	1	52
4	19	2	22
5	20	2	53
6	21	3	23
7	22	4	8
8	23	4	55
9	24	5	50
10	25	6	53
11	26	7	58
12	27	9	4
13	28	10	8
14	29	11	5
15	30	0	0

The Use of this Table is very
easy, as I shall instance in the foregoing Example.

Find the Moon's Age 27 under
(Moon's Age) and against it in the
Column of [Time] you have 9
Hours 4 Min. which added to 1
Hour 30 Min. the Time found a-
gainst *Berwick* in the Tide-Table,
the Sum 10 Hours 34 Min. is the
Time of High-Water required, differing from the former by only 32
Min. being not so considerable as
when near the Quarter, yet nearer
the Truth, because it allows for the
different Shifting of the Tides; but
either Way is exact enough for common Use.

Note; What is here said of the Tides is meant as to their general Motion; the Half Tides, Quarter Tides, and Currents, being a Thing that depends so much upon Experience, that it is in vain to think of them here.

A large Tide-Table after a New Method.

A	H. M.		A	H. M.
Berdeen	0 45		Briſtol *Key*	6 45
Army	1 30		Bridgwater	7 45
St. Andrews	2 15		*Cape* Blanco	9 45
Amſterdam	3 0		Buologne	11 0
Armentiere	3 0		*Race of* Blanquet	12 0
Arbroth	3 15			
Antwerp	6 0			
Archangel	6 0			
Abermorith	6 0			
Amazones *River in* South America	6 0			
Aldborough	9 45			

C	H. M.
CAPE Cantin *in* Barbary	0 0
Calais *without*	1 30
Camvere	1 30
Conquet	2 15
Cork *in* Ireland	4 30
Cape Clear *in* Ireland	4 30
Caldy	5 15
Carnarvan *Bay*	5 15
Cromer	7 0
Caſkets *without*	8 15
Cape Sierre-lion *in* Guiney	8 15
Chamberneſs	9 45
Cowes	10 30
Caen *in the* Foſs	10 30
Calais *Road*	10 30
Calſhot	11 15
Condado	12 0

B	H. M.
BEACHY	0 0
Bajador *in* Barbary	0 0
Blacktail Beacon	0 15
Blackneſs	1 30
Berwick	2 15
Bluet *without*	2 15
Britain South *Coaſt*	3 0
Biſcay *Coaſt*	3 0
Bourdeaux *River*	3 0
Buchaneſs	3 0
Bona Eſperanca	3 0
Beliſle	3 30
Breſt	3 15
Baſs *without*	3 45
Bridlington	4 0
Bourdeaux *River within*	3 45
Brovage *without*	3 45
Baltimore	4 30
Bree Sound	4 30
Bremen	6 0
Hackney	6 0
Briſtol	6 30

D	H. M.
Dunkirk	0 0
Dover *Pier*	0 0
Port Deſire *in* America	0 0
Downs	1 30
Dundee	2 15
Denby	2 15
Dort	3 0

	H. M.		H. M.
Dunbar	4 30	Guernsey	1 30
Dungarven	4 30	Goree	1 30
Dartmouth	6 0	Gravesend	1 30
St. David's Head	6 0	Galicia	3 0
Dublin *in* Ireland	8 15	Gascoigne	-3 0
Dunnose	9 45	Groyne	3 0
Diepe	10 30	Garonne *Mouth*	3 0
Dunwich	9 45		
Dungeness	9 45		
Dover	10 30		

H

		H. M.
H Ever		0 0
Hern		0 0
Holy Island		1 30
Hartlepool		3 0
Huntcliff *Foot*		3 45
Humber *Mouth*		5 15
Holms		6 0
Hull		6 0
Hamburgh		6 0
Hague		8 15
Harlem		9 0
Havre de Grace		9 0
St. Hellens		10 30
Harwich		11 0

E

	H. M.
E Emden	0 0
Eider	0 0
Elbe	0 0
Enchuysen	0 0
Edam	1 30
Edinburgh	4 30
Egmont	4 30
Exwater	7 30
Entrance of the Emes	7 30

I

	H. M.
I Utland *Isles*	0 0
Ireland W. *Coast*	3 0
Ireland South *Coast*	5 15
John de Luce	10 30

K

	H. M.
K Entish Knock	0 0
Killian	3 0
Kinsale *in* Ireland	4 30
Kilduyn	7 30

F

	H. M
F Landers *Coasts*	0 0
Flushing	0 45
Finmark *Coast*	1 30
Fountney *without*	2 15
Flamborough Head	4 0
Falmouth	5 15
Forn	5 15
Foy	5 30
Foulness	6 45
The Fly	7 30
Friesland *Coast*	7 30
Florida *in* Carolina	7 30
Foreland *North and South*	9 45
Firth	10 30
Fair Isle Roads	11 15

L

	H. M.
L Lisbon	2 15
St. Lucar	2 15
London	2 30
Leith	4 0
Lawreness	4 30
Lynn *without*	5 15

G

	H. M.
G Gibraltar *Road*	0 0
Graveling	0 0

Lunde

	H.	M.			H.	M.
Lundey	5	15		Porthus	3	0
Lynn	6	0		Pennes	3	45
Lanion	6	45		Plymouth	6	0
Land's End	7	30		St. Powls	6	0
Lizard	7	30		Podefcmefk *in* Ruffia	6	45
Lam *Bay*	8	15		Peterport	8	15
Leoflode	9	45		Portland	8	30
Lenow	9	45		Picardy *Coaft*	10	30

M		H.	M.	Q		H.	M.
MAZE *within*		0	45	**Q**Ueenborough		0	0
Malden		0	45	Quebec *in* Canada		6	0
St. Mark		2	15				
St. Matthew's *Point*		3	45				
Mount's *Bay*		4	30	R		H.	M.
Milford		5	15	**R**Ebdan		0	45
Moonlefs		5	15	Rochefter		0	45
St. Maloes		5	15	Rumney		1	30
Magnefs *Sound*		8	15	Ramkins		1	30
Mackwell's *Caftle*		8	15	Robin Hood's *Bay*		3	0
Ifle of Man		9	0	Rotterdam		3	0
Margate *Road*		11	15	Rouen		3	45
				Ruchel *without*		3	45
				Roan *River within*		3	45
N		H.	M.	Ramfey		5	15
NEwport *in the* Ifle of Wight		0	0	Rye		11	15
Nore Weft End		0	0	Rhodes		11	16
North C. Maggero		3	0				
Nantz *River without*		3	0				
Newcaftle		5	15	S		H.	M.
St. Nicholas *in* Ruffia		6	45	**S**HOE		0	0
Needles		9	45	Sheernefs		0	0
Normandy *Coaft*		10	30	Sleeve		0	0
Naze		11	15	Southampton		0	0
				Spits		0	0
				Shetland		3	0
O		H.	M.	Scilly		3	45
ORkney		9	0	Scarborough		3	45
Orwell		9	0	Sound		3	45
Oxfordnefs		9	45	Staples		3	45
				Seven Ifles		4	30
				Severn's *Mouth*		5	15
P		H.	M.	Stockton		5	15
POrtfmouth		0	0	Spurn		5	15
Poictou South *Coaft*		3	0	Salcomb		6	0

O 4   Start

	H.	M.
Start ———————	6	45
Sedmouth ——————	6	45
Shelberg ——————	9	0
Seven Cliffs —————	9	0
Shoreham —————	9	45
Seyn Head —————	10	30
Senegal —————	10	30

T	H.	M.
Terveer *within* —	0	45
Tenet —————	1	30
Terveer *without* ———	1	30
Tinmouth ————	3	0
Tees Mouth ————	3	0
Teneriff —————	3	0
Torbay —————	6	0
Texel —————	7	30
Tergon —————	9	45

U	H.	M.
Wreck —————	0	0
Ufe —————	3	0
Ulhant *without* ———	6	0
St. Vallery —————	10	30

W	H.	M.
Isle of Wight — —	0	0
Winchelsea ————	0	15
Weilands ——— —	1	30
Whitby ————	6	0
Waterford *in* Ireland —	0	0
Weymouth ——— —	6	0
Wells —————	6	0
Weymouth *Key* ———	6	45
Wieringham ——— —	7	0
Winterton —————	8	0

Y	H.	M.
Oughall *in* Ireland	4	30
Yarme —————	6	45
Yarmouth *Town* ———	10	0
Yarmouth *Pier* ———	10	0
Yarmouth *Road* ———	10	30

Z	H.	M.
Zeland *Coast* — —	1	30
Ziericke Sea ———	3	0

The Use of this Tide-Table is the same as in other Tide-Tables; for when you defire to know the Time of High-Water at any Place mentioned in the Table, you need only look in the Letter that the Name of the Place begins with, and there having found the Place you want, fee what Hour and Minute ftands againft it, which being added to the Hour and Minute of the Moon's coming to South, the Sum (abating twelve Hours, if it exceeds) is the Time of High-Water at the Place propofed that Day. —— *Example*; I defire to know what Time it will be High-Water at *Harwich*, *October* the 16th, 1770, I look in the Letter H, and find *Harwich*, and againft it 11 : 00, which added to the Moon's Southing that Day, the Sum is 20 : 36, from which caft away 12, the Remainder is the Time of High-Water: Of which fee more in the Explanation of the following Table of the Moon's Changing.

The

Year	January D. H.	Febru. D. H.	March D. H.	April D. H.	May D. H.	June D. H.
1769	8 1m	6 6a	8 6m	6 5a	6 1m	4 9m
1770	26 10a	25 4a	27 6m	25 5a	25 1m	23 9m
1771	15 10a	14 6a	16 noon	15 3m	14 3a	13 1m
1772	4 11a	2 8a	4 10m	3 5m	2 9a	{ 1 11m / 30 10a }
1773	23 8m	21 1m	23 5m	21 10f	21 3a	19 6a
1774	12 8m	11 1m	12 9m	11 1m	10 3a	9 7m
1775	{ 2 1m / 31 1a }	11 1a	{ 2 1m / 31 1a }	30 1m	29 noon	28 1m
1776	20 1m	18 1a	19 11a	18 11m	18 1m	16
1777	9 4a	8 3m	9 1a	9 1m	7 9m	5 7a
1778	28 6a	27 6m	28 3a	27 1m	26 8m	24 5a
1779	18 5m	16 7a	17 7m	15 4a	15 1m	14 8m
1780	6 8m	5 3m	6 6a	5 7m	4 5a	3 1m
1781	25 1m	23 10a	25 1a	25 7a	23 3a	22 1m
1782	14 1m	12 5a	14 4m	12 8a	13 1m	11 noon
1783	3 8m	1 11a	2 3a	1 8m	1 2m	29 7m

Years	July D. H.	August D. H.	Sep. D. H.	Oct. D. H.	November D. H.	December D. H.
1769	3 3a	{ 1 11a / 31 10m }	29 10a	29 2a	28 8m	28 3m
1770	22 4a	20 11a	19 8m	18 8a	17 10m	16 10a
1771	12 9m	10 4a	9 1m	8 8m	6 7a	6 8m
1772	30 7m	28 4a	27 1m	26 9m	24 8a	24 7m
1773	19 5a	18 5m	16 3a	16 2m	14 11m	13 11a
1774	8 9a	7 8a	6 3m	5 2a	4 2m	3 1a
1775	27 3m	26 1m	24 noon	23 6a	22 6m	22 6m
1776	16 5m	14 8a	13 1a	13 4m	11 5a	11 5m
1777	5 7m	3 9a	2 1a	2 6m	{ 1 1m / 30 3a }	30 7m
1778	24 2m	22 3a	21 6m	20 1m	19 6a	9 1a
1779	13 3a	10 5m	9 10m	9 1m	7 6a	7 1a
1780	2 7m	29 10a	28 8m	27 9a	26 1a	6 8a
1781	21 6m	19 2a	18 1m	17 8m	25 8a	15 11m
1782	10 10a	9 7m	7 3a	7 1m	5 9m	4 8a
1783	28 7a	27 6m	25 4a	25 1m	23 11m	22 8a

South about 12 Hours after Noon, which may more
properly be called next Morning, and confequently if
you abate 1 from her Age found when above 16 Days
old, the Southing found from thence is properly her
Southing for the Morning of the Day propoſed, &c.

SECT. VI.

How to keep a Reckoning.

HAVING thus learned to find the Variation, and
to work an Obſervation, and alſo to reckon your
Tides, with other Things neceſſary to be known, and
having as Maſter, Mate, or Pilot of a Ship, taken your
Charge for any Voyage, and having your
Log-Line, Glaſſes, and other Things in Rea-   *Fig.* 62.
dineſs, you muſt provide a Log-Board ruled
as you ſee in the Figure.

Then having ſet Sail, and got from the Shore, and
out of the Set of the Tide, obſerve how the Land bears
from you that you intend to take your Departure from;
and alſo as near as you can, compute the Diſtance you
are from it, and then that Courſe and Diſtance mark'd
off upon your Chart, gives the true Place of your Ship
when you begin your Reckoning.

This done, having ſettled your Watches, and ordered a Man to the Helm for two Hours, let him obſerve carefully what Courſe he ſteers by the Compaſs,
or if he be ordered to ſteer upon any given Courſe,
let him take Care to mind it, and at the End of two
Hours let the Mate heave the Log, having one attend-
ing him with the Half-Minute Glaſs, let the Mate over
haul off all the Line, (having firſt caſt the Log with a
few Fakes of Line into the Sea) till the Red Rag comes
to his Hand, and at that Inſtant cry *Turn*; the Man
with the Glaſs, juſt upon that Word, as the Red Rag
goes away, turning the Glaſs, and watching diligently:
When the Half Minute Glaſs is juſt out, the Man that
holds

holds the Glaſs, cries *Stop*; whereupon the Mate ſtops, and hauls in the Line, obſerving exactly how many Knots, half Knots, and Fathoms are gone out, and ſets it down upon the Log-board, againſt the Hour at which the Log was heaved, and thus proceed every two Hours during all the 24 Hours, and then the Log-board will be full the next Day at Noon.——Then having a Log-book with every Page ruled, like the Log-board, with the Day of the Month at the Top, take off the Log into your Book, and rub out the Chalks upon the Board, ſo is the Board ready for the next Day. I ſhall inſtance in a Voyage from a Place in Latitude 54 Deg. 8 Min. North, to another Place in Latitude 57 Deg. 50 Min. from whence we ſet ſail *Auguſt* 24, at Noon, and running that Day as by the Log following.

Note; Any 24 Hours Log, from Noon to Noon, is dated with the Day upon which it ends, and not upon the Day upon which it begins; then if we ſet Sail *Auguſt* 24, that 24 Hours Log muſt be called *Auguſt* the 25th, &c.

The Ship having run every two Hours as you ſee in the Log on the other Side, add up all the Knots together, the Sum 47 being doubled (becauſe you heave the Log but every two Hours) is 94, to which add the 4 half Knots, which in this Caſe are accounted as 4 whole Knots, becauſe they are alſo doubled : The Sum 98 Miles is the Diſtance run that Day.——The Courſe is N.E. but becauſe there is a Point Variation Weſt, the true Courſe is N.E. by N. therefore ſet down under the Log——made good 98 Miles N.E. by N.

Then find in the Table for working a Traverſe, your Northing and Eaſting for that Courſe and Diſtance, and ſet it in the blank Space on the Right-hand as you ſee done.

The Northing 81 : 5 Miles or Minutes added to the Latitude ſailed from 54 Deg. 8 Min. the Sum is the Latitude come to 55d. 29m. (the 5 which is placed after the 81 being but a Decimal, or half a Mile, need not be regarded) which ſet alſo down as you ſee.   For

For the Difference of Longitude you may find it by the Traverse Table, the Middle Latitude being 55 Deg. its Complement 35 Deg. being found, and under it in the Column of Departure, find the Departure, or Easting 54 : 4 or the nearest thereto; and against it in the Column of Distance you have 95 Min or 1d.—35m. for theDifference of Longitude made thatDay, all which set down as you see.

H.	K.	K.	P.	Course	Winds	Aug 15, 1770.
2	3			N.E.	West	Variation 1 Point West.
						Northing ————— 81.5
4	4					Easting ——————— 54.4
						Lat. come to — 55d. 29m.
6	4	1				Diff. Longitude— 1.35 E. ly.
8	5					
10	5					
12	5					
2	4	1				
4	4					
6	4					
8	3	1				
10	3	1				
12	3					

Made good 98 Miles North East by North.

Now

Now to find where your Ship is, in the firſt or laſt of the Charts before-mentioned, which are all made for this Voyage: And firſt, for the firſt particu-

Fig. 54. lar Chart; ſet off 98 Miles from A, upon a N. W. by N. Line, it will reach to the Point *a*, the Place of the Ship required; or take the Difference of Latitude made that Day 81 : 5 in one Pair of Compaſſes; and the Departure or Eaſting 54 : 4 in another Pair of Compaſſes; then with one Foot of each Pair of Compaſſes in A, run that in which you have the Difference of Latitude, along the Eaſt and Weſt Line AC, and that Pair in which you have the Departure along the North and South Line AB, the moveable Points will fall in *a* as before.

To find where the Ship is in the *Mercator*'s Chart, take the Latitude in one Pair of Compaſſes thus ; ſet one Foot upon the Line *a d*, in the Latitude

Fig. 55. 55 Deg. 29 Min. which is the Point *q*, and extend the other to any Parallel, as ſuppoſe the Line 55, 55; then after the ſame Manner take the Longitude in your Compaſſes, running them Parallel to any North and South Line, the Meeting of their moveable Points, *viz.* at *x*, is the Place of the Ship required.——*Note*; The Place of the Ship

Fig. 56. at *x*, in the third Chart is found by either of the two Methods by which it is found in the firſt.

How to keep Account of a Ship's Way upon a Wind.

A Ship is ſaid to ſail upon a Wind, when her Tacks are aboard, and her Yards ſharp braced, and is commonly ſuppoſed to be within ſix Points of the Wind with Lee-way, according to the Sail ſhe carries, &c.

All

All Sails set — 1 Point		Try under Mainsail
A Topsail in —— 2		only —— 5
Both Topsails in and	Points	Under Mizen only 6
a Sea —— 3		Lie a Hull, all Sails
Try under Mainsail		furled —— 7
and Mizen —— 4		

According to these Allowances of Lee-way, I shall
work the following Examples; nevertheless, in Practice Lee-way must be allowed according to Judgment,
for some Ships make more Lee-Way than others, with
the same Sails set, &c.

This Example, *August* 26, hath two Courses, which
may be reduced to one by the Rules for working a
Traverse by the Traverse Table.——The first Course
is N.E. by the Compass, but by the Variation allowed,
viz. one Point West, it is N.E. by N. but the Wind
being N.N.W. *viz* within six Points of the Course,
we must allow one Point Lee-way from the Wind,
which brings it again to N.E.

For the Distance, add up the first six Numbers of
Knots, during which Time she lay upon that Course,
the Sum 21 doubled, and the two half Knots added,
make in all 44 Miles N.E. for the first Course.

The second Distance by the same Rule, is 39 Miles,
the Course by the Compass N.W. but a Point Variation West makes it N.W. by W. and a Point Lee-way,
which is also Westerly, makes the Course made good to
be W.N.W. the Distance 39 Miles.

Course	Dist.	North	South	East	West
N.E.	44	31—1		31 1	
W.N.W.	39	14—9			36 0
		46—0			31 1
					4 9

By Tra-
verse Sail-
ing as per-
formed by
the Table,
you find the
Northing
46. 0. and
the

the Wefting 4.9. and by Cafe the fixth of Plane Sailing, by the Traverfe Table, the Diftance made good
is 46 Miles North, and Courfe 6 Degrees Weft; all
which fet down as in the firft Day's Work; then add
this Day's Northing to the Northing got before, the
Sum 127.5 is the Total Northing. Alfo fubtract this
Wefting 4.9 from the Fafting got before, the Remainder 49.5 is the Total Eafting, both which fet down as
you fee; alfo by adding 46 Minutes to the laft Day's
Latitude, the Sum 56 Deg. 15 Min.. is the Latitude
come to.

The Difference of Longitude is 9 Min. which becaufe it is Weft, it is to be fubtracted from the Longitude the Day before, and the Remainder 1d. 26m. is
the Longitude come to.

Now to find where the Ship is, upon the firft and
laft Charts, take the Total Northing in one Pair of
Compaffes, and the Total Eafting in another
Fig. 54. Pair, and proceed as before from A, the
Place failed from, it fhall produce the Point
e, for the Place arrived at: Alfo in the fecond, which
is a *Mercator's* Chart, find the Longitude and Latitude
come to, by the Method prefcribed in the
Fig. 55. firft Day's Work, and their Meeting or Interfection is at the Point *e,* the Place of the
Ship required.

Note; In the firft and laft Charts you may fet off one
Courfe and Diftance by another, as in Traverfe Sailing
Geometrical; but I fhall not infert that Way, left it too
much confufe the Draught.

H.	K.	½K	F.	Course	Winds	Aug. 26, 1770.
2	3			N.E.	NNW	Variation 1 Point West.
4	3	1				Northing ——— —— 45.7
6	4					Westing ——— —— 4.9
8	4					Total Northing —— 127.5
10	3	1				Total Easting ——— 45.5
12	4					Latitude come to — 56 15
2	3	1		N.W.	NNE	Diff. Longitude —— 0.9 W
4	3	1				Long. come to ——— 1.26
6	3	1				
8	3					
10	3					
12	3					

Made good 46 Miles, North 6 Degrees Westerly:

How to correct a Reckoning by an Observation.

IN the following Day's Work, *August* 27, there is made good 116 Miles; the Course with Variation allow'd, is N. E. by N. and the rest as you see found by the foregoing Method.

It is needless to give any more Examples of allowing Lee-way or Variation, the Rule being the same when Lee-way is 4, 5, 6, or 7 Points, as when it is but one

P   Point,

Point only minding to allow as much as it is, and the right Way. Thus if a Ship lie North with Wind at E.N.E. and 6 Points Lee-way, she makes her Way good at W.N.W. &c.

My Reckoning at Noon, *August* 27, brings me into Latitude 57, 51; but by Observation which I must prefer before the Dead Reckoning, I find I am in Latitude 57, 34; now to correct the Reckoning by an Observation, observe this General Rule.

If your Dif- ⎧ more ⎫ than the Depart. the ⎧ Log. or Dist,
ference of ⎨   ⎬   Fault is more like- ⎨
Latitude be ⎩ less ⎭ ly to be in the.   ⎩ Co. or Course

The Reason of the General Rule above, is evident from the Figure; for suppose a Ship at A sail *Fig. 63.* South Easterly, till by her Reckoning she is at C, or in the Parallel of Latitude BC, but by Observation she is in the Parallel of Latitude DEF: Now if we will suppose her computed Distance AC to be right, sweep an Arch with AC till it cut the Parallel DF in F, and make AF equal to AC: Now by this Means the Line AF is the Line described by the Ship's Motion, whereas we thought it had been the Line A C; but it is absurd to think that any Man should be so far mistaken in his Course as to steer from A to F, when he thought he had been steering from A to C; and therefore we must impute the Fault to the Distance, supposing that the Mistake lies there, and that when he thought he had steered from A to C, his Reckoning was a-head of the Ship, and that when he should have been at C he was only at E, the Space EC being but a tolerable Mistake in the whole Distance AC.

But in a more Easterly (or Westerly Course) suppose a Ship sails from A till her Course and Distance by Dead Reckoning is represented by the Line AI, and the Latitude come to, by the Parallel GI, but by Observation she is in the Parallel of Latitude H L K: Now if he was supposed to keep a right Account of his

H.	K.	K	F.	Courſe	Winds	Aug. 27, 1763.
2	3	1		N.E.	Weſt	Variation 1 Point Weſt.
4	4					Northing — — — 96.4
						Eaſting — — — 64.4
6	4	1		—		Total Northing — — 223.9
						Total Eaſting — — 113.9
8	5					Lat. come to — — 57.51
						Diff. Long. Eaſt — 1.58
10	5					Long. come to — — 3.24
12	5	1				Lat. by Obſer. — — 57.34
						Tot. Eaſt. correct. — 105.2
2	6					Tot. North. Cor. — 206.9
						Long. come to cor. — 3.8
4	6					
6	5	1			—	
8	4	1				
10	4	1				
12	4					

Made good 116 Miles North Eaſt by North.

Courſe and the Fault to be in the Diſtance, we muſt continue the Line A I till it cut the Parallel H K, and allow him to be miſtaken the whole Diſtance IK, in his Account, which is abſurd, and therefore muſt impute the Miſtake to the Courſe; and then with one foot of the Compaſſes in A, with the Extent A I deſcribe the Arch I L, and draw the Diſtance A L equal to A I, and ſuppoſe that when he thought he had ſailed along the Line A I, he had indeed ſailed along the Line A L, the Angle I A L being a much more tolerable

Fault

Fault in the Courfe, than the Diftance I K could be fuppofed to be in the Diftance.

Note; If you fail in a Current you may eafily be
grofly miftaken, either in Courfe or Diftance; but of
that fee more in the latter end of this Book.

Now in this Example the obferved Latitude, *Aug.*
27, differs from the Dead Reckoning Latitude by the
Quantity of 17 Miles or Minutes, and becaufe the Total Northing is more than the Total Eafting, I impute
the Fault to the Diftance by the Log, and then it is corrected by this Proportion.

As the Total Northing 223.9, to the Total Eafting
113.9, fo the Error in the Northing 17, to the Error
in the Eafting 8.7 neareft.

113.9		or 8 7 Miles neareft, to
17	223.9)1936.3(8⅐⅛⅒	be fubtracted from the
————	1451	Total Eafting 113.9 the
797.3		Remainder 105.2 is the
1139		Total Eafting corrected
————		
1936.3		

The Reafon of this Proportion is evident from the
Diagram; for In the Triangle ABC, as the Total
Northing by Reckoning AB, is to the Total Eafting by
reckoning BC, fo is the Error in the Northing E *q*,
to the Error in the Eafting *q* C; for the Sides of the
Triangle ABC, and E *q* C, are proportional; by *Eucl.
Lib.* 6. *Prob.* 4.

Then correct the Longitude; As the proper Difference between the Latitude by Reckoning, and the
Latitude by Obfervation, is to the Meridional Difference between the fame two Latitudes; fo is the Error
in Departure, to the Error in Longitude, that is, as 17,
is to 32; fo is 8.7, to the Error in Longitude.

```
   32        17)278(16   Minutes, (rejecting the Frac-
   8.7           108     tion, being but the Fraction
 _____          .6      of a Minute) to be subtracted
                         from the Longitude by Dead
  224                    Reckoning, the Remainder 3.8
  256.                   is the true Longitude Correct by
 _____                  Observation,
 278.4
```

But in Case your Course be far Easterly or Westerly,
and the observed Latitude differs from that found by
the Dead Reckoning, it is best to correct only the Latitude, reducing it to the observed Latitude, and not
to meddle at all with the Departure or Difference of
Longitude, because the Error is too small; as in the
foregoing Figure the true Departure H L differs so
little from the Departure found by Dead Reckoning
G I, that it is not worth while to correct it, unless you
have some certain Observation of Longitude to correct
it by.

When you have thus corrected your Latitude, Departure, and Difference of Longitude for this Day at
Noon, as here *Aug.* 27, then proceed to the next Day's
Work, add the next Day's Northing to the Northing
correcting this Day, to find the Northing next Day, and
not to the Northing found by Dead-Reckoning; and
do the same also with the Departure and Difference of
Longitude, not regarding that found by the Reckoning; and so proceed as before; and when you get another Observation to correct your reckoning by, add up
only the Northing and Easting made good since the last
Observation, and not of the whole Voyage, and so correct again according to this Example.

Now for setting off that Day's Work, *Aug.* 27, upon
each Chart, that Method is to be used which was before
directed to, and the Point Z in each Chart, shall represent the Place of the Ship, *Aug.* 27, at Noon.

Now if you desire to know the Course and Distance
from the Ship to the Place proposed, it may be done

P 5,   easily

easily and exactly upon the two last Charts, and chiefly
upon the last of all the three; for in it

Fig. 54 & 55. you need but set one Foot of the Compasses in the Point Z, where the Ship is,
and the other Foot in the Point B, the Place bound for;
that Extent applied to the graduated 'Line

Fig. 56. BD, accounting every Degree 60 Miles, and
every small Division 10 Miles, gives the true
Distance required 173 Miles.

In the second Chart, which is according to Mr.
Wright's Projection, commonly called *Mercator*'s, if you
use the Method which you have there prescribed, for
finding the Distance of any two Places upon a *Mercator's* Chart Geometrically, you will find it exactly a-
greeing with the former, *viz.* 173 Miles.

For finding the Course the Method is the same in
both; for the Rumbs are right Lines, equally divided,
or at equal Distances in both Charts; therefore if your
Charts have the Rhumbs upon them, observe to which
Rumb a Line drawn from the Ship to the Port would
be parallel, for that is the Course required; but if you
would still be more exact, you may find the true Course
and Distance from the Ship to the Port, by Case the
Sixth of *Mercator*'s Sailing Trigonometrical, thus, (having both Latitudes and Longitude given)

	Lat.	Long.
The Ship when correct by Observation is in	57.34	3 8
The Port bound for is in ——— ———	57.50	8 30

Proper Diff. Lat. 16, Merid. Diff. Lat. 30
Diff. Long. 5.22 or 322.

As Merid. Diff. of Latitude ·—— 30 ——	1.47712
To Radius ——————————— S.90 ——	10.00000
So Diff. of Longitude ——— 322 ——	2.50785
To Tang. of the Course ——— 84 41 ——	11.03073

Then for the Distances.

As Sine Comp. of Course —— 5.19 ——	8.96876
To proper Diff. of Latitude —— 16 ——	1.20412
So is Radius ———————————	10.00000
To the Distance ——— ——— 172 ——	2.23536

The

The Courfe is North 84.41 Eaft, and the Diftance 172 Miles.

But here we may fee the intolerable Error of the Plane Chart, the Ship being at z, and the Port bound for at the Point ☉, the Extent between them being applied to the graduated Line 54, 54, and allowing 60 Miles to each Degree, gives a Diftance vaftly too great for the Ship at z to the Port ☉, a moft infufferable Error.

I know it is objected by fome, that notwithftanding what can be faid againft the Plane Chart, it is ftill the moft ufeful and frequent among us, witnefs a great Part of the Charts and Waggoners that are now extant both in *Dutch* and *Englifh*, and if the Plane Chart was fo grofly falfe in fo fhort a Voyage as between the two Places before propofed, it would feem that thofe Plane Charts and Waggoners fhould not be fo much encouraged, or being ufed would not anfwer their End generally fo well as they do.

I anfwer, I take it for granted that thofe Charts are not grounded upon that Projection, that the Degrees of Longitude and Latitude are every where equal, as thofe commonly and properly called Plane Charts are, (for we find no fuch Thing as Degrees of Longitude upon them) but they are projected on the fame Ground that my Third Chart, before inferted, is grounded upon, *viz.* The true Courfe and Diftance from Place to Place is found, either by the Latitude and Longitude according to *Mercator's* Sailing, as I have directed in the Projection of that Chart, or elfe the Courfe and Diftance from Place to Place is found by Experience; thofe that have failed there obferving diligently what Courfe (with all proper Allowance for Variation, &c.) and alfo what Diftance carried them from one Place to another; and thus comparing their Obfervations of that Kind with the Obfervations of others, and correcting their Obfervations by their obferved Latitude, &c. And thus one obferving in one Place of the

World,

World, and another in another, and these Observations being compared, and the most agreeable chosen out and collected, may probably, have given rise to our large Waggoners now extant, which, though in Form of Plane Charts, yet in that Case must needs be true, and to be depended upon: For although the World is Globular, and not Plane, yet it is evident, that what is once the Course and Distance between any two Places, shall always be the Course and Distance between them. As for *Example*, Suppose, after all Allowance given for Variation, Lee-way, Currents, &c. I find that a N.E. by E. Course 240 Miles, or 80 Leagues, carries me from *Buchaness* to some known Place on the Coast of *Norway*; it is certain the same Course and Distance made good shall always be the same; and therefore if that Course and Distance was laid down upon any Chart between these two Places, I might safely depend upon that Chart for my next going there.

Indeed it may be objected, who can tell how to allow so exactly for Variation and Lee-way, but especially for unknown Currents, as to depend upon their reckoning for the true Course and Distance from Place to Place.

I answer, I agree to that; but yet when two, three, or more Ships sail upon the same Voyage, and find their Accounts nearly to agree, this may make them somewhat more confident of their Reckoning, and of the Truth of it; and this I take to be a Reason of the Improvements that are yet daily made in Charts and Waggoners; these Charts and Tables of Latitude and Longitude which are of a later Date, having doubtless attained nearer the Truth by a greater Confluence of Observations, and thereby differing more or less from these of more ancient Date, both in Latitude, but chiefly in Longitude of Places; which Difference we must impute to our attaining nearer to the Truth, and to Mistakes formerly committed, the Places themselves
remain

remain fixed and immoveable, as to their Situation upon the Surface of the Earth.

From what hath been said it will follow, that it is
not necessary to dissuade any Mariner from the Use of
those Charts and Waggoners now in Print, but it is rather an Encouragement to use them, and trust to them
as very good Helps; but let not this be an Inducement
to Persons that have had plentiful Experience of the
Truth and Sufficiency of those Charts and Waggoners,
presently to decline the Use of them, and fall to Work
to make Charts of their own, and trust to them, if
they have no other Way for it but that Hypothesis of
supposing the Degrees of Latitude and Longitude to be
every where equal (which I have sufficiently proved is
not the Ground of the Projection of those true Sea Charts
now in Use) the absurd and intolerable Falshood of which
Hypothesis aforesaid, I have sufficiently proved and
demonstrated.

Note; Whereas in the Examples before going, in the
three Days Work of keeping a Reckoning, I did for
the more Expedition find my Difference of Longitude by
the Traverse Table, by the Rule there delivered for that
Purpose; but yet if your Voyage be long, or near the
Pole, where the Degrees of Longitude grow much less,
you may, if you distrust the Truth of this Method,
examine your Longitude once in three or four Days, by
this Proposition following.

The Ship having sailed from Latitude 54.8, to Latitude 57.34, the proper Difference of Latitude is 206,
and the Meridional Difference of Latitude is 368, and
Departure made good is 105.2.

Co.   Ar.

Therefore As proper Diff. of Lat. — 206 —— 7.68614
To Merid. Diff. of Lat. — 368 —— 2.56584
So Departure ———— 105.2 —— 2.02201

To Diff. of Longitude — 188 —— 2.27399

Difference

Difference of Longitude 188 Min. or 3 Deg. 8 Min. The Longitude come to, being the very same as was found by the Traverse Table, although that is found at three Operations.

And now, seeing I have in the former Part of this Book laid down Methods for the Working of the several Cases and Questions in Navigation, and keeping a Reckoning both in Longitude and Latitude, not only by given Numbers, which I have called *Arithmetical Navigation*, but also without any Books, Tables, or Instruments, by a New Method never yet known or published; I shall instance in the first Day's Work in both Ways of Operation, it being too tedious to instance in all the three Day's Works, especially seeing there are sufficient Examples given of each Method elsewhere; and first by the given Numbers in Arithmetical Navigation.

The Course that Day is N.E. by N. 98 Miles.

The given Number for { Diff. Lat. is ——— 8315
three Points to find { Departure is ——— 5556

For Difference of Latitude.

        98
        83
    ———
       294
       784
    ———
    81|34 Diff. of Lat. 81.3.

It should be 81.5. but the Error is only $\frac{1}{10}$, and is occasioned by omitting the two last Figures of the given Number.

For the Departure.

        98 { The Depar-
       555 { ture or East
    ——— 54.39, or ra-
       490 ther 54.4.
       490
       490 {
    ——— {
    54|390 {

Difference of Latitude 81.3, or 1 Deg. 21 Min. hence the Latitude come to is 55 Deg. 29 Min.

For

For the Difference of Longitude, the Middle Latitude is 45.48, but the next greater whole Degree is 55, whose given Number in the second Table is 5735, but I shall only use the first two Figures 57.

Departure with Cyphers
Given Number or Divisor — 57)5440(95
  310
  ————
  25

The Diff. of Lon.ʳ 95m. or 1d. 35m. the Remainder being but a Fraction of a Minute, we reject it.

The same Day's Work cast up by a New Method.

The Course 33.45 or 33¼, the Distance 98 Miles.

For the Natural Radius for 33¼ Deg. by Method the Second, they that can work by cross Multiplication need not reduce the Fraction to a Decimal, but square it as follows;

  . 33¾
     33¾
  ————————
33 by 33 is — 1089
¼ of 33 is — 24¼
The same again 24¼
¼ of ¼ we reject.

  1138½
  3
  ————————
  3.415½
  57.3
  ————————
Nat. Rad. 60.7

As 60.7 to 98 :: 33¼ to Departure
     33¼
  ————
  294
  294½
  7
  ————
  3307

60.7)3307.0(54.4
     272.0
     ————
     398.0
          .492

For Diff. of Lat. by Rule the Third
Sum of the Sides 152.4
Diff. of the Sides 43.6

  9144
  4572
  6096
  ————————
Product — 6644.64(81.5 Square Root
        64   of the Product and Difference of Latitude required.
  ————
  161)244
      161
  ————
  (83)

  For

For Difference of Longitude, the Complement of Middle Latitude is 35, its Natural Radius is 61 : Therefore as 35, is to 54.4; so is 61, to the Difference of Longitude.

```
54.4
 61          35)3318(94¾; or 95 Minutes, or 1 Degree
----             168        35 Minutes, the Difference of
 54.4           ----        Longitude required.
3264             28
----
3318.4
```

And thus you see the Excellency and Usefulness of this New Method, by which, although by Stress of Weather or Cruelty of Enemies, you had lost all Charts, Books, Tables, and Instruments, yet you may without any of them keep as just an Account of your Ship's Way, both in Longitude and Latitude, as you can with them; as this and other Examples inserted elsewhere in this Book make manifest.

CHAP. VIII.

SECTION I.

How to make an Orthographic Projection of the Sphere, commonly called the Analemma; whereby, most of the necessary Questions in Astronomy may be resolved without Trigonometrical Calculation, only by the Help of Scale and Compasses.

ALTHOUGH I have inserted all the Astronomical Tables, ready calculated, that are of Use in the Practice of Navigation, and have also laid down a Method for finding the Variation of the Compass, without Azimuth or Amplitude, &c. yet for the Sake of

such

such as delight in Aſtronomical Operations, I ſhall
ſhew the Learner how to ſolve all neceſſary Aſtrono-
nomical Queſtions by the Analemma, or Orthographic
Projection upon the Plane of the Meridian ; and in this
Manner of Projection, the Eye is ſuppoſed to be perpendicular to that great Circle upon whoſe Plane the
Projection is made, and at an infinite Diſtance from
the ſaid Circle ; ſo that any Line let fall from the Eye
upon any Place within the ſaid Circle, ſhall be perpendicular to the ſaid Plane, and then will the Primitive
Circle (or Meridian of the Place, when the Projection
is upon the Plane of the Meridian; be a perfect Circle. All right Circles that divide the Projection into
two equal Parts are ſtrait Lines, or Diameters of the
Primitive Circle ; and all Oblique great Circles that divide the Primitive into two equal Parts, and yet touch
it at oppoſite Points are Semi-Elipſes ; and all leſſer Circles parallel to the Equator are projected into right
Lines, cutting the Primitive into two unequal Parts.
And Parts of Lines that ſerve for theSolution of any
Aſtronomical Queſtions, are Parts of ſome of thoſe Circles, or ſuppoſed to be ſo, and are meaſured or projected
cording to the following Directions.

The Primitive Circle being drawn with the Chord
of 60 Degrees (of any Radius large or ſmall) any Parts
of it are meaſured upon the Chords of the ſame Radius. Any Part of a Right Circle is meaſured upon
the Sines. Any Part of an Oblique Circle, or Elipſis
(when required to be meaſured, which is but ſeldom)
may be meaſured on the Chords, being firſt reduced to
the Primitive Circle thus; draw two Lines through the
two Points in the Oblique Circle, the Diſtance between
which is to be meaſured, and let the ſaid two Lines be
drawn parallel to that right Circle which cuts the ſaid
Oblique Circle at Right Angles, and mark where theſe
two Parallels cut the Primitive Circle, and the Diſtance
between theſe two Marks meaſured on the Chords, is
the

the Measure of that Part of the Oblique Circle required.

Note; There is no Way to project these semi-Ellipses, or Oblique Circles, but by finding a great Number of Points through which we are to pass, and so by a steady Hand or Help of a Bow, draw it through the said points; but the Operation being tedious, and of little Use, as it is seldom put in Practice: Nevertheless I shall hereafter shew how they are to be done, as also all the rest, as in the following Example.

An Orthographic Projection of the Sphere upon the Plane of the Meridian. For

Latitude of *London* —— 51d. —— 32m.⎫ N.
Sun's Declination North — 22d. —— 30m.⎭

First, With the Chord of 60, draw the Circle P E S Q, to represent the Meridian; then *Fig. 64.* Draw the Diameter H O to represent the Horizon, and at right Angles to it draw the Line ZN, to represent the Prime Vertical; then set off the Latitude 51 32 (taken off the Chords) from O to P, and from H to S, and draw the Line PRS to represent the Axis of the World, P the North Pole, and S the South Pole: Then at Right Angles with the Line P S, draw the Line ERQ to represent the Equator: Then set the Chord of the given Declination 22.30 from E to D, and from Q to C, and draw the Line D C the Parallel of the Sun's Declination: Set off the Chord of 23,29, the Sun's greatest Declination from E to T and from Q to K, and draw the Line T R K to represent the Ecliptic: set off also the Chord of 23.29 from P to *q* and *g*, and draw the Line *q g*, to represent the Artic Circle, and the same from S to *r* and *v*, and draw the Line *r v* to represent the Antartic Circle: Then from P, through the Point ☉ draw the Meridian P *b* ☉ B; thus, here is given in this Meridian, the Point P and the
Point

Point ☉; then to find the Point B, another Point through which this Meridian is to pass, divide the Line E R after the same Proportion that the Line D F is divided at the Point ☉, that is as F D, is to F ☉; so is R E, to R B, and make the Mark B, then by the same Means find the Point *b* in the Line *q g:* Then through the Points P *b* ☉ B with a Bow: Curving or Bending gradually, draw the dotted Curving Line P *b* ☉ B to represent a Part of another Meridian: All which being finished you may proceed to answer the following Problems.

1. *To find the Sun's Longitude or Distance from the next Equinoctial Point.*

The Sun's Longitude or Distance from the next Equinoctial Point, is an Arch of the Ecliptic, contained between the Center R (or Intersection of the Ecliptic and Equator) and the Point ☉, the Intersection of the Ecliptic and Parallel of Declination) so here R ☉ measured on the Sines, is the Sun's Distance from the next Equinoctial Point.

2. *Of the Sun's Right Ascension.*

It is an Arch of the Equator, contained between the Equinoctial Point R, and the Point where the Meridian drawn through ☉ cuts the Equator, as in B, so R B measured on the Sines, gives the Right Ascension required.

Note; If it be thought too much Trouble to draw these Meridians, all those Problems, which can be done by the Help of the Meridians, may be done without them, if you reduce the lesser Circles to the common Radius; therefore F ⋅ is the Sine of the Sun's Right Ascension to a lesser Radius F D; which may be reduced to the common Radius thus: Suppose I would find the Sun's Right Ascension, set F D from R to *d*, upon the Line R H, then with the Extent F ☉, and

one

one Foot in *d*, defcribe the Arch *e*; a Ruler laid from R, by the Extremity of that Arch will cut the Primitive Circle in *o*; then H *o* meafured on the Chords is equal to F ☉, the Sun's Right Afcenfion required.

Note; When any Arch of a leffer Circle is to be meafured, it muft be firft reduced to the common Radius, and meafured as before.

3. *To find the Hour of the Day.*

The Hour from Six o'Clock is an Arch of the Parallel of Declination, contained between the Line P S and the Point ☉, which reduced to the common Radius, and meafured as in *Problem* the fecond, gives the Degrees of a great Circle, which allowing 15 Degrees for an Hour, and one Degree to 4 Minutes of Time, gives the Hour from 6.

4. *To find the Time of the Sun's Rifing and Setting.*

The Arch F*y* reduced to the common Radius by *Prob.* II. and the Degrees reduced to the Time by *Prob.* III. gives the Hour that the Sun rifes before 6, or fets after 6.

Note; If the Sun had 22d. 30m. South Declination, it muft have been reprefented by the Parallel *f m*, and then the Arch *b i* reduced as before, would have given the Time of the Sun's rifing after, or fetting before 6 o'Clock.

5. *To find the Length of the Day or Night.*

That Part of the Parallel of Declination which is above the Horizon, reduced as in *Prob.* II. and III. and doubled gives the Length of the Day, and twice that Part which is below the Horizon gives the Length of the Night: Thus D*y* gives the Length of the Day, and *y c* the Length of the Night when the Sun hath 22d. 30m. North Declination, and *f b* gives the Length of the Day, and *b m* the Length of the Night in the fome South Declination.

6. To

6. *To find the Sun's Amplitude.*

The Sun's Amplitude is an Arch of the Horizon contained between the East or West Point, and the Point where the Sun rises or sets: Thus R y is the Amplitude in North, or R b in South Decination, which because it is an Arch of a great Circle is measured on the Sines.

Note; The Amplitude is always of the same Denomination with the Declination, whether North or South.

7. *To find what Hour the Sun will be due East or West.*

The Sun is East or West when in the Line ZN, and it is 6 o'Clock when the Sun is in the Line PS, therefore the Arch XF, reduced by *Prob.* II. and III. gives the Hour after 6 in the Morning, when the Sun is due East, or before 6 in the Evening, when it is due West.

8. *To find the Sun's Altitude when East or West.*

The Arch R X measured on the Sines, gives the Sun's Altitude when East or West.

9. *To find the Sun's Altitude at 6 o'Clock.*

The Arch FW equal to R a, measured on the Sines, gives the Sun's Altitude at 6.

10. *To find the Sun's Azimuth at 6.*

The Arch a F, reduced by *Prob.* II gives the Azimuth at 6 required, accounting the Azimuth from the East or West; but if from the North, the Arch F n, reduced as before, answers the *Problem.*

11. *To find the Sun's Azimuth at any given Hour; As suppose 5 in the Morning, or 7 at Night.*

I have thought fit to insert this *Problem,* not only for its Usefulness, but because it is performed by another

Q   Method

Method of reducing a leſſer Circle to the common Radius, which is the Inverſe of that Way of reducing, mentioned in *Prob.* II. As for Example, Becauſe 5 in the Morning is one Hour before 6, and every Hour of Time is equal to 15 Degrees of a greater (or leſſer) Circle; therefore ſet off 15 Degrees from F towards *y,* which is done thus.

Set off F C from R to *d* upon the Line R H; ſet off 15 of the Chords from A to *g,* and draw the pricked Line R *g*; the neareſt Diſtance from the Point *d* to that Line, ſet from F to *l,* is 15 Degrees, or an Hour of Time in that leſſer Circle, and the Point *l* is the Place of the Sun at 5 in the Morning, and the Arch *b l* reduced to the common Radius by *Prob.* II. gives the Sun's Azimuth from the Eaſt Northwards at 5 in the Morning, or from the Weſt Northwards at 7 in the Evening; and by this Method you may ſet off any Number of Degrees upon any Parallel or leſſer Circle, a Thing of great Uſe in Orthographic Projection.

S E C T. II.

Of the Stereographic Projection of the Sphere, upon the Plane of the Meridian.

STereographic Projection of the Sphere (as well as Orthographic) is a Branch of Perſpective, and ſheweth how to deſcribe, or transfer the *Fig.* 65. ſeveral Circles of the Sphere upon a Plane, which cutteth the ſaid Sphere in the Middle, the Eye being ſuppoſed to be placed in the Sphere's Superficies, perpendicular to the Center of the ſaid Plane; and this Projection hath its Excellency, that no other Perſpective Projection is capable of; for upon this all Circles of the Sphere greater or leſſer, (except right Circles) are alſo Circles upon the Plane of the Projection, and will anſwer the Appearance, and the Laws of Projecting and Meaſuring as well without the Primitive

Circle

Circle as within it, which makes this Projection more commodious and more practised than any other.

In this Kind of Projection there are four Sorts of Circles, *viz.* The Primitive Circle, a Right Circle, an Oblique Circle and a Parallel Circle.

1. The Primitive is a perfect round Circle, upon whose Plane the Projection is made; and in this Case it is the Meridian of the Place; as here the Circle ZHNO is the Primitive Circle.

2. A Right Circle is projected into a strait Line, or Diameter of the Primitive Circle which cuts the Primitive in two opposite Points, and divides it into two equal Parts, as HAO is the Horizon, or EAQ the Equator, &c.

3. An Oblique Circle cuts the Primitive at two opposite Points, but divides it into two unequal Parts, as Z ⊙ N, an Azimuth, or P ⊙ S, a Meridian.

4. A Parallel Circle is a lesser Circle, neither cutting the Primitive in opposite Points, nor dividing it into two equal Parts, as q ⊙ F, a Parallel of Declination, or B ⊙ P, a Parallel of Altitude.

P R O B. I. *To find the Center of any projected Circle.*

1. The Primitive Circle is always drawn with the Chord of 60 Degrees, upon the Center A.

2. A Right Circle (being a straight Line) hath no Center.

3. The Center of an Oblique Circle always falls in that Right Circle, which cuts the said Oblique Circle in two equal Parts; thus the Center of the Circle Z ⊙ N falls in the Right Circle H A O, and the Center of the Circle P O S falls in the Right Circle EAQ; and if you have three Points, as Z ⊙ N, through which the Circle is to pass, the Center is found by the sixth Geometrical *Problem* at the Beginning of this Book; but if you have given, the Angle that the Oblique Circle is to make with the Primitive Circle, set the Tangent of the given Angle from A in the Line HAO, or with the Secant of the said Angle, and one Foot in Z or N, with the other cross the Line H·A·O, either

Way

Way you shall find the Center of the Oblique Circle required.

4. For the Center of a Parallel Circle, having its Distance from the Right Circle, to which it is parallel, given, set off the Chord of that Distance upon the Primitive Circle from each End of the Right Circle, and the half Tangent of the said Distance from the Center A, upon that Right Circle, which cuts the other at Right Angles, and so you will have three Points, through which the Parallel is to pass, to find its Center by the sixth Geometrical Problem.

Or thus, Suppose I would find the Center of the Parallel of Declination *q* ☉ F, set the Tangent of the Complement of the Sun's Declination from *q* or F into the Line SAP, or the Secant of the same from A upon Line SAP continued, and that shall find the Center required.

PROB. II.   *To find the Pole of any Circle.*

Note; The Pole of any great Circle is always 90 Deg. from it's Periphery; Therefore,

1. The Pole of the Primitive is always at its Center A.

2. The Pole of a Right Circle is always in the Primitive, as the Pole of the Right Circle HAO is at Z, and the Poles of the Right Circle EAQ are at P and S, &c.

3. For the Pole of an Oblique Circle, as suppose the Circle Z ☉ N, observe where the said Oblique Circle cuts that Right Circle, which it intersects at Right Angles, as here in the Point G, lay a Ruler from Z to G, it cuts the Primitive in *y*; set off 90 of the Chords from *y* to K, a Ruler laid from Z to K, cuts the Circle HAO in *g*, the Pole of the Oblique Circle required.

4. The Pole of a Parallel Circle is always in the Pole of that Right Circle to which it is parallel; thus the Pole of the Parallel *q* ☉ F is in the Line AP, at the Point P, the Pole of the Right Circle EAQ, *viz.*

Note; An Arch of a Parallel Circle is never made a Side of a Spherical Triangle, they being all composed of Arches of Great Circles, whether Primitive, Right, or Oblique.

PROB.

PROB. III.   *To project any given Angle.*

1. To make an Angle of 23 29 at the Center of the Primitive with the Right Circle E Q.

Set 23 : 29 off the Chords from E to *q*, and from Q to W, and draw the Line or Right Circle *q* W, it makes (at A the Center) the Angle required.

2 To draw an Oblique Circle to make at Z an Angle of 48 Degrees with the Primitive.

The Center of this Circle will fall in the Right Circle HAO by *Prob.* I. Set the Tangent of 48 the given Angle from A towards O, it will reach to *r*, then upon *r*, as a Center with the Extent *r* Z or *r* N draw the Oblique Circle Z G N, it will make at Z the Angle required.

3 To draw an Oblique Circle to make a given Angle at the Primitive with another Oblique Circle, as suppose I would draw an Oblique Circle, to make at P an Angle of 48 Degrees with the Oblique Circle P *t* S.

Lay a Scale from P to *t* (the Point where the Oblique Circle cuts the Right Circle in which its Center falls) and it will cut the Primitive in *v*, set the given Angle 48 off the Chords from *v* to *x*, a Scale laid from P to *x* will cut the aforesaid Right Circle in *w*, draw the Circle P *w* S by Prob. I. it is the Obliqe Circle required.

4. To project any given Angle within the Primitive, but not at the Center.

We shall add this general Rule to what was laid down by Prob. I. *viz.* that as I said there the Centers of all Oblique Circles fall in that Right Circle, which cuts them at Right Angles, I shall now add, that the Centers of all Oblique Circles that pass through the same Point in the Right Circle, fall in a right Line drawn through the Center of the said Oblique Circle, and at Right Angles with the said Right Circle. *Example*, Because the Point *r* is the Center of the Oblique Circle ZGN, therefore the Centers of all Oblique Circles that pass through the Point G, shall fall in the Line *c d* continued, which paſſes through the Point *r*, and cuts the ſaid Circle at Right Angles, *&c.* which being well underſtood the Application is eaſy; as for Inſtance, it is required to draw an Oblique Circle through the Point G in the Right Circle HAO, to make with the ſaid Right Circle an Angle of 55 Degrees.

Lay a Scale from Z to G, it will cut the Primitive in *y*, the Arch H *y* meaſured on the Chords is 48 Degree, (the Meaſure of the Angle HZG) therefore ſet the Tangent of 48 from A to *r*, and through *r* at Right Angles to the Right Circles HAO draw *c d* continued, then with any Radius, and one Foot in G draw the Arch *f* R, and from the ſame Radius ſet off 35 Degrees (the Compliment of the Angle to be made at G) from *f* to *i*; a Scale, laid from G to *i* will cut *c d* in *m*, with the Extent *m* G, and one Foot in *m*, deſcribe the Oblique Circle SGP, it is the Circle required, and makes the given Angle 55 at G, with the Right Circle HAO.

Theſe Directions, with the Rules for meaſuring Arches of Circles and Angles, which follow in Order, will be ſufficient to inſtruct the Learner in projecting any Angle required to be made by a given Oblique Circle, with another Oblique Circle to be projected, and through a given Point in the ſaid Circle; but there is in this Caſe another Variety, which is,

To draw a Right Circle to make with an Oblique Circle any given Angle; in order to the Underſtanding of which, this general Rule may be neceſſary, *viz.*

That the Diſtance between the Poles of any two great Circles (meaſured upon a great Circle drawn between the ſaid Poles) is equal to the Angle made by the Interſection of the ſaid Circles, and conſequently to project the ſaid given Angle is but the Reverſe of the former; for having found the Pole of the given Oblique Circle, draw a Parallel Circle about the ſaid Pole diſtant from it, equal to the Quantity of the Angle propoſed, and where that Parallel cuts the Primitive ſhall be the Pole of the Right Circle, from which ſet off 90 De-

grees,

grees both Ways upon the Primitive, these Points, and
the Center shall be three Points in a Right Line, thro'
which the Right Circle is to be drawn, and wherever it
intersects the Oblique Circle, it makes the Angle required to be made, but that Intersection is confined to
that Point of Intersection only; but as we have not
yet taught how to draw a Parallel to an Oblique Circle
at any given Distance from it, that shall next follow.

P R O B. IV: *To draw a Parallel to an Oblique Circle at
any given Distance from it.*

It is required to draw a Circle parallel to the Oblique
Circle P *b* S, at 30 Degrees Distance from its Pole.

Find the Pole of the Oblique Circle by *Prob.* II. which
will be found at *f*, lay a Scale from S to *f*, it will cut
the Primitive in *l*, set 60 Degrees (the Parallel's Distance from the Pole of the Oblique Circle) both Ways
from *l* to P, and *v*, a Scale laid from S to P will cut
E Q in A, which happens at or near the Center of the
Circle, also a Scale laid from S to *v* will cut Q E (continued) in *a*; the Space *a* A being divided equally in
t, with the Extent *a t* or A *t*, and one Foot in *t*, describe
the Circle *a* Z A, it is the Parallel required, and *f* its
Pole is also the Pole of the Oblique Circle P *b* S to.
which it is parallel.

P R O B. V. *To set off any Number of Degrees, upon a
great Circle.*

In this Problem are three Varieties.

1. To set off any Number of Degrees upon the Primitive, suppose 51 deg. 32 min. from O upwards.

Take 51 32 off the Chords, and with that Extent and
one Foot in O, the other will reach to P, the Arch O P
being 51 32, as required.

2. To set off any Number of Degrees upon a Right
Circle; suppose 22 30 from the Center A towards P.

Take the Half Tangent of 22 30, and setting one
Foot in A the other will reach to *b*, and the Arch A *b* is
22 30 required, and *b* P its Complement to 90, *viz.* 67
30.

Note,

Note; If you set off any Number of Degrees upon a Right Circle from the Center, you must account from the Beginning of the half Tangents, and reckon forwards; but if you begin at the Primitive, you must then begin at 90 of the half Tangents and reckon backwards; thus if you had to set 67 30 from P, you must take 67 30 backwards, *viz.* from 90 to 22 30, which will reach to *b*, as before.

But if you have any Degrees to set off upon any intermediate Arch between the Center and the Periphery, it is something more difficult; but the following Rule is universal in all Varieties of this Problem.

Set the Chord of 22 30 from E towards Z, it will reach to *q*, lay a Scale from Q to *q* it will cut P S in *b*, then is A *b* 22 30, as was required.

P R O B. VI.   *How to measure any Side of a Triangle or Arch of a great Circle.*

1. Any Arch of the Primitive is always measured upon the Chords.

2. Any Arch of a Right Circle is always measured upon the half Tangents.

3. Any Arch of an Oblique Circle is measured thus;

Lay a Ruler from the Pole of the said Oblique Circle, over the Extremities of the said Arch, and observe where they cut the Primitive, and that Distance measured on the Chords gives the Quantity of the Arch required.

Example. Suppose I would measure the Arch ⊙ G of the Circle Z ⊙ N : Lay a Ruler from *g* (the Pole of the said Circle) to ⊙, it cuts the Primitive in B: Again, a Ruler laid from *g* to G cuts the Primitive in H ; the Distance ⸬H, measured on the Chords, gives the Quantity of the Arch ⊙ G required.

P R O B. VII.   *How to measure any Spherical Angle.*

This one Rule is universally useful in all Cases (whether the Sides including the Angle are Arches of the Primitive, Right or Oblique Circle) *viz.* Lay a Ruler

from

from the Angular Point, over the Poles of the two Circles, including the Angle, and obferve at what two Places the Ruler cuts the Primitive, for that Diftance meafured on the Chords, gives the Quantity of the Angle required.

Example. I would meafure the Angle Z ☉ P of the Triangle Z P ☉ : Lay a Ruler from ☉ to g (the Pole of one of the Circles including the Angle) it cuts the Primitive in &, and then a Ruler laid from ☉ to e (the Pole of the other Circle including the Angle) it cuts the Primitive in a; the Arch a & meafured on the Chords, gives the Quantity of the Angle Z ☉ P required.

By thefe Directions, and the Inverfe of them, you may fet off any Number of Degrees, or lay down any Angle required, as will be further illuftrated in the following Example, the Room affigned for this Subject not permitting me to enlarge upon it.

A Stereographic Projection of the Sphere, upon the Plane of the Meridian. For,

	d.	m.	
Latitude of *London* —— —— ——	51	32	} N.
Sun's Declination —— ——	22	30	

How to project the Sphere ftereographically upon the Plane of the Meridian.

Firft with the Chord of 60 Degrees, and one Foot of the Compaffes in the Center A, defcribe the Primitive Circle Z H N O to reprefent the Meridian; draw the Diameter HAO to reprefent the Horizon, and at Right Angles to it, draw the Diameter ZAN the Azimuth of Eaft and Weft : Then becaufe the Pole is 51d. 32m. above the Horizon, fet of 51d. 32m. off the Chords from O to P, and from H to S, then is P the North Pole, and S the South Pole : Draw the Diameter S P, the Axis of the World, and Hour Line of 6, and at Right Angles to it, the Diameter E Q the Equator;

Alfo

Alfo becaufe the Sun's Declination is 22d. 30m. draw the Parallel of Declination *q* ☉ F, at 22d. 30m. Diftance from the Equator Northward by *Prob.* I. hereof. Now the Sun being in the Parallel *q* ☉ F, and alfo in the Ecliptic *q* ☉ W, it muft be at the croffing of thofe Circles, *viz.* at the Point ☉, which is the true Place of the Sun in the Projection, which being found you have three Points, Z, ☉, N, through which to draw the Azimuth, Z ☉ N, and alfo three Points P, ☉, S, through which to draw the Meridian P ☉ S, both which are done by *Prob.* I. hereof; and then the Parallel *q*   F is the Parallel defcribed by the Sun's Motion, hence that Day the Sun's Rifing and Setting is at *g*, and at 6 o'Clock at *b*, and Eaft or Weft at *m*, and upon the Meridian at *q*, *&c.* which is fo intelligible, that it needs no further Illuftration.

For the Ecliptic, fet off 23 deg. 29 min. (the Sun's greateft Declination) from E to *q*, and from Q to W, and draw the Diameter *q* A W, which fhall reprefent the Ecliptic.

By *Prob.* I. hereof draw the Meridian P *m* S, and the Azimuth Z *b* N; and through the Point ☉ (where the Ecliptic cuts the Parallel of Declination) draw the Meridian P ☉ S.

1. *To find the Sun's Longitude or Diftance from the neareft Equinoctial Point.*

The Arch of the Equator A ☉, meafured on the half Tangents, and the Degrees and Minutes fo found, being reduced to Sines, Degrees and Minutes of the Ecliptic, allowing 30 Degrees to a Sine, *&c.* gives the Sun's Longitude required.

2. *To find the Sun's Right Afcenfion.*

The Arch of the Equator A *s* meafured on the half Tangents, gives the Sun's Right Afcenfion when he is in *Aries*, *Taurus*, or *Gemini*; but if in *Cancer*, *Leo*, or *Virgo*, the Degrees fo found fubtracted from 180, gives the Right Afcenfion; and when the Sun is in *Libra*, *Storpio*, or *Sagittary*, the Degrees fo found added to 180, gives the Right Afcenfion; but if in

Capricorn,

Capricorn, *Aquarius*, or *Pisces*, the Degrees so found subtracted from 360, shews the Right Ascension required.

Note; For finding the exact Degrees and Minutes of the Sun's Right Ascension, the same Rules of adding and subtracting the Degrees found, are to be used in the Orthographic Projection also.

3. *To find the Hour of the Day.*

Observe where the Meridian P O S cuts the Equator, as in *t*, the Arch A *t* of the Equator measured on the half Tangents, and reduced to Time, allowing 15 Deg. to an Hour, &c. gives the Hour past 6 in the Morning, or the Hours wanting of 6 at Night.

Or thus: In the Oblique Triangle Z P ☉, the Angle at P, measured by *Prob.* IV. and reduced to Time as above, gives the Hours wanting of 12, if in the Forenoon; or the Hours past 12, if in the Afternoon.

4. *To find the Time of the Sun's Rising and Setting.*

Through the Point *g* (where the Parallel of Declination cuts the Horizon) draw the Meridian P *g b* S, and observe where it cuts the Equator, as in *b*, the Arch A *b* measured on the half Tangents, and the Degrees reduced to Time, gives the Hour and Minutes that the Sun rises before 6 in the Morning, or sets after 6 in the Evening; or if in South Declination where the Parallel of Declination intersects the Horizon on the other Side of the Line PAS, or Hour Line of 6, it gives the Sun's Rising after 6, or setting before 6, &c.

5. *To find the Length of the Day or Night.*

The Hours and Minutes of the Sun's setting doubled gives the Length of the Day, the Time of the Sun's Rising doubled gives the Length of the Night.

6. *To find the Sun's Amplitude.*

The Arch of the Horizon A *g* (being an Arch contained between the Sun's Rising or Setting at *g*, and the due East or West Point in A) measured on the

half

half Tangents gives the true Amplitude from the East
or West as may be required.

7. *To find what Hour the Sun will be at East or West.*

The Sun is East or West in *m,* the Point where the
Parallel of Declination cuts the Prime Vertical, or East
and West Line Z A N ; therefore having drawn the
Meridian P *m* S, observe where it cuts the Equator, as
at *w,* then A *w* measured on the half Tangents, and
reduced to Time, gives the Hour after Six in the Morning that the Sun is due East, or the Hour before Six
in the Evening that the Sun is due West.

8. *To find the Sun's Altitude when East or West.*

The Arch A *m,* measured on the half Tangents, answers the Question, and gives the Sun's Altitude when
East or West.

9. *To find the Sun's Altitude at Six o'Clock.*

The Sun is in the Point *b* at Six o'Clock, therefore
the Arch *b e* measured by *Prob.* III. gives the Sun's
Altitude at 6 required.

10. *To find the Azimuth at Six o'Clock.*

The Arch of the Horizon A *e,* measured on the half
Tangents, gives the Sun's Azimuth from the East or
West as may be required, whose Complement to 90 *e o*
is the Azimuth from the North.

11. *To find the Sun's Azimuth at any Hour or Altitude, as*
supposed when in the Point ☉ in this Projection.

From Z to N, and through the Point ☉, draw the
Azimuth Circle Z ☉ N, as before directed, and observe
where it cuts the Horizon as in G, then the Arch A G,
measured on the half Tangents, gives the Azimuth from
the East or West as required.

Thus have I given you a Taste of all the most useful
Problems, relating to the Orthographic and Stereographic Projection of the Sphere; of which I shall add no
more at this Time, but proceed to Spherical Trigonometry, and its Application to Practical Astronomy, in
their proper Places.

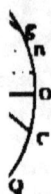

Fig. 64

Fig. 65

S E C T. III.

Of Spherical Trigonometry.

AS Plane Triangles confift of three right-lined Sides and three Angles, made by their interfect-ing each other, fo Spherical Triangles are made by the Interfection of the Arches of three great Circles of the Sphere, and of thofe there are three Sorts, *viz.* Right-angled, Oblique, and Quadrantal, but the two firft Kinds are moft in Ufe; and as the great Circles of the Sphere are either Meridians, Azimuths, &c. when they interfect each other, fo as to make a Spherical Triangle, the Solution of fuch Triangles is what is generally cal-led Aftronomy.

In Spherical Triangles obferve thefe Solutions.

1. Each Triangle hath 6 Parts, *viz.* 3 Sides and 3 Angles.

2. In Right-angled Triangles there are five, called Circular Parts, *viz.* the two Legs, the Hypothenufe, and 2 Acute Angles, the Right Angle being not look-ed upon as any Part in this Cafe.

3. Of thefe five there are always two given and one required (thefe two and Radius being fufficient to find out the fourth Term required) and of thefe three there is always one called Middle Part, and the other two are either Conjuncts, or Disjuncts, *viz.* either both next the Middle Part, or both feparated from it by a Part, not mentioned in the Queftion, as in the Scheme, fuppofing in the Pentagon ABCDE; if A, B and E were the Parts mentioned (whether given or required) A would be Middle Part, and B and E would be *Fig. 81.* Conjuncts, but if D, A and B were given, D would be the Middle Part, and A and B Dif-juncts, and if this be underftood of the Pentagon, it may be as readily underftood *Fig. 82.* of a Triangle; for fuppofe the three Sides,

B, C

B, C and E, and the two Angles A and D, all which are called Circular Parts, were any three of them the Parts given or required; as for Instance, if B, C and D, then C would be the Middle Part, and B and D Conjuncts, or if A, C and D were the Parts given or required, A would be the Middle Part, and C and D Disjuncts, because B is between A and C, and E is between A and D, &c.

4. Then the fundamental Proportion is, *As Radius is to the Tangent of one Conjunct; so is the Tangent of the other Conjunct, to the Sine of the Middle Part.*

And, *As Radius, is to the Sine Complement of one Disjunct; so is the Sine Complement of the other Disjunct, to the Sine of the Middle Part* ——— *Only* ———

5. The Hypothenuse and two Acute Angles, *viz.* A and D are called by their Compliments, that is, if any of them are Middle Parts, they are called Sine Complements instead of Sines; or if Conjuncts, they are called Tangent Complements instead of Tangents, and consequently if Disjuncts, they are called Sines, instead of Sine Compliments, *viz.*

6. As in the fundamental Proportion, the Middle Part is the last Term, it will follow that if the Middle Part be given, and a Conjunct required, the Proportion must be inverted, *viz. As the Tangent of the given Conjunct, is to Radius, so is the Sine of the Middle Part, to the Tangent of the required Conjunct,* &c. Understand the same of Disjuncts.

7. In all Spherical Triangels the Sines of the Sides are Proportional to the Sines of their opposite Angles, *viz.*

As the Sine of D, is to the Sine of B; so is
Fig. 82. the Sine of A, to the Sine of C; and so is
Radius, to the Sine of E, &c.

8. Oblique Triangles may be brought under the foregoing Rules by letting fall a Perpendicular, and working it as two Right-Angled Triangles, only observe to *let your Perpendicular fall from the End of a given Side, and opposite to a given Angle,* which when done, all the
twelve

twelve Cases in Oblique may be performed by these Rules, and the following Conclusions.

CONCLUSIONS.

1. The Co-fines of the Angles at the Base, are directly proportional to the Sines of the Vertical Angles.

2. The Sines of the Bases, and Tangents of the Angles at the Base are reciprocally proportional.

3. The Co-fines of the Segments of the Base, and Co-fines of the Hypothenufes are in direct Proportion.

4. The Tangents of the Hypothenufes are reciprocally proportional to the Co-fines of the Vertical Angles.

5. The Tangents of the Bases are proportional to the Tangents of the Vertical Angles.

6. In all Oblique Spherical Triangles, when a Perpendicular is let fall from the greatest Angle to the greatest Side, it will be, as the Tangent of half the Base, is to the Tangent of half the Sum of the other two Sides; so is the Tangent of half the Difference of the said Sides to the Tangent of half the Difference of the Segments of the Base; which half Difference being added to the half Base, the Sum is the greater Segment, but subtracted from it, the Remainder is the lesser Segment.

These are sufficient for all the 28 Cases, yet for Variety I shall insert another Operation where three Sides are given to find an Angle, *viz.*

Add the three Sides together, and from their half Sum, subtract the Side opposite to the Angle required; then to the Complement Arithmetical of the Log-Sines of the Sides containing the Angle required, add the Log-Sines of that half Sum, and Remainder, half the Total of these four Logarithms is the Sine Complement of half the Angle required.

These Directions well understood, being sufficient to solve all the 28 Cases of Spherical Triangles, I shall give a brief Instance of each Case, to help the Memory

as

as well as the Underftanding of the Learner, and fhall apply them more largely when I come to fpeak of Aftronomy itfelf.

Note; I fhall for Brevity fake put *S* for Sine, *S. C.* for Sine Complement, *T.* for Tangnt, *T. C.* for Tangent Complement, *R.* for Radius; and fhall only mention the Proportion here, but fhall perform the Calculations for the fame Reafons afterwards; *Note*, Mark the given Parts with a Dafh thus ', the required thus °.

C A S E I. *Fig.* 83.

Given $\left\{\begin{array}{l}\text{Angle A} \\ \text{Hyp. AD}\end{array}\right\}$ Required AB.

Here by the foregoing Directions, A is Middle Part, and AD and AB are Conjuncts, and though Middle Part is a Sine, yet being an Angle it becomes a Sine Complement, alfo AD being a Conjunct fhould be a Tangent; but being the Hypothenufe, it is a Tangent Complement, likewife becaufe Middle Part is given, the fundamental Proportion muft be inverted, all which being confidered, the Analogy is,

As *T.C.* AD .. *R.* : : *S.C.* A .. *T.* AB.

C A S E II. *Fig.* 84.

Given the Hypothenufe and an Angle, to find the Leg oppofite to the given Angle.

Given $\left\{\begin{array}{l}\text{The Angle A} \\ \text{Hypothenufe AD}\end{array}\right\}$ Required BD.

Here B D is Middle Part, and AD and A are Difjuncts, therefore the Proportion is,

As *Radius* .. *S.* AD : : *S.* A .. *S.* BD.

C A S E III. *Fig.* 85.

Given the Hypothenufe and an Angle, to find the other Angle.

Given $\left\{\begin{array}{l}\text{Hypothenufe AD} \\ \text{The Angle A}\end{array}\right\}$ Required the Angle D.

Here

Here A D is Middle Part, and the Angles A and D are Conjuncts, therefore the Proportion is,

As *T. C.* A ·· Radius : : *S. C.* AD. *T. C.*D.

CASE IV.   *Fig.* 86.

Given the Hypothenuse and a Leg, to find the Angle included between them.

Given $\begin{cases} \text{Hypot. AD} \\ \text{Leg AB.} \end{cases}$ Required the Angle A.

The Angle A is Middle Part, and AD and AB, are Conjuncts, therefore the Proportion is,

As Radius ·· *T.* AB : : *T.C.* AD ·· *S. C.* A.

CASE V.   *Fig.* 87:

Given the Hypothenuse and a Leg, to find the Angle opposite to the given Leg.

Given $\begin{cases} \text{Hypot. AD} \\ \text{Leg AB} \end{cases}$ Required the Angle D

Here AB is Middle Part, and AD and D are Disjuncts, therefore the Proportion is,

As *S.* AD ·· Radius : : *S.* AB ·· *S.* D.

CASE VI.   *Fig.* 88.

Given the Hypothenuse and a Leg, to find the other Leg.

Given $\begin{cases} \text{Hypot AD} \\ \text{Leg AB} \end{cases}$ Required BD.

Here A D is Middle Part, and A B and B D are Disjuncts, therefore the Proportion is,

As *S. C.* AB ·· Radius : : *S.C.* AD ·· *S. C.* BD.

R   CASE

CASE VII. *Fig.* 89.

Given an Angle and a Leg adjacent, to find the Leg opposite to the given Angle.

Given $\left\{\begin{array}{l}\text{Angle E}\\ \text{Leg EF}\end{array}\right\}$ Required the Leg FG.

In this Case EF is middle Part, and the other Parts are Conjuncts, therefore the Proportion is,

As *T. C.*E ·· Radius : : *S.* EF ·· *T.* FG.

CASE VIII. *Fig.* 90.

Given an Angle and a Leg adjacent, to find the Angle opposite to the given Leg:

Given $\left\{\begin{array}{l}\text{Angle F}\\ \text{Leg EF}\end{array}\right\}$ Required the Angle G.

Here the Angle G is middle Part, and E F and E are Disjuncts, therefore the Proportion is,

As Radius ·· *S. C.* EF : : *S.* E ·· *S. C.* G.

CASE IX. *Fig.* 91.

Given an Angle and a Leg adjacent, to find the Hypothenuse.

Given $\left\{\begin{array}{l}\text{Angle E}\\ \text{Leg EF}\end{array}\right\}$ Required EG.

The Angle E is middle Part, and EF and EG are Conjuncts, therefore the Proportion is,

As *T.* EF ·· Radius : : *S. C.* E. ·· *T. C.* EG.

CASE X. *Fig.* 92.

Given an Angle and a Leg opposite, to find the Leg adjacent to the given Angle.

Given $\left\{\begin{array}{l}\text{Angle E}\\ \text{Leg FG}\end{array}\right\}$ Required E F;

The

The Side EF is middle Part, and Angle E and Side FG, are Conjuncts, therefore the Proportion is,

As Radius : *T.* FG :: *T. C.* E -- *S.* EF:

CASE XI. *Fig.* 93.

Given an Angle and a Leg oppofite, to find the Angle adjacent to the given Leg.

Given $\left\{\begin{array}{l}\text{Angle E}\\ \text{Leg FG}\end{array}\right\}$ Required Angle G.

Here E is middle Part, and FG and Angle G are Disjuncts, therefore the Proportion is,

As *S. C.* FG -- Radius : : *S. C.* E -- *S.* G.

CASE XII. *Fig.* 94.

Given an Angle and a Leg oppofite, to find the Hypotbenufe.

Given $\left\{\begin{array}{l}\text{Angle E}\\ \text{Leg FG}\end{array}\right\}$ Required EG

Here FG is middle Part, and EG and Angle E are Disjuncts; therefore,

As *S.* E -- Radius : : FG -- *S.* EG.

CASE XIII. *Fig.* 95.

Given the Legs to find an Angle.

Given $\left\{\begin{array}{l}\text{Leg HI}\\ \text{Leg I K}\end{array}\right\}$ Required Angle H.

Hence H I is middle Part, and I K and H are Conjuncts, hence the Proportion is,

As *T.* I K   Radius :: *S.* H I -- *T. C.* H.

C A S E

C A S E XIV. *Fig.* 96.

Given the Legs to find the Hypothenuse.

Given $\begin{cases} \text{Leg H I} \\ \text{Leg I K} \end{cases}$ Required H K.

Hence H K is middle Part, and H I and I K are Disjuncts, therefore the Proportion is,

As Radius ·· *S. C.* H I : : *S. C.* I K. ·· *S. C.* HK.

C A S E XV. *Fig.* 97.

Given the Angles to find a Leg.

Given $\begin{cases} \text{Angle H} \\ \text{Angle K} \end{cases}$ Required H I.

Here the Angle K is middle Part, and H I and Angle H are Disjuncts, therefore the Proportion is,

As *S.* H ·· Radius : : *S. C.* K ·· *S. C.* H I.

C A S E XVI. *Fig.* 98.

Given the Angles to find the Hypothenuse:

Given $\begin{cases} \text{Angle H} \\ \text{Angle K} \end{cases}$ Required H K.

Here HK is middle Part, and the Angles H and K are Conjuncts, therefore the Proportion is,

As Radius ·· *T. C.* H : : *T. C.* K ·· *S. C.* HK.

These are the 16 Cases of Right-angled Triangles, and are all performed by the fundamental Proportion, commonly called my Lord *Napier's*, mentioned in Solution the 4th, by the Help of the Directions in Solution 2, 3, 5, and 6; the other 12 Cases are in Oblique, **and**

and performed by Solution 7 and 8, and the Conclusions following them.

Note; When we refer to any Case, we mean some of the 16 Cases of Right-angled Spherical Triangles before going.

Oblique Angles.

C A S E I. *Fig.* 99.

Two Sides and an Angle opposite to one of them being given, to find an Angle opposite to the other.

Note; It is to be known whether the Angle required be Acute or Obtuse; for if it be Acute, the Sine, Tangent, &c. that the Operation produceth, is the Sine Tangent, &c. of the Angle required; but if it is Obtuse, the Degrees found subtracted from 180, the Remainder is the Quantity of the Angle required.

Given $\begin{cases} \text{The Side LN} \\ \text{The Side NM} \\ \text{The Angle M} \end{cases}$ Required the Angle L.

The Solution of this depends upon Solution 7.

As S. LN ·· S.M : : S. NM ·· S. L.

C A S E II. *Fig.* 100.

Two Sides being given, with an Angle opposite to one of them, to find the Angle included between the given Sides.

Given $\begin{cases} \text{The Side L N} \\ \text{The Side N M} \\ \text{The Angle M} \end{cases}$ Required the Angle N.

Let fall the Perpendicular N P from N by Sol. 8. which falls within; then in the Right-angled Triangles MPN there are given the Side NM, and the An-

R 3   gle

gle M to find the Angle MNP by Cafe III. and then by Conclufion the fourth, the Propofition for finding the Angle LNP is,

As *T.* LN ·· *S. C.* MNP : : *T.* MN ·· *S. C.* LNP, and the Angle LNP being found and added to the Angle MNP, the Sum is the whole Angle LNM required.

CASE III. *Fig.* 101.

Given two Sides and an Angle oppofite to one of them, to find the third Side.

Given { The Side LN, The Side NM, The Angle M } Required the Side LM

Here as in Cafe II. becaufe the fame Things are given, the Perpendicular muft fall from the Angle N, upon the Side LM, and having let it fall to P by Sol. 8. you have given in the Right-angled Triangle MPN the Angle M and the Side MN, to find the Bafe MP by Cafe I. which being found, find the other Segment of the Bafe by Conclufion the third, *viz.*

As *S. C.* NM ·· *S C.* PM ∷ *S. C.* LN ·· *S. C.* LP.

Then MP, added to LP, is the Side LM required.

CASE IV. *Fig.* 102.

Two Angles and a Side oppofite to one of them being given, to find the Side oppofite to the other.

Note; Here as in Cafe I. it is required to be known whether the Side required be more or lefs than a Quadrant; for if it be lefs, the Sine or Tangent found anfwers the Cafe; but if it is more than a Quadrant, fubtract the Degrees and Minutes found from 180 Degrees, the Remainder is the Side required.

Given

$$\text{Given} \begin{cases} \text{The Angle N} \\ \text{The Angle M} \\ \text{The Side L N} \end{cases} \text{Required the Side L M.}$$

The Proportion in this Cafe deduced from Sol. 7. is,

As *S.* M ◦◦ *S.* LN : : *S.* N ◦◦ *S.* LM.

CASE V. *Fig.* 103.

Two Angles and a Side oppofite to one of them being given, to find the Side included between the given Angles.

$$\text{Given} \begin{cases} \text{The Angle L} \\ \text{The Angle M} \\ \text{The Side LN} \end{cases} \text{Required the Side LM.}$$

Here by Sol. 8. the Perpendicular is to fall from the Angle N upon the Side LM, and then in the Triangle LPN there is given the Angle L, and the Side LN to find the Bafe LP by Cafe I. of Right-angled Spherical Triangles, which being found, the other Segment PM is found by Conclufion the fecond, thus:

As *T.* M ◦◦ *S.* LP : : *T.* L ◦◦ *S.* MP.

The Segment MP added to LP makes the whole Side LM required.

CASE VI. *Fig.* 104.

Given two Angles and a Side oppofite to one of them, to find the third Angle.

$$\text{Given} \begin{cases} \text{The Angle L} \\ \text{The Angle M} \\ \text{The Side LN} \end{cases} \text{Required the Angle N.}$$

The Perpendicular falls from the Angle N by Sol. 8. then in the Right-angled Triangle LPN there is given the Angle L and Side LP, to find the Angle LNP by Cafe III. and then the other Angle MNP by Conclufion the firft, thus,

As *S.* C. L ◦◦ *S.* LNP ∷ *S.* C. M ◦◦ *S.* MNP.

R 4   Then

Then the Angle MNP added to LNP is the whole Angle N required.

CASE VII. *Fig.* 105.

Two Sides and an Angle included given, to find either of the other Angles.

Given $\left\{\begin{array}{l}\text{The Side XY}\\ \text{The Side YZ}\\ \text{The Angle XYZ}\end{array}\right\}$ Required the Angle Z.

Here by Sol. 8. the Perpendicular muſt fall either from X or Z, and it falls without, either upon the Side XY continued, or upon ZY continued; but where the Caſe is ſo ambiguous by Sol. 8, obſerve,

Let the Perpendicular fall from the End of a given Side, and oppoſite to a given Angle, and likewiſe oppoſite to the Angle required, if an Angle be required, or next to the Side required, if a Side be required.

By this Rule the Perpendicular muſt fall from the Angle X upon the Baſe Z Y continued to W, and then in the Right-angled Triangle YWX, there is given the Side X Y, and the Angle X Y W (being the Supplement of the given Angle XYZ to 180 Degrees) to find the Leg WY by Caſe I. of Right-angled Spherical Triangles, and adding WY to YZ the Sum is WZ; then by Concluſion the Second,

As *S.* WZ ·· *T.* XYW : : *S.* WY ·· *T.* XZW.

CASE VIII. *Fig.* 106.

Two Sides and an Angle included being given, to find the third Side.

Given $\left\{\begin{array}{l}\text{The Side XY}\\ \text{The Side YZ}\\ \text{The Angle XYZ}\end{array}\right\}$ Required the Side XZ.

In this Cafe the Perpendicular falls from the Angle X (by Sol. 8) without the Triangle, upon the Side Z Y continued to W, for the fame Reafons as in Cafe VII. or it might be let fall from the Angle Z upon the Side XW continued, but we fhall chufe the former, *viz.* to let it fall from the Angle X to the Side ZY continued to W, then as in Cafe 7, of Oblique Spherical Triangles you have given the Angle WYX, and the Hypothenufe YX to find WY by Cafe I. of Right-angled Spherical Triangles, then adding W Y to the given Side YZ, gives the whole Bafe WZ, then find the required Side XZ by Conclufion the third,

As S. C. WY ·· S. C. XY : : S.C. WZ ·· S. C. XZ.

C A S E IX. *Fig.* 107.

Two Angles and a Side included being given, to find either of the other Sides.

Given $\begin{cases} \text{The Angle B} \\ \text{The Angle G} \\ \text{The Side BG} \end{cases}$ Required the Side GN.

In this Cafe the Perpendicular falls from the Angle G upon the Side BN by Sol. 8. and falls within the Triangle, and then in the Triangle BGP, there is given the Side BG, and the Angle B, to find the Angle BGP by Cafe III. which Angle being found, fubtract it from the whole Angle BGN, the Remainder is the Vertical Angle PGN, then by Conclufion the fourth,

As S. C. PGN ·· T. BG : : S. C. BGP ·· T. GN, which is the Side required.

C A S E X. *Fig.* 108.

Two Angles and a Side comprehended being given, to find the other Angle.

Given

Given $\begin{cases} \text{The Angle B} \\ \text{The Angle G} \\ \text{The Side B G} \end{cases}$ Required the Angle N:

Here by Sol. 8. and the Reasons given in Case IX. the Perpendicular falls from the Angle G upon the Side BN, then in the Triangle BPG, there is given the Angle B, and the Side BG to find the Vertical Angle BGP by Case III. which being found and subtracted from the whole Vertical Angle BGN, there remains the Vertical Angle PGN, then find the required Angle PNG, by Conclusion the first, thus,

As S. BGP ·· S. C. GBP : : S. PGN ·· S. C. PNG.

CASE XI. *Fig.* 109.

Three Sides given to find an Angle.

Given $\begin{cases} \text{The Side BG} \\ \text{The Side GN} \\ \text{The Side BN} \end{cases}$ Required the Angle GBN

This Case may be performed by various Methods, or by Conclusion 6, but being the required Angle is one of the Angles at the Base, it may be more commodiously done (a Perpendicular being let fall) by Conclusion the sixth, that is, as Tangent of half the Base BN, is to the Tangent of half the Sum of the Sides BG and GN; so is the Tangent of half the Difference of the said Sides, to the Tangent of half the Difference of the Segments of the Base.

Then in the Triangle BPG there is given the Hypothenuse BG, and Leg BP, to find the Angle required by Case IV. of Right angled Spherical Triangles, *viz.*

As Radius ·· *T.* BP : : *T. C.* BG ·· *S. C.* PBG, the Angle required.

Note; When the greatest Angle is required it is better to Work it by Conclusion the sixth.

CASE

C A S E XII. *Fig.* 110.

Given the Angles to find a Side.

Take the Complement of the greateſt Angle to 180
Degrees, and then the Angles will be turned into Sides;
and are to be calculated in all Reſpects as in Caſe the
Eleventh, which need not be repeated in this Place,
but ſhall be more large when we come to the Application hereof to Aſtronomy.

S E C T. IV.

*The Demonſtration of the firſt Five of the Conclu-
ſions, by which Oblique Spherical Triangles are
ſolved by the Help of a Perpendicular.*

OBſerve all Spherical Triangles which require a Perpendicular to be let fall according to Sol. 8.
that the Perpendicular divides the Oblique Triangle
into two Right-angled ones, and is common to both, if
both the Angles at the Baſe be acute, for then it falls
within; or if it falls without, the Rules are evidently
the ſame if we look upon the Oblique Triangles to be
folded in the Perpendicular, the leſſer unfolded out of
the greater, the ſame Perpendicular is common to both,
and the Demonſtration is the ſame.

The Demonſtration is deduced from the fundamental
Proportion Sol. 4. provided we make Radius, and the
Perpendicular (which is common to both Triangles)
the two firſt Terms in both Proportions; for they being deſtroyed, the two laſt Terms in both bear the ſame
Proportion to each other, as the firſt and laſt in each
did, as for *Example.*

Çoncluſion I. The Co-ſines of the Angles at the
Baſe are directly proportional to the Sines of the vertical Angles.

In

In the Triangle
$$\left\{\begin{array}{l} \text{LNP : As } \textit{Radius} \cdot \text{S. C. NP} : : \text{S. LNP} \cdot \cdot \\ \text{S. C. L.} \\ \text{MNP : As } \textit{Radius} \cdot \text{S. C. NP} : : \text{S. MNP} \cdot \cdot \\ \text{S.C.M.} \end{array}\right.$$

Deſtroy the two firſt Terms in each, it will be,

As S. LNP ·· S. C. L : : S. MNP ·· S. C. M.

That is, as the Sine of one Vertical Angle, is to S. C. of its Angle at the Baſe ; ſo is the Sine of the other Vertical Angle, to the S. C. of its Angle at the Baſe. *Which was to be proved.*

Conclufion II. The Sines of the Baſes are reciprocally proportional to the Tangents of the Angles at the Baſe.

In the Triangle LNM, let fall the Per- *Fig.* 104. pendicular NP by Sol. 8, and then by the Directions above, the Baſes and the Angles at the Baſe are the Parts concerned in the Queſtion, to make (with Radius and the common Perpendicular, the four Terms in the Queſtion, and then by Cafe X. of Right-angled Spherical Triangles, it is,

In the Triangle
$$\left\{\begin{array}{l} \text{LNP : As } \textit{Radius} \cdot \text{T. NP} : : \text{T. C. L} \cdot \cdot \text{S.} \\ \text{LP.} \\ \text{MNP : As } \textit{Radius} \cdot \text{T. NP} : : \text{T. C. M} \cdot \cdot \text{S.} \\ \text{MP.} \end{array}\right.$$

Deſtroy the two firſt Terms in each (being both the ſame) and it will be,

As T. C. L ·· S. LP : : T. C. M ·· S. MP.

Alternately, as T. C. L ·· T. C. M : : S. LP ·· S. MP ; but as T. C. of any Arch, is to the T. C. of any other Arch; ſo is T. of the latter, to the T. of the former *Supplement to* Barrow's *Euclid*) therefore ;

As

As T. M $\cdot\cdot$ S. LP : : T. L $\cdot\cdot$ S. MP.

Conclusion the Third. The Co-sines of the Segments of the Base are directly proportional to the Co-sines of the Hypothenuses.

In the Triangle LNM let fall the Perpendicular NP to divide it into two Right-   *Fig.* 104. angled ones LNP and MNP by Sol. 8. then as the Bases and Hypothenuses are the Parts concerned with Radius and the common Perpendicular NP, it will be by Case the XIV.

In the
Triangle
$\begin{cases} \text{LNP : As } \textit{Radius} \cdot\cdot S. C. \text{NP} : : S. C. \text{LP :} \\ \quad S. C. \text{LN.} \\ \text{MNP : As } \textit{Radius} \cdot\cdot S. C. \text{NP} \cdot\cdot S. C. \text{MP :} \\ \quad S. C. \text{MN} \end{cases}$

Destroy the two first Terms in each, it is,

As S. C. LP $\cdot\cdot$ S. C. LN : : S. C. MP $\cdot\cdot$ S. C. MN.

Conclusion the Fourth. The Tangents of the Hypothenuses are reciprocally proportional to the Co-sines of the Vertical Angles.

Let fall the Perpendicular NP, as before, and then in the Triangle LNP the Side LN,   *Fig.* 104. and Angle N, and in the Triangle MNP, the Side MN, and the Angle N are the Parts concerned with Radius, and the common Perpendicular NP, then by Case the IV. it is,

In the
Triangle
$\begin{cases} \text{LNP : As } \textit{Radius} \cdot\cdot T. \text{NP} : : T. C. \text{LN} \cdot\cdot \\ \quad S. C. \text{N.} \\ \text{MNP : As } \textit{Radius} \cdot\cdot T. \text{NP} : : T. C. \text{MN} \cdot\cdot \\ \quad S. C. \text{N.} \end{cases}$

Destroy the two first Terms in each, it is,

As

As *T. C.* LN ·· *S. C.* LNP :: *T. C.* MN ·· *S. C.* MNP.
or by mutually changing Tang. Comp. for Tangent,
as *T.* MN ·· *S.C.* LNP : : *T.* LN ·· *S. C.* MNP.

Conclusion the Fifth. The Tangents of the Bases are
directly proportional to the Tangents of the Vertical
Angles.

In the Oblique. Triangle LNM divided
Fig. 104. into two Right-angled ones by the Perpendicular NP, as before, the Parts concerned
are the Bases and Vertical Angles, *viz.* in the Triangle
LPN, are the Angle LNP, and Sides LP and NP, and
in the Triangle MPN, are the Angle NMP, and the
Sides PM and PN, but because PN is middle Part, it
cannot come in with Radius to possess the first and second
Places in the Proportion, without Transposition or Alteration, which may be thus done—First, by Case the
Tenth:

In the
Triangle
$$\begin{cases} \text{LPN : As } Rad. \cdots T. \text{ LP. :: } T. \text{ } C. \text{ LNP } \cdots \\ S. \text{ NP.} \\ \text{MNP : As } Rad. \cdots T. \text{ BM. :: } T. \text{ } C. \text{ MNP } \cdots \\ S. \text{ NP.} \end{cases}$$

Therefore (*by Supplement to* Barrow's *Euclid*)

In the
Triangle
$$\begin{cases} \text{LNP : As } T. \text{ } C. \text{ PL : } Rad. \text{ :: } T. \text{ } C. \text{ LNP } \cdots \\ S. \text{ NP.} \\ \text{MPN : As } T.C. \text{ PM } \cdots Rad. \text{ :: } T.C. \text{ MNP } \cdots \\ S. \text{ NP.} \end{cases}$$

By Permutation, *Euclid Lib.* 5. *Def.* 12.

In the
Triangle
$$\begin{cases} \text{LPN : As } T. \text{ } C. \text{ LP } \cdots T.C. \text{ LNP :: } Rad. \cdots \\ S. \text{ NP.} \\ \text{MNP : As } T.C. \text{ PM } \cdots T.C. \text{ MNP :: } Rad. \cdots \\ S. \text{ NP.} \end{cases}$$

Destroy the two last Terms, being both the same in
each, and the remaining Proportion will be,

As

As $T. C. LP \cdot T. C. LNP :: T. C. PM \cdot T. C. MNP.$

But as the Tangent Complement of any Arch is to the Tangent Complement of any other Arch, so the Tangent of the latter Arch to the Tangent of the former, (because every Tangent multiplied by its Complement produceth the Square of Radius) therefore it is;

As $T. C. LP \cdot T. C. LNP :: T. LNP \cdot T. LP.$

And for the same Reason,

As $T. C. PM \cdot T. C. MNP :: T. PM \cdot T. MNP.$

Therefore in an inverse Proportion,

As $T. LNP \cdot T. LP :: T. LMP \cdot T. MP.$

That is, as the Tangent of one Segment of the Base is to the Tangent of its Vertical Angle; so is the Tangent of the other Segment of the Base, to the Tangent of its Vertical Angle. *Which was to be demonstrated.*

These Conclusions are sufficient in all Cases of Oblique Spherical Triangles, and I thought it more proper to demonstrate them here, than to interrupt the Learner with a Demonstration in the Practice of the 28 Cases of Spherical Triangles.

Of *Quadrantal Triangles.*

As Right-angled Triangles are so called, because they have one Angle, that is a Quadrant or 90 Degrees; so the Quadrantal one takes that Name from their having one Side of a Quadrant or 90 Degrees, as in the Triangle A⊙P, the Side A P being a Quadrant or 90 Degrees, or in Astronomical Terms (which we shall explain hereafter) the Side A ⊙ is the Sun's *Fig. 65.* Longitude, the Sine ⊙P is the Complement of the Sun's Declination, and AP is a Quadrant, or the Pole's Distance from the Equinoctial, &c.

These

These Quadrantal Triangles are solved by the same universal Proportion, as Right-angled ones are, if we use the Sides of the Quadrantals as we do the Angles of the Right-angled, *viz.* as in Right-angled Triangles, the Right Angle is excepted, and the other two Angles and the three Sides are called the five circular Parts, so in Quadrantal, the Quadrantal Side is excepted, and the other two Sides and the three Angles make the five circular Parts; but all the Parts whether Sides or Angles (except the Quadrantal Side) are called Conjuncts, Disjuncts and Middle Part, as Right-angled Triangles are, and with the same Limitation, *viz.* that the two Angles next the Quadrantal Side, are called by their Complements, *viz.* if they are Conjuncts they are called Tangent Complements, and Sines if they are Disjuncts; and if Middle Part they are Sine Complements, and in all other Cases they are performed as in Solution 4, with the Caution in Solution 5 and 6, as one Example will sufficiently explain.

Suppose in the Triangle AP⊙ there is given *Fig.* 65. its Sides A⊙ and ⊙P, to find an Angle at ⊙, here A⊙ and ⊙P are Conjuncts, and the Angle ⊙ is Middle Part, and because the Parts given and required are remote from the Quadrantal Side, they are called by their Complements; and hence the Proportion is:

As Rad. ·· *T. C.* A ⊙ : : *T. C.* ⊙ P : : *S. C.* A⊙P.

It will be needless to repeat any more Examples, it being scarce possible, after these Directions, for any that understands Right-angled Spherical Triangles, to mistake in solving the Quadrantal ones.

A

A
TABLE
OF
DIFFERENCE
OF
Latitude and *Departure*
IN
MINUTES and TENTH PARTS,
to every *Degree* and *Quarter-Point*
OF THE
COMPASS,
For the exact working of a
TRAVERSE,
And readily finding the
LONGITUDE
BY
INSPECTION,
According to Middle Latitude.

S

Dist.	1 Deg.		2 Deg.		¼ Point		3 Deg.		4 Deg.		5 Deg.		Dist.
	Lat	Dep	Lat	Dep	Lat	Dep	Lat	Dep	Lat	Dep	Lat	Dep	
1	01.0	00.0	01.0	00.0	01.0	00.0	01.0	00.1	01.0	00.1	01.0	00.1	1
2	02.0	00.0	02.0	00.1	02.0	00.1	02.0	00.1	02.0	00.1	02.0	00.2	2
3	03.0	00.1	03.0	00.1	03.0	00.1	03.0	00.2	03.0	00.1	03.0	00.3	3
4	04.0	00.1	04.0	00.1	04.0	00.2	04.0	00.2	04.0	00.1	04.0	00.3	4
5	05.0	00.1	05.0	00.2	05.0	00.2	05.0	00.3	05.0	00.1	05.0	00.4	5
6	06.0	00.1	06.0	00.2	06.0	00.2	06.0	00.3	06.0	00.4	06.0	00.5	6
7	07.0	00.1	07.0	00.2	07.0	00.1	07.0	00.4	07.0	00.5	07.0	00.6	7
8	08.0	00.1	08.0	00.3	08.0	00.4	08.0	00.4	08.0	00.6	08.0	00.7	8
9	09.0	00.1	09.0	00.3	09.0	00.4	09.0	00.5	09.0	00.6	09.0	00.8	9
10	10.0	00.2	10.0	00.4	10.0	00.5	10.0	00.5	10.0	00.7	10.0	00.9	10
11	11.0	00.2	11.0	00.4	11.0	00.5	11.0	00.6	11.0	00.8	11.0	01.0	11
12	12.0	00.2	12.0	00.4	12.0	00.6	12.0	00.6	12.0	00.8	12.0	01.0	12
13	13.0	00.2	13.0	00.5	13.0	00.6	13.0	00.7	13.0	00.9	12.9	01.1	13
14	14.0	00.2	14.0	00.3	14.0	00.7	14.0	00.7	14.0	01.0	13.9	01.2	14
15	15.0	00.3	15.0	00.5	15.0	00.7	15.0	00.8	15.0	01.0	14.9	01.3	15
16	16.0	00.3	16.0	00.6	16.0	00.8	16.0	00.8	16.0	01.1	15.9	01.4	16
17	17.0	00.3	17.0	00.6	17.0	00.8	17.0	00.9	17.0	01.2	16.9	01.5	17
18	18.0	00.3	18.0	00.6	18.0	00.9	18.0	00.9	18.0	01.1	17.9	01.6	18
19	19.0	00.3	19.0	00.7	19.0	00.9	19.0	01.0	19.0	01.3	18.9	01.7	19
20	20.0	00.4	20.0	00.7	20.0	01.0	20.0	01.0	20.0	01.4	19.9	01.7	20
21	21.0	00.4	21.0	00.7	21.0	01.0	21.0	01.1	20.9	01.5	20.9	01.8	21
22	22.0	00.4	22.0	00.8	22.0	01.1	22.0	01.1	21.9	01.5	21.9	01.9	22
23	23.0	00.4	23.0	00.8	23.0	01.1	23.0	01.2	22.9	01.6	22.9	02.0	23
24	24.0	00.4	24.0	00.8	24.0	01.2	24.0	01.3	23.9	01.7	23.9	02.1	24
25	25.0	00.4	25.0	00.9	25.0	01.2	25.0	01.3	24.9	01.7	24.9	02.2	25
26	26.0	00.5	26.0	00.9	26.0	01.3	26.0	01.4	25.9	01.8	25.9	02.3	26
27	27.0	00.5	27.0	00.9	27.0	01.7	27.0	01.4	26.9	01.9	26.9	02.4	27
28	28.0	00.5	28.0	01.0	28.0	01.4	28.0	01.5	27.9	02.0	27.9	02.4	28
29	29.0	00.5	29.0	01.0	29.0	01.4	29.0	01.5	28.9	02.0	28.9	02.5	29
30	30.0	00.5	30.0	01.2	30.0	01.5	30.0	01.6	29.9	02.1	29.9	02.6	30
31	31.0	00.5	31.0	01.1	31.0	01.5	31.0	01.6	30.9	02.1	30.9	02.7	31
32	32.0	00.6	32.0	01.1	32.0	01.6	32.0	01.7	31.9	02.2	31.9	02.8	32
33	33.0	00.6	33.0	01.2	33.0	01.6	33.0	01.7	32.9	02.3	32.9	02.9	33
34	34.0	00.6	34.0	01.2	34.0	01.7	34.0	01.8	33.9	02.4	33.9	03.0	34
35	35.0	00.6	35.0	01.2	35.0	01.7	35.0	01.8	34.9	02.4	34.9	03.1	35
36	36.0	00.6	36.0	01.3	36.0	01.8	36.9	01.9	35.9	02.5	35.9	03.1	36
37	37.0	00.6	37.0	01.3	37.0	01.9	36.9	01.9	36.9	02.6	36.9	03.2	37
38	38.0	00.7	38.0	01.3	38.0	01.9	37.9	02.0	37.9	02.7	37.9	03.3	38
39	39.0	00.7	39.0	01.4	39.0	01.9	38.9	02.0	38.9	02.7	38.9	03.4	39
40	40.0	00.7	40.0	01.4	40.0	02.0	39.9	02.1	39.9	02.8	39.9	03.5	40
41	41.0	00.7	41.0	01.4	41.0	02.1	41.0	02.1	40.9	02.9	40.9	03.6	41
42	42.0	00.7	42.0	01.5	41.9	02.1	41.9	02.2	41.9	02.9	41.9	03.7	42
43	43.0	00.8	43.0	01.5	42.9	02.1	42.9	02.2	42.9	03.0	42.8	03.8	43
44	44.0	00.8	44.0	01.5	43.9	02.2	43.9	02.3	43.9	03.1	43.8	03.8	44
45	44.0	00.8	44.0	01.6	44.9	02.2	44.9	02.4	44.9	03.1	44.8	03.9	45
46	46.0	00.8	46.0	01.6	45.9	02.3	45.9	02.4	45.9	03.2	45.8	04.0	46
47	47.0	00.8	47.0	01.6	46.9	02.3	46.9	02.5	46.9	03.3	46.8	04.1	47
48	48.0	00.8	48.0	01.7	47.9	02.4	47.9	02.5	47.9	03.3	47.8	04.2	48
49	49.0	00.8	49.0	01.7	48.9	02.4	48.9	02.6	48.9	03.4	48.8	04.3	49
50	50.0	00.9	50.0	01.7	49.9	02.5	49.9	02.6	49.9	03.5	49.8	04.4	50
Dist.	Dep	Lat	Dep	Lat	Dep	Lat	Dep	Lat	Dep	Lat	Dep	Lat	Dist.
	89 Deg.		88 Deg.		7¾ Point		87 Deg		86 Deg.		85 Deg.		

Dist.	1 Deg.		2 Deg.		¼ Point		3 Deg.		4 Point		5 Deg.		Dist.
	Lat	Dep	Lat	Dep	Lat	Dep	Lat	Dep	Lat	Dep	Lat	Dep	
51	51.0	00.9	51.0	01.8	50.9	02.5	50.9	02.7	50.9	03.6	50.8	04.4	51
52	52.0	00.9	52.0	01.8	51.9	02.6	51.9	02.7	51.9	03.6	51.8	04.5	52
53	53.0	00.9	53.0	01.8	52.9	02.6	52.9	02.8	52.9	03.7	52.8	04.6	53
54	54.0	00.9	54.0	01.9	53.9	02.7	53.9	02.8	53.9	03.8	53.8	04.7	54
55	55.0	00.9	55.0	01.9	54.9	02.7	54.9	02.8	54.9	03.8	54.8	04.8	55
56	56.0	01.0	56.0	02.0	55.9	02.7	55.9	02.9	55.9	03.9	55.8	04.9	56
57	57.0	01.0	57.0	02.0	56.9	02.8	56.9	03.0	56.9	04.0	56.8	05.0	57
58	58.0	01.0	58.0	02.0	57.9	02.8	57.9	03.0	57.9	04.1	57.8	05.1	58
59	59.0	01.0	59.0	02.1	58.9	02.9	58.9	03.1	58.9	04.1	58.8	05.1	59
60	60.0	01.0	60.0	02.1	59.9	02.9	59.9	03.1	59.9	04.2	59.8	05.2	60
61	61.0	01.1	61.0	02.1	60.9	03.0	60.9	03.2	60.9	04.3	60.8	05.3	61
62	62.0	01.1	62.0	02.2	61.9	03.0	61.9	03.2	61.9	04.4	61.8	05.4	62
63	63.0	01.1	63.0	02.2	62.9	03.1	62.9	03.3	62.9	04.4	62.8	05.5	63
64	64.0	01.1	64.0	02.2	63.9	03.1	63.9	03.3	63.9	04.5	63.8	05.6	64
65	65.0	01.1	65.0	02.3	64.9	03.2	64.9	03.4	64.9	04.5	64.8	05.7	65
66	66.0	01.2	66.0	02.3	65.9	03.2	65.9	03.6	65.9	04.6	65.7	05.8	66
67	67.0	01.2	67.0	02.3	66.9	03.3	66.9	03.5	66.8	04.7	66.7	05.8	67
68	68.0	01.2	68.0	02.4	67.9	03.3	67.9	03.6	67.8	04.7	67.7	05.9	68
69	69.0	01.2	69.0	02.4	68.9	03.4	68.9	03.6	68.8	04.8	68.7	06.0	69
70	70.0	01.2	70.0	02.4	69.9	03.4	69.9	03.7	69.8	04.9	69.7	06.1	70
71	71.0	01.2	71.0	02.5	70.9	03.5	70.9	03.7	70.8	05.0	70.7	06.2	71
72	72.0	01.3	72.0	02.5	71.9	03.5	71.9	03.8	71.8	05.0	71.7	06.3	72
73	73.0	01.3	73.0	02.5	72.9	03.6	72.9	03.8	72.8	05.1	72.7	06.4	73
74	74.0	01.3	74.0	02.6	73.9	03.6	73.9	03.9	73.8	05.2	73.7	06.5	74
75	75.0	01.3	75.0	02.6	74.9	03.7	74.9	03.9	74.8	05.2	74.7	06.5	75
76	76.0	01.3	76.0	02.7	75.9	03.7	75.9	04.0	75.8	05.3	75.7	06.6	76
77	77.0	01.3	77.0	02.7	76.9	03.8	76.9	04.0	76.8	05.4	76.7	06.7	77
78	78.0	01.4	78.0	02.7	77.9	03.8	77.9	04.1	77.7	05.4	77.7	06.8	78
79	79.0	01.4	79.0	02.8	78.9	03.9	78.9	04.1	78.8	05.5	78.7	06.9	79
80	80.0	01.4	80.0	02.8	79.9	03.9	79.9	04.2	79.8	05.6	79.7	07.0	80
81	81.0	01.4	81.0	02.8	80.9	04.0	80.9	04.2	80.8	05.7	80.7	07.1	81
82	82.0	01.4	82.0	02.9	81.9	04.0	81.9	04.3	81.8	05.7	81.7	07.2	82
83	83.0	01.5	83.0	02.9	82.9	04.1	82.9	04.3	82.8	05.8	82.7	07.2	83
84	84.0	01.5	84.0	02.9	83.9	04.1	83.9	04.4	83.8	05.9	83.7	07.3	84
85	85.0	01.5	84.9	03.0	84.9	04.2	84.9	04.4	84.8	05.9	84.7	07.4	85
86	86.0	01.5	85.9	03.0	85.9	04.2	85.9	04.5	85.8	06.0	85.7	07.5	86
87	87.0	01.5	86.9	03.0	86.9	04.3	86.9	04.6	86.8	06.1	86.7	07.6	87
88	88.0	01.5	87.9	03.1	87.9	04.3	87.9	04.6	87.8	06.1	87.7	07.7	88
89	89.0	01.6	88.9	03.1	88.9	04.4	88.9	04.7	88.8	06.2	88.7	07.8	89
90	90.0	01.6	89.9	03.1	89.9	04.4	89.9	04.7	89.8	06.3	89.7	07.8	90
91	91.0	01.6	90.9	03.2	90.9	04.5	90.9	04.8	90.8	06.4	90.7	07.9	91
92	92.0	01.6	91.9	03.2	91.9	04.5	91.9	04.8	91.8	06.4	91.6	08.0	92
93	93.0	01.6	92.9	03.2	92.9	04.6	92.9	04.9	92.8	06.5	92.6	08.1	93
94	94.0	01.6	93.9	03.3	93.9	04.6	93.9	04.9	93.8	06.6	93.6	08.2	94
95	95.0	01.7	94.9	03.3	94.9	04.7	94.9	05.0	94.8	06.6	94.6	08.3	95
96	96.0	01.7	95.9	03.4	95.9	04.7	95.9	05.0	95.8	06.7	95.6	08.4	96
97	97.0	01.7	96.9	03.4	96.9	04.8	96.9	05.1	96.8	06.8	96.6	08.5	97
98	98.0	01.7	97.9	03.4	97.9	04.8	97.9	05.1	97.8	06.8	97.6	08.5	98
99	99.0	01.7	98.9	03.5	98.9	04.9	98.9	05.2	98.8	06.9	98.6	08.6	99
100	100.0	01.7	99.9	03.5	99.9	04.9	99.9	05.2	99.8	07.0	99.6	08.7	100
	Dep	Lat	Dep	Lat	Dep	Lat	Dep	Lat	Dep	Lat	Dep	Lat	Dist.
	89 Deg.		88 Deg.		7¾ Point		87 Deg.		86 Point		85 Deg.		

Dist.	½ Point Lat	½ Point Dep	6 Deg. Lat	6 Deg. Dep	7 Deg. Lat	7 Deg. Dep	8 Deg. Lat	8 Deg. Dep	¼ Point Lat	¼ Point Dep	9 Deg. Lat	9 Deg. Dep	Dist.
1	01.0	00.1	01.0	00.1	01.0	00.1	01.0	00.1	01.0	00.1	01.0	00.2	1
2	02.0	00.2	02.0	00.2	02.0	00.2	02.0	00.3	02.0	00.3	02.0	00.3	2
3	03.0	00.3	03.0	00.3	03.0	00.4	03.0	00.4	03.0	00.4	03.0	00.5	3
4	04.0	00.4	04.0	00.4	04.0	00.5	04.0	00.6	04.0	00.6	03.9	00.6	4
5	05.0	00.5	05.0	00.5	05.0	00.6	05.0	00.7	04.9	00.7	04.9	00.8	5
6	06.0	00.6	06.0	00.6	06.0	00.7	05.9	00.8	05.9	00.9	05.9	00.9	6
7	07.0	00.7	07.0	00.7	06.9	00.9	06.9	01.0	06.9	01.0	06.9	01.1	7
8	08.0	00.8	08.0	00.8	07.9	01.0	07.9	01.1	07.9	01.2	07.9	01.3	8
9	09.0	00.9	08.9	00.9	08.9	01.1	08.9	01.2	08.9	01.3	08.9	01.4	9
10	10.0	01.0	09.9	01.0	09.9	01.2	09.9	01.4	09.9	01.5	09.9	01.6	10
11	10.9	01.1	10.9	01.1	10.9	01.3	10.9	01.5	10.9	01.6	10.9	01.7	11
12	11.9	01.2	11.9	01.3	11.9	01.5	11.9	01.7	11.9	01.8	11.9	01.9	12
13	12.9	01.3	12.9	01.4	12.9	01.6	12.9	01.8	12.9	01.9	12.8	02.0	13
14	13.9	01.4	13.9	01.5	13.9	01.7	13.9	01.9	13.8	02.1	13.8	02.2	14
15	14.9	01.5	14.9	01.6	14.9	01.8	14.9	02.1	14.8	02.2	14.8	02.3	15
16	15.9	01.6	15.9	01.7	15.9	01.9	15.8	02.2	15.8	0.23	15.8	02.5	16
17	16.9	01.7	16.9	01.8	16.9	02.1	16.8	02.4	16.8	0.25	16.8	02.7	17
18	17.9	01.8	17.9	01.9	17.9	02.2	17.8	02.5	17.8	0.26	17.8	02.8	18
19	18.9	01.9	18.9	02.0	18.9	02.3	18.8	02.6	18.8	0.28	18.8	03.0	19
20	19.9	02.0	19.9	02.1	19.8	02.4	19.8	02.8	19.8	0.29	19.8	03.1	20
21	20.9	02.1	20.9	02.2	20.8	02.6	20.8	02.9	20.8	03.1	20.7	03.3	21
22	21.9	02.2	21.9	02.3	21.8	02.7	21.8	03.1	21.8	03.2	21.7	03.4	22
23	22.9	02.3	22.9	02.4	22.8	02.8	22.8	03.2	22.8	03.4	22.7	03.6	23
24	23.9	02.4	23.9	02.5	23.8	02.9	23.8	03.3	23.7	03.5	23.7	03.8	24
25	24.9	02.4	24.9	02.6	24.8	03.0	24.8	03.5	24.7	03.7	24.7	03.9	25
26	25.9	02.5	25.9	02.7	25.8	03.2	25.7	03.6	25.7	03.8	25.7	04.1	26
27	26.9	02.6	26.9	02.8	26.8	03.3	26.7	03.8	26.7	04.0	26.7	04.2	27
28	27.9	02.7	27.8	02.9	27.8	03.4	27.7	03.9	27.7	04.1	27.7	04.4	28
29	28.9	02.8	28.8	03.0	28.8	03.5	28.7	04.0	28.7	04.3	28.6	04.5	29
30	29.9	02.9	29.8	03.1	29.8	03.7	29.7	04.2	29.7	04.4	29.6	04.7	30
31	30.8	03.0	30.8	03.2	30.8	03.8	30.7	04.3	30.7	04.5	30.6	04.9	31
32	31.8	03.1	31.8	03.3	31.8	03.9	31.7	04.5	31.7	04.7	31.6	05.1	32
33	32.8	03.2	32.8	03.4	32.8	04.0	32.7	04.6	32.6	04.8	32.6	05.2	33
34	33.8	03.3	33.8	03.6	33.7	04.1	33.7	04.7	33.6	05.0	33.6	05.4	34
35	34.8	03.4	34.8	03.7	34.7	04.3	34.7	04.9	34.6	05.1	34.6	05.5	35
36	35.8	03.5	35.8	03.8	35.7	04.4	35.6	05.0	35.6	05.3	35.6	05.6	36
37	36.8	03.6	36.8	03.9	36.7	04.5	36.6	05.1	36.6	05.4	36.5	05.8	37
38	37.8	03.7	37.8	04.0	37.7	04.6	37.6	05.3	37.6	05.6	37.5	05.9	38
39	38.8	03.8	38.8	04.1	38.7	04.8	38.6	05.4	38.6	05.7	38.5	06.1	39
40	39.8	03.9	39.8	04.2	39.7	04.9	39.6	05.6	39.6	05.9	39.5	06.3	40
41	40.8	04.0	40.8	04.3	40.7	05.0	40.6	05.7	40.6	06.0	40.5	06.4	41
42	41.8	04.1	41.8	04.4	41.7	05.1	41.6	05.8	41.5	06.2	41.5	06.6	42
43	42.8	04.2	42.8	04.5	42.7	05.2	42.6	06.0	42.5	06.3	42.5	06.7	43
44	43.8	04.3	43.8	04.6	43.7	05.4	43.6	06.1	43.5	06.5	43.5	06.9	44
45	44.8	04.4	44.8	04.7	44.7	05.5	44.6	06.3	44.5	06.6	44.4	07.0	45
46	45.8	04.5	45.7	04.8	45.7	05.6	45.6	06.4	45.5	06.7	45.4	07.2	46
47	46.8	04.6	46.7	04.9	46.6	05.7	46.5	06.6	46.5	06.9	46.4	07.3	47
48	47.8	04.7	47.7	05.0	47.6	05.9	47.5	06.7	47.5	07.0	47.4	07.5	48
49	48.8	04.8	48.7	05.1	48.6	06.0	48.5	06.8	48.5	07.2	48.4	07.7	49
50	49.8	04.9	49.7	05.2	49.6	06.1	49.5	07.0	49.5	07.3	49.4	07.8	50
Dist.	Dep	Lat	Dep	Lat	Dep	Lat	Dep	Lat	Dep	Lat	Dep	Lat	Dist.
	7½ Points		84 Deg.		83 Deg.		82 Deg.		7½ Point		81 Deg.		

Dist.	¼ Point		6 Deg.		7 Deg.		8 Deg.		¼ Point		9 Deg.		Dist.
	Lat	Dep	Lat	Dep	Lat	Dep	Lat	Dep	Lat	Dep	Lat	ep	
51	50.8	05.0	50.7	05.3	50.6	06.2	50.5	07.1	50.6	07.5	50.4	08.0	51
52	51.7	05.1	51.7	05.4	51.6	06.3	51.5	07.2	51.4	07.6	51.1	08.1	52
53	52.7	05.2	52.7	05.5	52.6	06.5	52.5	07.4	52.4	07.8	52.3	08.3	53
54	53.7	05.3	53.7	05.6	53.6	06.6	53.5	07.5	53.4	07.9	53.3	08.4	54
55	54.7	05.4	54.7	05.7	54.6	06.7	54.5	07.7	54.4	08.1	54.1	08.6	55
56	55.7	05.5	55.7	05.9	55.6	06.8	55.5	07.8	55.4	08.2	55.3	08.8	56
57	56.7	05.6	56.7	06.0	56.6	06.9	56.4	07.9	56.4	08.4	56.3	08.9	57
58	57.7	05.7	57.7	06.1	57.6	07.1	57.4	08.1	57.4	08.5	57.5	09.1	58
59	58.7	05.8	58.7	06.2	58.6	07.2	58.4	08.2	58.4	08.7	58.2	09.2	59
60	59.7	05.9	59.7	06.3	59.3	07.3	59.4	08.4	59.4	08.8	59.1	09.4	60
61	60.7	06.0	60.7	06.4	60.5	07.4	60.4	08.5	60.3	08.9	60.1	09.5	61
62	61.7	06.1	61.7	06.5	61.5	07.6	61.4	08.6	61.3	09.1	61.1	09.7	62
63	62.7	06.2	62.7	06.6	62.5	07.7	62.4	08.8	62.3	09.2	62.1	09.9	63
64	63.7	06.3	63.6	06.7	63.5	07.8	63.4	08.9	63.3	09.4	63.2	10.0	64
65	64.7	06.4	64.6	06.8	64.5	07.9	64.4	09.0	64.3	09.5	64.1	10.2	65
66	65.6	06.5	65.6	06.9	65.4	08.1	65.4	09.2	65.3	09.7	65.2	10.3	66
67	66.7	06.6	66.6	07.0	66.5	08.2	66.3	09.3	66.3	09.8	66.2	10.5	67
68	67.7	06.7	67.6	07.1	67.5	08.3	67.3	09.5	67.3	10.0	67.8	10.6	68
69	68.7	06.8	68.6	07.2	68.5	08.4	68.3	09.6	68.3	10.1	68.2	10.8	69
70	69.7	06.9	69.6	07.3	69.5	08.5	69.3	09.7	69.2	10.3	69.1	10.9	70
71	70.7	07.0	70.6	07.4	70.5	08.7	70.3	09.9	70.2	10.4	70.1	11.1	71
72	71.7	07.1	71.6	07.5	71.5	08.8	71.3	10.0	71.2	10.6	71.1	11.3	72
73	72.6	07.2	72.6	07.6	72.5	08.9	72.3	10.2	72.2	10.7	72.1	11.5	73
74	73.6	07.3	73.6	07.7	73.4	09.0	73.3	10.3	73.2	10.8	73.1	11.6	74
75	74.6	07.3	74.6	07.8	74.4	09.1	74.3	10.4	74.2	11.0	74.1	11.7	75
76	75.6	07.4	75.6	07.9	75.4	09.3	75.3	10.6	75.2	11.1	75.1	11.9	76
77	76.6	07.5	76.6	08.0	76.4	09.4	76.3	10.7	76.2	11.3	76.1	12.0	77
78	77.6	07.6	77.6	08.1	77.4	09.5	77.2	10.9	77.2	11.4	77.1	12.2	78
79	78.6	07.7	78.6	08.2	78.4	09.6	78.2	11.0	78.1	11.6	78.1	12.4	79
80	79.6	07.8	79.6	08.4	79.4	09.7	79.2	11.1	79.1	11.7	79.1	12.5	80
81	80.6	07.9	80.6	08.5	80.4	09.9	80.2	11.3	80.1	11.9	80.0	12.7	81
82	81.6	08.0	81.5	08.6	81.4	10.0	81.2	11.4	81.1	12.0	81.0	12.8	82
83	82.6	08.1	82.5	08.7	82.3	10.1	82.1	11.5	82.1	12.2	82.0	13.0	83
84	83.6	08.2	83.5	08.8	83.4	10.2	83.1	11.7	83.1	12.3	83.0	13.1	84
85	84.6	08.3	84.5	08.9	84.4	10.4	84.2	11.8	84.1	12.5	84.0	13.3	85
86	85.6	08.4	85.5	09.0	85.4	10.5	85.2	12.0	85.1	12.6	84.5	13.4	86
87	86.6	08.5	86.5	09.1	86.1	10.6	86.0	12.1	86.0	12.8	85.9	13.6	87
88	87.6	08.6	87.5	09.2	87.1	10.7	87.1	12.2	87.0	12.9	86.9	13.8	88
89	88.6	08.7	88.5	09.3	88.1	10.8	88.1	12.4	88.0	13.1	87.9	13.9	89
90	89.6	08.8	89.5	09.4	89.1	11.0	89.1	12.5	89.0	13.2	88.5	14.1	90
91	90.6	08.9	90.5	09.5	90.1	11.1	90.0	12.6	91.0	13.4	89.5	14.2	91
92	91.6	09.0	91.5	09.6	91.1	11.2	91.1	12.8	91.0	13.5	90.5	14.0	92
93	92.6	09.1	92.5	09.7	92.1	11.3	92.1	12.9	92.0	13.6	91.9	14.5	93
94	93.5	09.2	93.5	09.8	93.1	11.5	93.1	13.1	93.0	13.8	92.8	14.7	94
95	94.5	09.1	94.5	09.9	94.3	11.6	94.1	13.2	94.0	13.9	93.8	14.9	95
96	95.5	09.4	95.5	10.0	95.1	11.7	95.1	13.4	95.0	14.1	94.8	15.0	96
97	96.5	09.5	96.5	10.1	96.1	11.8	96.0	13.5	96.0	14.2	95.8	15.2	97
98	97.5	09.6	97.5	10.2	97.1	11.9	97.0	13.6	96.9	14.4	96.8	15.3	98
99	98.5	09.7	98.5	10.3	98.1	12.1	98.0	13.8	97.9	14.5	97.8	15.5	99
100	99.5	09.8	99.4	10.4	99.2	12.2	99.0	13.9	98.9	14.7	98.8	15.6	100
Diff.	Dep	Lat	Dep	Lat	Dep	Lat	Dep	Lat	Dep	Lat	Lat		Diff.

Diff.	10 Deg.		11 Deg.		1 Point		12 Deg.		13 Deg.		14 Deg.		Diff.
	Lat	Dep	Lat	Dep	Lat	Dep	Lat	Dep	Lat	Dep	Lat	Dep	
1	01.0	00.2	01.0	00.2	01.0	00.2	01.0	00.2	01.0	00.2	01.0	00.2	1
2	02.0	00.3	02.0	00.4	02.0	00.4	02.0	00.4	01.9	00.4	01.9	00.5	2
3	03.0	00.5	02.9	00.6	02.9	00.6	02.9	00.6	02.9	00.7	02.9	00.7	3
4	03.9	00.7	03.9	00.8	03.9	00.8	03.9	00.8	03.9	00.9	03.9	01.0	4
5	04.9	00.9	04.9	01.0	04.9	01.0	04.9	01.0	04.9	01.1	04.9	01.2	5
6	05.9	01.0	05.9	01.1	05.9	01.2	05.9	01.2	05.8	01.3	05.8	01.4	6
7	06.9	01.2	06.9	01.3	06.9	01.4	06.8	01.5	06.8	01.6	06.8	01.7	7
8	07.9	01.4	07.9	01.5	07.8	01.6	07.8	01.7	07.8	01.8	07.8	01.9	8
9	08.9	01.6	08.8	01.7	08.8	01.9	08.8	02.0	08.8	02.0	08.7	02.2	9
10	09.8	01.7	09.8	01.9	09.8	02.0	09.8	02.1	09.7	02.2	09.7	02.4	10
11	10.8	01.9	10.8	02.1	10.8	02.1	10.8	02.3	10.7	02.5	10.7	02.7	11
12	11.8	02.1	11.8	02.3	11.8	02.3	11.7	02.5	11.7	02.7	11.6	02.9	12
13	12.8	02.3	12.8	02.5	12.7	02.7	12.7	02.8	12.7	02.9	12.6	03.1	13
14	13.8	02.4	13.7	02.7	13.7	02.7	13.7	02.9	13.6	03.1	13.6	03.4	14
15	14.8	02.6	14.7	02.9	14.7	02.9	14.7	03.1	14.6	03.4	14.6	03.6	15
16	15.8	02.8	15.7	03.1	15.7	03.1	15.6	03.3	15.6	03.6	15.5	03.9	16
17	16.7	03.0	16.7	03.2	16.7	03.3	16.6	03.5	16.6	03.8	16.5	04.1	17
18	17.7	03.1	17.7	03.4	17.7	03.5	17.6	03.7	17.5	04.0	17.5	04.4	18
19	18.7	03.3	18.6	03.6	18.6	03.6	18.6	03.9	18.5	04.3	18.4	04.6	19
20	19.7	03.5	19.6	03.8	19.6	03.9	19.6	04.2	19.5	04.5	19.4	04.8	20
21	20.7	03.6	20.6	04.0	20.6	04.1	20.5	04.4	20.5	04.7	20.4	05.1	21
22	21.7	03.8	21.6	04.2	21.6	04.3	21.5	04.6	21.4	04.9	21.3	05.3	22
23	22.6	04.0	22.6	04.4	22.6	04.4	22.5	04.7	22.4	05.2	22.3	05.6	23
24	23.6	04.2	23.6	04.6	23.5	04.7	23.5	05.0	23.4	05.4	23.3	05.8	24
25	24.6	04.3	24.5	04.8	24.5	04.9	24.5	05.2	24.4	05.6	24.3	06.0	25
26	25.6	04.5	25.5	05.0	25.5	05.1	25.4	05.4	25.3	05.8	25.2	06.3	26
27	26.6	04.7	26.5	05.2	26.5	05.3	26.4	05.6	26.3	06.1	26.2	06.5	27
28	27.6	04.9	27.5	05.3	27.5	05.5	27.4	05.8	27.3	06.3	27.2	06.8	28
29	28.6	05.0	28.5	05.5	28.4	05.7	28.4	06.0	28.3	06.5	28.1	07.0	29
30	29.5	05.2	29.4	05.7	29.4	05.9	29.3	06.2	29.2	06.7	29.1	07.3	30
31	30.5	05.4	30.4	05.9	30.4	06.0	30.3	06.4	30.2	07.0	30.1	07.5	31
32	31.5	05.6	31.4	06.1	31.4	06.2	31.3	06.7	31.2	07.2	31.0	07.7	32
33	32.5	05.7	32.4	06.3	32.4	06.4	32.3	06.9	32.2	07.4	32.0	08.0	33
34	33.5	05.9	33.4	06.5	33.3	06.6	33.3	07.1	33.1	07.6	33.0	08.2	34
35	34.5	06.1	34.4	06.7	34.3	06.8	34.2	07.3	34.1	07.9	34.0	08.5	35
36	35.4	06.3	35.3	06.9	35.3	07.0	35.2	07.5	35.1	08.1	34.9	08.7	36
37	36.4	06.4	36.3	07.1	36.2	07.2	36.2	07.7	36.1	08.3	35.9	09.0	37
38	37.4	06.6	37.3	07.2	37.3	07.4	37.2	07.9	37.0	08.5	36.9	09.2	38
39	38.4	06.8	38.3	07.4	38.3	07.6	38.1	08.1	38.0	08.8	37.8	09.4	39
40	39.4	06.9	39.3	07.6	39.2	07.8	39.1	08.3	39.0	09.0	38.8	09.7	40
41	40.4	07.1	40.2	07.8	40.2	08.0	40.1	08.5	39.9	09.2	39.8	09.9	41
42	41.4	07.3	41.2	08.0	41.2	08.2	41.1	08.7	40.9	09.4	40.8	10.2	42
43	42.3	07.5	42.2	08.2	42.2	08.4	42.1	08.9	41.9	09.7	41.7	10.4	43
44	43.3	07.6	43.2	08.4	43.2	08.4	43.0	09.1	42.9	09.9	42.7	10.6	44
45	44.3	07.8	44.2	08.6	44.1	08.8	44.0	09.4	43.8	10.1	43.7	10.9	45
46	45.3	08.0	45.2	08.8	45.1	09.0	45.0	09.6	44.8	10.2	44.6	11.1	46
47	46.3	08.2	46.1	09.0	46.1	09.2	46.0	09.8	45.8	10.6	45.6	11.4	47
48	47.3	08.3	47.1	09.2	47.1	09.4	47.0	10.2	46.8	10.8	46.6	11.6	48
49	48.3	08.5	48.1	09.3	48.1	09.6	47.9	10.2	47.7	11.0	47.5	11.9	49
50	49.2	08.7	49.1	09.5	49.0	09.8	48.9	10.4	48.7	11.2	48.5	12.1	50
Diff.	Dep	Lat	Dep	Lat	Dep	Lat	Dep	Lat	Dep	Lat	Dep	Lat	Diff.
	80 Deg.		79 Deg.		7 Points		78 Deg.		77 Deg.		76 Deg.		

Dist	10 Deg.		11 Deg.		1 Point		12 Deg.		13 Deg.		14 Deg.		Dist
	Lat	Dep	Lat	Dep	Lat	Dep	Lat	Dep	Lat	Dep	Lat	Dep	
51	50.3	08.9	50.1	09.7	50.0	10.0	49.9	10.6	49.7	11.5	49.5	12.3	51
52	51.2	09.0	51.0	09.9	51.0	10.1	50.9	10.8	50.7	11.7	50.5	12.6	52
53	52.2	09.2	52.0	10.1	52.0	10.3	51.8	11.0	51.6	11.9	51.4	12.8	53
54	53.2	09.4	53.0	10.3	53.0	10.5	52.8	11.2	52.6	12.1	52.4	13.1	54
55	54.2	09.5	54.0	10.5	54.0	10.7	53.8	11.4	53.6	12.4	53.4	13.3	55
56	55.1	09.7	55.0	10.7	54.9	10.9	54.8	11.6	54.6	12.6	54.3	13.5	56
57	56.1	09.9	56.0	10.9	55.9	11.1	55.8	11.8	55.5	12.8	55.3	13.8	57
58	57.1	10.1	56.9	11.1	56.8	11.3	56.7	12.0	56.5	13.0	56.3	14.0	58
59	58.1	10.2	57.9	11.3	57.9	11.5	57.7	12.2	57.5	13.3	57.2	14.3	59
60	59.1	10.4	58.9	11.4	58.8	11.7	58.7	12.5	58.5	13.5	58.2	14.5	60
61	60.1	10.6	59.9	11.6	59.8	11.9	59.7	12.7	59.4	13.7	59.2	14.8	61
62	61.1	10.8	60.9	11.8	60.8	12.1	60.6	12.9	60.4	13.9	60.2	15.0	62
63	62.0	10.9	61.8	12.0	61.8	12.3	61.6	13.1	61.4	14.2	61.1	15.2	63
64	63.0	11.1	62.8	12.2	62.7	12.5	62.6	13.3	62.4	14.4	62.1	15.5	64
65	64.0	11.3	63.8	12.4	63.7	12.7	63.6	13.5	63.3	14.6	63.1	15.7	65
66	65.0	11.5	64.8	12.6	64.7	12.9	64.6	13.7	64.3	14.8	64.0	16.0	66
67	66.0	11.6	65.8	12.8	65.7	13.1	65.5	13.9	65.3	15.1	65.0	16.2	67
68	67.0	11.8	66.8	13.0	66.7	13.3	66.5	14.1	66.3	15.3	66.0	16.4	68
69	68.0	12.0	67.7	13.2	67.7	13.5	67.5	14.3	67.2	15.5	66.9	16.7	69
70	68.9	12.2	68.7	13.4	68.7	13.7	68.5	14.5	68.2	15.7	67.9	16.9	70
71	69.9	12.3	69.7	13.5	69.6	13.9	69.4	14.8	69.2	16.0	68.9	17.2	71
72	70.9	12.5	70.7	13.7	70.6	14.0	70.4	15.0	70.2	16.2	69.9	17.4	72
73	71.9	12.7	71.7	13.9	71.6	14.2	71.4	15.2	71.1	16.4	70.8	17.7	73
74	72.9	12.8	72.6	14.1	72.6	14.4	72.4	15.4	72.1	16.7	71.8	17.9	74
75	73.9	13.0	73.6	14.3	73.6	14.6	73.3	15.6	73.1	16.9	72.8	18.1	75
76	74.8	13.2	74.6	14.5	74.5	14.8	74.3	15.8	74.1	17.1	73.8	18.4	76
77	75.8	13.4	75.6	14.7	75.5	15.0	75.3	16.0	75.0	17.3	74.7	18.6	77
78	76.8	13.5	76.6	14.9	76.5	15.2	76.3	16.2	76.0	17.5	75.7	18.9	78
79	77.8	13.7	77.5	15.1	77.5	15.4	77.2	16.4	77.0	17.8	76.7	19.1	79
80	78.8	13.9	78.5	15.3	78.5	15.6	78.2	16.6	78.0	18.0	77.6	19.4	80
81	79.8	14.1	79.5	15.5	79.4	15.8	79.2	16.8	78.9	18.2	78.6	19.6	81
82	80.8	14.2	80.5	15.6	80.4	16.0	80.1	17.0	79.9	18.4	79.6	19.7	82
83	81.7	14.4	81.5	15.8	81.4	16.2	81.1	17.3	80.9	18.7	80.5	20.1	83
84	82.7	14.6	82.5	16.0	82.4	16.4	82.1	17.5	81.8	18.9	81.5	20.3	84
85	83.7	14.8	83.4	16.2	83.4	16.6	83.1	17.7	82.8	19.1	82.5	20.6	85
86	84.7	14.9	84.4	16.4	84.3	16.8	84.1	17.9	83.8	19.3	83.4	20.8	86
87	85.7	15.1	85.4	16.6	85.3	17.0	85.1	18.1	84.8	19.6	84.4	21.0	87
88	86.7	15.3	86.4	16.8	86.3	17.2	86.1	18.3	85.7	19.8	85.4	21.3	88
89	87.6	15.4	87.4	17.0	87.4	17.4	87.1	18.5	86.8	20.0	86.4	21.5	89
90	88.6	15.6	88.3	17.2	88.3	17.6	88.0	18.7	87.7	20.2	87.3	21.8	90
91	89.6	15.8	89.3	17.4	89.3	17.8	89.0	18.9	88.7	20.5	88.3	22.0	91
92	90.6	16.0	90.3	17.6	90.2	17.9	90.0	19.1	89.6	20.7	89.3	22.3	92
93	91.6	16.1	91.3	17.7	91.2	18.1	91.0	19.3	90.6	20.9	90.2	22.5	93
94	92.6	16.3	92.3	17.9	92.2	18.3	91.9	19.5	91.6	21.1	91.2	22.7	94
95	93.6	16.5	93.3	18.1	93.2	18.5	92.9	19.7	92.6	21.4	92.2	23.0	95
96	94.5	16.7	94.2	18.3	94.2	18.7	93.9	20.0	93.5	21.6	93.1	23.2	96
97	95.5	16.8	95.2	18.5	95.1	18.9	94.9	20.2	94.5	21.8	94.1	23.5	97
98	96.5	17.0	96.2	18.7	96.1	19.1	95.9	20.4	95.1	22.0	95.1	23.7	98
99	97.5	17.2	97.2	18.9	97.1	19.3	96.8	20.6	96.5	22.3	96.1	23.9	99
100	98.5	17.4	98.2	19.1	98.1	19.5	97.8	20.8	97.4	22.5	97.0	24.2	100
Dist	Dep	Lat	Dep	Lat	Dep	Lat	Dep	Lat	Dep	Lat	Dep	Lat	Dist
	80 Deg.		79 Deg.		7 Points		78 Deg.		77 Deg.		76 Deg.		

Diff	1¼ Point		15 Deg.		16 Deg.		1⅜ Point		17 Deg.		18 Deg		Diff
	Lat	Dep	Lat	Dep	Lat	Dep	Lat	Dep	Lat	Dep	Lat	De	
1	01.0	00.2	01.0	00.3	01.0	00.3	01.0	00.3	01.0	00.3	01.0	00.3	1
2	01.9	00.5	01.9	00.5	01.9	00.6	01.9	00.6	01.9	00.6	01.9	00.6	2
3	02.9	00.7	02.9	00.8	02.5	00.8	02.9	00.9	02.9	00.9	02.9	00.9	3
4	03.9	01.0	03.9	01.0	03.8	01.1	03.8	01.2	03.8	01.2	03.8	01.2	4
5	04.8	01.2	04.8	01.3	04.8	01.4	04.8	01.5	04.8	01.5	04.8	01.5	5
6	05.8	01.5	05.8	01.6	05.6	01.7	05.7	01.7	05.7	01.8	05.7	01.y	6
7	06.8	01.7	06.8	01.8	06.7	01.9	06.7	02.0	06.7	02.0	06.7	02.2	7
8	07.8	01.9	07.7	02.1	07.7	02.2	07.7	02.3	07.6	02.3	07.6	02.5	8
9	08.7	02.2	08.7	02.3	08.7	02.5	08.6	02.6	08.6	02.6	08.6	02.8	9
10	09.7	02.4	09.7	02.6	09.6	02.8	09.6	02.9	09.6	02.9	09.5	03.1	10
11	10.7	02.7	10.6	02.8	10.6	03.0	10.5	03.1	10.5	03.2	10.5	03.4	11
12	11.6	02.9	11.6	03.1	11.5	03.3	11.5	03.5	11.5	03.5	11.4	03.7	12
13	12.6	03.2	12.6	03.4	11.5	03.6	12.4	03.8	11.4	03.8	12.4	04.0	13
14	13.6	03.4	13.5	03.6	13.5	03.9	13.4	04.1	13.4	04.1	13.3	04.3	14
15	14.5	03.6	14.5	03.9	14.4	04.1	14.4	04.4	14.3	04.4	14.3	04.6	15
16	15.5	03.9	15.5	04.1	15.4	04.4	15.2	04.6	15.3	04.7	15.2	04.9	16
17	16.5	04.1	16.4	04.4	16.3	04.7	16.3	04.9	16.3	05.0	16.2	05.3	17
18	17.5	04.4	17.4	04.7	17.3	05.0	17.2	05.2	17.2	05.3	17.1	05.6	18
19	18.4	04.6	18.4	04.9	18.3	05.2	18.2	05.5	18.2	05.6	18.1	05.9	19
20	19.4	04.9	19.3	05.2	19.2	05.5	19.1	05.8	19.1	05.8	19.0	06.2	20
21	20.4	05.1	20.3	05.4	20.2	05.8	20.1	06.1	20.1	06.1	20.0	06.5	21
22	21.3	05.3	21.2	05.7	21.1	06.1	21.1	06.4	21.0	06.4	20.9	06.8	22
23	22.3	05.6	22.2	06.0	22.1	06.3	22.0	06.7	22.0	06.7	21.9	07.1	23
24	23.3	05.8	23.2	06.2	23.1	06.6	23.0	07.0	22.9	07.0	22.8	07.4	24
25	24.2	06.1	24.1	06.5	24.0	06.9	23.9	07.3	23.9	07.3	23.8	07.7	25
26	25.2	06.3	25.1	06.7	25.0	07.2	24.9	07.5	24.9	07.6	24.7	08.0	26
27	26.2	06.6	26.1	07.0	26.0	07.4	25.8	07.8	25.8	07.9	25.7	08.3	27
28	27.2	06.8	27.0	07.2	26.9	07.7	26.8	08.1	26.8	08.2	26.6	08.7	28
29	28.1	07.0	28.0	07.5	27.9	08.0	27.8	08.4	27.7	08.5	27.6	09.0	29
30	29.1	07.3	29.0	07.8	28.8	08.3	28.7	08.7	28.7	08.7	28.5	09.3	30
31	30.1	07.5	29.9	08.0	29.8	08.5	29.7	09.0	29.6	09.1	29.5	09.6	31
32	31.0	07.8	30.9	08.3	30.8	08.8	30.6	09.3	30.6	09.4	30.4	10.0	32
33	32.0	08.0	31.9	08.5	31.7	09.1	31.6	09.6	31.6	09.6	31.4	10.3	33
34	33.0	08.3	32.8	08.8	32.7	09.4	32.5	09.9	32.5	09.9	32.3	10.5	34
35	33.9	08.5	33.8	09.1	33.6	09.6	33.5	10.2	33.5	10.2	33.3	10.8	35
36	34.9	08.7	34.6	09.3	34.6	09.9	34.4	10.4	34.4	10.5	34.2	11.1	36
37	35.9	09.0	35.7	09.6	35.6	10.2	35.4	10.7	35.2	11.5	35.1	11.5	37
38	36.9	09.2	36.7	09.8	36.5	10.5	36.4	11.0	36.3	11.1	36.1	11.7	38
39	37.8	09.5	37.7	10.1	37.5	10.7	37.3	11.3	37.3	11.4	37.1	12.0	39
40	38.8	09.7	38.6	10.4	38.5	11.0	38.3	11.6	38.3	11.7	38.0	12.4	40
41	39.8	10.0	39.5	10.6	39.4	11.3	39.2	11.9	39.2	12.0	39.0	12.7	41
42	40.7	10.2	40.6	10.9	40.4	11.6	40.2	12.2	40.2	12.3	39.9	13.0	42
43	41.7	10.4	41.6	11.1	41.3	11.8	41.1	12.5	41.1	12.6	40.9	13.3	43
44	42.7	10.7	42.5	11.4	42.3	12.1	42.1	12.8	41.1	12.9	41.8	13.6	44
45	43.6	10.9	43.5	11.6	43.3	12.4	43.1	13.1	43.0	13.2	42.8	13.9	45
46	44.6	11.2	44.4	11.9	44.2	12.7	44.0	13.4	44.0	13.4	43.7	14.2	46
47	45.6	11.4	45.4	12.1	45.2	13.0	45.0	13.6	44.9	13.7	44.7	14.5	47
48	46.6	11.7	46.4	12.4	46.1	13.2	45.9	13.9	45.9	14.0	45.7	14.8	48
49	47.5	11.9	47.4	12.7	47.1	13.5	46.9	14.2	46.9	14.3	46.6	15.1	49
50	48.5	12.1	48.3	12.9	48.1	13.8	47.8	14.5	47.8	14.6	47.6	15.4	50
Diff	Dep	Lat	Dep	Lat	Dep	Lat	Dep	Lat	Dep	Lat	Dep	Lat	Diff
	6¾ Point		75 Dég.		74 Deg		6⅝ Point		73 Deg.		72 Deg.		

Dist.	1¼ Point Lat	Dep	15 Deg. Lat	Dep	16 Deg. Lat	Dep	1⅜ Point Lat	Dep	17 Deg. Lat	Dep	18 Deg. Lat	Dep	Dist.
51	49.5	12.4	49.3	11.2	49.0	14.1	48.8	14.8	48.8	14.9	48.5	15.8	51
52	50.4	12.6	50.2	11.5	50.0	14.3	49.7	15.1	49.7	15.2	49.4	16.1	52
53	51.4	12.9	51.1	11.7	50.9	14.6	50.7	15.3	50.7	15.5	50.4	16.4	53
54	52.4	13.1	52.2	14.0	51.9	14.9	51.7	15.7	51.6	15.8	51.3	16.7	54
55	53.3	13.4	53.1	14.2	52.9	15.2	52.6	16.0	52.6	16.1	52.3	17.0	55
56	54.3	13.6	54.1	14.5	53.8	15.4	54.1	16.3	53.5	16.4	53.3	17.1	56
57	55.3	13.9	55.1	14.8	54.8	15.7	54.1	16.5	54.1	16.7	54.2	17.6	57
58	56.3	14.1	56.0	15.0	55.8	16.0	55.5	16.8	55.5	17.0	55.2	17.9	58
59	57.2	14.3	57.0	15.3	56.7	16.3	56.5	17.1	56.4	17.5	56.1	18.2	59
60	58.2	14.0	58.0	15.5	57.7	16.5	57.4	17.4	57.4	17.8	57.1	18.5	60
61	59.2	14.8	58.9	15.8	58.6	16.8	58.4	17.7	58.3	17.8	58.0	18.8	61
62	60.1	15.1	59.9	16.0	59.6	17.1	59.3	18.0	59.1	18.1	59.0	19.2	62
63	61.1	15.1	60.9	16.3	60.6	17.4	60.3	18.3	60.2	18.4	59.9	19.5	63
64	62.1	15.0	61.8	16.6	61.3	17.6	61.2	18.6	61.2	18.7	60.9	19.8	64
65	63.0	15.1	62.8	16.8	62.5	17.9	62.2	18.9	62.2	19.0	61.8	20.1	65
66	64.0	16.0	63.7	17.1	63.4	18.2	63.2	19.2	63.1	19.3	62.8	20.4	66
67	65.0	16.3	64.7	17.3	64.4	18.5	64.1	19.4	64.1	19.6	63.7	20.7	67
68	66.0	16.3	65.7	17.6	65.4	18.7	65.1	19.7	65.0	19.9	64.7	21.0	68
69	66.9	16.8	66.6	17.9	66.3	19.0	66.0	20.0	66.0	20.3	65.6	21.3	69
70	67.9	17.0	67.6	18.1	67.3	19.3	67.0	20.3	66.9	20.5	66.6	21.6	70
71	68.9	17.3	68.6	18.4	68.3	19.6	67.9	20.6	67.9	20.8	67.5	21.9	71
72	69.8	17.5	69.5	18.6	69.2	19.8	68.9	20.9	68.8	21.1	68.5	22.2	72
73	70.8	17.7	70.5	18.9	70.2	20.1	69.8	21.2	69.8	21.4	69.4	22.6	73
74	71.8	18.0	71.5	19.2	71.1	20.4	70.8	21.5	70.8	21.6	70.4	22.9	74
75	72.7	18.2	72.4	19.4	72.1	20.7	71.8	21.8	71.7	21.9	71.3	23.2	75
76	73.7	18.5	73.4	19.7	73.1	20.9	72.7	22.1	72.7	22.2	72.3	23.5	76
77	74.7	18.7	74.4	19.9	74.0	21.2	73.6	22.4	73.6	22.5	73.2	23.8	77
78	75.7	18.9	75.3	20.2	75.0	21.5	74.6	22.6	74.6	22.8	74.2	24.1	78
79	76.6	19.2	76.3	20.4	75.9	21.8	75.6	22.9	75.5	23.1	75.1	24.4	79
80	77.6	19.4	77.3	22.0	76.9	22.0	76.6	23.2	76.5	23.4	76.1	24.7	80
81	78.6	19.7	78.2	21.0	77.9	22.3	77.5	23.5	77.5	23.7	77.0	25.0	81
82	79.5	15.9	79.2	21.2	78.8	22.6	78.5	23.8	78.4	24.0	78.0	25.3	82
83	80.5	20.2	80.2	21.5	79.8	22.9	79.4	24.1	79.4	24.3	78.9	25.6	83
84	81.5	20.4	81.1	21.7	80.8	25.1	80.4	24.3	80.3	24.5	79.9	26.0	84
85	82.4	20.7	82.1	22.0	81.7	23.4	81.3	24.7	81.3	24.8	80.8	26.3	85
86	83.4	20.9	83.1	22.3	82.7	23.7	82.3	25.0	82.2	25.1	81.8	26.6	86
87	84.4	21.1	84.0	22.5	83.6	24.0	83.3	25.3	83.2	25.4	82.7	26.9	87
88	85.4	21.4	85.0	22.8	84.6	24.3	84.1	25.5	84.2	25.7	83.7	27.2	88
89	86.3	21.6	86.0	23.0	85.5	24.5	85.2	25.8	85.1	26.0	84.6	27.5	89
90	87.3	21.9	86.9	23.3	86.5	24.8	86.1	26.1	86.1	26.3	85.6	27.8	90
91	88.3	22.1	87.9	23.5	87.5	25.1	87.1	26.4	87.0	26.6	86.5	28.1	91
92	89.2	22.4	88.9	23.8	88.4	25.4	88.0	26.7	88.0	26.9	87.5	28.4	92
93	90.2	22.6	89.8	24.1	89.4	25.6	89.0	27.0	88.9	27.2	88.4	28.7	93
94	91.2	22.8	90.8	24.3	90.4	25.9	90.0	27.3	89.9	27.5	89.4	29.0	94
95	92.1	23.1	91.8	24.6	91.3	26.2	90.9	27.6	90.8	27.8	90.4	29.4	95
96	93.1	23.3	92.7	24.8	92.3	26.5	91.9	27.9	91.8	28.1	91.3	29.7	96
97	94.1	23.6	93.7	25.1	93.2	26.7	92.8	28.2	92.8	28.4	92.3	30.0	97
98	95.1	23.8	94.7	25.4	94.2	27.0	93.8	28.5	93.7	28.7	93.2	30.3	98
99	96.0	24.1	95.6	25.6	95.2	27.3	94.7	28.7	94.7	28.9	94.1	30.6	99
100	97.0	24.3	96.6	25.9	96.1	27.6	95.7	29.0	95.6	29.2	95.1	30.9	100

Dep	Lat	Dep	Lat	Dep	Lat	Dep	Lat	Dep	Lat	Dep	Lat	
6¼ Point		75 Deg.		74 Deg.		6⅜ Point		73 Deg.		72 Deg.		

Dist.	19 Deg.		1¼ Point		20 Deg.		21 Deg.		22 Deg.		2 Point		Dist.
	Lat	Dep	Lat	Dep	Lat	Dep	Lat	Dep	Lat	Dep	Lat	Dep	
1	00.9	00.3	00.9	00.3	00.9	00.3	00.9	00.4	00.9	00.4	00.9	00.4	1
2	01.9	00.7	01.9	00.7	01.9	00.7	01.9	00.7	01.9	00.7	91.8	00.8	2
3	02.8	01.0	02.8	01.0	02.8	01.0	02.8	01.1	02.8	01.1	02.8	01.1	3
4	03.8	01.3	03.8	01.3	03.8	01.4	03.7	01.4	03.7	01.5	03.7	01.5	4
5	04.7	01.6	04.7	01.7	04.7	01.7	04.7	01.8	04.6	01.9	04.6	01.9	5
6	05.7	02.0	05.6	01.0	05.6	02.1	05.6	02.1	05.6	02.2	05.5	02.3	6
7	06.6	02.3	06.6	02.4	06.6	02.4	06.5	02.5	06.5	02.6	06.5	02.7	7
8	07.6	02.6	07.5	02.7	07.5	02.7	07.5	02.9	07.4	03.0	07.4	03.1	8
9	08.5	02.9	08.5	03.0	08.5	03.1	08.4	03.2	08.3	03.3	08.3	03.4	9
10	09.5	03.1	09.4	03.4	09.4	03.4	09.3	03.6	09.3	03.7	09.2	03.8	10
11	10.4	03.6	10.4	03.7	10.3	03.8	10.3	03.9	10.2	04.1	10.2	04.2	11
12	11.3	03.9	11.3	04.0	11.3	04.1	11.2	04.3	11.1	04.5	11.1	04.6	12
13	12.3	04.2	12.2	04.4	12.2	04.4	12.2	04.7	12.1	04.9	12.0	05.0	13
14	13.2	04.6	13.2	04.7	13.2	04.8	13.1	05.0	13.0	05.2	12.9	05.4	14
15	14.2	04.9	14.1	05.1	14.1	05.1	14.0	05.4	13.9	05.6	13.9	05.7	15
16	15.1	05.2	15.1	05.4	15.0	05.5	14.9	05.7	14.8	06.0	14.8	06.1	16
17	16.1	05.5	16.0	05.7	16.0	05.8	15.9	06.1	15.8	06.4	15.7	06.5	17
18	17.0	05.9	16.9	06.1	16.9	06.2	16.8	06.5	16.7	06.7	16.6	06.9	18
19	18.0	06.2	17.9	06.4	17.9	06.5	17.8	06.8	17.6	07.1	17.6	07.3	19
20	18.9	06.5	18.8	06.7	18.8	06.8	18.7	07.2	18.5	07.5	18.5	07.7	20
21	19.9	06.8	19.8	07.1	19.7	07.2	19.6	07.5	19.5	07.9	19.4	08.0	21
22	20.8	07.1	20.7	07.4	20.7	07.5	20.5	07.9	20.4	08.2	20.3	08.4	22
23	21.7	07.5	21.7	07.7	21.6	07.9	21.5	08.2	21.3	08.6	21.2	08.8	23
24	22.7	07.8	22.6	08.1	22.6	08.2	22.4	08.6	22.3	09.0	22.2	09.2	24
25	23.6	08.1	23.1	08.4	23.5	08.5	23.3	09.0	23.2	09.4	23.1	09.6	25
26	24.6	08.5	24.5	08.8	24.4	08.9	24.3	09.3	24.1	09.7	24.0	09.9	26
27	25.5	08.8	25.4	09.1	25.4	09.2	25.2	09.7	25.0	10.1	24.9	10.2	27
28	26.5	09.1	26.4	09.4	26.3	09.6	26.1	10.0	26.0	10.5	25.9	10.7	28
29	27.4	09.4	27.3	09.8	27.3	09.9	27.1	10.4	26.9	10.9	26.8	11.1	29
30	28.4	09.5	28.2	10.1	28.2	10.3	28.0	10.8	27.8	11.2	27.7	11.5	30
31	29.3	10.1	29.2	10.4	29.1	10.6	28.9	11.1	28.7	11.6	28.6	11.9	31
32	30.3	10.4	30.1	10.8	30.1	10.9	29.9	11.5	29.7	12.0	29.6	12.3	32
33	31.2	10.7	31.1	11.1	31.0	11.3	30.8	11.8	30.6	12.4	30.5	12.6	33
34	32.1	11.1	32.0	11.5	31.9	11.6	31.7	12.2	31.5	12.7	31.4	13.0	34
35	33.1	11.4	33.0	11.8	32.9	12.0	32.7	12.5	32.5	13.1	32.3	13.4	35
36	34.0	11.7	33.9	12.1	33.8	12.3	33.6	12.9	33.4	13.5	33.3	13.8	36
37	35.0	12.0	34.8	12.5	34.8	12.7	34.5	13.3	34.3	13.9	34.2	14.2	37
38	35.9	12.4	35.8	12.8	35.7	13.0	35.5	13.6	35.2	14.2	35.1	14.5	38
39	36.9	12.7	36.7	13.1	36.6	13.3	36.4	14.0	36.2	14.6	36.0	14.9	39
40	37.8	13.0	37.7	13.5	37.6	13.7	37.4	14.3	37.1	15.0	37.0	15.3	40
41	38.8	13.3	38.6	13.8	38.5	14.0	38.3	14.7	38.0	15.4	37.9	15.7	41
42	39.7	13.7	39.5	14.1	39.5	14.4	39.2	15.1	38.9	15.7	38.8	16.1	42
43	40.6	14.0	40.5	14.5	40.4	14.7	40.1	15.4	39.9	16.1	39.7	16.5	43
44	41.6	14.3	41.4	14.8	41.3	15.0	41.1	15.8	40.8	16.5	40.7	16.8	44
45	42.5	14.7	42.4	15.2	42.3	15.4	42.0	16.1	41.7	16.9	41.6	17.2	45
46	43.5	15.0	43.3	15.5	43.2	15.7	42.9	16.5	42.7	17.2	42.5	17.6	46
47	44.4	15.3	44.2	15.8	44.2	16.1	43.9	16.8	43.6	17.6	43.4	18.0	47
48	45.4	15.6	45.2	16.2	45.1	16.4	44.8	17.2	44.5	18.0	44.3	18.4	48
49	46.3	16.0	46.1	16.5	46.0	16.8	45.7	17.6	45.4	18.4	45.3	18.8	49
50	47.3	16.3	47.1	16.8	47.0	17.1	46.7	17.9	46.3	18.7	46.2	19.1	50
Dist.	Dep	Lat	Dep	Lat	Dep	Lat	Dep	Lat	Dep	Lat	Dep	Lat	Dist.
	71 Deg.		6¾ Point		70 Deg.		69 Deg.		68 Deg.		6 Point		

Dist	19 Deg.		1¾ Point		20 Deg.		21 Deg.		22 Deg.		2 Points		Dist
	Lat	Dep	Lat	Dep	Lat	Dep	Lat	Dep	Lat	Dep	Lat	Dep	
50	48.2	16.6	48.0	17.1	47.9	17.2	47.6	18.3	47.3	19.1	47.1	19.5	50
51	49.2	16.9	49.0	17.5	48.9	17.8	48.5	18.6	48.2	19.5	48.9	19.9	52
52	50.2	17.1	49.9	17.9	49.8	18.1	49.5	19.0	49.1	19.9	49.0	20.3	53
53	51.0	17.6	50.8	18.2	50.7	18.5	50.4	19.4	50.1	20.2	49.9	20.7	54
54	52.0	17.9	51.8	18.5	51.7	18.8	51.3	19.7	51.0	20.6	50.8	21.0	55
55	52.9	18.2	52.7	18.9	52.6	19.2	52.3	20.1	51.9	21.0	51.7	21.4	56
56	53.9	18.6	53.7	19.2	53.6	19.5	53.2	20.4	52.8	21.0	52.7	21.8	57
57	54.8	18.9	54.6	19.5	54.5	19.8	54.1	20.8	53.8	21.7	53.6	22.2	58
58	55.8	19.2	55.5	19.9	55.4	20.2	55.1	21.1	54.7	22.1	54.5	22.6	59
60	56.7	19.5	56.5	20.2	56.4	20.5	56.0	21.5	55.6	22.5	55.4	23.0	60
61	57.7	19.9	57.4	20.6	57.3	20.9	56.9	21.9	56.5	22.8	56.4	23.3	61
62	58.6	20.1	58.4	20.9	58.3	21.2	57.9	22.2	57.5	23.2	57.1	23.7	62
63	59.6	20.5	59.3	21.2	59.2	21.5	58.8	22.6	58.4	23.6	58.2	24.1	63
64	60.5	20.8	60.3	21.6	60.1	21.9	59.7	22.9	59.3	24.0	59.1	24.5	64
65	61.5	21.2	61.2	21.9	61.1	22.2	60.7	23.3	60.3	24.3	60.1	24.9	65
66	62.4	21.5	62.1	22.2	62.0	22.6	61.6	23.7	61.2	24.7	61.0	25.2	66
67	63.3	21.8	63.1	22.6	63.0	22.9	62.6	24.0	62.1	25.1	61.9	25.6	67
68	64.3	22.1	64.0	22.9	63.9	23.3	63.5	24.4	63.1	25.5	62.8	26.0	68
69	65.2	22.5	65.0	23.2	64.8	23.6	64.4	24.7	64.0	25.8	63.7	26.4	69
70	66.2	22.8	65.9	23.6	65.8	23.9	65.4	25.1	64.9	26.2	64.7	26.8	70
71	67.1	23.1	66.8	23.9	66.7	24.3	66.3	25.4	65.8	26.6	65.6	27.2	71
72	68.1	23.4	67.8	24.3	67.7	24.6	67.2	25.8	66.8	27.0	66.5	27.6	72
73	69.0	23.8	68.7	24.6	68.6	25.0	68.2	26.2	67.7	27.3	67.4	27.9	73
74	70.0	24.1	69.7	24.9	69.5	25.3	69.1	26.5	68.6	27.7	68.4	28.3	74
75	70.9	24.4	70.6	25.3	70.5	25.6	70.0	26.9	69.5	28.1	69.3	28.7	75
76	71.9	24.7	71.6	25.6	71.4	26.0	71.0	27.2	70.5	28.5	70.2	29.1	76
77	72.8	25.1	72.5	25.9	72.4	26.3	71.9	27.6	71.4	28.8	71.1	29.5	77
78	73.7	25.4	73.4	26.3	73.3	26.7	72.8	28.0	72.3	29.2	72.1	29.8	78
79	74.7	25.7	74.4	26.6	74.2	27.0	73.8	28.3	73.3	29.6	73.0	30.2	79
80	75.6	26.0	75.3	27.0	75.2	27.4	74.7	28.7	74.2	30.0	73.9	30.6	80
81	76.6	26.4	76.3	27.3	76.1	27.7	75.6	29.0	75.1	30.3	74.8	31.0	81
82	77.5	26.7	77.2	27.6	77.1	28.0	76.6	29.4	76.0	30.7	75.8	31.4	82
83	78.5	27.0	78.2	28.0	78.0	28.4	77.5	29.7	77.0	31.1	76.7	31.8	83
84	79.4	27.3	79.1	28.3	78.9	28.7	78.4	30.1	77.9	31.5	77.6	32.1	84
85	80.4	27.7	80.0	28.6	79.9	29.1	79.4	30.5	78.8	31.8	78.5	32.5	85
86	81.3	28.0	81.0	29.0	80.8	29.4	80.3	30.8	79.7	32.2	79.5	32.9	86
87	82.3	28.3	81.9	29.3	81.8	29.8	81.2	31.2	80.7	32.6	80.4	33.3	87
88	83.2	28.7	82.9	29.6	82.7	30.1	82.2	31.5	81.6	33.0	81.3	33.7	88
89	84.1	29.0	83.8	30.0	83.6	30.4	83.1	31.9	82.5	33.3	82.2	34.1	89
90	85.1	29.3	84.7	30.3	84.6	30.8	84.0	32.2	83.4	33.7	83.2	34.4	90
91	86.0	29.6	85.7	30.7	85.5	31.1	85.0	32.6	84.4	34.1	84.1	34.8	91
92	87.0	30.0	86.6	31.0	86.5	31.5	85.9	33.0	85.3	34.5	85.0	35.2	92
93	87.9	30.3	87.6	31.3	87.4	31.8	86.8	33.3	86.2	34.8	85.9	35.6	93
94	88.9	30.6	88.5	31.7	88.3	32.1	87.8	33.7	87.1	35.2	86.8	36.0	94
95	89.8	30.9	89.4	32.0	89.3	32.5	88.7	34.0	88.1	35.6	87.8	36.4	95
96	90.8	31.3	90.4	32.3	90.2	32.8	89.6	34.4	89.0	36.0	88.7	36.7	96
97	91.7	31.6	91.3	32.7	91.1	33.1	90.6	34.8	89.9	36.3	89.6	37.1	97
98	92.7	31.9	92.3	33.0	92.1	33.5	91.5	35.1	90.9	36.7	90.5	37.5	98
99	93.6	32.3	93.2	33.4	93.0	33.9	92.4	35.5	91.8	37.1	91.5	37.9	99
100	94.5	32.6	94.2	33.7	94.0	34.2	93.4	35.8	92.7	37.5	92.4	38.1	100

Dist.	23 Deg. Lat	23 Deg. Dep	24 Deg. Lat	24 Deg. Dep	25 Deg. Lat	25 Deg. Dep	2¼ Point Lat	2¼ Point Dep	26 Deg. Lat	26 Deg. Dep	27 Deg. Lat	27 Deg. Dep	Dist.
1	00.5	00.4	00.9	00.4	00.9	00.4	00.9	00.4	00.9	00.4	00.9	00.5	1
2	01.8	00.8	01.8	00.8	01.8	00.8	01.8	00.9	01.8	00.9	01.8	00.9	2
3	02.8	01.2	02.7	01.2	02.7	01.3	02.7	01.3	02.7	01.3	02.7	01.4	3
4	03.7	01.6	03.6	01.6	03.6	01.7	03.6	01.7	03.6	01.8	03.6	01.8	4
5	04.6	02.0	04.6	02.0	04.5	02.1	04.5	02.1	04.5	02.2	04.5	02.3	5
6	05.5	02.3	05.5	02.4	05.4	02.5	05.4	02.6	05.4	02.6	05.3	02.7	6
7	06.4	02.7	06.4	02.8	06.3	03.0	06.3	03.0	06.3	03.1	06.2	03.2	7
8	07.4	03.1	07.3	03.3	07.2	03.4	07.2	03.4	07.2	03.5	07.1	03.6	8
9	08.3	03.5	08.2	03.7	08.1	03.8	08.1	03.8	08.1	03.9	08.0	04.1	9
10	09.2	03.9	09.1	04.1	09.1	04.2	09.0	04.3	09.0	04.4	08.9	04.5	10
11	10.1	04.3	10.0	04.5	10.0	04.6	09.9	04.7	09.9	04.8	09.8	05.0	11
12	11.0	04.7	11.0	04.9	10.9	05.1	10.8	05.1	10.8	05.3	10.7	05.4	12
13	11.9	05.1	11.9	05.3	11.8	05.5	11.8	05.6	11.7	05.7	11.6	05.9	13
14	12.9	05.5	12.8	05.7	12.7	05.9	12.7	06.0	12.6	06.1	12.5	06.4	14
15	13.8	05.9	13.7	06.1	13.6	06.3	13.6	06.4	13.5	06.6	13.4	06.8	15
16	14.7	06.2	14.6	06.5	14.5	06.3	14.5	05.8	14.4	06.9	14.3	07.3	16
17	15.6	06.6	15.5	06.9	15.4	07.2	15.4	07.3	15.3	07.5	15.1	07.7	17
18	16.6	07.0	16.4	07.3	16.3	07.6	16.3	07.7	16.2	07.9	16.0	08.2	18
19	17.5	07.4	17.4	07.7	17.2	08.0	17.2	08.1	17.1	08.3	16.9	08.6	19
20	18.4	07.8	18.3	08.1	18.1	08.5	18.1	08.6	18.0	08.8	17.8	09.1	20
21	19.3	08.2	19.2	08.5	19.0	08.9	19.0	09.0	18.9	09.2	18.7	09.5	21
22	20.3	08.6	20.1	08.9	19.9	09.3	19.9	09.4	19.8	09.6	19.6	10.0	22
23	21.2	09.0	21.0	09.4	20.8	09.8	20.8	09.8	20.7	10.1	20.5	10.4	23
24	22.1	09.4	21.9	09.8	21.8	10.1	21.7	10.3	21.6	10.5	21.4	10.9	24
25	23.0	09.8	22.8	10.2	22.7	10.6	22.6	10.7	21.5	11.0	22.3	11.3	25
26	23.9	10.2	23.8	10.6	23.6	11.0	23.5	11.1	23.4	11.4	23.2	11.8	26
27	24.9	10.5	24.7	11.0	24.5	11.4	24.4	11.5	24.3	11.8	24.1	12.3	27
28	25.8	10.9	25.6	11.4	25.4	11.8	25.4	12.0	25.2	12.3	24.9	12.7	28
29	26.7	11.3	26.5	11.8	26.3	12.3	26.2	12.4	26.1	12.7	25.8	13.2	29
30	27.6	11.7	27.4	12.2	27.2	12.7	27.1	12.8	27.0	13.2	26.7	13.6	30
31	28.5	12.1	28.4	12.6	28.1	13.1	28.0	13.3	27.9	13.6	27.6	14.1	31
32	29.5	12.5	29.2	13.0	29.0	13.5	28.9	13.7	28.8	14.0	28.5	14.5	32
33	30.4	12.9	30.1	13.4	29.9	13.9	29.8	14.1	29.7	14.5	29.4	15.0	33
34	31.3	13.3	31.1	13.8	30.8	14.4	30.6	14.5	30.6	14.9	30.3	15.4	34
35	32.3	13.7	32.0	14.2	31.7	14.8	31.6	15.0	31.5	15.3	31.2	15.9	35
36	33.1	14.1	32.9	14.6	32.6	15.2	32.5	15.4	32.4	15.8	32.1	16.3	36
37	34.1	14.5	33.8	15.0	33.5	15.6	33.4	15.8	34.3	16.2	33.0	16.8	37
38	35.0	14.8	34.7	15.5	34.4	16.1	34.4	16.3	34.2	16.7	33.9	17.3	38
39	35.9	15.2	35.6	15.9	35.3	16.5	35.1	16.7	35.1	17.1	34.7	17.7	39
40	36.8	15.6	36.5	16.1	36.3	16.9	36.2	17.1	36.0	17.5	35.6	18.2	40
41	37.7	16.0	37.5	16.7	37.2	17.1	37.1	17.5	36.8	17.9	36.5	18.6	41
42	38.7	16.4	38.4	17.1	38.1	17.7	38.0	18.0	37.7	18.4	37.4	19.1	42
43	39.6	16.8	39.3	17.5	39.0	18.2	38.9	18.4	38.6	18.9	38.3	19.5	43
44	40.5	17.2	40.2	17.9	39.9	18.6	39.8	18.8	39.5	19.3	39.2	20.0	44
45	41.4	17.6	41.1	18.3	40.8	19.0	40.7	19.2	40.4	19.7	40.1	20.4	45
46	42.3	18.0	42.0	18.7	41.7	19.4	41.6	19.7	41.3	20.2	41.0	20.9	46
47	43.3	18.4	42.9	19.1	42.6	19.9	42.3	20.1	42.2	20.6	41.9	21.3	47
48	44.2	18.8	43.8	19.5	43.5	20.3	43.4	20.5	43.1	21.0	42.8	21.8	48
49	45.1	19.2	44.8	19.9	44.4	20.7	44.3	20.9	44.0	21.5	43.7	22.2	49
50	46.0	19.5	45.7	20.3	45.3	21.1	45.2	21.4	44.9	21.9	44.6	22.7	50
	Dep	Lat	Dep	Lat	Dep	Lat	Dep	Lat	Dep	Lat	Dep	Lat	
	67 Deg.		66 Deg.		65 Deg.		5¼ Point		64 Deg.		63 Deg.		

Dift.	23 Deg.		24 Deg.		25 Deg.		2½ Point		26 Deg.		27 Deg.		Dift.
	Lat	Dep	Lat	Dep	Lat	Dep	Lat	Dep	Lat	Dep	Lat	Dep	
51	46.9	19.9	46.6	20.7	46.1	21.6	46.1	21.8	45.8	22.4	45.4	23.2	51
52	47.9	20.3	47.5	21.1	47.1	22.0	47.0	22.2	46.7	22.8	46.1	23.6	52
53	48.8	20.7	48.4	21.6	48.0	22.4	47.9	21.7	47.6	23.2	47.2	24.1	53
54	49.7	21.1	49.3	22.0	48.9	22.8	48.8	23.1	48.5	23.7	48.1	24.5	54
55	50.6	21.5	50.2	22.4	49.8	23.2	49.7	23.5	49.4	24.1	49.0	25.0	55
56	51.5	21.9	51.2	22.8	50.8	23.7	50.6	23.9	50.3	24.5	49.5	25.4	56
57	52.5	22.3	52.1	23.2	51.7	24.1	51.5	24.4	51.2	25.0	50.4	25.4	57
58	53.4	22.7	53.0	23.6	52.6	24.5	52.4	24.8	52.1	25.4	51.2	26.3	58
59	54.3	23.1	53.9	24.0	53.5	24.9	53.2	25.2	53.0	25.9	52.6	26.8	59
60	55.2	23.4	54.8	24.4	54.4	25.4	54.2	25.7	53.9	26.3	53.1	27.2	60
61	56.1	23.8	55.7	24.8	55.3	25.8	55.1	26.1	54.8	26.7	54.0	27.7	61
62	57.1	24.2	56.6	25.2	56.2	26.2	56.0	26.5	55.7	27.1	55.1	28.1	62
63	58.0	24.6	57.5	25.6	57.1	26.6	57.0	26.9	56.6	27.6	56.1	28.6	63
64	58.9	25.0	58.5	26.0	58.0	27.0	57.9	27.4	57.5	28.1	57.0	29.1	64
65	59.8	25.4	59.4	26.4	58.9	27.5	58.8	27.8	58.4	28.5	57.5	29.5	65
66	60.8	25.8	60.3	26.8	59.8	27.9	59.7	28.2	59.3	28.9	58.1	30.0	66
67	61.7	26.2	61.2	27.2	60.7	28.3	60.6	28.6	60.2	29.4	59.7	30.4	67
68	62.6	26.6	62.1	27.7	61.6	28.7	61.5	29.1	61.1	29.8	60.1	30.9	68
69	63.5	27.0	63.0	28.1	62.5	29.2	62.4	29.5	62.0	30.2	61.3	31.3	69
70	64.4	27.3	63.9	28.5	63.4	29.6	63.3	29.9	62.9	30.7	62.4	31.8	70
71	65.4	27.7	64.8	28.9	64.3	30.0	64.2	30.4	63.8	31.1	63.1	32.2	71
72	66.3	28.1	65.8	29.3	65.3	30.4	65.1	30.8	64.7	31.6	64.7	32.7	72
73	67.2	28.5	66.7	29.7	66.2	30.8	66.0	31.2	65.6	32.0	65.0	33.1	73
74	68.1	28.9	67.6	30.1	67.1	31.3	66.9	31.6	66.5	32.4	65.8	33.6	74
75	69.0	29.3	68.5	30.5	68.0	31.7	67.8	32.1	67.4	32.9	66.2	34.1	75
76	70.0	29.7	69.4	30.9	68.9	32.1	68.7	32.5	68.3	33.3	67.2	34.5	76
77	70.9	30.1	70.3	31.3	69.8	32.5	69.6	32.9	69.2	33.8	68.6	35.0	77
78	71.8	30.5	71.3	31.7	70.7	33.0	70.5	33.3	70.1	34.2	69.6	35.4	78
79	72.7	30.9	72.2	32.1	71.6	33.4	71.4	33.8	71.0	34.6	70.4	35.9	79
80	73.6	31.3	73.1	32.5	72.5	33.8	72.3	34.2	71.9	35.1	71.3	36.3	80
81	74.6	31.6	74.0	32.9	73.4	34.2	73.2	34.6	72.8	35.5	72.2	36.8	81
82	75.5	32.0	74.9	33.3	74.3	34.7	74.1	35.1	73.7	35.9	73.1	37.2	82
83	76.4	32.4	75.8	33.7	75.2	35.1	75.0	35.5	74.6	36.4	74.0	37.7	83
84	77.3	32.8	76.7	34.1	76.1	35.5	75.9	35.9	75.5	36.8	74.8	38.1	84
85	78.2	33.2	77.6	34.6	77.0	35.9	76.8	36.1	76.4	37.3	75.7	38.6	85
86	79.2	33.6	78.6	35.0	77.9	36.3	77.7	36.8	77.3	37.7	76.4	39.0	86
87	80.1	34.0	79.5	35.4	78.8	37.2	78.6	37.2	78.2	38.1	77.5	39.5	87
88	81.0	34.4	80.4	35.8	79.8	37.2	79.6	37.6	79.1	38.6	78.4	40.0	88
89	81.9	34.8	81.3	36.2	80.7	37.6	80.5	38.1	80.0	39.0	79.3	40.4	89
90	82.8	35.2	82.2	36.6	81.6	38.0	81.4	38.5	80.9	39.5	80.2	40.9	90
91	83.8	35.6	83.1	37.0	82.5	38.5	82.3	38.9	81.8	39.9	81.1	41.3	91
92	84.7	35.9	84.0	37.4	83.4	38.9	83.2	39.3	82.7	40.3	82.0	41.8	92
93	85.6	36.3	85.0	37.8	84.3	39.3	84.1	39.8	83.6	40.8	82.5	42.2	93
94	86.5	36.7	85.9	38.2	85.2	39.7	85.0	40.2	84.5	41.2	83.8	42.7	94
95	87.4	37.1	86.8	38.6	86.1	40.1	85.9	40.6	85.4	41.6	84.6	43.1	95
96	88.3	37.5	87.7	39.0	87.0	40.6	86.8	41.0	86.3	42.1	85.5	43.6	96
97	89.3	37.9	88.6	39.4	87.9	41.0	87.7	41.5	87.2	42.5	86.4	44.0	97
98	90.2	38.4	89.5	39.9	88.8	41.4	88.6	41.9	88.1	43.0	87.3	44.5	98
99	91.2	38.7	90.4	40.3	89.7	41.8	89.5	42.3	89.0	43.4	88.2	44.9	99
100	92.0	39.1	91.4	40.7	90.6	42.3	90.4	42.8	89.9	43.8	89.1	45.4	100
Dift.	Dep	Lat	Dep	Lat	Dep	Lat	Dep	Lat	Dep	Lat	Dep	Lat	Dift.
	67 Deg.		66 Deg.		65 Deg.		5¼ Point		64 Deg.		63 Deg.		

Dist	28 Deg.		2½ Point		29 Deg.		30 Deg.		2¾ Point		31 Deg.		Dist
	Lat	Dep	Lat	Dep	Lat	Dep	Lat	Dep	Lat	Dep	Lat	Dep	
1	00,9	00,5	00,9	00,5	00,9	00,5	00,9	00,5	00,9	00,5	00,9	00,5	1
2	01,8	00,9	01,8	00,9	01,7	01,0	01,7	01,0	01,7	01,0	01,7	01,0	2
3	02,6	01,4	02,6	01,4	02,6	01,5	02,6	01,5	02,6	01,5	02,6	01,5	3
4	03,5	01,9	03,5	01,9	03,5	01,9	03,5	02,0	03,4	02,1	03,4	02,1	4
5	04,4	02,3	04,4	02,4	04,4	02,4	04,3	02,5	04,3	02,6	04,3	02,6	5
6	05,3	02,8	05,3	02,8	05,2	02,9	05,2	03,0	05,1	03,1	05,1	03,1	6
7	06,2	03,3	06,2	03,3	06,1	03,4	06,1	03,5	06,0	03,6	06,0	03,6	7
8	07,1	03,8	07,1	03,8	07,0	03,9	06,9	04,0	06,9	04,1	06,9	04,1	8
9	07,9	04,2	07,9	04,2	07,9	04,4	07,8	04,5	07,7	04,6	07,7	04,6	9
10	08,8	04,7	08,8	04,7	08,7	04,8	08,7	05,0	08,6	05,1	08,6	05,1	10
11	09,7	05,2	09,7	05,2	09,6	05,3	09,5	05,5	09,4	05,7	09,4	05,7	11
12	10,6	05,6	10,6	05,7	10,5	05,8	10,4	06,0	10,3	06,2	10,3	06,2	12
13	11,5	06,1	11,5	06,1	11,4	06,3	11,3	06,5	11,1	06,7	11,1	06,7	13
14	12,4	06,6	12,3	06,6	12,2	06,8	12,1	07,0	12,0	07,2	12,0	07,2	14
15	13,2	07,0	13,2	07,1	13,1	07,3	13,0	07,5	12,9	07,7	12,9	07,7	15
16	14,1	07,5	14,1	07,5	14,0	07,8	13,9	08,0	13,7	08,2	13,7	08,2	16
17	15,0	08,0	15,0	08,0	14,9	08,2	14,7	08,5	14,6	08,7	14,6	08,8	17
18	15,9	08,5	15,9	08,5	15,7	08,7	15,6	09,0	15,4	09,3	15,4	09,3	18
19	16,8	08,9	16,8	09,0	16,6	09,2	16,5	09,5	16,3	09,8	16,3	09,8	19
20	17,7	09,4	17,6	09,4	17,5	09,7	17,3	10,0	17,2	10,3	17,2	10,3	20
21	18,5	09,9	18,5	09,9	18,4	10,2	18,2	10,5	18,0	10,8	18,0	10,8	21
22	19,4	10,3	19,4	10,4	19,2	10,7	19,1	11,0	18,9	11,3	18,9	11,3	22
23	20,3	10,8	20,3	10,8	20,1	11,1	19,9	11,5	19,7	11,8	19,7	11,8	23
24	21,2	11,3	21,2	11,3	21,0	11,6	20,8	12,0	20,6	12,3	20,6	12,3	24
25	22,1	11,7	22,0	11,8	21,9	12,1	21,6	12,5	21,4	12,9	21,4	12,9	25
26	23,0	12,2	22,9	12,3	22,7	12,6	22,5	13,0	22,3	13,4	22,3	13,4	26
27	23,8	12,7	23,8	12,7	23,6	13,1	23,4	13,5	23,1	13,9	23,1	13,9	27
28	24,7	13,1	24,7	13,2	24,5	13,6	24,0	14,0	24,0	14,4	24,0	14,4	28
29	25,6	13,6	25,6	13,7	25,4	14,1	25,1	14,5	24,9	14,9	24,9	14,9	29
30	26,5	14,1	26,5	14,1	26,2	14,5	26,0	15,0	25,7	15,4	25,7	15,4	30
31	27,4	14,6	27,3	14,6	27,1	15,0	26,8	15,5	26,6	15,9	26,6	16,0	31
32	28,3	15,0	28,2	15,1	28,0	15,5	27,7	16,0	27,4	16,4	27,4	16,5	32
33	29,1	15,5	29,1	15,6	28,9	16,0	28,6	16,5	28,3	17,0	28,3	17,0	33
34	30,0	16,0	30,0	16,0	29,7	16,5	29,4	17,0	29,2	17,5	29,1	17,5	34
35	30,9	16,4	30,9	16,5	30,6	17,0	30,3	17,5	30,0	18,0	30,0	18,0	35
36	31,8	16,9	31,7	17,0	31,5	17,5	31,2	18,0	30,9	18,5	30,9	18,5	36
37	32,7	17,4	32,6	17,4	32,4	17,9	32,0	18,5	31,7	19,0	31,7	19,1	37
38	33,5	17,8	33,5	17,9	33,2	18,4	32,9	19,0	32,6	19,5	32,6	19,6	38
39	34,4	18,3	34,4	18,4	34,1	18,9	33,8	19,5	33,4	20,0	33,4	20,1	39
40	35,3	18,8	35,3	18,9	35,0	19,4	34,6	20,0	34,3	20,6	34,3	20,6	40
41	36,2	19,2	36,2	19,3	35,9	19,9	35,5	20,5	35,2	21,1	35,1	21,1	41
42	37,1	19,7	37,0	19,8	36,7	20,4	36,4	21,0	36,0	21,6	36,0	21,6	42
43	38,0	20,2	37,9	20,3	37,6	20,8	37,2	21,5	36,9	22,1	36,9	22,1	43
44	38,8	20,7	38,8	20,7	38,5	21,3	38,1	22,0	37,7	22,6	37,7	22,7	44
45	39,7	21,1	39,7	21,2	39,4	21,8	39,0	22,5	38,6	23,1	38,6	23,2	45
46	40,6	21,6	40,6	21,7	40,3	22,3	39,8	23,0	39,5	23,6	39,4	23,7	46
47	41,5	22,1	41,4	22,2	41,1	22,8	40,7	23,5	40,3	24,2	40,3	24,2	47
48	42,4	22,6	42,3	22,6	42,0	23,3	41,6	24,0	41,2	24,7	41,1	24,7	48
49	43,3	23,0	43,2	23,1	42,9	23,8	42,4	24,5	42,0	25,2	42,0	25,2	49
50	44,1	23,5	44,1	23,6	43,6	24,2	43,3	25,0	42,9	25,7	42,9	25,7	50
Dist	Dep	Lat	Dep	Lat	Dep	Lat	Dep	Lat	Dep	Lat	Dep	Lat	Dist
	62 Deg.		5¾ Points		61 Deg.		60 Deg.		5½ Point		59 Deg.		

Dist	28 Deg.		2¾ Point		29 Deg.		30 Deg.		2¾ Point		31 Deg.		Dist
	Lat	Dep	Lat	Dep	Lat	Dep	Lat	Dep	Lat	Dep	Lat	Dep	
51	45.0	23.9	45.0	24.0	44.6	24.7	44.2	25.5	43.7	26.2	43.7	26.3	51
52	45.9	24.4	45.9	24.5	45.5	25.2	45.0	26.0	44.6	26.7	44.6	26.8	52
53	46.8	24.9	46.7	25.0	46.4	25.7	45.9	26.5	45.5	27.2	45.4	27.3	53
54	47.7	25.4	47.6	25.5	47.2	26.2	46.8	27.0	46.3	27.8	46.3	27.8	54
55	48.6	25.8	48.5	25.9	48.1	26.7	47.6	27.5	47.2	28.3	47.1	28.3	55
56	49.4	26.3	49.4	26.4	49.0	27.1	48.5	28.0	48.0	28.8	48.0	28.8	56
57	50.3	26.8	50.3	26.9	49.9	27.6	49.4	28.5	48.9	29.3	48.9	29.4	57
58	51.2	27.2	51.2	27.3	50.7	28.1	50.2	29.0	49.7	29.8	49.7	29.9	58
59	52.1	27.7	52.0	27.8	51.6	28.6	51.1	29.5	50.6	30.3	50.6	30.4	59
60	53.0	28.2	52.9	28.3	52.5	29.1	52.0	30.0	51.5	30.8	51.4	30.9	60
61	53.9	28.6	53.8	28.8	53.3	29.6	52.8	30.5	52.3	31.4	52.3	31.4	61
62	54.7	29.1	54.7	29.2	54.2	30.1	53.7	31.0	53.2	31.9	53.1	31.9	62
63	55.6	29.6	55.6	29.7	55.1	30.5	54.6	31.5	54.0	32.4	54.0	32.4	63
64	56.5	30.0	56.4	30.2	56.0	31.0	55.4	32.0	54.9	32.9	54.9	33.0	64
65	57.4	30.5	57.3	30.6	56.8	31.5	56.3	32.5	55.7	33.4	55.7	33.5	65
66	58.3	31.0	58.2	31.1	57.7	32.0	57.2	33.0	56.6	33.9	56.6	34.0	66
67	59.2	31.5	59.1	31.6	58.6	32.5	58.0	33.5	57.5	34.4	57.4	34.5	67
68	60.0	31.9	60.0	32.1	59.5	33.0	58.9	34.0	58.3	35.0	58.3	35.0	68
69	60.9	32.4	60.9	32.5	60.3	33.5	59.8	34.5	59.2	35.5	59.1	35.5	69
70	61.8	32.9	61.7	33.0	61.2	33.9	60.6	35.0	60.0	36.0	60.0	36.0	70
71	62.7	33.3	62.6	33.5	62.1	34.4	61.5	35.5	60.9	36.5	60.9	36.6	71
72	63.6	33.8	63.5	33.9	63.0	34.9	62.4	36.0	61.8	37.0	61.7	37.1	72
73	64.5	34.3	64.4	34.4	63.8	35.4	63.2	36.5	62.6	37.5	61.6	37.6	73
74	65.3	34.7	65.3	34.9	64.7	35.9	64.0	37.0	63.5	38.0	63.4	38.1	74
75	66.2	35.2	66.1	35.4	65.6	36.4	64.9	37.5	64.3	38.6	64.3	38.6	75
76	67.1	35.7	67.0	35.8	66.5	36.8	65.8	38.0	65.2	39.1	65.1	39.1	76
77	68.0	36.2	67.9	36.3	67.3	37.3	66.7	38.5	66.0	39.6	66.0	39.7	77
78	68.9	36.6	68.8	36.8	68.2	37.8	67.5	39.0	66.9	40.1	66.9	40.2	78
79	69.7	37.1	69.7	37.2	69.1	38.3	68.4	39.5	67.8	40.6	67.7	40.7	79
80	70.6	37.6	70.6	37.7	70.0	38.8	69.3	40.0	68.6	41.2	68.6	41.2	80
81	71.5	38.0	71.4	38.2	70.8	39.3	70.1	40.5	69.5	41.6	69.4	41.7	81
82	72.4	38.5	72.3	38.6	71.7	39.8	70.3	41.0	70.3	42.2	70.3	42.2	82
83	73.3	39.0	73.2	39.1	72.6	40.3	71.9	41.5	71.2	42.7	71.1	42.7	83
84	74.2	39.4	74.1	39.6	73.5	40.7	72.7	42.0	72.0	43.2	72.0	43.3	84
85	75.0	39.9	75.0	40.1	74.3	41.2	73.6	42.5	72.9	43.7	72.9	43.8	85
86	75.9	40.4	75.8	40.5	75.2	41.7	74.5	43.0	73.8	44.3	73.7	44.3	86
87	76.8	40.8	76.7	41.0	76.1	42.2	75.3	43.5	74.6	44.7	74.6	44.8	87
88	77.7	41.3	77.6	41.5	77.0	42.7	76.2	44.0	75.5	45.2	75.4	45.3	88
89	78.6	41.8	78.5	42.0	77.8	43.1	77.1	44.5	76.3	45.8	76.3	45.8	89
90	79.5	42.3	79.4	42.4	78.7	43.6	77.9	45.0	77.2	46.3	77.1	46.3	90
91	80.3	42.7	80.3	42.9	79.6	44.1	78.8	45.5	78.1	46.8	78.0	46.9	91
92	81.2	43.2	81.1	43.4	80.5	44.6	79.7	46.0	78.9	47.3	79.7	47.4	92
93	82.1	43.7	82.0	43.8	81.3	45.1	80.5	46.5	79.8	47.8	79.7	47.9	93
94	83.0	44.1	82.9	44.3	82.2	45.6	81.4	47.0	80.6	48.3	80.6	48.4	94
95	83.9	44.6	83.8	44.8	83.1	46.1	82.3	47.5	81.5	48.8	81.4	48.9	95
96	84.8	45.1	84.7	45.2	84.0	46.5	83.1	48.0	82.3	49.3	82.3	49.4	96
97	85.6	45.5	85.5	45.7	84.8	47.0	84.0	48.5	83.2	49.9	83.1	50.0	97
98	86.5	46.0	86.4	46.2	85.7	47.5	84.9	49.0	84.1	50.4	84.0	50.5	98
99	87.4	46.5	87.3	46.7	86.6	48.0	85.7	49.5	84.9	50.9	84.9	51.0	99
100	88.3	46.9	88.2	47.1	87.5	48.5	86.6	50.0	85.8	51.4	85.7	51.5	100

Dist		Dep	Lat	Dep	Lat	Dep	Lat	Dep	Lat	Dep	Lat	Dep	Lat	Dist
	62 Deg.		5¼ Point		61 Deg.		60 Deg.		5¼ Point		59 Deg.			

Dist.	32 Deg.		33 Deg.		3 Points		34 Deg.		35 Deg.		36 Deg.		Dist.
	Lat	Dep	Lat	Dep	Lat	Dep	Lat	Dep	Lat	Dep	Lat	Dep	
1	00.8	00.5	00.8	00.5	00.8	00.6	00.8	00.6	00.8	00.6	00.8	00.6	1
2	01.7	01.1	01.7	01.1	01.7	01.1	01.6	01.1	01.6	01.1	01.6	01.2	2
3	01.5	01.6	02.5	01.6	02.5	01.7	02.5	01.7	02.5	01.7	02.4	01.8	3
4	03.4	02.1	03.4	02.2	03.5	02.2	03.3	02.2	03.3	02.3	03.2	02.4	4
5	04.2	02.6	04.2	02.7	04.2	02.8	04.1	02.8	04.1	02.9	04.0	02.9	5
6	05.1	03.2	05.0	03.3	05.0	03.3	05.0	03.4	04.9	03.4	04.9	03.5	6
7	05.9	03.7	05.9	03.8	05.8	03.9	05.8	03.9	05.7	04.0	05.7	04.1	7
8	06.8	04.2	06.7	04.4	06.6	04.4	06.6	04.5	06.6	04.6	06.5	04.7	8
9	07.6	04.8	07.5	04.9	07.5	05.0	07.5	05.0	07.4	05.2	07.3	05.3	9
10	08.5	05.3	08.4	05.4	08.3	05.6	08.3	05.6	08.2	05.7	08.1	05.9	10
11	09.3	05.8	09.2	06.0	09.1	06.1	09.1	06.2	09.0	06.3	08.9	06.5	11
12	10.2	06.4	10.1	06.5	10.0	06.7	09.9	06.7	09.8	06.9	09.7	07.0	12
13	11.0	06.9	10.9	07.1	10.8	07.2	10.8	07.3	10.6	07.5	10.5	07.6	13
14	11.9	07.4	11.7	07.6	11.6	07.8	11.6	07.8	11.5	08.0	11.3	08.2	14
15	12.7	07.9	12.6	08.2	12.5	08.3	12.4	08.3	12.3	08.6	12.1	08.8	15
16	13.6	08.5	13.4	08.7	13.3	08.9	13.3	08.9	13.1	09.2	12.9	09.4	16
17	14.4	09.0	14.3	09.3	14.1	09.4	14.1	09.5	13.9	09.8	13.8	10.0	17
18	15.3	09.5	15.1	09.8	15.0	10.0	14.9	10.1	14.7	10.3	14.6	10.6	18
19	16.1	10.1	15.9	10.3	15.8	10.6	15.8	10.6	15.6	10.9	15.4	11.2	19
20	17.0	10.6	16.8	10.9	16.6	11.1	16.6	11.2	16.4	11.5	16.2	11.8	20
21	17.8	11.1	17.6	11.4	17.5	11.7	17.4	11.7	17.2	12.0	17.0	12.3	21
22	18.7	11.7	18.5	12.0	18.3	12.2	18.2	12.3	18.0	12.6	17.8	12.9	22
23	19.5	12.2	19.3	12.5	19.1	12.8	19.1	12.9	18.8	13.2	18.6	13.5	23
24	20.4	12.7	20.1	13.1	20.0	13.3	19.9	13.4	19.7	13.8	19.4	14.1	24
25	21.2	13.2	21.0	13.6	20.8	13.9	20.7	14.0	20.5	14.3	20.2	14.7	25
26	22.0	13.8	21.8	14.2	21.6	14.4	21.6	14.5	21.3	14.9	21.0	15.3	26
27	22.9	14.3	22.6	14.7	22.4	15.0	22.4	15.0	22.1	15.5	21.8	15.9	27
28	23.7	14.8	23.6	15.2	23.3	15.6	23.2	15.7	22.9	16.1	22.7	16.5	28
29	24.6	15.4	24.3	15.8	24.1	16.1	24.0	16.2	23.8	16.6	23.5	17.0	29
30	25.4	15.9	25.2	16.3	24.9	16.7	24.9	16.8	24.6	17.2	24.3	17.6	30
31	26.3	16.4	26.0	16.9	25.8	17.2	25.7	17.3	25.4	17.8	25.1	18.2	31
32	27.1	17.0	26.8	17.4	26.6	17.8	26.5	17.9	26.2	18.4	25.9	18.8	32
33	28.0	17.5	27.7	18.0	27.4	18.3	27.4	18.5	27.0	18.9	26.7	19.4	33
34	28.8	18.0	28.5	18.5	28.3	18.9	28.2	19.0	27.9	19.5	27.5	20.0	34
35	29.7	18.5	29.4	19.1	29.1	19.4	29.0	19.6	28.7	20.1	28.3	20.6	35
36	30.5	19.1	30.2	19.6	29.9	20.0	29.8	20.3	29.5	20.6	29.1	21.2	36
37	31.4	19.6	31.0	20.1	30.8	20.6	30.7	20.6	30.3	21.2	29.9	21.7	37
38	32.2	20.1	31.9	20.7	31.6	21.1	31.5	21.2	31.1	21.8	30.7	22.3	38
39	33.1	20.7	32.7	21.2	32.4	21.7	32.3	21.8	32.0	22.4	31.6	22.9	39
40	33.9	21.2	33.6	21.8	33.3	22.2	33.2	22.4	32.8	22.9	32.4	23.5	40
41	34.8	21.7	34.4	22.3	34.1	22.8	34.0	22.9	33.6	23.5	33.2	24.1	41
42	35.6	22.3	35.2	22.9	34.9	23.3	34.8	23.5	34.4	24.1	34.0	24.7	42
43	36.5	22.8	36.1	23.4	35.8	23.9	35.6	24.0	35.2	24.7	34.8	25.3	43
44	37.3	23.3	36.9	24.0	36.6	24.4	36.5	24.6	36.0	25.2	35.6	25.9	44
45	38.2	23.8	37.7	24.5	37.4	25.0	37.3	25.2	36.9	25.8	36.4	26.5	45
46	39.0	24.4	38.6	25.1	38.2	25.6	38.1	25.7	37.7	26.4	37.2	27.0	46
47	39.9	24.9	39.4	25.6	39.1	26.1	39.0	26.3	38.5	27.0	38.0	27.6	47
48	40.7	25.4	40.3	26.1	39.9	26.7	39.8	26.8	39.3	27.5	38.8	28.2	48
49	41.6	26.0	41.1	26.7	40.7	27.2	40.6	27.4	40.1	28.1	39.6	28.8	49
50	42.4	26.5	41.9	27.2	41.6	27.8	41.4	28.0	41.0	28.7	40.4	29.4	50
Dist.	Dep	Lat	Dep	Lat	Dep	Lat	Dep	Lat	Dep	Lat	Dep	Lat	Dist.
	58 Deg.		57 Deg.		5 Points		56 Deg.		55 Deg.		54 Deg.		

Dist	32 Deg. Lat	Dep	33 Deg. Lat	Dep	3 Points Lat	Dep	34 Deg. Lat	Dep	35 Deg. Lat	Dep	36 Deg. Lat	Dep	Dist
51	43.2	27.0	42.8	27.8	42.4	28.3	42.3	28.5	41.8	29.3	41.3	30.0	51
52	44.1	27.6	43.6	28.3	43.2	28.9	43.1	29.1	42.6	29.8	42.1	30.6	52
53	44.9	28.1	44.5	28.9	44.1	29.4	43.9	29.6	43.4	30.4	42.9	31.2	53
54	45.8	28.6	45.3	29.4	44.9	30.0	44.6	30.2	44.2	31.0	41.7	31.7	54
55	46.6	29.1	46.1	30.0	45.7	30.6	45.6	30.8	45.1	31.5	44.5	32.3	55
56	47.5	29.7	47.0	30.6	46.6	31.1	46.4	31.3	45.9	32.1	45.3	32.9	56
57	48.3	30.2	47.8	31.0	47.4	31.7	47.2	31.9	46.7	32.7	46.1	33.5	57
58	49.2	30.7	48.6	31.6	48.2	32.2	48.1	32.4	47.5	33.2	46.9	34.1	58
59	50.0	31.3	49.5	32.1	49.0	32.8	48.9	33.0	48.3	33.8	47.7	34.7	59
60	50.9	31.8	50.3	32.7	49.9	33.3	49.7	33.6	49.2	34.4	48.5	35.3	60
61	51.7	32.3	51.2	33.2	50.7	33.9	50.6	34.1	50.0	35.0	49.3	35.9	61
62	52.6	32.8	52.0	33.8	51.6	34.4	51.4	34.7	50.8	35.6	50.2	36.4	62
63	53.4	33.4	52.8	34.3	52.4	35.0	52.2	35.2	51.6	36.1	51.0	37.0	63
64	54.3	33.9	53.7	34.9	53.2	35.6	53.1	35.8	52.4	36.7	51.8	37.6	64
65	55.1	34.4	54.5	35.4	54.0	36.1	53.9	36.3	53.2	37.3	52.6	38.2	65
66	56.0	35.0	55.4	35.9	54.9	36.7	54.7	36.9	54.1	37.9	53.4	38.8	66
67	56.8	35.5	56.2	36.5	55.7	37.2	55.5	37.5	54.9	38.4	54.2	39.4	67
68	57.7	36.0	57.0	37.0	56.5	37.8	56.4	38.0	55.7	39.0	55.0	40.0	68
69	58.5	36.6	57.9	37.6	57.4	38.3	57.2	38.6	56.5	39.6	55.8	40.6	69
70	59.4	37.1	58.7	38.1	58.2	38.9	58.0	39.1	57.3	40.2	56.6	41.1	70
71	60.2	37.6	59.5	38.7	59.0	39.4	58.9	39.7	58.2	40.7	57.4	41.7	71
72	61.1	38.2	60.4	39.2	59.9	40.0	59.7	40.3	59.0	41.3	58.2	42.3	72
73	61.9	38.7	61.2	39.8	60.7	40.6	60.5	40.8	59.8	41.9	59.1	42.9	73
74	62.8	39.2	62.1	40.3	61.5	41.1	61.3	41.4	60.6	42.4	59.5	43.5	74
75	63.6	39.7	62.9	40.8	62.4	41.7	62.2	41.9	61.4	43.0	60.7	44.1	75
76	64.4	40.3	63.7	41.4	63.2	42.2	63.0	42.5	62.3	43.6	61.5	44.7	76
77	65.3	40.8	64.6	41.9	64.0	42.8	63.8	43.1	63.1	44.2	62.3	45.3	77
78	66.1	41.3	65.4	42.5	64.9	43.3	64.7	43.6	63.9	44.7	63.1	45.8	78
79	67.0	41.9	66.3	43.0	65.7	43.9	65.5	44.2	64.7	45.3	63.9	46.4	79
80	67.8	42.4	67.1	43.6	66.5	44.4	66.3	44.7	65.5	45.9	64.7	47.0	80
81	68.7	42.9	67.9	44.1	67.4	45.0	67.1	45.3	66.4	46.5	65.5	47.6	81
82	69.5	43.4	68.8	44.7	68.2	45.6	68.0	43.9	67.2	47.0	66.3	48.2	82
83	70.4	44.0	69.6	45.2	69.0	46.1	68.8	46.4	68.0	47.6	67.1	48.8	83
84	71.2	44.5	70.5	45.7	69.8	46.7	69.6	47.0	68.8	48.2	68.0	49.4	84
85	72.1	45.0	71.3	46.3	70.7	47.2	70.5	47.5	69.6	48.8	68.8	50.0	85
86	72.9	45.6	72.1	46.8	71.5	47.8	71.3	48.1	70.4	49.3	69.6	50.5	86
87	73.8	46.1	73.0	47.4	72.3	48.3	72.1	48.6	71.3	49.9	70.4	51.1	87
88	74.6	46.6	73.8	47.9	73.2	48.9	73.0	49.2	72.1	50.5	71.2	51.7	88
89	75.5	47.2	74.6	48.5	74.0	49.4	73.8	49.8	72.9	51.0	72.0	52.3	89
90	76.3	47.7	75.5	49.0	74.8	50.0	74.6	50.3	73.7	51.6	72.8	52.9	90
91	77.2	48.2	76.3	49.6	75.7	50.5	75.4	50.9	74.5	52.2	73.6	53.5	91
92	78.0	48.7	77.2	50.1	76.5	51.1	76.3	51.4	75.4	52.8	74.4	54.1	92
93	78.9	49.3	78.0	50.6	77.3	51.6	77.1	52.0	76.2	53.3	75.2	54.7	93
94	79.7	49.8	78.8	51.2	78.2	52.2	77.9	52.5	77.0	53.9	76.0	55.2	94
95	80.6	50.3	79.7	51.7	79.0	52.8	78.8	53.1	77.8	54.5	76.9	55.8	95
96	81.4	50.9	80.5	52.3	79.8	53.3	79.6	53.7	78.6	55.1	77.7	56.4	96
97	82.3	51.4	81.4	52.8	80.7	53.9	80.4	54.2	79.5	55.6	78.5	57.0	97
98	83.1	51.9	82.2	53.4	81.5	54.4	81.2	54.8	80.3	56.2	79.3	57.6	98
99	84.0	52.5	83.1	53.9	82.3	55.0	82.1	55.4	81.1	56.8	80.1	58.2	99
100	84.8	53.0	83.9	54.5	83.1	55.5	82.9	55.9	81.9	57.4	80.9	58.8	100
Dist	Dep	Lat	Dep	Lat	Dep	Lat	Dep	Lat	Dep	Lat	Dep	Lat	Dist
	58 Deg.		57 Deg.		5 Points		56 Deg.		55 Deg.		54 Deg.		

Diff	34 Point		37 Deg		38 Deg		39 Deg		34 Point		40 Deg		Diff
	Lat	Dep	Lat	Dep	Lat	Dep	Lat	Dep	Lat	Dep	Lat	Dep	
1	00,8	00,6	00,8	00,6	00,8	00,6	00,8	00,6	00,8	00,6	00,8	00,6	1
2	01,6	01,2	01,6	01,2	01,6	01,2	01,6	01,3	01,5	01,3	01,5	01,3	2
3	02,4	01,8	02,4	01,8	02,4	01,8	02,3	01,9	02,3	01,9	02,3	01,9	3
4	03,2	02,4	03,2	02,4	03,8	02,5	03,1	02,5	03,1	02,5	03,1	02,6	4
5	04,0	03,0	04,0	03,0	03,9	03,1	03,9	03,1	03,9	03,2	03,8	03,2	5
6	04,8	03,6	04,8	03,6	04,7	03,7	04,7	03,8	04,6	03,8	04,6	03,8	6
7	05,6	04,2	05,6	04,2	05,5	04,3	05,4	04,4	05,4	04,4	05,4	04,5	7
8	06,4	04,8	06,4	04,8	06,3	04,9	06,2	05,0	06,2	05,1	06,1	05,1	8
9	07,2	05,4	07,2	05,4	07,1	05,5	07,0	05,7	07,0	05,7	06,9	05,8	9
10	08,0	06,0	03,0	06,0	07,9	06,2	07,8	06,3	07,7	06,3	07,7	06,4	10
11	08,8	06,6	08,8	06,6	08,7	06,8	08,5	06,9	08,5	07,0	08,4	07,1	11
12	09,6	07,1	09,6	07,2	09,5	07,4	09,3	07,6	09,3	07,6	09,2	07,7	12
13	10,4	07,7	10,4	07,8	10,2	08,0	10,1	08,2	10,0	08,2	10,0	08,4	13
14	11,2	08,3	11,2	08,4	11,0	08,6	10,9	08,8	10,8	08,9	10,7	09,0	14
15	12,0	08,9	12,0	09,0	11,8	09,2	11,7	09,4	11,6	09,5	11,5	09,6	15
16	11,9	09,5	12,8	09,6	12,6	09,9	12,4	10,1	12,4	10,1	12,3	12,3	16
17	13,7	10,1	13,6	10,2	13,4	10,5	13,2	10,7	13,1	10,8	13,0	10,9	17
18	14,5	10,7	14,4	10,8	14,2	11,1	14,0	11,3	13,9	11,4	13,8	11,6	18
19	15,3	11,3	15,2	11,4	15,0	11,7	14,8	12,0	14,7	12,1	14,6	12,2	19
20	16,1	11,9	16,0	12,0	15,8	12,3	15,5	12,6	15,5	12,7	15,3	12,9	20
21	16,9	12,5	16,8	12,6	16,5	12,9	16,3	13,2	16,2	13,3	16,1	13,5	21
22	17,7	13,1	17,6	13,2	17,3	13,5	17,1	13,8	17,0	14,0	16,9	14,1	22
23	18,5	13,7	18,4	13,8	18,1	14,2	17,9	14,5	17,8	14,6	17,6	14,8	23
24	19,3	14,3	19,2	14,4	18,9	14,8	18,6	15,2	18,6	15,2	19,4	15,4	24
25	20,1	14,9	20,0	15,0	19,7	15,4	19,4	15,7	19,3	15,9	19,1	16,1	25
26	20,9	15,5	20,8	15,6	20,5	16,0	20,2	16,4	20,1	16,5	19,9	16,7	26
27	21,7	16,1	21,6	16,2	21,3	16,6	21,0	17,0	20,9	17,1	20,7	17,4	27
28	22,5	16,7	22,4	16,8	22,1	17,2	21,8	17,6	21,6	17,6	21,4	18,0	28
29	23,3	17,3	23,2	17,5	22,9	17,9	22,5	18,2	22,4	18,4	21,8	18,6	29
30	24,1	17,9	24,0	18,1	23,6	18,5	23,3	18,9	23,2	19,0	23,0	19,3	30
31	24,9	18,5	24,8	18,7	24,4	19,1	24,1	19,5	24,0	19,7	23,7	19,9	31
32	25,7	19,1	25,6	19,3	25,2	19,7	24,9	20,2	24,7	20,3	24,5	20,6	32
33	26,5	19,7	26,4	19,9	26,0	20,3	25,6	20,8	25,5	20,9	25,3	21,2	33
34	27,3	20,3	27,2	20,5	26,8	20,9	26,4	21,4	26,3	21,6	26,0	21,9	34
35	28,1	20,8	28,0	21,1	27,6	21,5	27,2	22,0	27,1	22,2	26,8	22,5	35
36	28,9	21,4	28,7	21,7	28,4	22,2	28,0	22,7	27,8	22,8	27,6	23,1	36
37	29,7	22,0	29,5	22,3	29,2	22,8	28,8	23,3	28,6	23,5	28,3	23,8	37
38	30,5	22,6	30,3	22,9	29,9	23,4	29,5	23,9	29,4	24,1	29,1	24,4	38
39	31,3	23,2	31,1	23,5	30,7	24,0	30,3	24,5	30,1	24,7	29,9	25,1	39
40	32,1	23,8	31,9	24,1	31,5	24,6	31,1	25,2	30,9	25,4	30,6	25,7	40
41	32,9	24,4	32,7	24,7	32,3	25,2	31,9	25,8	31,7	26,0	31,4	26,4	41
42	33,7	25,0	33,5	25,3	33,1	25,9	32,6	26,4	32,5	26,6	32,2	27,0	42
43	34,5	25,6	34,3	25,9	33,9	26,5	33,4	17,1	33,2	27,3	32,9	27,6	43
44	35,3	26,2	35,2	26,5	34,7	27,1	34,2	27,7	34,0	27,9	33,7	28,3	44
45	36,1	26,8	35,9	27,2	35,5	27,7	35,0	28,3	34,8	28,5	34,5	28,9	45
46	36,9	27,4	36,7	27,9	36,2	28,3	35,7	28,9	35,6	29,2	35,2	29,6	46
47	37,7	28,0	37,5	28,3	37,0	28,9	36,5	29,6	36,3	29,8	36,0	30,2	47
48	38,6	28,6	38,3	28,9	37,8	29,6	37,3	30,2	37,1	30,5	36,8	30,9	48
49	39,4	29,2	39,1	29,5	38,6	30,2	38,1	30,8	37,9	31,1	37,5	31,5	49
50	40,2	29,8	39,9	30,1	39,4	30,8	38,9	31,5	38,6	31,7	38,3	32,1	50
Diff	Dep	Lat	Dep	Lat	Dep	Lat	Dep	Lat	Dep	Lat	Dep	Lat	Diff
	34 Point		53 Deg		52 Deg		51 Deg		44 Point		50 Deg		

Dist	1½ Point		37 Deg.		38 Deg.		39 Deg.		1½ Point		40 Deg.		Dist
	Lat	Dep	Lat	Dep	Lat	Dep	Lat	Dep	Lat	Dep	Lat	Dep	
51	41,0	30.4	40.7	30.7	40.2	31.4	39.6	32.1	39.4	32.4	39.1	32.8	51
52	41.8	31.0	41.5	31.3	41.0	32.0	40.4	32.7	40.2	33.0	39.8	33.2	52
53	42.6	31.6	42.3	31.9	41.8	32.6	42.2	33.4	41.0	31.6	40.6	34.1	53
54	43.4	32.2	43.1	32.5	42.6	33.2	42.0	34.0	41.7	34.3	41.4	34.7	54
55	44.2	32.8	43.9	33.1	43.3	31.9	42.7	34.6	42.3	34.9	42.1	35.4	55
56	45.0	33.4	44.7	33.7	44.1	34.5	43.5	36.2	43.3	35.5	41.9	36.0	56
57	45.8	34.0	45.5	34.3	44.9	35.1	44.3	35.9	44.1	36.1	43.7	36.6	57
58	46.6	34.5	46.3	34.9	45.7	35.7	45.1	36.5	44.8	36.8	44.4	37.3	58
59	47.4	35.1	47.1	35.5	46.5	36.3	45.8	37.1	45.6	37.4	45.2	37.9	59
60	48.2	35.7	47.9	36.1	47.3	36.9	46.6	37.8	46.4	38.1	46.0	38.6	60
61	49.0	36.3	48.7	36.7	48.1	37.6	47.4	38.4	47.2	38.7	46.7	39.2	61
62	49.8	36.9	49.5	37.3	48.9	38.2	48.1	39.0	47.9	39.3	47.5	49.9	62
63	50.6	37.5	50.3	37.9	49.6	38.8	49.0	39.6	48.7	40.0	48.3	40.5	63
64	51.4	28.1	51.1	38.5	50.4	39.4	49.7	40.3	49.6	40.6	49.0	41.1	64
65	52.2	28.7	51.9	39.1	51.2	40.0	50.5	40.9	50.2	41.3	49.8	41.8	65
66	53.0	39.3	52.7	39.7	52.0	40.6	51.3	41.5	51.0	41.9	50.6	42.4	66
67	53.8	39.9	53.5	40.3	52.8	41.3	52.2	42.2	51.8	42.5	51.3	43.2	67
68	54.6	40.5	54.3	40.9	53.6	41.9	52.8	42.8	52.6	43.1	52.1	41.7	68
69	55.4	41.1	55.1	41.5	54.4	42.5	53.6	43.4	53.3	43.7	52.9	44.4	69
70	56.2	41.7	55.9	42.1	55.2	43.1	54.4	44.1	54.1	44.4	53.6	45.0	70
71	57.0	42.3	56.7	42.7	55.9	43.7	55.2	44.7	54.9	45.0	54.4	45.6	71
72	57.8	42.9	57.5	43.3	56.7	44.3	56.0	45.3	55.7	45.7	55.2	46.3	72
73	58.6	43.5	58.3	43.9	57.5	44.9	56.7	46.6	56.4	46.3	55.9	46.9	73
74	59.4	44.1	59.1	44.5	58.3	45.6	57.5	46.6	57.2	46.9	56.7	47.6	74
75	60.2	44.7	59.9	45.1	59.1	46.2	58.3	47.2	58.0	47.6	57.5	48.2	75
76	61.0	45.3	60.7	45.7	59.9	46.8	59.1	47.8	58.7	48.2	58.2	48.9	76
77	61.8	45.9	61.5	46.3	60.7	47.4	59.8	48.5	59.5	48.8	59.0	49.5	77
78	62.6	46.5	62.3	46.9	61.5	48.0	60.6	49.1	60.3	49.5	59.7	50.2	78
79	63.5	47.1	63.1	47.5	62.3	48.6	61.4	49.7	61.1	50.1	60.5	50.8	79
80	64.3	47.7	63.9	48.1	63.0	49.2	62.2	50.3	61.8	50.8	61.3	51.4	80
81	65.1	48.3	64.7	48.7	63.8	49.9	62.9	51.0	62.6	51.4	61.0	52.1	81
82	65.9	48.8	65.5	49.3	64.6	50.5	63.7	51.6	63.4	52.0	62.8	52.7	82
83	66.7	49.4	66.3	49.9	65.4	51.1	64.5	52.2	64.2	52.7	63.6	53.4	83
84	67.5	50.0	67.1	50.6	66.2	51.7	65.3	52.9	64.9	53.3	64.3	54.0	84
85	68.3	50.6	67.9	51.2	67.0	52.3	66.1	53.5	65.7	53.9	65.1	54.6	85
86	69.1	51.2	68.7	51.8	67.8	52.9	66.8	54.1	66.5	54.6	65.9	55.3	86
87	69.9	51.8	69.5	52.4	68.6	53.5	67.6	54.7	67.3	55.2	66.6	55.9	87
88	70.7	52.4	70.3	53.0	69.3	54.1	68.4	55.4	68.0	55.8	67.4	56.6	88
89	71.5	53.0	71.1	53.6	70.1	54.8	69.2	56.0	68.8	56.5	68.2	57.2	89
90	72.3	53.6	71.9	54.2	70.9	55.4	69.9	56.6	69.6	57.1	68.9	57.9	90
91	73.1	54.2	72.7	54.8	71.7	56.0	70.7	57.3	70.3	57.7	69.7	58.5	91
92	73.9	54.8	73.5	55.4	72.5	56.6	71.5	57.9	71.1	58.4	70.5	59.1	92
93	74.7	55.4	74.3	56.0	73.3	57.3	72.1	58.5	71.9	59.0	71.2	59.8	93
94	75.5	56.0	75.1	56.6	74.1	57.9	73.0	59.2	72.7	59.6	72.0	60.4	94
95	76.3	56.6	75.9	57.2	74.9	58.5	73.8	59.8	73.4	60.3	72.8	61.1	95
96	77.1	57.2	76.7	57.8	75.6	59.1	74.6	60.4	74.2	60.9	73.5	61.7	96
97	77.9	57.8	77.5	58.4	76.4	59.7	75.4	61.0	75.0	61.5	74.3	62.4	97
98	78.7	58.4	78.3	59.0	77.2	60.3	76.2	61.7	75.8	62.2	75.1	63.0	98
99	79.5	59.0	79.1	59.6	78.0	61.0	76.9	62.3	76.5	62.8	75.8	63.6	99
100	80.2	59.6	79.9	60.2	78.8	61.6	77.7	62.9	77.3	63.4	76.6	64.3	100
Dist	Dep	Lat	Dep	Lat	Dep	Lat	Dep	Lat	Dep	Lat	Dep	Lat	Dist
	4½ Point		53 Deg.		52 Deg.		51 Deg.		4½ Point		50 Deg.		

Diff	41 Deg.		42 Deg.		3¾ Point		43 Deg.		44 Deg.		4 Points		Diff
	Lat	Dep	Lat	Dep	Lat	Dep	Lat	Dep	Lat	Dep	Lat	Dep	
1	00.6	00.7	00.7	00.7	00.7	00.7	00.7	00.7	00.7	00.7	00.7	00.7	1
2	01.5	01.3	01.5	01.3	01.5	01.3	01.4	01.4	01.4	01.4	01.4	01.4	2
3	02.3	02.0	02.2	02.0	02.2	02.0	02.2	02.0	02.2	02.1	02.1	02.1	3
4	03.0	02.6	03.0	02.7	03.0	02.7	03.0	02.7	02.9	02.8	02.8	02.8	4
5	03.8	03.3	03.7	03.3	03.7	03.3	03.7	03.4	03.6	03.5	03.5	03.5	5
6	04.5	03.9	04.5	04.0	04.4	04.0	04.4	04.1	04.3	04.2	04.2	04.2	6
7	05.3	04.6	05.2	04.7	05.2	04.7	05.1	04.8	05.0	04.9	04.9	04.9	7
8	06.0	05.2	05.9	05.4	05.9	05.4	05.9	05.5	05.8	05.6	05.7	05.7	8
9	06.8	05.9	06.7	06.0	06.7	06.0	06.6	06.1	06.5	06.3	06.3	06.4	9
10	07.5	06.6	07.4	06.7	07.4	06.7	07.3	06.8	07.2	06.9	07.1	07.1	10
11	08.3	07.2	08.2	07.4	08.2	07.4	08.0	07.5	07.9	07.6	07.8	07.8	11
12	09.1	07.9	08.9	08.0	08.9	08.1	08.8	08.2	08.6	08.3	08.5	08.5	12
13	09.8	08.5	09.7	08.7	09.6	08.7	09.5	08.9	09.3	09.0	09.2	09.2	13
14	10.6	09.2	10.4	09.4	10.4	09.4	10.2	09.5	10.1	09.7	09.9	09.9	14
15	11.3	09.8	11.1	10.0	11.1	10.1	11.0	10.2	10.8	10.4	10.6	10.6	15
16	12.1	10.5	11.9	10.7	11.9	10.7	11.7	10.9	11.5	11.1	11.3	11.9	16
17	12.8	11.2	12.6	11.4	12.6	11.4	12.4	11.6	12.2	11.8	12.0	12.0	17
18	13.6	11.8	13.4	12.0	13.3	12.1	13.2	12.3	12.9	12.5	12.7	12.7	18
19	14.3	12.5	14.1	12.7	14.1	12.8	13.9	13.0	13.7	13.2	13.4	13.4	19
20	15.1	13.1	14.9	13.4	14.8	13.4	14.6	13.6	14.3	13.9	14.1	14.1	20
21	15.8	13.8	15.6	14.0	15.6	14.1	15.4	14.3	15.1	14.6	14.8	14.8	21
22	16.6	14.4	16.3	14.7	16.3	14.8	16.1	15.0	15.8	13.3	15.6	15.6	22
23	17.4	15.1	17.1	15.4	17.0	15.4	16.8	15.7	16.5	16.0	16.3	16.3	23
24	18.1	15.7	17.8	16.1	17.8	16.1	17.6	16.4	17.3	16.7	17.0	17.0	24
25	18.9	16.4	18.6	16.7	18.5	16.8	18.3	17.1	18.0	17.4	17.7	17.7	25
26	19.6	17.1	19.3	17.4	19.3	17.5	19.0	17.7	18.7	18.1	18.4	18.4	26
27	20.3	17.7	20.1	18.1	20.0	18.1	19.7	18.4	19.4	18.8	19.1	19.1	27
28	21.1	18.4	20.8	18.7	20.7	18.8	20.5	19.1	20.1	19.5	19.8	19.8	28
29	21.9	19.0	21.5	19.4	21.5	19.5	21.2	19.8	20.9	20.1	20.5	20.5	29
30	22.6	19.7	22.3	20.1	22.2	20.1	21.9	20.5	21.6	20.8	21.2	21.2	30
31	23.4	20.3	23.0	20.7	23.0	20.8	22.7	21.1	22.3	21.5	21.9	21.9	31
32	24.1	21.0	23.8	21.4	23.7	21.5	23.4	21.8	23.0	22.2	22.6	22.6	32
33	24.9	21.7	24.5	22.1	24.5	22.2	24.1	22.5	23.7	22.9	23.3	23.3	33
34	25.7	22.3	25.3	22.7	25.2	22.8	24.9	23.2	24.5	23.6	24.0	24.0	34
35	26.4	23.0	26.0	23.4	25.9	23.5	25.6	23.9	25.2	24.3	24.7	24.7	35
36	27.2	23.6	26.8	24.1	26.7	24.2	26.3	24.6	25.9	25.0	25.5	25.5	36
37	27.9	24.3	27.5	24.8	27.4	24.8	27.0	25.3	26.6	25.7	26.2	26.2	37
38	28.7	24.9	28.3	25.4	28.2	25.5	27.8	25.9	27.4	26.4	26.9	26.9	38
39	29.4	25.6	29.0	26.1	28.9	26.2	28.5	26.6	28.1	27.1	27.6	27.6	39
40	30.2	26.2	29.7	26.8	29.6	26.9	29.3	27.3	28.8	27.8	28.3	28.3	40
41	30.9	26.9	30.5	27.4	30.4	27.5	30.0	28.0	29.5	28.5	29.0	29.0	41
42	31.7	27.6	31.2	28.1	31.1	28.2	30.7	28.6	30.2	29.2	29.7	29.7	42
43	32.5	28.2	32.0	28.8	31.8	28.9	31.4	29.3	30.9	29.9	30.4	30.4	43
44	33.2	28.9	32.7	29.4	32.6	29.6	32.2	30.0	31.6	30.6	31.1	31.1	44
45	34.0	29.5	33.4	30.1	33.3	30.2	32.9	30.7	32.4	31.3	31.8	31.8	45
46	34.7	30.2	34.2	30.8	34.1	30.9	33.6	31.4	33.1	32.0	32.5	32.5	46
47	35.5	30.8	34.9	31.4	34.8	31.6	34.4	32.1	33.8	32.6	33.2	33.2	47
48	36.2	31.5	35.7	32.1	35.6	32.2	35.1	32.7	34.5	33.3	33.9	33.9	48
49	37.0	32.1	36.2	32.8	36.3	32.9	35.8	33.4	35.2	34.0	34.6	34.6	49
50	37.7	32.8	37.2	33.5	37.0	33.6	36.6	34.1	36.0	34.7	35.4	35.4	50

	Dep	Lat	Dep	Lat	Dep	Lat	Dep	Lat	Dep	Lat	Dep	Lat	Diff
	49 Deg.		48 Deg.		4¼ Point		47 Deg.		46 Deg.		4 Points		

Dist.	41 Deg. Lat	41 Deg. Dep	42 Deg. Lat	42 Deg. Dep	3¾ Point Lat	3¾ Point Dep	43 Deg. Lat	43 Deg. Dep	44 Deg. Lat	44 Deg. Dep	4 Points Lat	4 Points Dep	Dist.
51	38.5	31.5	37.9	34.2	37.8	34.3	37.3	34.8	36.7	35.4	36.1	36.1	51
52	39.2	34.2	38.6	34.8	38.5	34.9	38.0	35.5	37.4	36.2	36.8	36.8	52
53	40.0	34.8	39.4	35.5	39.3	35.6	38.8	36.2	38.1	36.8	37.5	37.5	53
54	40.8	35.4	40.1	36.1	40.0	36.3	39.5	36.8	38.8	37.5	38.2	38.2	54
55	41.5	36.1	40.9	36.8	40.8	36.9	40.2	37.5	39.6	38.2	38.9	38.9	55
56	42.3	36.7	41.6	37.5	41.5	37.6	41.0	38.2	40.3	38.9	39.6	39.6	56
57	43.0	37.4	42.4	38.1	42.3	38.3	41.7	38.9	41.0	39.6	40.3	40.3	57
58	43.8	38.1	43.1	38.8	43.0	38.0	42.4	39.6	41.7	40.3	41.0	41.0	58
59	44.5	38.7	43.8	39.5	43.7	39.6	43.2	40.2	42.4	41.0	41.7	41.7	59
60	45.3	39.4	44.6	40.2	44.5	40.3	43.9	40.9	43.2	41.7	42.4	42.4	60
61	46.0	40.0	45.3	40.8	45.2	41.0	44.6	41.6	43.9	42.4	43.1	43.1	61
62	46.8	40.7	46.1	41.5	45.9	41.6	45.3	42.3	44.6	43.1	43.8	43.8	62
63	47.5	41.3	46.8	42.2	46.7	42.3	46.1	43.0	45.3	43.8	44.5	44.5	63
64	48.3	42.0	47.6	42.8	47.4	43.0	46.8	43.6	46.0	44.5	45.3	45.3	64
65	49.1	42.6	48.3	43.5	48.2	43.7	47.5	44.3	46.8	45.2	46.0	46.0	65
66	49.8	43.3	49.0	44.2	48.9	44.3	48.3	45.0	47.5	45.8	46.7	46.7	66
67	50.6	44.0	49.8	44.8	49.6	45.0	49.0	45.7	48.2	46.5	47.4	47.4	67
68	51.3	44.6	50.5	45.5	50.4	45.7	49.7	46.4	48.9	47.2	48.1	48.1	68
69	52.1	45.3	51.3	46.2	51.1	46.3	50.5	47.1	49.6	47.9	48.8	48.8	69
70	52.8	45.9	52.0	46.8	51.9	47.0	51.2	47.7	50.4	48.6	49.5	49.5	70
71	53.6	46.6	52.8	47.5	52.6	47.7	51.9	48.4	51.1	49.3	50.2	50.2	71
72	54.3	47.2	53.5	48.2	53.4	48.4	52.7	49.1	51.8	50.0	50.9	50.9	72
73	55.1	47.9	54.2	48.8	54.1	49.0	53.4	49.8	52.5	50.7	51.6	51.6	73
74	55.8	48.6	55.0	49.5	54.8	49.7	54.2	50.5	53.2	51.4	52.3	52.3	74
75	56.6	49.2	55.7	50.2	55.6	50.4	54.9	51.2	53.9	52.1	53.0	53.0	75
76	57.4	49.9	56.5	50.9	56.3	51.0	55.6	51.8	54.7	52.8	53.7	53.7	76
77	58.1	50.5	57.2	51.5	57.1	51.7	56.3	52.5	55.4	53.5	54.4	54.4	77
78	58.9	51.2	58.0	52.2	57.8	52.4	57.0	53.2	56.1	54.2	55.1	55.1	78
79	59.6	51.8	58.7	52.9	58.5	53.1	57.8	53.9	56.8	54.9	55.9	55.9	79
80	60.4	52.5	59.4	53.5	59.3	53.7	58.5	54.6	57.5	55.6	56.6	56.6	80
81	61.1	53.1	60.2	54.2	60.0	54.4	59.2	55.2	58.3	56.3	57.3	57.3	81
82	61.9	53.8	60.9	54.9	60.8	55.1	60.0	55.9	59.0	57.0	58.0	58.0	82
83	62.6	54.5	61.7	55.5	61.5	55.7	60.7	56.6	59.7	57.7	58.7	58.7	83
84	63.4	55.1	62.4	56.2	62.2	56.4	61.4	57.3	60.4	58.4	59.4	59.4	84
85	64.1	55.8	63.2	56.9	63.0	57.1	62.2	58.0	61.1	59.0	60.1	60.1	85
86	64.9	56.4	63.9	57.5	63.7	57.8	62.9	58.7	61.9	59.7	60.8	60.8	86
87	65.7	57.1	64.6	58.2	64.5	58.4	63.6	59.1	62.6	60.4	61.5	61.5	87
88	66.4	57.7	65.4	58.9	65.2	59.1	64.4	60.0	63.3	61.1	62.2	62.2	88
89	67.2	58.4	66.1	59.6	65.9	59.8	65.1	60.7	64.0	61.8	62.9	62.9	89
90	67.9	59.0	66.9	60.2	66.7	60.4	65.8	61.4	64.7	62.5	63.6	63.6	90
91	68.7	59.7	67.6	60.9	67.4	61.1	66.6	62.1	65.3	63.2	64.3	64.3	91
92	69.4	60.4	68.4	61.6	68.2	61.8	67.3	62.7	66.2	63.9	65.1	65.1	92
93	70.2	61.0	69.1	62.2	68.9	62.5	68.0	63.4	66.9	64.6	65.8	65.8	93
94	70.9	61.7	69.9	62.9	69.6	63.1	68.8	64.1	67.6	65.3	66.5	66.5	94
95	71.7	62.3	70.6	63.6	70.4	63.8	69.5	64.8	68.3	66.0	67.2	67.2	95
96	72.5	63.0	71.3	64.2	71.1	64.5	70.2	65.5	69.1	66.7	67.9	67.9	96
97	73.2	63.6	72.1	64.9	71.9	65.2	70.9	66.2	69.8	67.4	68.6	68.6	97
98	74.0	64.3	72.8	65.6	72.6	65.8	71.7	66.9	70.5	68.1	69.3	69.3	98
99	74.7	65.0	73.6	66.2	73.4	66.5	72.4	67.5	71.2	68.8	70.0	70.0	99
100	75.5	65.6	74.3	66.9	74.1	67.2	73.1	68.2	71.9	69.5	70.7	70.7	100
Dist.	Dep	Lat	Dep	Lat	Dep	Lat	Dep	Lat	Dep	Lat	Dep	Lat	Dist.

| | 49 Deg. | | 48 Deg. | | 4¼ Point | | 47 Deg. | | 46 Deg. | | 4 Points | |

A

TABLE

OF

Meridional Parts,

TO EVERY

DEGREE and MINUTE

OF THE

MERIDIAN,

For performing feveral Cafes in

NAVIGATION,

According to

MERCATOR,

Or, more properly,

Mr. *WRIGHT*'s Projection.

T 4

M	o d	1 d	2 d	3 d	4 d	5 d	6 d	7 d	8 d	9 d	10 d	M
0	0	60	120	180	240	300	361	421	481	541	601	0
1	1	61	121	181	241	301	362	422	482	542	602	1
2	2	62	122	182	242	302	363	423	483	543	603	2
3	3	63	123	183	243	303	364	424	484	544	604	3
4	4	64	124	184	244	304	365	425	485	545	605	4
5	5	65	125	185	245	305	366	426	486	546	606	5
6	6	66	126	186	246	306	367	427	487	547	607	6
7	7	67	127	187	247	307	368	428	488	548	608	7
8	8	68	128	188	248	308	369	429	489	549	609	8
9	9	69	129	189	249	309	370	430	490	550	610	9
10	10	70	130	190	250	310	371	431	491	551	611	10
11	11	71	131	191	251	311	372	432	492	552	612	11
12	12	72	132	192	252	312	373	433	493	553	613	12
13	13	73	133	193	253	313	374	434	494	554	614	13
14	14	74	134	194	254	314	375	435	495	555	615	14
15	15	75	135	195	255	315	376	436	496	556	616	15
16	16	76	136	196	256	316	377	437	497	557	617	16
17	17	77	137	197	257	317	378	438	498	558	618	17
18	18	78	138	198	258	318	379	439	499	559	619	18
19	19	79	139	199	259	319	380	440	500	560	620	19
20	20	80	140	200	260	320	381	441	501	561	621	20
21	21	81	141	201	261	321	382	442	502	562	622	21
22	22	82	142	202	262	322	383	443	503	563	623	22
23	23	83	143	203	263	323	384	444	504	564	624	23
24	24	84	144	204	264	324	385	445	505	565	625	24
25	25	85	145	205	265	325	386	446	506	566	626	25
26	26	86	146	206	266	326	387	447	507	567	627	26
27	27	87	147	207	267	327	388	448	508	568	628	27
28	28	88	148	208	268	328	389	449	509	569	629	28
29	29	89	149	209	269	329	390	450	510	570	630	29
30	30	90	150	210	270	330	391	451	511	571	631	30
31	31	91	151	211	271	331	392	452	512	572	632	31
32	32	92	152	212	272	332	393	453	513	573	633	32
33	33	93	153	213	273	333	394	454	514	574	634	33
34	34	94	154	214	274	334	395	455	515	575	635	34
35	35	95	155	215	275	335	396	456	516	576	636	35
36	36	96	156	216	276	336	397	457	517	577	637	36
37	37	97	157	217	277	337	398	458	518	578	638	37
38	38	98	158	218	278	338	399	459	519	579	639	38
39	39	99	159	219	279	339	400	460	520	580	640	39
40	40	100	160	220	280	340	401	461	521	581	641	40
41	41	101	161	221	281	341	402	462	522	582	642	41
42	42	102	162	222	282	342	403	463	523	583	643	42
43	43	103	163	223	283	343	404	464	524	584	644	43
44	44	104	164	224	284	344	405	465	525	585	645	44
45	45	105	165	225	285	345	406	466	526	586	646	45
46	46	106	166	226	286	346	407	467	527	587	647	46
47	47	107	167	227	287	347	408	468	528	588	648	47
48	48	108	168	228	288	348	409	469	529	589	649	48
49	49	109	169	229	289	349	410	470	530	590	650	49
50	50	110	170	230	290	350	411	471	531	591	651	50
51	51	111	171	231	291	351	412	472	532	592	652	51
52	52	112	172	232	292	352	413	473	533	593	653	52
53	53	113	173	233	293	353	414	474	534	594	654	53
54	54	114	174	234	294	354	415	475	535	595	655	54
55	55	115	175	235	295	355	416	476	536	596	656	55
56	56	116	176	236	296	356	417	477	537	597	657	56
57	57	117	177	237	297	357	418	478	538	598	658	57
58	58	118	178	238	298	358	419	479	539	599	659	58
59	59	119	179	239	299	359	420	480	540	600	660	59

M	11 d	12 d	13 d	14 d	15 d	16 d	17 d	18 d	19 d	20 d	21 d	M
0	664	725	787	848	910	973	1035	1098	1161	1225	1289	0
1	665	726	788	849	911	974	36	99	63	26	90	1
2	666	727	789	851	913	975	37	1100	64	27	92	2
3	667	728	790	852	914	976	38	01	65	28	92	3
4	668	729	791	853	915	977	39	42	66	29	93	4
5	669	730	792	854	916	978	41	03	67	30	95	5
6	670	731	793	855	917	979	42	05	68	32	96	6
7	671	732	794	856	918	980	43	06	69	33	97	7
8	672	733	795	857	919	981	44	07	70	34	98	8
9	673	735	796	858	920	982	45	08	71	35	99	9
10	674	736	797	859	921	983	1046	1109	1176	1236	1300	10
11	675	737	798	860	922	984	47	10	73	37	01	11
12	676	738	799	861	923	985	48	11	74	38	02	12
13	677	739	800	862	914	986	49	12	75	39	03	13
14	678	740	801	863	925	987	50	13	76	40	04	14
15	679	741	802	864	926	988	51	14	77	41	05	15
16	680	742	803	865	927	989	52	15	78	42	06	16
17	681	743	804	866	928	990	53	16	79	43	07	17
18	682	744	805	867	929	991	54	17	81	44	08	18
19	683	745	806	868	930	992	55	18	82	45	10	19
20	684	746	807	869	931	994	1056	1119	1181	1246	1311	20
21	685	747	808	870	912	995	57	20	84	48	12	21
22	686	748	809	871	933	996	58	21	47	49	13	22
23	688	749	810	872	934	997	59	22	86	50	14	23
24	689	750	811	873	935	998	60	23	87	51	15	24
25	690	751	812	874	936	999	61	25	88	52	16	25
26	691	752	813	875	937	1000	63	26	89	53	17	26
27	692	753	814	876	938	1001	64	27	50	54	18	27
28	693	754	816	877	939	1002	65	23	91	55	19	28
29	694	755	817	878	941	1003	66	29	92	56	20	29
30	695	756	818	879	942	1004	1067	1130	1193	1257	1321	30
31	696	757	819	880	943	05	68	31	94	58	22	31
32	697	758	820	881	944	06	69	32	95	59	24	32
33	698	759	821	882	945	07	70	33	96	60	25	33
34	699	760	822	884	946	08	71	34	97	61	85	34
35	720	761	823	885	947	09	72	35	99	62	27	35
36	701	762	824	886	948	10	73	36	1200	64	48	36
37	702	763	825	887	949	11	74	37	01	65	29	37
38	703	764	826	828	950	12	75	38	02	66	30	38
39	704	765	827	889	951	13	76	39	03	67	31	39
40	705	766	828	890	952	1014	1077	1140	1204	1268	1332	40
41	706	767	829	891	953	15	78	41	05	69	33	41
42	707	768	830	892	954	16	79	42	07	70	34	42
43	708	769	831	893	955	18	80	44	07	71	35	43
44	709	770	832	894	956	19	81	45	08	72	36	44
45	710	771	833	895	957	20	82	46	09	73	18	45
46	711	772	834	896	958	21	83	47	10	74	39	46
47	712	773	835	897	959	22	85	48	11	75	40	47
48	713	774	836	898	560	23	86	49	13	76	41	48
49	714	775	837	899	961	24	87	50	13	77	42	49
50	715	776	838	900	962	1025	1088	1151	1214	1278	1343	50
51	716	777	839	901	963	26	89	52	16	80	44	51
52	717	779	840	902	964	27	90	53	17	81	45	52
53	718	780	841	903	965	28	91	54	18	82	46	53
54	719	781	842	904	966	29	92	55	19	83	47	54
55	720	782	843	905	968	30	93	56	20	84	48	55
56	721	783	844	906	969	31	94	57	21	85	49	56
57	722	784	845	907	970	32	95	58	22	86	50	57
58	723	785	846	908	971	33	96	59	23	87	52	58
59	724	786	847	909	972	34	87	60	24	88	53	59

M	22 d	23 d	24 d	25 d	26 d	27 d	28 d	29 d	30 d	31 d	32 d	M
0	1356	1419	1464	1580	1616	1684	1751	1819	1889	1958	2028	0
1	55	10	85	51	18	85	52	21	50	59	10	1
2	56	21	86	52	19	86	53	22	91	60	11	2
3	57	22	87	53	20	87	55	23	92	61	32	3
4	58	23	88	54	21	88	56	24	93	63	33	4
5	59	24	90	56	22	89	57	25	94	64	34	5
6	60	25	91	57	23	90	58	25	95	65	35	6
7	61	26	92	58	24	91	59	27	96	66	37	7
8	62	27	93	59	25	93	60	29	98	67	38	8
9	63	28	94	60	26	94	61	30	99	69	39	9
10	1364	1419	1495	1561	1628	1655	1762	1831	1900	1970	22,0	10
11	66	31	96	62	29	96	64	32	01	71	41	11
12	67	32	97	63	30	97	65	33	02	72	43	12
13	68	33	98	64	31	98	66	34	03	73	44	13
14	69	34	59	65	32	99	67	35	05	74	45	14
15	70	35	1500	67	33	1200	68	37	06	76	45	15
16	71	36	02	68	34	01	69	38	07	77	47	16
17	72	37	03	69	35	03	70	39	08	78	48	17
18	73	38	04	70	37	04	72	40	09	79	50	18
19	74	39	05	71	38	05	73	41	10	80	51	19
20	1375	1440	1506	1572	1639	1706	1774	1842	1911	1981	20,2	20
21	76	41	07	73	40	07	75	43	13	83	53	21
22	77	43	08	74	41	08	76	45	14	84	54	22
23	78	44	09	75	42	09	77	46	15	85	56	23
24	80	45	10	77	43	10	78	47	16	86	57	24
25	81	46	11	78	44	12	80	48	17	87	58	25
26	82	47	11	79	45	13	81	49	18	88	59	26
27	83	48	14	80	47	14	82	50	20	90	60	27
28	84	49	15	81	48	15	83	52	21	91	61	28
29	85	50	16	82	49	16	84	53	22	92	63	29
30	1386	1451	1517	1583	1650	1717	1785	1854	1923	1993	2064	30
31	87	52	18	84	51	18	86	55	24	94	65	31
32	88	53	19	85	52	20	87	56	25	95	66	32
33	89	55	20	86	53	21	88	57	27	97	67	33
34	90	56	21	88	54	22	90	58	28	98	69	34
35	91	57	22	89	56	23	91	60	29	99	70	35
36	93	58	24	90	57	24	91	61	30	2000	71	36
37	94	59	25	91	58	25	93	62	01	01	72	37
38	95	60	26	92	59	26	94	63	32	02	73	38
39	96	61	27	93	60	27	95	64	34	04	75	39
40	1397	1462	1528	1594	1661	1729	1797	1865	1935	2005	2076	40
41	98	63	29	95	62	30	98	6	06	06	77	41
42	59	64	30	96	63	31	99	68	37	07	78	42
43	1400	65	31	98	64	31	1800	69	38	08	79	43
44	01	67	32	99	66	33	01	70	39	10	80	44
45	02	68	33	1600	67	34	02	71	41	11	81	45
46	03	69	35	01	68	35	03	72	42	12	83	46
47	05	70	36	02	69	36	05	73	43	13	84	47
48	06	71	37	03	70	38	06	75	44	14	85	48
49	07	72	38	04	71	39	07	76	45	15	86	49
50	1408	1473	1519	1603	1672	1740	1808	1877	1946	2017	2088	50
51	09	74	40	06	73	41	09	78	48	18	89	51
52	10	75	41	08	75	42	10	79	49	19	50	52
53	11	76	42	09	76	43	11	80	50	20	91	53
54	12	77	43	10	77	44	13	81	51	21	92	54
55	13	79	44	11	78	45	14	83	52	23	94	55
56	14	80	46	12	79	47	15	84	53	24	95	56
57	15	81	47	13	80	48	16	85	55	25	96	57
58	16	82	48	14	81	49	17	86	56	26	97	58
59	18	83	49	15	82	50	18	87	57	27	98	59

66	30	11	01	84
27	60	34	09	63
88	61	34	10	86
90	62	37	11	87
91	64	38	11	89
92	65	39	14	60
51	66	40	15	91
94	6	42	16	92
2196	2167	2343	2418	2454
97	70	44	19	95
98	71	45	20	56
99	72	46	21	58
2100	74	48	23	99
02	75	49	26	2500
03	76	50	25	01
04	77	51	27	03
05	78	53	28	04
07	80	54	29	05
2208	2281	2355	2430	2506
09	82	56	32	08
10	83	58	33	09
11	85	59	34	10
12	86	60	35	12
14	87	61	37	13
15	88	63	38	14
16	50	64	39	15
17	91	65	40	17
19	92	66	42	18
2220	2393	2368	2443	2519
21	95	69	44	21
22	96	70	45	22
24	97	71	47	23
25	58	73	46	24
26	90	74	45	26

M	44 d	45 d	46 d	47 d	48 d	49 d	50 d	51 d	52 d	53 d	54 d	M/c
0	2946	3010	3116	3203	3292	3382	3474	3569	3665	3764	3865	c
1	47	11	17	04	93	84	76	70	67	65	66	1
2	49	13	18	06	95	85	78	71	68	67	68	2
3	50	14	20	07	96	87	79	74	70	69	70	3
4	51	36	21	09	98	88	81	75	72	70	71	4
5	53	17	123	10	99	90	82	77	73	72	73	5
6	54	18	24	12	3301	91	84	78	75	74	75	6
7	56	40	26	13	02	93	85	80	77	75	77	7
8	57	27	14	03	94	87	82	78	77	78	8	
9	58	41	29	16	05	96	88	83	80	79	80	9
10	2960	3044	3130	3217	3306	3397	3490	3585	3682	3780	3882	10
11	61	46	31	19	08	99	92	86	83	82	83	11
12	63	47	33	20	3400	91	88	85	84	85	12	
13	64	48	34	22	11	02	95	90	86	85	87	13
14	65	50	36	23	12	03	96	91	88	87	89	14
15	67	51	37	25	14	05	98	93	90	89	90	15
16	68	53	39	26	16	07	99	94	91	90	92	16
17	70	54	40	28	17	08	3501	96	93	92	94	17
18	71	55	42	29	19	10	03	98	95	94	95	18
19	72	57	43	13	20	11	04	99	96	95	97	19
20	2974	3058	3144	3232	3222	3413	3506	3601	3698	3797	99	20
21	75	60	46	14	23	14	07	02	99	99	3901	21
22	76	61	47	35	25	16	09	04	3701	3800	02	22
23	78	63	49	37	26	17	10	06	03	02	04	23
24	79	64	50	38	28	19	12	07	04	04	06	24
25	81	65	52	40	29	20	14	09	06	06	07	25
26	82	67	53	41	31	22	15	10	08	07	09	26
27	83	68	55	42	32	23	17	12	09	09	11	27
28	85	70	56	44	34	25	18	14	11	11	13	28
29	86	71	57	45	35	27	20	15	13	12	14	29
30	2988	3073	3159	3247	3317	3428	3521	3617	3714	3814	3916	30
31	89	74	60	48	38	30	23	18	16	16	18	31
32	91	75	62	50	40	31	25	20	17	17	19	32
33	92	77	63	51	41	33	26	22	19	19	21	33
34	93	78	65	53	43	34	28	23	21	21	23	34
35	95	80	66	54	44	36	29	25	22	23	25	35
36	96	81	68	56	46	37	31	26	24	24	26	36
37	98	83	69	57	47	39	32	28	26	26	28	37
38	99	84	71	59	49	40	34	30	27	27	30	38
39	3000	85	72	60	50	42	36	31	29	29	32	39
40	3002	3087	3173	3262	3352	3443	3537	3633	3731	3831	3933	40
41	03	88	75	63	53	45	39	34	32	32	35	41
42	05	90	76	65	55	47	40	36	34	34	37	42
43	06	91	78	66	56	48	42	38	36	36	38	43
44	07	93	79	68	58	50	43	39	37	38	40	44
45	09	94	81	69	59	51	45	41	39	39	42	45
46	10	95	82	71	61	53	47	43	41	41	44	46
47	12	97	84	72	62	54	48	44	42	43	45	47
48	13	98	85	74	64	56	49	46	44	45	47	48
49	14	3100	87	75	65	57	50	47	46	46	49	49
50	3016	3101	3188	3277	3167	3459	3553	3649	3747	3848	3951	50
51	17	03	90	78	68	60	55	51	49	49	52	51
52	19	04	91	80	70	62	56	52	50	51	54	52
53	20	05	92	81	71	64	58	54	52	53	56	53
54	21	07	94	83	73	65	59	55	54	55	58	54
55	23	08	95	84	74	67	60	57	55	56	59	55
56	24	10	97	86	76	68	62	59	57	58	61	56
57	26	11	98	87	78	70	64	60	59	60	63	57
58	27	13	3200	89	79	71	65	62	60	61	64	58
59	28	14	01	90	86	73	67	64	62	63	66	59

M	55 d	56 d	57 d	58 d	59 d	60 d	61 d	62 d	63 d	64 d	65 d	M
0	3568	4074	4183	4254	4109	4527	4649	4775	4905	5019	5179	0
1	70	76	84	96	11	29	51	77	07	42	81	1
2	71	77	86	4198	13	31	33	79	09	44	84	2
3	73	79	88	00	15	33	55	81	12	46	86	3
4	75	81	90	02	17	35	57	60	14	49	88	4
5	77	83	92	04	19	37	60	86	16	51	91	5
6	78	85	94	06	21	39	62	88	18	53	93	6
7	80	86	95	08	23	41	65	90	20	55	95	7
8	82	88	97	09	25	43	66	92	23	58	98	8
9	84	90	99	11	27	45	68	94	25	60	5200	9
10	3585	4092	4201	4313	4129	4547	4670	4796	4927	5062	5203	10
11	87	94	03	15	31	49	72	98	29	65	03	11
12	89	95	05	17	33	51	74	4801	31	67	07	12
13	91	97	07	19	34	53	76	03	34	69	10	13
14	92	99	08	21	36	55	78	05	36	71	12	14
15	94	5101	10	23	38	57	80	07	38	74	14	15
16	96	03	12	25	40	59	82	09	40	76	17	16
17	98	04	14	27	42	62	84	11	43	78	19	17
18	99	06	16	28	44	64	87	14	45	81	22	18
19	4601	08	18	30	46	66	89	16	47	83	24	19
20	4603	4110	4220	4332	4148	4568	4691	4818	4949	5085	5226	20
21	05	12	21	34	50	70	93	20	51	88	29	21
22	06	13	23	36	52	72	95	22	54	90	31	22
23	08	15	25	38	54	74	97	24	56	92	34	23
24	10	17	27	40	56	76	99	26	58	95	36	24
25	12	19	29	42	58	78	4701	29	60	97	38	25
26	14	21	31	44	60	80	03	31	63	99	41	26
27	15	22	32	46	62	82	05	33	65	5102	43	27
28	17	24	34	47	64	84	07	35	67	04	46	28
29	19	26	36	49	66	86	10	37	69	06	48	29
30	4621	4128	4238	4351	4168	4588	4712	4839	4972	5108	5250	30
31	22	30	40	53	70	90	14	42	74	11	53	31
32	24	31	42	55	72	92	16	44	76	13	55	32
33	26	33	44	57	74	94	18	46	78	15	58	33
34	28	35	46	59	61	96	20	48	81	18	60	34
35	29	37	47	61	78	98	22	70	83	20	63	35
36	31	39	49	63	80	4600	24	52	85	22	65	36
37	33	41	51	65	82	02	26	55	87	25	67	37
38	35	42	53	67	84	04	28	57	90	27	70	38
39	37	44	55	69	86	06	31	59	92	29	72	39
40	4638	4146	4257	4370	4188	4608	4733	4861	4994	5132	5275	40
41	40	48	59	72	90	10	13	63	96	36	77	41
42	41	50	60	74	92	12	17	65	99	36	80	42
43	44	52	62	76	94	14	19	68	5001	19	82	43
44	45	53	64	78	95	16	41	70	03	41	84	44
45	47	55	65	80	97	18	41	72	05	43	87	45
46	49	57	68	82	99	20	43	74	08	46	89	46
47	51	59	70	84	4501	23	47	76	10	48	92	47
48	52	61	72	86	03	25	50	79	12	51	94	48
49	54	62	74	88	05	27	52	81	14	11	97	49
50	4656	4164	4275	4390	4507	4629	5754	4883	5017	5155	5259	50
51	58	66	77	92	09	31	56	85	19	31	5301	51
52	60	68	79	94	11	33	58	87	21	60	04	52
53	61	70	81	56	11	35	60	90	23	61	06	53
54	63	72	83	98	16	37	62	92	26	63	09	54
55	65	73	85	99	17	19	64	94	28	67	11	55
56	67	75	87	01	19	41	66	96	30	69	14	56
57	69	77	89	03	21	43	69	98	33	72	16	57
58	70	79	91	05	23	45	71	4501	35	74	19	58
59	72	81	92	07	25	47	73	03	37	76	21	59

M.	66 d	67 d	68 d	69 d	70 d	71 d	72 d	73 d	74 d	75 d	76 d	M
0	5324	5474	5631	5758	5906	6114	5335	6532	5746	6570	7210	0
1	26	77	33	97	69	49	38	38	49	74	14	1
2	28	79	36	5800	72	52	41	41	53	78	18	2
3	31	82	39	03	75	55	45	45	57	82	22	3
4	33	84	41	06	78	58	48	49	60	86	27	4
5	36	87	44	09	81	61	51	52	64	90	31	5
6	38	89	47	11	84	64	54	55	68	94	35	6
7	41	92	50	14	86	67	58	57	71	97	39	7
8	43	95	52	17	89	61	61	62	75	7001	43	8
9	46	97	55	20	92	73	64	65	79	05	47	9
10	5148	5500	5658	5823	5995	6177	5367	6369	5783	7009	7252	10
11	51	02	60	25	98	60	71	72	86	13	56	11
12	53	05	63	28	6001	83	74	76	90	17	60	12
13	56	07	66	31	04	86	77	76	93	21	64	13
14	58	10	68	34	07	89	80	83	97	25	68	14
15	61	13	71	37	10	92	84	8.	5801	29	73	15
16	63	15	74	39	13	95	87	96	03	33	77	16
17	66	18	76	42	16	98	90	93	08	37	81	17
18	68	20	79	45	19	6201	94	9.	12	41	85	18
19	71	23	81	48	22	05	97	6600	15	45	89	19
20	5373	5526	5685	5851	6025	6208	6400	6603	5819	7048	7254	20
21	76	28	87	54	28	11	01	07	23	52	98	21
22	78	31	90	56	31	14	07	10	26	56	7302	22
23	80	33	93	59	34	17	10	14	30	60	06	23
24	83	36	95	62	37	20	13	17	34	64	11	24
25	85	39	98	65	40	23	17	21	38	68	15	25
26	88	41	5701	68	43	26	20	24	41	72	19	26
27	90	44	04	71	46	30	23	28	45	76	23	27
28	93	46	06	74	49	33	27	31	49	80	28	28
29	95	49	09	76	52	36	30	35	53	84	32	29
30	5398	5552	5712	5879	6055	6239	6411	6619	6856	7088	7316	30
31	5401	54	15	82	58	42	37	42	60	92	41	31
32	03	57	17	85	61	45	46	64	96	45	32
33	06	59	20	88	64	49	43	49	68	7100	49	33
34	08	62	23	91	67	52	47	51	71	04	53	34
35	11	65	25	94	70	55	50	56	75	08	5.	35
36	13	67	28	96	73	58	53	60	79	12	62	36
37	16	70	31	99	76	61	57	63	83	16	66	37
38	18	73	34	5902	79	64	60	67	86	20	71	38
39	21	75	36	05	82	68	63	70	90	24	75	39
40	5423	5578	5739	5908	6085	6271	6467	6674	6894	7128	7379	40
41	26	80	42	11	88	74	70	77	98	32	84	41
42	28	83	45	14	91	77	73	81	6901	36	88	42
43	31	86	47	17	94	80	77	85	05	40	92	43
44	33	88	50	19	97	83	80	88	09	45	97	44
45	36	91	53	22	6100	87	83	92	13	49	7401	45
46	38	94	56	25	03	90	87	95	17	53	06	46
47	41	96	58	28	06	93	90	99	20	57	10	47
48	43	99	61	31	09	96	94	6702	24	61	14	48
49	46	5602	64	34	12	99	97	06	28	65	19	49
50	5448	5604	5767	5937	6115	6303	6500	6710	6932	7169	7423	50
51	51	07	70	40	18	06	04	13	36	73	27	51
52	54	10	72	43	21	09	07	17	40	77	32	52
53	56	12	75	46	24	12	11	20	43	81	36	53
54	59	15	78	48	27	15	14	24	47	85	41	54
55	61	17	81	51	30	19	17	28	51	89	45	55
56	64	20	84	54	33	22	21	31	55	94	54	56
57	66	23	86	57	36	25	24	35	58	98	49	57
58	69	25	89	60	40	28	28	38	61	7202	58	58
59	71	28	92	63	43	32	31	42	66	06	63	59

M	77 d	78 d	79 d	80 d	81 d	82 d	83 d	84 d	85 d	M
0	7467	7745	8046	8375	8739	9145	9606	10137	10765	0
1	72	49	51	81	45	51	14	147	770	1
2	76	54	56	87	51	60	22	157	788	2
3	81	59	61	93	58	67	31	165	799	3
4	85	64	67	58	65	74	39	175	811	4
5	90	69	72	8404	71	82	47	185	812	5
6	94	74	77	10	78	89	55	194	824	6
7	98	78	83	16	84	96	64	205	845	7
8	7503	83	88	22	91	9101	72	214	858	8
9	07	88	93	27	97	11	80	224	869	9
10	7512	7793	8099	8433	8804	9118	9689	10234	881	10
11	16	98	8104	39	10	25	97	244	893	11
12	21	7803	09	45	17	33	5706	254	906	12
13	25	08	15	51	23	40	14	264	917	13
14	30	13	20	57	30	48	23	273	929	14
15	35	17	25	63	36	55	31	283	941	15
16	39	22	31	69	43	61	40	293	953	16
17	44	27	36	74	49	70	48	303	965	17
18	48	32	41	80	56	77	57	314	978	18
19	53	37	47	86	61	85	65	324	990	19
20	7557	7842	8152	8492	8869	9191	9774	10334	11003	20
21	62	47	58	98	76	9300	83	344	014	21
22	66	52	63	8504	82	07	91	354	027	22
23	71	57	68	10	89	15	9800	364	039	23
24	76	62	74	16	96	22	09	374	052	24
25	80	67	79	22	8903	30	17	385	064	25
26	85	72	85	26	09	37	26	395	077	26
27	89	77	90	34	16	45	35	405	089	27
28	94	82	96	40	23	53	44	416	102	28
29	99	87	8201	46	30	60	52	426	114	29
30	7603	7892	8207	8552	8936	9168	9861	10437	11127	30
31	08	97	12	58	43	76	70	447	140	31
32	12	7902	18	65	50	81	79	457	153	32
33	17	07	23	71	57	91	88	468	166	33
34	22	12	29	77	63	99	97	478	179	34
35	26	17	34	83	70	9407	9906	489	191	35
36	31	22	40	89	77	14	15	500	205	36
37	36	27	45	95	84	22	24	510	218	37
38	40	32	51	8601	91	30	33	521	211	38
39	45	37	56	07	98	38	42	522	244	39
40	7650	7942	8262	8614	9005	9445	9951	10512	11257	40
41	54	48	67	20	12	53	59	533	270	41
42	59	53	71	26	18	61	69	564	283	42
43	64	58	76	32	25	69	78	575	297	43
44	68	63	84	38	32	77	87	586	310	44
45	73	68	90	44	39	85	9595	597	324	45
46	78	73	95	51	46	93	10005	608	337	46
47	83	78	8301	57	53	9501	10015	619	351	47
48	87	83	07	63	60	09	10024	630	365	48
49	92	89	12	69	67	17	10033	641	378	49
50	7697	7954	8318	8676	9074	9525	10043	10652	11392	50
51	7702	99	24	82	81	33	10052	663	406	51
52	06	8004	29	88	88	41	10061	674	420	52
53	11	09	35	95	96	49	10071	685	434	53
54	16	14	41	8701	9103	57	10080	695	448	54
55	21	20	47	07	10	65	10090	708	452	55
56	25	25	52	14	17	73	10099	719	476	56
57	30	30	58	20	24	81	10109	730	470	57
58	35	35	64	26	31	89	10118	742	504	58
59	40	40	69	33	38	98	10127	7.3	518	59

M	86 d	87 d	88 d	89 d	M
0	11355	12522	13916	16500	0
1	547	541	948	357	1
2	561	561	974	416	2
3	576	580	14003	476	3
4	590	599	033	537	4
5	605	619	063	599	5
6	620	639	093	662	6
7	634	659	123	726	7
8	649	679	154	792	8
9	664	699	184	838	9
10	11679	12719	14216	16926	10
11	694	739	247	996	11
12	709	759	279	17067	12
13	724	780	311	139	13
14	739	801	343	213	14
15	755	821	376	289	15
16	770	842	408	366	16
17	785	863	441	445	17
18	801	884	475	526	18
19	816	906	509	609	19
20	11832	12927	14543	17693	20
21	848	949	578	781	21
22	863	970	613	870	22
23	879	992	648	952	23
24	895	13014	684	18056	24
25	911	036	720	153	25
26	927	059	756	252	26
27	943	081	793	355	27
28	959	104	830	461	28
29	976	126	868	570	29
30	11992	13149	14906	18682	30
31	12008	172	943	799	31
32	025	195	983	920	32
33	041	219	15022	19045	33
34	058	242	062	174	34
35	075	266	102	309	35
36	092	290	143	450	36
37	109	314	184	596	37
38	126	338	226	749	38
39	143	362	268	909	39
40	12160	13386	15311	20076	40
41	177	411	354	253	41
42	194	437	398	439	42
43	212	461	442	635	43
44	229	486	487	843	44
45	247	511	532	21063	45
46	264	537	579	303	46
47	282	563	625	557	47
48	300	589	673	832	48
49	318	615	721	22132	49
50	12336	13641	15770	22459	50
51	354	668	819	21821	51
52	373	695	869	23226	52
53	391	721	920	23685	53
54	409	749	972	24215	54
55	428	776	16024	24842	55
56	447	804	078	25609	56
57	465	832	132	26598	57
58	484	860	187	27992	58
59	503	888	243	30375	59

Neceſſary

TABLES

OF THE

SUN's DECLINATION,

AND

Latitude and Longitude of Places,

WITH

A TABLE of the Magnitudes,
Right Aſcenſions and Declinations
of ſome of the Principal

FIXED STARS:

ALSO,

A TABLE

OF THE

SUN's RIGHT ASCENSION, &c.

U

A Table of the Sun's Declination.
First after Leap-Year,
1769, 1773, 1777, 1781, 1785.

Days	Jan. South		Feb. South		March South		April North		May North		June North	
	D	M	D	M	D	M	D	M	D	M	D	M
1	22	59	16	57	07	24	04	44	15	14	22	09
2	22	54	16	40	07	01	05	07	15	32	22	17
3	22	48	16	22	06	38	05	31	15	50	22	24
4	22	43	16	04	06	14	05	54	16	07	22	31
5	22	35	15	56	05	51	06	17	16	24	22	38
6	22	27	15	27	05	28	06	39	16	41	22	44
7	22	19	15	08	05	05	07	02	16	58	22	50
8	22	11	14	49	04	41	07	24	17	14	22	55
9	22	03	14	30	04	18	07	46	17	30	23	00
10	21	54	14	11	03	54	08	08	17	45	23	05
11	21	45	13	52	03	30	08	30	18	01	23	09
12	21	35	13	32	03	06	08	52	18	16	23	13
13	21	25	13	12	02	42	09	14	18	31	23	17
14	21	15	12	51	02	19	09	35	18	45	23	20
15	21	05	12	30	01	55	09	57	19	00	23	22
16	20	54	12	09	01	32	10	18	19	14	23	24
17	20	42	11	48	01	08	10	39	19	27	23	26
18	20	29	11	27	00	44	11	00	19	40	23	27
19	20	16	11	05	00	21	11	21	19	53	23	28
20	20	02	10	43	No.	02	11	41	20	06	23	29
21	19	48	10	22	00	26	12	02	20	18	23	29
22	19	34	10	00	00	59	12	22	20	30	23	29
23	19	20	09	38	01	14	12	42	20	41	23	28
24	19	05	09	16	01	37	13	02	20	52	23	27
25	18	50	08	54	02	01	13	21	21	03	23	25
26	18	35	08	31	02	24	13	40	21	14	23	23
27	18	20	08	09	02	48	14	00	21	24	23	21
28	18	04	07	46	03	11	14	19	21	34	23	18
29	17	48			03	35	14	38	21	43	23	15
30	17	31			03	58	14	56	21	52	23	11
31	17	14			04	21			22	01		

A Table of the Sun's Declination.
First after Leap-Year.

1769, 1773, 1777, 1781, 1785

Days	July North		August North		Sept. North		Octob. South		Nov. South		Decem. South	
	D	M	D	M	D	M	D	M	D	M	D	M
1	23	07	17	58	08	09	03	23	14	37	21	55
2	23	03	17	43	07	47	03	46	14	56	22	04
3	22	59	17	27	07	25	04	09	15	15	22	13
4	22	54	17	11	07	03	04	32	15	34	22	21
5	22	48	16	55	06	41	04	55	15	52	22	29
6	22	42	16	38	06	19	05	18	16	10	22	36
7	22	35	16	21	05	56	05	41	16	28	22	43
8	22	28	16	04	05	33	06	04	16	45	22	49
9	22	21	15	47	05	10	06	27	17	02	22	55
10	22	14	15	29	04	47	06	49	17	19	23	00
11	22	06	15	11	04	24	07	12	17	36	23	05
12	21	58	14	53	04	01	07	35	17	52	23	10
13	21	49	14	35	03	38	07	58	18	08	23	14
14	21	40	14	17	03	15	08	20	18	24	23	17
15	21	31	13	58	02	52	08	42	18	39	23	20
16	21	21	13	39	02	29	09	04	18	54	23	23
17	21	11	13	20	02	06	09	26	19	09	23	25
18	21	00	13	00	01	43	09	48	19	24	23	27
19	20	49	12	41	01	20	10	10	19	38	23	28
20	20	38	12	21	00	56	10	32	19	51	23	29
21	20	27	12	01	00	33	10	54	20	04	23	29
22	20	15	11	41	00	09	11	15	20	17	23	29
23	20	03	11	21	Son	14	11	36	20	30	23	28
24	19	50	11	00	00	38	11	57	20	42	23	27
25	19	37	10	39	01	01	12	18	20	54	23	26
26	19	24	10	18	01	25	12	38	21	05	23	23
27	19	10	09	57	01	48	12	59	21	16	23	20
28	18	56	09	36	02	12	13	29	21	26	23	17
29	18	42	09	15	02	36	13	39	21	36	23	14
30	18	28	08	53	03	00	13	58	21	46	23	10
31	18	13	08	31			14	17			23	06

A Table of the Sun's Declination.
Second after Leap-Year,
1770, 1774, 1778, 1782, 1786.

Days	Jan. South		Feb. South		March South		April North		May North		June North	
	D	M	D	M	D	M	D	M	D	M	D	M
1	23	00	17	00	07	28	04	39	15	09	22	06
2	22	54	16	43	07	05	05	02	15	27	22	14
3	22	48	16	25	06	42	05	25	15	45	22	21
4	22	42	16	07	06	19	05	48	16	03	22	29
5	22	35	15	49	05	56	06	11	16	20	22	35
6	22	28	15	31	05	32	06	34	16	37	22	42
7	22	21	15	12	05	09	06	56	16	53	22	47
8	22	13	14	53	04	46	07	19	17	10	22	53
9	22	04	14	34	04	25	07	41	17	26	22	58
10	21	55	14	14	03	59	08	03	17	41	23	03
11	21	46	13	55	03	35	08	25	17	57	23	07
12	21	36	13	35	03	12	08	47	18	12	23	11
13	21	26	13	15	02	48	09	09	18	27	23	15
14	21	15	12	54	02	24	09	30	18	42	23	18
15	21	04	12	34	02	01	09	52	18	56	23	20
16	20	53	12	13	01	37	10	13	19	10	23	23
17	20	41	11	52	01	13	10	34	19	23	23	25
18	20	29	11	31	00	50	10	55	19	37	23	26
19	20	16	11	09	00	26	11	16	19	50	23	27
20	20	03	10	48	00	02	11	37	20	02	23	28
21	19	50	10	26	No.	21	11	57	20	15	23	28
22	19	36	10	04	00	45	12	17	20	26	23	28
23	19	22	09	42	01	09	12	37	20	38	23	27
24	19	08	09	20	01	32	12	57	20	49	23	26
25	18	53	08	58	01	56	13	17	21	00	23	25
26	18	38	08	36	02	19	13	36	21	11	23	23
27	18	22	08	13	02	43	13	55	21	21	23	21
28	18	07	07	50	03	06	14	14	21	31	23	18
29	17	50			03	30	14	33	21	40	23	15
30	17	34			03	53	14	51	21	49	23	11
31	17	07			04	10			21	58		

A Table of the Sun's Declination.
Second after Leap-Year.

1770, 1774, 1778, 1782, 1786

Days	July North		Avguſt North		Sept. North		Octob. South		Nov. South		Decem. South	
	D	M	D	M	D	M	D	M	D	M	D	M
1	23	08	18	00	08	13	03	17	14	33	21	53
2	23	03	17	45	07	51	03	41	14	52	22	02
3	22	59	17	29	07	29	04	04	15	11	22	1J
5	22	54	17	13	07	07	04	27	15	29	22	19
5	22	48	16	57	06	45	04	50	15	48	22	26
6	22	42	16	40	06	23	05	13	16	06	22	34
7	22	36	16	24	06	00	05	36	16	23	22	40
8	22	29	16	07	05	37	05	59	16	41	22	47
9	22	22	15	50	05	15	06	22	16	58	22	53
10	22	15	15	32	04	52	06	45	17	15	22	58
11	22	07	15	14	04	29	07	08	17	32	23	03
12	21	59	14	56	04	06	07	30	17	48	23	08
13	21	50	14	38	03	43	07	53	18	04	23	12
14	21	41	14	20	03	20	08	15	18	20	23	16
15	21	32	14	01	02	57	08	38	18	36	23	19
16	21	22	13	42	02	34	09	00	18	51	23	21
17	21	12	13	23	02	10	09	22	19	05	23	24
18	21	01	13	04	01	47	09	44	19	20	23	25
19	20	51	12	44	01	24	10	06	19	34	23	27
20	20	40	12	24	01	00	10	27	19	47	23	28
21	20	28	12	04	00	37	10	49	20	01	23	28
22	20	16	11	44	00	14	11	10	20	14	23	28
23	20	04	11	24	Sou 10		11	31	20	26	23	27
24	19	52	11	03	00	33	11	52	20	39	23	26
25	19	39	10	43	00	57	12	13	20	50	23	25
26	19	26	10	22	01	20	12	34	21	02	23	23
27	19	12	10	01	01	44	12	54	21	13	23	20
28	18	58	09	40	02	07	13	14	21	24	23	17
29	18	44	09	18	02	30	13	34	21	34	23	14
30	18	30	08	57	02	54	13	54	21	44	23	11
31	18	15	08	36			14	14			23	07

U 3

A Table of the Sun's Declination.
Third after Leap-Year.

1771, 1775, 1779, 1783, 1787.

Days	Jan. South		Feb. South		March South		April North		May North		June North	
	D	M	D	M	D	M	D	M	D	M	D	M
1	23	02	17	05	07	31	04	33	15	05	22	04
2	22	57	16	48	07	11	04	56	15	23	22	12
3	22	51	16	31	06	48	05	19	15	41	22	20
4	22	45	16	13	05	25	05	42	15	59	22	28
5	22	38	15	55	06	02	06	05	16	16	22	35
6	22	31	15	36	05	39	06	28	16	33	22	41
7	22	23	15	18	05	16	06	51	16	50	22	47
8	22	15	14	59	04	52	07	13	17	06	22	53
9	22	07	14	40	04	29	07	35	17	22	22	58
10	21	59	14	20	04	05	07	57	17	38	23	03
11	21	50	14	00	03	42	08	20	17	54	23	07
12	21	40	13	40	03	18	08	42	18	09	23	11
13	21	30	13	20	02	55	09	04	18	24	23	14
14	21	19	13	00	02	31	09	25	18	38	23	17
15	21	08	12	40	02	07	09	47	18	53	23	20
16	20	57	12	19	01	43	10	08	19	07	23	23
17	20	45	11	58	01	20	10	29	19	20	23	25
18	20	33	11	37	00	56	10	50	19	33	23	27
19	20	20	11	16	00	33	11	11	19	46	23	28
20	20	07	10	54	00	09	11	31	19	59	23	29
21	19	54	10	32	Nor.	14	11	52	20	12	23	29
22	19	40	10	10	00	38	12	12	20	24	23	29
23	19	26	09	48	01	02	12	32	20	36	23	28
24	19	12	09	26	01	26	12	52	20	47	23	27
25	18	58	09	04	01	50	13	12	20	58	23	26
26	18	43	08	42	02	13	13	31	21	09	23	24
27	18	27	08	20	02	37	13	50	21	19	23	22
28	18	11	07	57	03	00	14	09	21	29	23	20
29	17	55			03	24	14	28	21	39	23	17
30	17	39			03	47	14	47	21	48	23	15
31	17	22			04	10			21	56		

A Table of the Sun's Declination. Third after Leap-Year,

1771, 1775, 1779, 1783, 1787.

Days	July North		August North		Sept. North		Octob. South		Nov. South		Decem. South	
	D	M	D	M	D	M	D	M	D	M	D	M
1	23	09	18	05	08	20	03	10	14	27	21	51
2	23	05	17	50	07	58	03	34	14	47	22	00
3	23	01	17	34	07	36	03	57	15	06	22	09
4	22	56	17	18	07	13	04	20	15	25	22	17
5	22	51	17	01	06	51	04	41	15	43	22	25
6	22	45	16	46	06	29	05	07	16	01	22	32
7	22	39	16	29	06	07	05	30	16	19	22	39
8	22	32	16	12	05	44	05	53	16	37	22	46
9	22	25	15	55	05	22	06	16	16	54	22	52
10	22	18	15	38	04	59	06	39	17	11	22	58
11	22	10	15	20	04	36	07	02	17	28	23	03
12	22	02	15	02	04	13	07	25	17	44	23	08
13	21	53	14	44	03	50	07	48	18	00	23	12
14	21	44	14	26	03	27	08	10	18	16	23	16
15	21	35	14	07	03	04	08	32	18	32	23	19
16	21	26	13	48	02	40	08	54	18	47	23	22
17	21	16	13	29	02	17	09	16	19	02	23	24
18	21	05	13	10	01	53	09	38	19	16	23	26
19	20	54	12	50	01	30	10	00	19	30	23	28
20	20	43	12	30	01	07	10	22	19	44	23	29
21	20	32	12	10	00	44	10	43	19	58	23	29
22	20	20	11	51	00	20	11	04	20	11	23	29
23	20	08	11	30	Sou.	03	11	25	20	24	23	28
24	19	56	11	09	00	26	11	46	20	36	23	27
25	19	43	10	49	00	50	12	07	20	48	23	26
26	19	30	10	28	01	13	12	28	20	59	23	24
27	19	17	10	07	01	37	12	49	21	10	23	22
28	19	03	09	46	02	00	13	09	21	21	23	19
29	18	49	09	25	02	23	13	29	21	31	23	16
30	18	35	09	04	02	46	13	49	21	41	23	12
31	18	20	08	42			14	08			23	08

A Table of the Sun's Declination. Leap-Year.

1772, 1776, 1780, 1784, 1788.

Days	Jan. South		Feb. South		March South		April North		May North		June North	
	D	M	D	M	D	M	D	M	D	M	D	M
1	23	03	17	09	07	16	04	50	15	19	22	10
2	22	57	16	52	06	53	05	13	15	37	22	18
3	22	51	16	35	06	30	05	36	15	54	22	25
4	22	45	16	17	06	07	05	59	16	11	22	32
5	22	39	15	59	05	44	06	22	16	28	22	39
6	22	33	15	41	05	21	06	45	16	45	22	45
7	22	26	15	23	04	58	07	08	17	02	22	51
8	22	18	15	04	04	34	07	30	17	18	22	56
9	22	10	14	45	04	11	07	52	17	34	23	01
10	22	01	14	25	03	47	08	14	17	50	23	06
11	21	52	14	05	03	24	08	36	18	05	23	10
12	21	42	13	45	03	00	08	58	18	20	23	13
13	21	32	13	25	02	37	09	20	18	35	23	17
14	21	22	13	05	02	13	09	42	18	49	23	20
15	21	11	12	45	01	49	10	03	19	03	23	21
16	21	00	12	24	01	25	10	24	19	17	23	25
17	20	48	12	03	01	01	10	45	19	31	23	27
18	20	36	11	42	00	38	11	06	19	44	23	28
19	20	23	11	21	00	14	11	27	19	57	23	29
20	20	10	10	59	Nor.	09	11	47	20	09	23	29
21	19	57	10	38	00	33	12	07	20	21	23	29
22	19	44	10	16	00	56	12	27	20	33	23	29
23	19	30	09	54	01	20	12	47	20	44	23	28
24	19	16	09	31	01	44	13	07	20	55	23	27
25	19	01	09	09	02	08	13	27	21	06	23	25
26	18	46	08	47	02	31	13	46	21	16	23	23
27	18	31	08	25	02	54	14	05	21	26	23	20
28	18	15	08	02	03	17	14	24	21	36	23	17
29	17	59	07	39	03	41	14	43	21	45	23	14
30	17	43			04	04	15	01	21	54	23	11
31	17	26			04	27			22	02		

A Table of the Sun's Declination. Leap-Year,

1772, 1776, 1780, 1784, 1788.

Day	July North		Auguſt North		Sept. North		Octob. South		Nov. South		Decem South	
	D	M	D	M	D	M	D	M	D	M	D	M
1	23	07	17	53	08	03	03	28	14	42	21	58
2	23	02	17	38	07	41	03	52	15	01	22	07
3	22	57	17	22	07	19	04	15	15	20	22	15
4	22	52	17	06	06	57	04	38	15	39	22	23
5	22	46	16	50	06	35	05	01	15	57	22	31
6	22	40	16	33	06	12	05	24	16	15	22	38
7	22	34	16	16	05	50	05	47	16	33	22	45
8	22	27	15	59	05	27	06	10	16	50	22	51
9	22	20	15	42	05	04	06	33	17	07	22	56
10	22	12	15	24	04	41	06	56	17	24	23	01
11	22	04	15	06	04	18	07	19	17	40	23	06
12	21	55	14	48	03	55	07	41	17	56	23	10
13	21	46	14	30	03	32	08	04	18	12	23	13
14	21	37	14	11	03	09	08	26	18	28	23	16
15	21	28	13	53	02	46	08	48	18	43	23	21
16	21	18	13	34	02	23	09	10	18	58	23	23
17	21	08	13	15	02	.00	09	32	19	13	23	25
18	20	57	12	55	01	36	09	54	19	27	23	27
19	20	46	12	36	01	13	10	16	19	41	23	28
20	20	35	12	16	00	49	10	39	19	54	23	29
21	20	24	11	56	00	26	10	59	20	07	23	30
22	20	12	11	35	00	02	11	20	20	20	23	30
23	19	59	11	15	Sou.	21	11	41	20	32	23	28
24	19	46	10	54	00	44	12	02	20	41	23	26
25	19	33	10	33	01	08	12	23	20	56	23	24
26	19	20	10	12	01	31	12	44	21	09	23	22
27	19	06	09	51	01	55	13	04	21	19	23	20
28	18	52	09	30	02	18	13	24	21	33	23	17
29	18	38	09	09	02	41	13	44	21	42	23	13
30	18	23	08	47	03	05	14	04	21	47	23	09
31	18	08	08	25			14	23			23	04

A TABLE of the Latitudes and Longitudes of the Principal Harbours, Headlands and Iflands in the WORLD: Corrected by the lateft and beft Obfervations: Accounting the Longitude from the Meridian of L O N D O N.

Note, *When the Latitude and Longitude of an Ifland is given, the middle of the Ifland is meant, except fome particular Parts of it be expreffed.*

Places Names.	Latitude N. or S.		Longitude E. or W.	
	D.	M.	D.	M.
B Erwick	55	50	1	44 W
Newcaftle	55	10	1	30 W
Stockton	54	33	1	25 W
Flambro'-head	54	8	0	10 E
Spurn	53	45	0	13 E
Yarmouth	52	45	1	40 E
Orfordnefs	52	15	1	12 E
L O N D O N	51	32	0	0
North Foreland	51	28	1	20 E
South Foreland	51	12	1	20 E
Beachy	50	48	0	25 E
Dunnofe Ifle of Wight	50	38	1	23
Portland	50	30	2	40
Tepfham	50	37	3	25
Start Point	50	9	3	45
Lizard	49	55	5	14
Land's-end	50	6	6	0
St. Mary's Scilly	49	57	6	45
Hartland Point	51	8	4	35
Lundy Ifle	51	20	4	40
Milford	51	44	5	00
Briftol	51	32	2	35

The Coaft of England. — North Latitude. — Weft Longitude.

Places Names.	Latitude N. or S.		Longitude E. or W.	
	D.	M.	D.	M.
The Coast of England.				
St. David's Head	51	55	05	22
Barsey Isle	52	46	04	58
Holy Head	53	23	04	40
Liverpool	53	20	02	58
Isle of Man S.W. End	53	45	05	00
White-haven	54	20	03	20
Carlisle	54	45	02	35
The Coast of Scotland.				
Glasgow	55	53	04	05
North part of Sky Island	57	50	05	35
N. part of Lewis Island	58	20	07	15
St. Kilda	58	10	09	45
Farra Head	58	33	05	05
N.ermost Isles of Orkney	59	13	03	30
Fair Isle	59	30	03	00
Shetland South Point	59	52	01	30
Buchaness	57	45	01	20
Aberdeen	57	24	01	37
Dundee	56	30	02	36
Edinburgh	55	57	02	59
The Sea Coasts about the Islands of Ireland.				
Grimes Hole, or Geuberman's Rock	66	23	29	30
Garmart Isles or Gille	65	48	27	30
Westmania Isles	63	37	23	17
Rock Point	64	00	26	13
Snow Hill	6:	11	27	14
Fair Foreland	66	00	26	17
Marza or Largerness	66	08	24	00
Grimsa Isle	67	22	22	44
Lange Ness	66	56	13	00
Scilly or Papey Island	64	50	12	10
Horn Bay	64	42	12	10
Merchants Foreland	63	25	17	06
Portland	64	2	21	05

Dublin

Places Names.	Latitude N. or S.		Longitude E. or W.	
	D.	M.	D.	M.
The Coast of Ireland.				
Dublin	53	12	06	56
Wexford	52	13	07	25
Waterford	52	07	07	52
Old Head of Kingſale	51	35	08	55
Cork	51	49	09	30
Cape Clear	51	10	10	30
Cow and Calf	51	22	10	36
Limrick	52	21	09	48
Galway	53	10	10	03
Sline-head	53	20	11	15
Londonderry	54	55	08	00
Belfaſt	54	36	06	40
The Coast of Holland and Flanders.				
Scaw	57	26	10	10
Helighland	54	28	08	35
Hambrough	53	41	10	24
Embden	53	05	08	50
The Fly	53	18	05	35
The Texel	53	15	05	09
Amſterdam	52	21	04	59
Rotterdam	51	55	04	25
The Brill	51	56	04	06
Oſtend	51	12	02	57
Sluyce	51	19	03	50
Calais	50	57	02	00
The Coast of France and Portugal.				
Bulloign	50	45	01	38
Diep	49	56	01	06
Cape de Hague	49	46	02	06
Caſkets	49	50	02	26
Guernſey	49	36	02	40
Morlaix	48	37	03	50
Iſle Baſs	49	00	04	00
Uſhant	48	30	05	05

Places Names.	Latitude N. or S.		Longitude E. or W.	
	D.	M.	D.	M.
The Coast of France and Portugal.				
Brest	48	23	04	32
Penmark	47	48	04	15
Bell-Isle	47	20	03	10
Nantz	47	14	01	42
Island Dieu	46	36	02	15
Island Ree	46	13	01	30
Rochel	46	10	01	14
Oleroon	46	00	00	54
Bourdeaux	44	50	00	28
Bilboa	43	30	03	00
Cape Pinas	43	51	06	00
Cape Ortegal	44	02	07	48
Cape Finisterre	43	10	09	40
Port a Port	41	16	09	20
Burlings	39	39	09	20
Rock of Lisbon	38	54	09	40
Lisbon	38	42	08	53
Cape St. Vincent	37	00	08	52
Cadiz	36	33	06	02
Cape Trefalgar	36	10	05	48
The Coast of the Main Continent within the Straits.				
Gibralter	36	11	05	03 W
Malaga	36	43	03	50 W
Cape de Gat	36	38	02	01 W
Cape Paul	37	50	00	15
Alicant	38	34	00	10
Cape Martin	36	46	00	28
Tortosa	40	54	01	10
Barcelona	41	30	02	21
Marseilles	43	20	05	19
Toulon	43	06	05	59
Geniva	44	25	08	40
Leghorn	43	30	10	29
Civita Vechia	42	10	12	03
				Rome

North Latitude. West Longitude.

North Latitude. East Longitude.

Places Names.	Latitude N. or S.		Longitude E. or W.	
	D.	M.	D.	M.
Rome ———————	41	54	12	45
Naples ———————	40	51	14	45
Cape Spartavanto ——	38	00	16	58
Cape Collone ———	38	56	18	06
Gallipoli ———————	40	08	18	39
Cape St. Maria ———	39	56	19	00
Ancona ———————	43	40	14	30
Venice ———————	45	25	12	10
Ragufa ———————	42	40	20	00
La Valona ———————	40	56	21	04
Lepanto ———————	38	20	22	15
C. Matapan ———————	36	35	22	30
C. St. Angelo ———	36	31	23	40
Cape Colona ———	37	40	24	44
Athens ———————	38	00	24	09
Cape Martelo S. P. of Negropont ———	38	00	25	45
Cape Monte Sancto —	40	25	25	05
Gallipoli ———————	40	30	27	20
Constantinople ———	41	00	29	00
Smyrna ———————	38	30	27	27
Ephefus ———————	37	54	27	55
Antiochetta ———	36	30	32	47
Scandaroon ———	36	34	36	08
Antiochia———————	36	11	36	25
Tripoli———————	34	40	36	10
Joppa———————	32	32	36	00
Cairo ———————	30	05	34	20
Alexandria———————	31	07	30	20
Cape Rufato ———	32	53	21	05
Cape Miferato ———	32	21	16	23
Tripoli ———————	32	55	13	00
Cape Bona —— ——	37	05	10	00
Bona ———————	37	00	07	20
				Tunis

(left margin, vertical): The Sea Coasts on the Main Continent within the Straits.

(right of latitude column, vertical): North Latitude.

(right of longitude column, vertical): East Longitude.

Places Names.	Latitude N. or S.		Longitude E. or W.	
	D.	M.	D.	M.
C. on the Main-Continent within the Straits.				
Tunis	36	50	10	16
Algiers	36	45	03	15
Oran	35	46	00	26
Cape de Tres forcas	35	34	02	04
Tetuan	35	28	05	06
Ceuta	35	50	04	45
Tangier	35	55	05	45
Islands within the Straits.				
Alboran	36	00	02	27 W
Formentura	38	34	01	31
Yvica	38	55	01	30
Majorca City	39	48	02	49
Port Mahon Minorca	40	42	04	19
Gallitta	37	40	09	03
Sardinia South End	38	54	09	12
Corsica North End	42	56	09	45
Gorgon	43	34	09	38
Capria	43	06	10	15
Eliboa	42	44	10	45
Messina	38	12	16	25
Palermo	38	12	13	48
Maritimo	38	04	12	25
Syracuse	37	05	15	55
Cape Passero	36	38	15	40
Malta	35	54	14	34
Limosa	36	08	13	11
Limpadosa	35	34	12	45
Sematto	35	46	14	15
Corfu	39	45	20	06
Cephalonia	38	15	21	00
Zant	37	47	21	14
Madon or Morea	36	52	21	32
Lemnos	39	59	25	37
Scio	38	20	26	12

Negro

Places Names.	Latitude N. or S.		Longitude E. or W.	
	D.	M.	D.	M.
Iſlands within the Straits. Negropont	38	36	24	05
CapeSt. JohnW. End of Candy	35	20	23	57
Cape Solomon, E. End of Candy	35	00	27	06
Rhodes City	36	40	28	00
Weſt End of Cyprus	35	40	32	23
Eaſt End of Cyprus	35	35	35	08
The Coaſt of Barbary and Guiney. Cape Spartel	35	46	05	55
Salle	33	43	06	30
Cape Cantin	32	26	09	10
Cape de Geer	30	24	10	06
Cape Non	28	15	11	04
Cape Bajador	26	12	14	30
Cape Olerado	23	41	15	50
Cape Blanco	20	35	17	23
Senegal	15	28	16	20
Cape de Verde	14	43	17	05
River Gambia	13	08	15	20
Serralion	08	36	12	15
Cape Monte	06	23	10	45
Miſerado	06	55	09	35
River Seſter	05	28	08	13
Cape Palmas	04	18	05	56
River St. Andrew	05	00	03	45
Jaque Jaque	04	16	02	45
Aſlene	04	15	02	20
Cape Three Points	04	28	01	21
River Volto	05	50	03	20
River Formoſa	07	00	07	40
Cape Formoſa	04	22	06	40
New Callabar	04	42	08	33
Old Callabar	04	18	09	45 River

Places Names.	Latitude N. or S.		Longitude E. or W.	
	D.	M.	D.	M.
The Coast of Barbary and Guiney.				
Safmons River —	3	10 N	10	8
River Camarones —	3	25 N	10	10
River de Angre —	0	49 N	10	1
Island Cabos —	0	40	11	36
Lopas —	0	55	9	30
River Congo —	5	45	15	25
Angola —	8	51	15	56
Cape Negro —	16	8	12	31
Cape St. Thomas —	24	10	14	43
Secos —	28	56	15	56
Cape Bona Esperance —	34	6	18	35
Western Islands.				
Corvo —	39	54	30	55
Flores —	39	32	30	54
Fyal —	38	53	28	16
Pico —	38	40	27	20
St. George —	38	52	26	03
Tercera —	38	57	25	34
St. Michael —	38	06	23	36
St. Maries —	37	co	23	38
Canary Islands.				
Ferro —	28	00	17	45
Palma —	28	40	17	36
Gomero —	28	08	17	06
Teneriff —	28	20	16	28
Madeira West-end —	32	23	17	26
Porto Sancto —	32	58	15 ·	54
Canaria —	27	52	15	10
Forteventura —	28	4	13	34
Lancerota —	29	2	12	44

(South Latitude · East Longitude for the Coast of Barbary and Guiney; North Latitude · West Longitude for the Western and Canary Islands.)

X Cape

Places Names.	Latitude N. or S.		Longitude E. or W.	
	D.	M.	D.	M.
Cape de Verde Islands				
St. Antonio	17	35	24	40
St. Vincent	17	15	24	26
St. Lucia	17	07	24	20
St. Nicholas	17	00	23	38
Brava	14	28	23	54
Fuego	14	52	23	30
Jago	15	08	22	50
Isle of May	15	16½	22	12
Isle Sal	16	45	22	04?
Bonavista	16	05	22	05
Southern Islands				
St. Matthews	01	30 S	06	11 W
Ascension	07	40 S	13	45 W
St. Helena	06	00 S	06	04 W
Fornandepo	02	50 N	10	40 E
Princes's Island	01	50 N	09	15 E
St. Thomas	00	00—	08	20 E
Annabona	02	10 S	05	35 E
Coasts of the Main Continent in the East-Indies.				
Cape Bona Esperance	34	06	18	35
Cape Lagullas	34	55	21	20
River St. Lucia	28	25	32	15
Cape Corientes	23	42	36	15
Misambique	15	05	41	10
Cape Falso	09	00	39	10
Tongon	04	50	39	7
Mombaso	03	55	38	30
Melinde	03	00	39	40
River Lamas	01	20	40	13
River de Fugor	00	00	42	5
Magadoxo	02	21 N	44	50
Cape Bassos	04	10 N	47	38
Cape Guardafoy	11	50 N	51	20
Cape Rosalgat	22	41 N	59	45
Cape Muca, or Muskat	23	36 N	57	40

Bussera

Places Names.	Latitude N or S.		Longitude E. or W.	
	D.	M.	D.	M.
Buffera	29	50	49	05
Gambaroon	27	30	56	36
Surrat	21	08	72	25
Goa	15	32	73	50
Calecute	11	17	75	34
Cochin	09	58	76	05
Cape Comarine	07	50	78	15
Columba in Zeloan	07	07	79	35
Fort St. George	13	08	80	42
Dew Point	15	59	81	25
Vifegapatam	17	40	84	07
Pondy	18	45	85	20
Cape Palmiris	20	45	88	00
Ballafore	21	10	87	50
Piply	21	25	87	58
Bengal	22	17	92	29
Cape Negrais	16	23	93	25
Malacca	02	11	102	10
Formofa	01	55	101	40
Siam Entrance	13	10	101	01
Cambodia Entrance	10	30	105	00
Cape Avarilla	13	20	108	03
Cochin, or Chinchen	14	05	107	58
Canton	23	08	113	08
Amoy or Quemöy	24	35	116	55
Lampo	30	10	120	25
Ifland Chufan	30	05	120	35
Nanqun	33	15	120	05
Madagafcar or ⎫ S. Eaft	25	47	46	10
St. Lawrence ⎭ N. Eaft	12	10	50	50
St. John de Lifbon	25	24	53	30
Mayetta	12	10	45	45
Mohilla	12	05	44	41

The Coafts of the Main Continent in the Eaft-Indies. North Latitude. Eaft Longitude.

Iflands in the Eaft-Indies. S. Latitude. Eaft Longit.

X 2   Co.

Places Names.	Latitude N. or S.		Longitude E. or W.	
	D.	M.	D.	M.
Comero	11	50	43	44
St. Juan de Nova	17	30	43	05
Mauritius	20	05	52	55
Diegoroes	19	45	61	35
Romiras de Caste Limos	20	00	67	16
Amsterdam	38	50	74	24
St. Brandon	16	38	64	30
Diego Gratiofa	08	42	68	25
Quabella	03	49	52	39
Baffos de Chagos	06	45	68	44
Yas de Diego Roys	00	10	72	05
Maldivia { N. End	07	15	73	05
Maldivia { S. End	00	40	76	15
Malique	09	00	73	05
Saccatra	12	21	54	07
Abdeleur	12	00	53	03
C. Gallo in Zeloan	06	07	80	45
Yas de Amber	00	00	52	30
Andaman the middle	12	40	93	19
Nicobar	7	05	93	45
Sumatra N.W. End	5	28	94	45
Virkins Island	2	22	95	07
Naffau Island	2	54	99	32
Hencola	3	50	104	08
Sumatra S.E. End	5	22	105	14
Engano	5	50	101	53
Selam	8	20	102	13
Princes Island	6	30	104	02
Bantam in Java	6	11	105	55
Batavia	6	16	106	37
Jova E. End	8	35	113	37
Straits of Sunda	6	02	105	38
Banco S. End	3	25	106	57
				Borneo

Islands in the East-Indies.

Places Names.	Latitude N. or S.		Longitude E. or W.	
	D.	M.	D.	M.
Islands in the East-Indies.				
Borneo S. Point ———	4	04	112	52
Bandy Isles ———	4	55	127	17
Celebes { South End —	5	40	119	07
{ North End—	1	40	121	00
Mindano W. Point —	6	40	119	35
Borneo N. Point ———	7	10	112	55
Luconia { S. Point —	13	30	120	10
{ N.E. Point	18	55	119	45
Aynian { W. Point ——	19	30	107	6
{ E. Point——	19	55	109	56
Formosa { S. Point —	22	00	119	56
{ N. Point—	25	32	120	45
Piscadore Isles———	23	30	118	35
Islands Chusan ———	30	28	121	15
Japan { S.E. Point —	35	30	140	30
{ S.W. Point—	35	00	128	30
The Coast of America in the South-Sea from California to Cape Horn				
Cape St. Sebastian ———	42	45	127	55
Cape St. Lucas ———	23	20	111	46
Cape Corientes ———	19	40	110	30
Aquapulco ———	17	00	104	18
Aquatulco ———	15	27	101	3
Guatimala ———	14	25	97	0
Panama ———	8	50	81	18
Bay Bonaventuro ——	3	18	79	06
Islands of Gallopega —	0	0	90	10
Cape del Ajugo ———	6	38	83	50
Lima ———	12	30	77	20
Arica ———	18	17	73	10
La Serena ———	29	00	76	32
Island Juan Fernando—	33	20	85	48
Baldivia ———	39	35	81	20
Port Stevens ———	49	50	82	36
Cape Victory ———	52	05	83	10
Cape Horn ———	57	34	79	55

Places Names.	Latitude N. or S.		Longitude E. or W.	
	D.	M.	D.	M.
C. of Brazil in S. America from Cape Horn to Cape Roque.				
Magellan E. Entrance	52	00	75	10
River Julian	48	40	74	32
Cape Blanco near Riv. Camarones	46	50	72	10
Buenos Ayres R. Plata	35	40	58	0
River Grand	31	35	51	50
St. Catherines	28	0	48	50
Cape Frio	22	55	42	26
Spirito Sancto	20	10	42	0
P. Segura	16	32	40	30
Bay Todos Sanctos	12	50	40	30
River St. Francisco	10	45	37	46
Cape St. Augustine	8	30	35	28
Cape Roque	5	0	35	42
Tristian de Cunha	37	5	13	20
Trinidada	20	30	30	0
(South Latitude)			*(West Longitude)*	
The Coast of the Main Continent in the West-Indies.				
River Amazones Ent.	0	0	49	20
North Cape	2	5	49	25
Suranam	6	25	55	35
Oronoque	8	15	59	25
C. Coquipaco	12	48	70	40
Carthagena	10	28	75	20
Scots Settlement	8	30	78	45
Nicaraque Entrance	11	25	84	5
Cape Catoche	21	10	86	15
Camphecha	19	30	92	5
La Vera Cruz	19	10	97	55
Cape Florida	24	58	80	35
(North Latitude)			*(West Longitude)*	

Tri-

Places Names.	Latitude N. or S.		Longitude E. or W.	
	D.	M.	D.	M.
Trinidada	10	34	60	25
Tobago	11	10	59	10
Granada	11	57	60	20
Barbadoes	12	58	58	50
St. Vincent	13	10	60	10
St. Lucia	13	56	60	07
Martinico	14	48	60	55
Dominico	15	25	60	27
Marigalante	15	58	60	20
Guardalupe	16	10	61	15
Monserat	16	45	61	15
Antigua	17	5	61	44
Nevis	17	6	62	32
St. Christophers	17	20	62	38
Barbuda	17	53	60	42
St. Bartholomew	17	54	62	5
St. Martins	18	5	62	9
Anguilla	18	16	62	15
Virgins	18	28	63	25
St. Cruz	17	48	63	25
Bieque	18	0	63	15
Port Rico St. John's	18	33	65	30
St. Domingo Hispan.	18	25	69	30
Port Royal Jamaica	17	40	76	33
East End of Cuba	20	20	74	05
Havana	22	40	82	50
Bay Hondy	22	45	83	40
Cape St. Antonia	21	50	85	42
Bermudas	32	25	63	40
Bahama Bank N. point	27	50	78	45
Bahama Islands	26	45	79	08
Abacco S. point	26	00	77	01
Harbour Island	25	40	76	45
Andross N. point	25	15	79	0
Providence	25	0	77	15

The Caribbee Islands. — North Latitude. — West Longitude.

Bahama Islands — North Latitude. — West Longitude.

Places Names.	Latitude N. or S.		Longitude E. or W.	
	D.	M.	D.	M.
Bahama Islands.				
Illuthera S. point	24	30	75	52
Cat Island	24	25	75	10
Watling Island	24	7	74	40
Rum Key	23	45	74	55
Exuma	23	25	75	57
Crooked Isle N. point	22	58	74	10
Atkins Key	22	10	74	05
Morapervouz	21	57	74	40
Atwoods Key	22	59	73	30
French Keys	22	40	73	35
Mayaguana	22	35	72	50
Hog Styel	21	15	73	50
Hyneago W. End	20	57	73	20
Caicos Bank N. point	21	40	71	55
Turks Island	21	35	70	8
Abrolho N. point	21	40	69	10
Plate rack	20	10	68	20
The Coast of Carolina, Virginia, Maryland, Pensilvania, New-England, and Newfoundland.				
Charles Town upon Ashly River	33	05	78	50
Cape Hatteras	35	15	74	20
Cape Henry	37	0	75	30
Cape Charles	37	14	74	15
Cape Hinlopen	38	54	74	55
Long Island	40	50	72	40
New York	40	58	73	53
Cape Cod	42	12	68	55
Boston	42	30	69	23
Cape Sable	43	50	64	58
Island Sable	44	20	59	5
Cape Britain	46	10	58	25
Querbeck	47	00	60	50
Bay of Breft	52	10	56	57
Bell Island	52	5	55	30

Places Names.	Latitude N. or S.		Longitude E. or W.	
	D.	M.	D.	M.
Cape St. John	50	19	52	45
Cape Bonavista	49	12	52	13
Trinity Bay	48	47	52	15
Conception Bay	48	13	52	10
St. John's Harbour	48	00	51	35
Bay of Bulls	47	50	51	26
Cape Race	46	40	51	52
Cape St. Mary	47	10	53	20
Placentia	47	47	53	58
Cape Roy	48	01	57	35
Buttons Isle	60	25	66	30
C. Charles	62	05	75	30
C. Walsingham	62	35	77	50
Mansfield Isle	61	42	80	30
C. Jones	55	3	79	05
Ruperts River	51	26	79	25
Albany River	52	32	84	47
The Cubbs	54	15	82	40
C. Henrietta Maria	55	5	84	22
Port Nelso	57	5	93	57
C. Churchill	59	0	95	20
C. Southampton	61	57	86	48
Shark Point	64	27	82	55
Nottingham Isle	63	38	79	47
Queen Ann's Foreland	63	42	74	47
Resolution Isle	61	55	65	10
Cape Farewell	59	40	46	45

Coast of Carolina, Virginia, Maryland, Pensilvania, New-England, &c. — North Latitude. West Longitude.

The Coast of Hudson's-Bay and Straits. — North Latitude. West Longitude.

Sound

Places Names.	Latitude N. or S.		Longitude E. or W.	
	D.	M.	D.	M.
Sound Royal ———	66	22	24	33
Bargazar Point ———	66	20	16	35
Whales Back ———	65	27	20	35
Merchants Foreland —	63	25	17	5
Halliford ———	64	20	34	43
Fair Foreland ———	66	10	26	25
Grims Island ———	67	15	22	35
Weſtmania Iſles———	63	35	22	50
Feio Iſles ———	62	6	5	0
Beerenberg, or John Mayn's Iſles ———	71	45	4	30
Point Lookout ———	76	30	15	35
Horn Sound ———	76	43	13	40
Fair Foreland———	79	18	10	50
Hacluits Headland———	79	55	11	0
Helies Sound ———	78	55	21	45
Lees Foreland ———	78	5	23	25
Whales Head———	77	18	21	30
Hope Iſland ———	76	18	23	45
Cherry or Bear Iſle —	74	35	18	5
Admiralty Iſland ———	75	5	55	50
Fretum Burrough ———	69	58	61	20
Cape Candenole ———	69	07	42	35
Catſnoſe———	65	43	35	14
Archangel bar ———	64	30	40	30
Croſs Iſland ———	66	31	36	13
Sweetnoſe ———	68	8	34	42
Kilduyn———	69	30	31	20
North Cape ———	71	25	23	2
Surroy———	71	5	18	40
Tromſound ———	70	20	16	0
Leefort SW. print ———	68	15	9	45
Dronten———	63	40	10	40
Stadland———	62	10	4	38
North-bergen ———	60	10	5	40
Naze of Norway —	57	45	7	42

(The Coaſt of Ireland, Greenland, Nova Zembla, and the Northern Iſles.)

North Latitude. *Weſt Longitude.*

Maer-

Places Names.	Latitude N. or S.		Longitude E. or W.	
	D.	M.	D.	M.
Maerden	58	19	9	5
Larwick	58	54	9	20
Chriſtiana	59	40	9	55
Macſterland	57	58	11	45
Gottenberg	57	43	12	15
Elſineur	56	20	12	42
Copenhagen	55	44	12	45
Valſterborn	55	28	12	55
Kalmer	56	40	16	40
Stockholm	59	20	19	25
Wyburg	60	55	29	26
Peterſburg	60	00	30	20
Narve	59	08	28	44
Revel	59	20	24	55
Riga	56	59	24	56
Derwinda	57	10	22	10
Conningſberg	54	40	20	40
Dantzick	54	22	18	40
Wiſbuy in Gotland	57	37	18	40
Bornholm	55	17	14	53
Straelſound	54	25	13	20
Lubeck	54	05	11	5
Anout, or Anholt	56	42	11	18
Leſow	57	6	10	40
Scaw	57	30	10	30

The Sea-Coaſt in the Sound and Baltick Sea.

North Latitude.

Eaſt Longitude.

A

A TABLE of the Sun's Right Ascension.

Days Month	Jan.		Feb.		March		April		May		June	
	H.	M.	H.	M.	H.	M.	H.	M.	H.	M.	H.	M.
1	18	50	21	02	22	51	0	44	2	36	4	38
2	18	54	21	6	22	54	0	48	2	40	4	42
3	18	58	21	10	22	58	0	51	2	44	4	46
4	19	2	21	14	23	02	0	55	2	48	4	50
5	19	7	21	18	23	5	0	59	2	51	4	54
6	19	11	21	22	23	9	1	3	2	55	4	58
7	19	16	21	26	23	13	1	6	2	59	5	2
8	19	20	21	30	23	16	1	10	3	3	5	6
9	19	24	21	34	23	20	1	14	3	7	5	11
10	19	29	21	38	23	24	1	17	3	10	5	15
11	19	33	21	42	23	28	1	21	3	14	5	19
12	19	38	21	46	23	31	1	25	3	18	5	23
13	19	42	21	50	23	35	1	29	3	22	5	27
14	19	46	21	54	23	39	1	33	3	26	5	32
15	19	51	21	58	23	42	1	36	3	30	5	36
16	19	55	22	02	23	46	1	40	3	34	5	40
17	20	00	22	5	23	50	1	44	3	38	5	44
18	20	4	22	9	23	53	1	47	3	42	5	48
19	20	8	22	13	23	57	1	51	3	46	5	52
20	20	12	22	17	0	1	1	54	3	50	5	56
21	20	16	22	21	0	4	1	58	3	54	6	1
22	20	20	22	24	0	8	2	2	3	58	6	5
23	20	25	22	28	0	11	2	6	4	2	6	9
24	20	29	22	32	0	15	2	10	4	6	6	13
25	20	33	22	36	0	19	2	13	4	10	6	17
26	20	37	22	40	0	22	2	17	4	14	6	22
27	20	41	22	43	0	26	2	21	4	18	6	26
28	20	45	22	47	0	29	2	25	4	22	6	30
29	20	50			0	33	2	29	4	26	6	34
30	20	54			0	37	2	32	4	30	6	38
31	20	58			0	40			4	34		

A TABLE of the Sun's Right Ascension.

Days Month	July		August		Sept.		Oct.		Nov.		Dec.	
	H.	M.	H.	M.	H.	M.	H.	M.	H.	M.	H.	M.
1	6	42	8	47	10	43	12	31	14	28	16	32
2	6	46	8	51	10	46	12	35	14	32	16	36
3	6	51	8	55	10	50	12	38	14	36	16	40
4	6	55	8	59	10	54	12	42	14	39	16	45
5	6	59	9	3	10	57	12	46	14	43	16	49
6	7	3	9	7	11	1	12	49	14	47	16	54
7	7	7	9	11	11	5	12	53	14	51	16	58
8	7	11	9	14	11	8	12	57	14	55	17	2
9	7	15	9	18	11	12	13	1	14	59	17	7
10	7	19	9	22	11	15	13	4	15	3	17	11
11	7	23	9	26	11	19	13	8	15	7	17	15
12	7	27	9	29	11	23	13	12	15	11	17	20
13	7	31	9	33	11	26	13	15	15	15	17	24
14	7	36	9	37	11	30	13	19	15	19	17	29
15	7	40	9	40	11	33	13	22	15	24	17	33
16	7	44	9	44	11	37	13	26	15	28	17	38
17	7	48	9	48	11	41	13	30	15	32	17	42
18	7	52	9	51	11	44	13	34	15	36	17	47
19	7	56	9	55	11	48	13	38	15	40	17	51
20	8	0	9	59	11	51	13	41	15	45	17	55
21	8	4	10	3	11	55	13	45	15	49	17	59
22	8	8	10	7	11	59	13	49	15	53	18	4
23	8	12	10	10	12	2	13	51	15	57	18	9
24	8	16	10	14	12	6	13	57	16	2	18	13
25	8	20	10	17	12	9	14	0	16	6	18	18
26	8	24	10	21	12	13	14	4	16	11	18	22
27	8	28	10	25	12	17	14	8	16	15	18	27
28	8	31	10	28	12	20	14	12	16	19	18	31
29	8	35	10	32	12	24	14	16	16	21	18	35
30	8	39	10	36	12	27	14	20	16	28	18	40
31	8	43	10	39			14	24			18	44

A

A TABLE of the Declination, Right-Ascension, and Magnitude of the Principal fixed Stars.

Names of the Stars.	Magnitude.	Right Ascen. H. M.	Declination. D. M.
Pegasus's lower Wing	2	0 1	13 52 N
Pole Star	2	0 45	88 02 N
Girdle of Andromeda	2	0 56	34 18 N
Achernar	1	1 28	58 27 S
Bright * of ♈	2	1 54	22 19 N
Medusa's Head	2	2 52	40 01 N
Perseus Right Side	2	3 07	49 00 N
Brightest of 7 *'s	3	3 33	23 31 N
Bull's Eye Aldebaran	1	4 22	16 01 N
Goat Capella	1	4 59	45 44 N
Orion's Left Foot	1	5 53	08 29 S
Orion's Left Shoulder	2	5 12	06 06 N
Middle Star in Orion's Belt	3	5 24	01 22 S
Orion's Right Shoulder	2	5 42	07 21 N
Waggoner	2	5 41	44 53 N
Syrius the Great Dog	1	6 35	16 24 S
Castor	2	7 19	32 23 N
Procyon the Little Dog	1	7 27	05 49 N
Pollux	2	7 31	28 35 N
South Foot, or Claw of the Crab	3	8 03	09 54 N
Hydraea Heart	2	9 16	07 38 S
Southern in the Lyon's Neck	1	9 54	17 55 N
Lyon's Heart	1	9 55	13 08 N
Upper Pointer	2	10 48	63 06 N
Lyon's Tail	1	11 37	15 54 N
Virgin's Girdle	3	12 44	04 42 N
Virgin's Spike	1	13 13	09 54 S
Arcturus	1	14 05	20 26 N
South Ballance	2	14 38	15 02 S
North Ballance	3	15 04	08 29 S
Scorpion's Heart	1	16 15	25 53 S
Head of Ophiucus	2	17 24	12 45 N
Brightest in the Harp	1	18 29	38 34 N
Vulture's Heart	2	19 39	08 15 N
Swan's Tail	2	20 33	44 26 N
Pegasus's Mouth	3	21 32	08 46 N
Fomelhaut	1	22 44	30 53 S
Marchab	2	22 53	13 56 N

The

The Use of the Table of Fixed-Star's, and Sun's Right Afcenfion.

FIRST, Suppofe you fee a Star upon the Meridian, and know not what Star it is, fee for the Sun's Right Afcenfion the Day propofed, and to that add the Hour of the Night when the Star is upon the Meridian, cafting away 24, if it exceeds; and for that Sum look in the Table of Fixed Stars, under [Right Afcenfion,] and the Star againft which it ftands is the Star that you faw upon the Meridian.

Example. January 11, and at 11 at Night, I fee a Star upon the Meridian, which I fuppofe to be of the Firft Magnitude, *viz.* one of the biggeft of the Fixed Stars; I find the Sun's Right Afcenfion that Day 19 : 33, to which add the Hour of the Night 11, the Sum is 30 : 33, but cafting away 24, there refts 6 : 33, which I look for in the Table of Fixed Stars, under [Right Afcenfion] and find it to ftand againft *Syrius,* the *Great Dog,* fo I know it is the *Great Dog* that I faw upon the Meridian, which is a Star of the Firft Magnitude; and fo in others.

But if you know a Star, and defire to know what Time it will be upon the Meridian, fubtract the Right Afcenfion of the Sun that Day from the Right Afcenfion of the Star mentioned in the Table (borrowing 24, if need be) the Remainder is the Time of the Star's coming to the Meridian. *Example;* I defire to know what Time the *Great Dog*

Dog comes to South *February* the 12th ; the Sun's Right Afcenfion that Day is 21 : 46, which fubtracted from 06 : 35 the Right Afcenfion of the *Great Dog* added to 24, becaufe it cannot be done otherwife, the Remainder 8 : 49 is the Hour of the Night that the *Great Dog* comes to the South.

Again, if you fee a known Star upon the Meridian, and defire thereby to know the Time of the Night, you may prefently know what Hour it is by the abovefaid Rule ; and in common Cafes, if you know the Time of a known Star's coming to the Meridian, you may nearly compute it, by allowing 4 Minutes to every Day. *Example,* I have found as above, that the *Great Dog* Souths *February* the 12th, at 49 Minutes paft 8 at Night ; and feeing the fame Star upon the Meridian *February* the 20th, I defire to know the Hour of the Night ; now from *February* 12 to *February* 20, is 8 Days, and allowing 4 Minutes a Day, the amount is 32 Minutes, which fubtracted from 8 Hours 49 Minutes, leaves the Time of the *Great-Dog's* Southing *February* the 20th.   This Allowance of 4 Minutes a Day may ferve for common Ufe, but if you would be exact, work by the preceding Rules.

In the laft Column of the Table, you have the Declination of the Fixed Stars, which is of Ufe in taking an Obfervation ; of which fee more in Chapter VII. Section IV. of this Treatife.

The Magnitude is of Ufe to know the Bignefs of a Star ; thofe of the firft Magnitude being the biggeft ; the fixth the leaft, and the reft of Bignefs proportionable to their Magnitudes exprefs'd in the Table.

Chap.

CHAP. IX.

Containing feveral entertaining and ufeful
QUESTIONS.

SECT. I. *Of Currents.*

IN failing in a Current, it is evident that a Ship doth not make her Way good according to the Courfe fteered by the Compafs, and the Diftance run by the Log, becaufe at the fame Time being privately carried by a Current, her true Courfe and Diftance is compounded of the Courfe, and Diftance failed, and of the Direction and Motion of the Current, therefore where both thefe are given, you muft firft lay down the Courfe and Diftance failed, and then the Courfe and Motion of the Current; to this laft Spot thus found, draw a Line from the Place failed from, the fame will fhew the true Courfe and Diftance made good, as in the firft Cafe following: A Ship fails S. E. 100 Miles from C to A, and a Current fets in the fame Time Weft 30 Miles; now if there was no Current, the true Courfe and Diftance of the Ship would be reprefented by the Line CA, and the Point A fhould reprefent the Place failed to, but becaufe a Current fets Weft 30 Miles in the fame Time, I fet off 30 Miles W. from A to D, and the Point D reprefents the Place failed to, and the Line CD is the true Courfe and Diftance made good; and

hence

mand what is the true Courſe and Diſtance made good
in that 24 Hours Time. With the Courſe South Eaſt,
and Diſtance 100 lay it down as in Plane Sailing Geometrical, and then it will appear as in the Triangle ABC, that the Ship ſhould be at A, *Fig.* 66.
but becauſe the Current ſets 30 Miles Weſt in
the ſame Time, ſet off 30 Miles from A to D, becauſe
from A to D is Weſt, then is the true Place of the
Ship at D; therefore draw the Line C D, then is the
Triangle CBD the true Projection of the Queſtion, with
Allowance for the Current, in which Angle B C D
29.56 is the true Courſe made good, the Hypothenuſe
CD 81.6 the Diſtance, the Leg C B 70.7 is the Difference of Latitude, and B D 40.7 is the Departure, as
may be found by meaſuring them Geometrically.

Arithmetical Calculation.

By Caſe the firſt of Plane Sailing

As Radius	——	90	0	10.00000
To the Diſtance	——	100	0	2.00000
So Sine Comp. Courſe	—	45	0	9.84948
To the Difference of Latitude	70	7	——	1.84948

The Departure is alſo 70.7 equal to the Difference of
Latitude, becauſe the Courſe is South Eaſt, *viz.* at an
Angle of 45 Degrees.

But becauſe the Current hath ſet her 30 Miles Weſt
in the ſame Time, therefore ſubtract 30 from the Departure found 70.7 there remains 40.7, the true Departure
with Allowance for the Current.

Then you have the true Difference of Latitude and
Departure given to find her Courſe and Diſtance, by Caſe
the Sixth of Plane Sailing.

As Difference of Latitude	70	7	——	1.84948
To Radius	——	90	0	10.00000
So true Departure	——	40	7	1.60959
To Tangent of the Courſe	29	56	——	9.76011

Y 2   Then

Then for the Diſtance.

As Sine of the Courſe ——	29	56	—— 9.69809
To the Departure ——	40	7	—— 1.60959
So is Radius ————	90	0	—— 10.00000

To the Diſtance made good 81 6 —— 1.91150

Or the Diſtance may be found thus, by Caſe the fourth of Oblique Plane Triangles.

In the Oblique Triangle ACD there is given the Side AC 100, and the Side AD 30, and the Angle between them 45 Degrees: Then by Caſe the fourth of Oblique Plane Triangles.

Co. Ar.

As the Sum of the Sides ——	130	—— 7.88606
To the Difference of the Sides	70	—— 1.84509
So the Tang. of half the Sum } of the unknown Angles }	67.30	— 10.38277

To the Tang. of half their Diff. 52.26 — 10.11392

Hence the Angle ADC is 119.56, and the Angle ACD is 15.4.

Then for the Side CD, which is the true Diſtance made good.

As Sine of ACD ——	15d. 4m.	Co.Ar. 0.58513
To the Side AD ——	30	—— 1.47712
So is the Sine of CAD ——	45 0	—— 9.84948

To the Side CD 81.6 the Diſtance required 1.91173

But ſuppoſe the Current had ſet upon an Oblique Courſe, and not due Eaſt, Weſt, North, or South, the Operation had been more difficult: As for Example,

A Ship

A Ship sails S.E. 100 Miles a Day, in a Current that sets N.N.W. 30 Miles a Day; I demand the Course, Distance, Difference of Latitude, and Departure made good in one Day.

Geometrical Construction.

Lay down the Triangle ABC, as in the foregoing Question, with the Course S. E. viz. an *Fig. 67.* Angle of 45 Degrees, the Distance 100, the Difference of Latitude 70.7, and the Departure the same as found before; then should the Ship be at A; but because the Current in that Time hath set the Ship 30 Miles N.N.W. therefore set off 30 Miles N.N.W. from A to *d*, which may be done by the Rule laid down in Traverse Sailing Geometrical; for seeing the Line CA is South East from C to A, it must needs be North West from A to C; and then seeing the Current sets N.N.W. which is 2 Points to the Northward of North West; therefore with the Chord of 60, and one Foot in A, draw the Arch *g d* upon which set off 2 Points from *g* to *d*, and draw A *d*, which is a N.N.W. Line, upon which set off 30 (the Current's Race) from A to *d*, and then is the true Place of the Ship at *d*, her true Distance, C *d* 73.2 Miles, her Difference of Latitude C *q* 43.0 Miles, and her Departure *q d* 59.2 Miles, and the Arch *k b* measures the Angle at C 54d. 0m. the Course required.

Arithmetical Calculation.

In the Triangle A S *d* Right-angle at S, you have given the Hypothenuse A *d* 30, and the Angle at A 67. 30 (because A *d* is a N.N.W. Line) to find AS and S *d* by the first of Right-angled Plane Triangles.

Y 3   A*s*

As Radius —— —— ——	90	00	——	10.00000
To the Hypothenuse A *d* ——		30	——	1.47712
So is the Sine of the Angle at A 67	30	——	9.96561

| To the Leg S *d* —— —— | 27 | 7 | —— | 1.44273 |

As Radius —— —— ——	90	00	——	10.00000
To the Hypothenuse A *d* —		30	——	1.47712
So is the Sine of the Angle at *d* 22	30	——	9.58284

| To the Leg AS —— —— | 11 | 5 | —— | 1.05996 |

The Leg S *d* 27.7 equal to *q* B, subtracted from the whole Difference of Latitude CB 70.7, leaves C *q* the true Difference of Latitude 43 0; and the Leg SA 11.5, subtracted from the whole Departure 70.7, leaves 59.2, the true Departure: By which you may find the true Course and Distance by Case the sixth of Plane Sailing.

As Difference of Latitude —	43	0	——	1.63346
To Radius —— ——	90	0	——	10.00000
So is the Departure ——	59	2	——	1.77232

| To the Tangent of the Course | 54 | 0 | —— | 10.13886 |

As Sine Comp. Course ——	36	0	——	9.76921
To Difference of Latitude —	43	0	——	1.63346
So is Radius —— ——	90	0	——	10.00000

| To the Distance —— —— | 73 | 2 | —— | 1.86425 |

The true Course is 54 Degrees from the South Eastward, or South East near three Quarters East, and the Distance is 73.2 Miles.

CASE

C A S E II.   *Courſe and Diſtance ſailed, and Courſe and Diſtance made good, given, to find the Courſe and Motion of the Current.*

A Ship ſails (by the Compaſs) S. by E. 36 Miles, and then arrives at a Place which is known to bear from the Place ſailed from S.E. by S. 54 Miles, (having been deceived by an unknown Current) I demand which Way the Current ſets, and how faſt, ſuppoſing the Ship to ſail by the Log 4 Miles an Hour.

Geometrical Conſtruction.

In this and all other Caſes of Currents, as well as Plane Sailing, Traverſe, &c. draw the   *Fig.* 68. North and South Line A B, and ſet off the Courſe ſteered S. by E. and Diſtance 36 Miles from A to C; then ſet off alſo the Courſe and Diſtance made good 54 Miles S.E. by S. from A to D: Then becauſe by the Courſe ſteer'd by the Compaſs, and Diſtance run by the Log, the Ship ſhould have been at C, but is found at D, therefore I am ſure there is ſome Current hath ſet me in the ſame Time from C to D, therefore draw the Line CD for the Set of the Current, which meaſured will be found to be 25 Miles. And the Angle ACD accounted from the N. by W. Point (becauſe the Line CA is a N. by W. Line) will be found to be 11 Points from the N. by W. Eaſtwards, *viz.* E.S.E. od. 9m. Southerly, for the true Courſe of the Current by the Rule laid down in Traverſe Sailing Geometrical for laying down Courſes by Diſtance of Points.

Arithmetical Calculation.

In the Oblique Triangle ACD, you have given the Side AC 36 Miles, and the Side AD 54 Miles, and the Angle between them 2 Points, or 22d. 30m. (being the Diſtance between S. by E. and S.E. by S.) to find the

Y 4   Angle

Angle ACD, and the Side CD, by Case the fourth of
Oblique Plane Triangles.   Co. Ar.
As the Sum of the given Sides ——— 90 — 8.04576
To their Difference ——— ——— 18 — 1.25527
So Tang. of ½ Sum of unknown Ang. 78.45 10.70134

To Tang. of ½ their Difference ——— 49.9 10.03237

The half Difference of the Angles — 45d. 9m.
Added to the half Sum ——— 78   45

The Sum is the Angle ACD ——— 123   54
which reduced to Points of the Compass, is 11 Points,
o Deg. 9 Min. and that accounted from N. by W. finds
E.S.E. o deg. 9 min. Southerly, for the true Course of
the Current.
       Then for the Side CD the Current's Race in that Time:

                                                     Co. Ar.
As Sine of the Angle ACD   123d. 54m: — 0.08092
To the Side AD ———   54   ——— 1.73239
So is Sine of the Angle CAD 22   30   ——— 9.58283

To the Side CD required — 25 ——— 1.39614

    So that the Current sets 25 Miles E.S.E. 9m. Sou-
therly, in the Time that the Ship sail'd by the Log, 36
Miles S. by E. and supposing the Ship to sail 4 Miles
an Hour, she would sail 36 Miles in 9 Hours, in which
Time the Current sets 25 Miles, therefore divide 25 by
9, the Quotient 2⅓ Miles is the hourly Motion of the
Current.

    CASE III. Course and Distance made good by the
Ship, and Course and Motion of the Current given, to
find the Course and Distance sail'd: Or, more properly
thus, having the Bearing and Distance between two
Ports or Islands given, and having also the Course and
Motion of a Current that lies between them given, to
find what Course to steer by the Compass, or how much
                                                          to

to Windward of your true Courſe to ſteer, ſo that the Compound Motion of the Ship may juſt ſet her to the deſired Port.

There are two Iſlands A and B; the Courſe from B to A is South 40 Degrees Weſterly, diſtance 80 Miles: The Current ſets Eaſt 2½ Miles an Hour, a Ship ſails 4½ Miles an Hour, I demand what Courſe ſhe muſt ſteer from B to A, and how far ſhe muſt ſail by the Log before ſhe arrives at A, the Port deſired?

Geometrical Conſtruction.

Draw the N. and S. Line BD, and ſet off the Courſe and Diſtance from B to A South   *Fig.* 69. 40 Degrees Weſt, 80 Miles from B to A: Then becauſe the Current ſets Eaſt, draw the Eaſt and Weſt Line AC at Pleaſure, by the Rule laid down in Traverſe Sailing *Geometrical*, then is the Side AB 80 Miles, and the Angle BAC 130d. 00m. given, but you have no other Side nor Angle given in the Oblique Triangle ABC, but you have the Proportion of the two Sides CA and CB, for AC repreſents the Motion of the Current 2½ Miles an Hour, and BC repreſents the Motion of the Ship thro' the Water 4½ Miles an Hour: Therefore find the Angle ABC by Caſe the Second of Oblique Plane Triangles.

As the Side BC 4½, or in Decimals 4.5 *Co. Ar.* 9.34679
To the Sine of the Angle BAC 130.0   —— 9.88425
So the Side AC 2½, or   ——   2.5   —— 0.39794

To the Sine of the Angle ABC 25.11   —— 9.61898

The Angle ABC is 25d —11m. therefore make the Angle ABC, 25.11, and draw the Line BC to cut AC in C, and then is the Projection finiſhed, and the Angle ABC 25.11 added to the Angle ABD 40d.—0m. the Sum 65d. 11m. from the South Weſterly, or W. S. W. almoſt ⅓ Southerly, is the Courſe that the Ship muſt ſteer to gain the Port with Allowance for the Current.

Then

Then for the Side BC, the Distance sailed by the Log, subtract the Sum of the two Angles A and B 155d.—11m. from 180.—om. the Remainder 24d.—49m. is the Angle ACB.   Then,

As Sine of the Angle ACB — 24.49 *Co. Ar.* 0.37705
To its opposite Side BA ——— 80 ——— 1.90309
So Sine of the Angle BAC — 130—0 ——— 9.88429

To the Side opposite BC — 146 ——— 2.16439

The Distance sailed by the Log, is 146 Miles, and the Rate of Sailing is 4½ Miles an Hour, therefore divide 146 by 4½, the Quotient 32⅘ is the Hours that the Ship will be in Sailing from B to A.

Now if you would prove the Work, multiply 32⅘, the Hours that the Ship is in sailing by 2½ the Miles that the Current sets in one Hour, the Quotient 81½, is the Miles that the Current sets in that Time, represented by the Side AC, which you will also find to be true by the following Canon.

As Sine of BAC ——— ——— 130.0 *Co. Ar.* 0.11575
To Side opposite BC ——— 146 ——— 2.16435
So Sine of ABC ——— ——— 25.11 ——— 9.62891

To Side opposite AC ——— 81.1 ——— 1.90901

Note, It is necessary in this Case to know how fast the Ship sails, for the faster she sails, the less she need lie to Windward of her true Course against the Current.

Case IV. *or* Question IV.

There are other Varieties in sailing in a Current, some of which I shall instance for the Learner's Improvement and Diversion.

A Cur-

A Current ſets 32 Miles a Day E.N.E. *a Ship ſailing thercin ſteers* S.S.E. *by the Compaſs, and finds that in 24 Hours ſhe is 70 Miles diſtant from the Place ſailed from: I demand upon what Point ſhe hath made her Way good, and how far ſhe hath ſailed by the Log ?*

Geometrical Conſtruction.

Draw the N. and S. Line AB, and ſet off the Ship's Courſe ſteer'd S.S.E. 22.30, and *Fig.* 70. draw the Line A C continued ; then any where upon that Line, as at C, draw an E.N.E. Line (as the Line DC) by the Rule laid down in Traverſe Sailing Geometrical to repreſent the Set of the Current, upon which ſet off 32 Miles, the Current's Motion in 24 Hours, from C to D, then at the neareſt Diſtance from D to the Line AC, (which here happens to be the Length of the Line CD, becauſe CD is Perpendicular to AC) draw the Parallel *b* D, then with 70 Miles (the Diſtance made good) in your Compaſſes, and one Foot in A deſcribe the Arch B *g* continued, and where it cuts the Parallel *b* D as in *g*, begin the Line *g* E, drawing it parallel to the Line DC : Then is A *g* 70 Miles, and Diſtance made good, E *g* 32 Miles, the Set of the Current E.N.E. The Line AE 62.3 Miles, the Diſtance ſail'd by the Log, and the Angle EA *g* 27d. 12m. added to the Angle B A C 22d. 30m. the Sum 49d. 42m. is the Courſe made good from the South Eaſtward, or the Arch *k m* meaſured on the Rumbs, gives SE. 4 Degrees, 42 Minutes Eaſterly.

Arithmetical Calculation.

In the Triangle A E *g* you have given the Diſtance made good A *g* 70 Miles,. and the Motion of the Current E *g* 32 Miles, and the Angle A E *g* (which happens here to be a Right-angle) 90 deg. to find the Angle EA *g* thus :

As

As Side A g ———— —— 70 —— 1.84509
To the Sine of the Angle oppoſite } 90 0 — 10.00000
   A E g ———— ———— }
So Side E g ———— —— 32 —— 1.50515

To the Sine of the Angle oppo. E A g. 27 12 — 9.66006

The Angle E A g 27d. 12m. added to the Angle BAC 22.30, the Sum 49.42 is the true Courſe made good from the South Eaſtwards, *viz.* S. E. 4d. 42m, Eaſterly:

Then for the Side AE the Diſtance ſailed by the Log.

As Sine AE g ———— — 90d. 0m — 10.00000
To Side oppoſite A g —— 70 ———— — 1.84509
So Sine A g E —— —— 62   48 —— 9.94910

To Side oppoſite AE —— 62.3 ———— 1.79419

The Courſe made good is S.E. 4d. 42m. Eaſterly, and the Diſtance ſailed by the Log is 62.3 Miles.

And if you would prove the Work by inverting the Queſtion, and propoſing it in the firſt Caſe of Current Sailing, thus,

A Ship ſails S.S.E. 62.3 Miles in a Current that ſets E.N.E. 32 Miles in the ſame Time; you will find the Anſwer produces 70 Miles S.E. 4d. 42m. Eaſterly, for the Courſe and Diſtance made good.

Note; Although in this Caſe the Triangle A E g is Right-angled, becauſe the Current's Race E. N. E. makes a Right-angle with the Ship's Courſe S.S.E. yet in any other Caſe it would have been an Oblique Triangle; but the Rules both for Projection and Calculation would have been the ſame.

Queſtion the Fifth.

A Ship ſails 72 Miles a Day by the Log, in a Current that ſets Eaſt 12 Miles a Day, and then finds that ſhe hath made her Way good South Eaſt, I demand what Courſe ſhe hath ſteer'd by the Compaſs, and what Diſtance ſhe has made good?

Geome-

Geometrical Conſtruction.

Draw AB, and with an Angle of 45 Deg.
the Courſe made good, draw the Line A C   *Fig.* 71.
continued, then with 72 the Diſtance ſailed
by the Log, ſweep the Arch B *d*; then any where upon
the Line AB, as at B, make an Eaſt and Weſt Line,
becauſe the Current ſets Eaſt; then take 12 Miles in
your Compaſſes, the Currents Motion; and ſet it parallel to BC, and ſo that it may juſt extend from the
Arch B *d* to the Line A C, as here from *e* to *g*, and
draw the Line *e g*, and 'tis done—or if you think this
Method too mechanical for laying down the Line *g e*,
you may do it thus: You foreſee that in the Triangle
A *e g* when laid down, there will be given the Side A *e*
72, and the Angle oppoſite to it 45 deg. and the Side
eg 12, to find the Angle *e* A *g*, thus,

As Side A *e* ——— ——— 72 *Co. Ar.* 8.14267
To Sine of theAngle oppoſite A *g e* 45.0 —— 9.84948
So Sine *eg* ———   ——— 12 —— 1.07918

To Sine of Angle oppoſite *g* A *e* 6.46 — 9.07133

which added to the Angle A *g e* 45 deg. the Sum 51.46
ſubtracted from 180, leaves 128.14, the Angle A *eg*;
therefore having ſet off 72 from A, upon the Line A *e*
from A to *e*, at *e* draw the Line *e g*, to make an Angle
of 128.14 with the Line A *e*, this Line if carefully
drawn will juſt contain 12 ſuch Parts, whereof the Line
A *e* contains 72, by that Time it is extended to cut the
Line A *g*.

Then for the Side A *g* the Diſtance made good.

As Sine *e* A *g* ——— ——— —— 6.46 *Co. Ar.* 0.92876
To Side *eg* ——— ——— —— 12 ——— 1.07918
So Sine of A *eg* ——— ——— 128.14 — 9.89514

To Side A*g* theDiſtance made good 80 ——— 1.90308
'*t* here-

Therefore for the Courſe ſteer'd, ſubtract the Angle *e* A *g* 6.46 from the whole Angle given *b* A *g* 45, the Remainder 38.14 is the Angle *b* A *e* the Courſe ſteer'd, and the Side A *g*, found to be 80, is the Diſtance made good.

Note, In finding the Sine of any Angle above 90 Degrees you muſt ſubtract the Angle (whoſe Sine is required) from 180 Degrees, the Sine of the Remainder is the Sine of the Angle required, as in the Example above, where the Sine of 128.14 is required, ſubtract 128.14 from 180, the Remainder 51d. 46m. ſought in the Table of Sines, the Sine anſwering to it is 9.89514, which is alſo the Sine of 128.14 which was required; the Reaſon of which is evident from Plate 1ſt, Fig. 7. of this Book, in which if you account 128d. 14m. from *k* upon the Arch *k* 90 S, it will reach to the Point *x*, and the Perpendicular *x z* is equal to a Perpendicular let fall from 51d. 46m. as you may obſerve by the Degrees numbered from S upwards towards 90, the ſame Degree and Minute which anſwers to 128.14, if numbered from *k*, anſwers to 51.46, if numbered from S, &c.

And thus much for Plane Sailing in a Current; many more Queſtions might be invented from other Data's, but I would ſtudy Brevity, that the Book may not be too chargeable to the Buyer, ſuppoſing that by a Right Underſtanding of theſe Rules, the Ingenious will be able to project and anſwer any other Caſe or Queſtion of this Kind; and as for Traverſe Sailing in a Current, although I thought to have placed it in a Section or Chapter by itſelf, yet I find it altogether needleſs; for the Courſes being firſt reduced to one, by the Rules laid down in Traverſe Sailing, the Operation for allowing for known Currents, or finding the Courſe and Motion of unknown Currents, is the very ſame with the Rules here laid down.

O F

O F

Turning to WINDWARD

IN A

C U R R E N T.

THIS may also be divided into several Cases, of which I shall speak in Order; but the most useful is, where the Course and Motion of the Current is given, with the Course and Distance between the Place sailed from, and the Place bound for, and from what Point the Wind blows, and how near the Wind the Ship will make her Way good, (for these four Things are commonly given or known) to find how long she must lie upon each Tack to gain her Port, supposing her Rate of running, or Miles sailed in an Hour by the Log, be also given or known.

Example

Example, Queftion the Firft.

There are two Iflands A B, A is diftant from B 90
Leagues due North: A Current fets from A towards
B South 2 Miles an Hour, a Ship at A intending for
B, meets with the Wind at South, gets her Starboard
Tacks aboard, and makes her Way good within 72
Degrees of the Wind, and fails four Miles an Hour by
the Log, I demand how long fhe muft lie upon each
Tack to gain her Port, and what Courfe fhe makes
good?

Geometrical Conftruction.

Draw AB reprefenting the Bearing and Dif-
Fig. 72. tance of the two Iflands, *viz.* N. and S. 90
Leagues, upon the Middle of which erect the
Perpendicular *d* C, then draw the Line AC, to make an
Angle of 72 Degrees, with the Line HB, and continue
the Line AC till it cut the Perpendicular *d* C in C, and
draw the Line CB, fo fhall the Ifofceles Triangle ABC
reprefent the two Iflands, and the Ship's Way to them
without any Allowance for the Current, the Line AC
reprefenting the Ship's Way, with her Starboard Tacks
aboard 72 deg. from the Wind, and the Line CB her
Way, with her Larboard Tacks aboard to fetch the
Ifland; but becaufe every Hour while the Ship fails
four Miles by the Log, the Current fets her two Miles
to the Southward, therefore find how many Hours the
Ship is in failing from A to C by the Log; in order to
which you muft find the Side A C by dividing the
Ifofceles Triangle ABC into two Right-angled Triangles
A*d*C and B*d*C, then in the Triangle A*d*C you have
given the Leg A*d* 45 (being half the whole Line A B
which is 90) and the Angle *d*A C 72 Deg. confequently the Angle AC*d* 18 deg. to find the Hypothenufe
A C, by Cafe the fourth, of Right angled Plane
Triangles.

As

As Sine of AC *d* —— —— 18d. 0m: 9.48998
To Side oppoſite A *d* —— 45 —— 1.65321
So is Radius A *d* C —— —— 90.0 —— 10.00000

To Hypothenuſe AC —— —— 145¼ —— 2.16323

Hence AC is 145¼ Leagues, or 437 Miles, which divided by 4 the Miles ſailed in one Hour, the Quotient 109¼ is the Hours, in which the Ship ſails from A to C; but the Current ſetting South two Miles an Hour, 'tis plain that in 109¼ Hours it hath ſet South 218 Miles, or 72¼ Leagues nearly; therefore draw the South Line C *g* and thereon lay 72¼ Leagues, becauſe whilſt ſhe ſail'd by the Log from A to C, the Current hath ſet her from C to *g*, then draw the Line A *g*, which repreſent the true compound Motion of the Ship, or Courſe made good; but becauſe ſhe is to lie upon that Tack, but only till ſhe be got half Way to the Port ſail'd for, obſerve where the Line A *g* cuts the Perpendicular *d* C in *e*, and then is the Angle *d* A *e* 49d. 39m. the true Courſe made good by Reaſon of the Current, and the Line A *e* the true Diſtance ſail'd upon the Starboard Tack, *viz.* 69.5 Leagues, and *e* B the Diſtance ſailed upon Larboard Tack being alſo 69.5 Leagues. But to know how far ſhe will have ſailed by the Log, by that Time that ſhe will be at *e* by the Help of the Current, draw *e k* parallel to *g* C, this Parallel ſhall cut AC and A *g* proportionably by *Euclid Lib.* 6. *Prop.* 2. *viz.* As A *g* to A *e*, ſo AC to A *k*, hence A *k* meaſured will be found to be 55.5 Leagues, or 166.7 Miles ſailed by the Log, which divided by four, the Miles ſailed by the Log in one Hour, gives 41.775 Hours, the Time to ſtand upon the Starboard Tack, and the ſame upon the Larboard, to fall in with the Iſland at B, *&c.*

Arithmetical Calculation.

In the Oblique Triangle AC *g*, there is given the Side AC 437 Miles, and the Side C *g* 218 Miles, and the

Angle

Angle included A C *g* 108d. om. to find the Angle
C A *g*, by Caſe the Fourth of Oblique Angled Plane
Triangles.

As Sum of the Sides ———   655 Co. *Ar.* 7.18376
To Diff. of the Sides ———   219 ——— 2.34044
So Tang. of ½ Sum of unknown Ang. 36.0 ——— 9.86126

To Tang. of half their Difference 13.39 ——— 9.38546

    The half Difference 13d. 39m. ſubtracted from the
half Sum 36d. om. reſts 22d. 21m. the Angle C A *g*,
which ſubtracted from the whole Angle *d* A C 72d. om.
leaves the Angle *d*A*e* 49d. 39m. the Courſe made good.
    Then for the true Diſtance made good A *e*: In the
Triangle A *d e* you have given A *d* 45 Leagues, and the
Angle *d* A *e* 49d. 39m. conſequently the Angle A *e d*
40d. 21m. to find A *e* by Caſe the Second of Right-
angled Plane Triangles.

As Sine of *e d* —— —— 40d. 21m. — 9.81121
To Side oppoſite A *d* — 45 —— — 1.65321
So Radius —— —— — 90   o — 10.00000

To Hypothenuſe A *e* — 69   5 — 1.84200

    Then for the Diſtance ſail'd by the Log A *k* (while by
the Help of the Current ſhe is carried to *e*.)
    In the Triangle A*k e*, you have given A *e* 69.5, and
you have given all the three Angles, *viz.* A *k e*, equal
to AC*g* 108d. om. and A*ek* equal to A *g* C equal to *d*A*e*,
49d. 39m. and *k* A *e* 22d. 21m. equal to A C, to
find the Side A *k* by Caſe the firſt of Oblique-angled
Plane Triangles.

As Sine of A *k e* — — 108d. om. *Co. Ar.* 0.02179
To Side oppoſite A *e* — 69.5 — — 1.84198
So Sine of A *e k* —— 49.5 39 — 9.88201

To Side oppoſite A *k*. — 55.7 —— ═ 1.74578
                               The

The Side A *k*, the Diſtance ſailed by the Log, is 55.7 Leagues, or 167.1 Miles, which divided by 4, the Miles ſailed in one Hour by the Log, the Quotient 41.775 is the Hours that ſhe muſt lie upon the Starboard Tack, in which Time ſhe is carried away by the Help of the Current from A to *e*, and then lying the ſame Time upon the Larboard-Tack, ſhe will arrive at B, the deſired Port; ſo that in 83,55 Hours, ſhe performs the Voyage of 90 Leagues, although upon a Wind, by the Help of the Current, which had there been no Current, would have required 218 Hours, or 9 Days and 2 Hours

Queſtion the Second.

There are two Iſlands diſtant 400 Miles N. and S. from each other, ſuppoſe A and B, A being the Northermoſt: A Ship at A intending for B, meets with the Wind at South, ſhe gets her Starboard Tacks aboard, and makes her Way good through the Water, within 72 Degrees of the Wind, a Current at the ſame Time ſetting South 24 Miles a Day, ſhe ſtood four Days upon each Tack, and then arrived at her Port at B: I demand her Courſe and Diſtance made good upon each Tack?

Fig. 73.

Geometrical Conſtruction.

Draw the Line AB 400, and at an Angle of 72 Degrees, draw AD and BD, to cut each other in D, and from that Interſection let fall the Perpendicular DC, which will fall upon the Middle of the Line AB, then is the Projection finiſhed without Allowance for the Current; but becauſe in that four Days that ſhe had her Starboard Tacks aboard, the Current had ſet her 96 Miles, ſet off 96 from C to *e* and *b*, and draw *b k* and *e g* Parallel to CD, and where theſe Parallels cut the Lines AD and BD, as at *g* and *k*, draw the Line *g k* to cut

Z 2

cut the Perpendicular CD in *m,* and then to the Interſection at *m,* draw the Lines A *m* and B *m,* for theſe Lines ſhall repreſent the true compound Motion of the Ship or Courſe and Diſtance made good, *viz.* The Angle CA *m* 58d. is the Courſe made good from the Meridian, and the Line A *m* equal to *m* B is the true Diſtance ſail'd upon each Tack, *viz.* 377.4:

Arithmetical Calculation.

In the Triangle A *e g* Right angled at *e,* you have given the Leg A *e* 104, and the Angle *e* A *g* 72d. conſequently the Angle A *g e* 18d. to find the Hypothenuſe A *g.*

As Sine of A *g e* ———— 18d. om. — 9.48998
To Side oppoſite A *e* ——— 104 ——— 2.01703
So Radius ————————— 90.0 ———— 10.00000

To the Hypothenuſe A *g* 336.5 ——— 2.52705

Then in the Oblique Triangle A *g m* you have given the Side Ag 336 5 and the Side *g m* 96, becauſe equal to C *e,* and the Angle included A *g m* 108d. om. to find the Side A *m.*

                                                 Co. Ar.

As Sum of the Sides ——— ——— 432 5—7.36402
To Difference of the Sides ——— 240.5—2.38111
So Tang. of half Sum of unknown Angles 36.0—9.86126

To Tangent of half their Difference — 22 0—9.60639

Hence the Angle *g* A *m* is 14d. oom. which ſubtracted from the whole Angle 72, leaves the Courſe made good, CA *m* 58d. oom. Then for the Diſtance made good.

                                                 Co. Ar.

As Sine of *g* A *m* ——— —— 14d. oom. — 0.61632
To Side oppoſite *g m* ——— 96 ——— —— 1.98227
So Sine A *g m* ——— ——— 108.0 ——— 9 97820

To Side oppoſite A *m* —— 377.4 ——— 2.57679
                                                 The

The Diſtance made good upon each Tack is 377.4 Miles; the Diſtance ſailed by the Log upon each Tack is A g and B k 336.5, the Courſe made good 58d. 0m. from the Meridian.

Queſtion the Third.

A Maſter of a Ship having taken Freight from an Iſland at A, to another Iſland at B, and having his Son aboard along with him to gain Experience, he requires of his Father, whether there was not a Current ſetting upon ſome Point between the two Iſlands: His Father refuſing to tell him, bids him obſerve well the Dead Reckoning Outward-bound, and alſo Home, and ſee if he could not, by comparing them, find the true Courſe and Motion of the Current: Whereupon the Captain, who knew very well which Way the Current ſet, and how the Iſlands bore upon each other, cauſed them to ſteer away South, and in Running 270 Miles upon that Courſe, they arrived at B, and having done his Buſineſs there, they ſteered North Weſt, and Running 380 Miles upon that Courſe, they arrived at A. Now I demand how the Iſlands bear from each other, and how far diſtant; alſo which Way the Current ſets, and how faſt, ſuppoſing the Ship ſailed always at the Rate of 5 Miles an Hour by the Log?

Geometrical Conſtructions:

Draw the North and South Line A C 270 Miles, the firſt Courſe and Diſtance ; then becauſe her Courſe Home was North Weſt to A, draw the Line A D at the Angle of 45 degrees from AC (becauſe if DA be North Weſt, AD muſt be South Eaſt, and conſequently make an Angle of 45 Degrees with the Line AC) and ſet off 380 from A to D, and draw the Line CD, which repreſents the Motion of the Current during the Ship's ſailing from A to C, and from D to A; for finding of which you have given the Side AC 270, and the Side AD 380, and the

Fig. 74:

contain'd

contain'd Angle 45 deg. to find the Angle A C D by
Caſe the fourth, of Oblique Plane Triangles.

As Sum of the Sides ——— 650 *Co. Ar.* 7.18709
To the Difference of the Sides 110 ——— 2.04139
So Tang. of ½ Sum of unkn. Angles 67.30 — 10.38277

To Tang. of ½ their Difference 22.13 — 9.61125

The half Sum added to the half Difference, makes
the bigger Angle ACD 89.43.

Then for the Side CD.

As Sine of ACD ——— ——— 89.43 *Co. Ar.* 0.00001
To Side oppoſite AD ——— 380 ——— 2.57978
So Sine of CAD ——— ——— 45.0 ——— 9.84948

To Side oppoſite CD ——— 268.7 ——— 2.42927

The Side CD 268.7 Miles is the Currents Race during the whole Voyage: Now to find the Bearing and
Diſtance of the Iſlands, we ſee that in the whole Voyage,
out and home, the Ship runs 650 Miles, which at 5
Miles an Hour, requires 130 Hours, in which Time the
Current ſets 268.7 Miles, but her Voyage outward being but 270 Miles, which at 5 Miles an Hour requires
but 54 Hours, therefore ſay by the Rule of Three, if in
130 Hours the Current ſets 268.7 Miles, how far will it
ſet in 54 Hours?

$$268.7$$
$$54$$
——————
$$10748$$
$$13435$$
——————
3|0) 1450|9.8 (111.6

The Current's Drift while the Ship ſails by the Log.
from A to C is 111.6, which ſet upon the Line C D
from

from C to B, and draw AB which reprefents the true Courfe and Diftance between the two Iflands, for the finding of which by Calculation, you have given, in the Triangle A B C, the Side A C 270, and the Side C B 111.6, and the contained Angle ACB 89.43, to find the Angle CAB, by Cafe the Fourth, of Oblique Plane Triangles,

			Co. Ar.
As Sum of Sides	——	381 6	7.41840
To the Difference of the Sides	—	158.4	2.19975
So Tang. of halfSum unknown Angles	45.8	—10.00202

| To Tang. of half their Difference | 22.38 | 9.62017 |

Hence the Angle CAB is 22d. 30m. the true Courfe from A to B South Eafterly, which is South South Eaft exactly.

Then for the Side AB the true Diftance between the Iflands.

			Co. Ar.
As Sine of C A B	— —	22.30	0.41716
To Side oppofite C B	— 111.6	—	2.04766
So Sine of ACB	— —	89.43	9.99999

| To Side oppofite AB | — 291.6 | 2.46481 |

The true Diftance between the two Iflands is 291.6 Miles.

The Courfe of the Current is the Angle ACD, found by the firft Operation 89d. 43m. from the North Eafterly, which is Eaft 17 Minutes Northerly, and as for the Rate or Motion of the Current's running, it is found by dividing the whole of the Current 268.7, by the whole Number of Hours that the Ship was under Sail, viz. 130, the Quotient is the Miles that the Current fets in one Hour, as appears by the Operation.

$$130 \overline{)\ 268.700\ (\ 2.066}$$
$$\underline{8.70}$$
$$\underline{9.00}$$
$$\overline{120}$$

Hence in Anſwer to what was demanded ; the Courſe from A to B is S.S.E. ——The Diſtance 291.6 Miles ; the Courſe of the Current is Eaſt od. 17m. Northerly, and its Rate or Motion is 2.066 Miles an Hour.

Now if you would prove the Truth of the Operation you may eaſily do it, by inverting the Queſtion, and ſtating it as in the firſt Caſe of *Current Sailing*, thus,

A Ship ſails South 5 Miles an Hour in a Current that ſets Eaſt 17 Min. Northerly 2 $\frac{12}{1000}$ or 2.066 Miles an Hour, I demand the Courſe and Diſtance made good, and you will find it produce S.S.E. for the Courſe, and 291.6 Miles for the Diſtance; but I ſhall leave the Operation for the Reader's Practice.

Uſeful

Pleaſant and Uſeful

QUESTIONS

I Thought to have added other Varieties of Queſtions concerning Currents; but this being but an Appendix to a ſmall Treatiſe, I fear if I ſhould proceed, I ſhould exceed my intended Bounds, and therefore ſhall add no more, ſuppoſing that by what is laid down the ingenious Student may eaſily learn to project and anſwer any other neceſſary Queſtion in that Kind; I ſhall now proceed according to my Promiſe, to inſert ſome other Pleaſant and Uſeful Queſtions for the Reader's Improvement and Diverſion.

Queſtion the Firſt.

There is a round Sea or Lake of an unknown Diameter, upon which the Wind is obſerved to blow always one Way, but upon what Point of the Compaſs is unknown; upon the Coaſt thereof are two Ports A and B, diametrically oppoſite to each other, ſo ſituate that the Trade Wind blowing there, will be ſomewhere upon the Larboard Quarter, during their ſailing from A to B, but how many Points abaft the Beam is unknown, the Courſe and Diſtance between the two Ports is alſo unknown.———A Ship at A intending for B, was no ſooner got out of the Harbour, but ſhe was

aſſaulted

*assaulted by Pyrates, who after a long Conflict carried away
her Main Mast, Mizen Mast, Fore Top Mast, and also her
Rudder, and came aboard to Plunder, taking away her Car-
goe, the Master's Books, Instruments, Charts, Compasses,
&c. being left in this Distress he sets his Fore Sail, and
having no Rudder, nor any more Sail to command her with,
lets her drive right before the Wind, 'till at last she drove
a-shore in a little Creek, altogether unknown by the Master,
nor did he know the Course or Distance from this Creek, either
to the Port sailed from, or to the Port bound for: However here he got such Masts and Sails as he had occasion for,
with a Rudder, and all that was necessary to the working of
a Ship, but could get no Books, Charts, Instruments, or
Compasses: I demand what Means he must use to guide him-
self to the Port at B?*

Answer.

Although at first Sight this Question seems impossible
to be answered from what is given, yet with a little Consideration it is very easy; the Solution of it is grounded altogether upon the Thirty-first Proposition of the Third
Book of *Euclid*, where it is prov'd that an Angle in a Semicircle, is a Right-angle, &c.   Now the two Ports A
and B being proposed opposite to each other, or in a Semicircle, it will necessary follow that to what Place soever
the Ship was blown from A, right before the Wind,
and put a Shore, the same Wind upon the Beam would
carry her to the Port at B, as appears by the Figure
annexed.

The Angles ACB and ADB and AEB are
Fig. 75. all Right-angles by the Proposition above
named; and therefore if from A the Ship is
blown before the Wind to C, 'tis plain that the same
Wind continuing, (as is here supposed) the Wind must
needs be upon the Beam from C to B, because A C B
is just Square or a Right-angle.

Or, suppose the Wind blow from A to D, and the
Ship driven before the Wind, is blown a Shore at D, it

Is plain that (the Angle ADB being a Right-angle) the same Wind muft be upon the Beam from D to B——and for the fame Reafon, if blown before the Wind from A to E, the Wind will be upon the Beam from E to B; *See the Demonftration in the Book itfelf, viz. Euclid.*

Queftion the Second.

There is a round Lake, whofe Diameter is 360 Miles, upon the Coaft of which lie two Ports, A and B, whofe nearest Diftance in a ftrait Line is 120 Miles, but their Bearing or Situation is not known: A Ship at A fails away Weft a certain Diftance, and then arrives at another unknown Port at L, whofe Bearing and Diftance both from A and B are unknown, and at the Port at L, the Mafter takes in Goods for B, I demand what Courfe he muft fleer to find the Port at B?

First, Draw the Circle K L M N to reprefent the Lake whofe Diameter K M or L N *Fig. 76.* is 360 Miles, then affume any Point in the Circumference of the Circle, as at A, to reprefent the Port at A ; then with 120 in your Compaffes, the Diftance of the Ports in a ftrait Line, and one Foot in A, the other will reach to B, biffect the Arch AB at N, and thro' N and the Center *o*, draw the Diameter N *o* L, and at Right-Angles to it the Diameter KM, and draw the Line AB, then is the Line A *q* B the Chord of the Arch ANB, and B *q* is the Sine of the Arch BN to the Radius K *o* equal to 180 Miles, the Semidiameter of the Lake. Therefore,

As K *o* equal to B *o* — 180 — 2.25527
To Radius — — 90 00 — 10.00000
So B *q* equal to half B *q* A — 60 — — 1.77815

To Sine of the Angle B *o q* — 19.28 —— 9.52238

The Angle B *o q* 19.28 doubled is 38.56, the whole Angle B *o* A (becaufe the Angles B *o q* and *q o* A, are equal.) Now fuppofe a Ship at A fails Weft to an unknown

known Port at L, I have already proved the Angle AOB to be 38.56, and therefore the Angle ALB is 19.28, because an Angle at the Center of a Circle is doubled to an Angle at the Circumference, by *Euclid, Lib.* 3. *Prob.* 20; and therefore if from A to L be *West*, it is plain that from L to B is *East* 19 Deg. 28 Min. Northerly, and that is his Course from L to B.

Or, for Variety, if you suppose the Port at S to be the Port failed to, and that from A to S is *West*, then from S to B will be *East* 19 Deg. 28 Min. *Northerly* as before, for the Angles ALB and ASB are equal, as is sufficiently demonstrated in *Euclid's* Elements in the forementioned Proposition.

Question the Third.

A Fleet of Ships with four Men of War, with Wind at W. by S. sails away S.E. 3 Miles an Hour, till bearing of some Pyrates to the Northward from them, the Commodore sends out a Man of War a Cruising, who sails due North 7 Miles an Hour, for the Space of 11 Hours, and finding the Pyrate, he (after an Hours Conflict) took him, the Fleet all the while sailing S E. 3 Miles an Hour. Now I demand what Course the Man of War and the Pyrates must steer, that they must just fall in with the Fleet, without altering their Course, suppose they sail 7 Miles an Hour.

Fig. 77.

Geometrical Construction.

First draw the *North* and *South* Line CA continued to B, then from A (the Place where the Fleet was when they heard of the Pyrates) set off the S.E. Line AE; then because the Cruising Man of War failed 7 Miles an Hour North, for 11 Hours, *viz.* 77 Miles: Set off 77 Miles from A to B, also because the Commodore failed 3 Miles an Hour in 11 Hours that the Man of War was in Chase, and the one Hour that he was engaged, in all 12 Hours, which at 3 Miles an Hour is 36 Miles S.E. which set from A to D, then is the Commodore

modore at D, and the Cruizer at B, when the Cruiser
set sail after the Commodore: Now because the Line
AE is South East, or 45 Degrees, the Angle BAD is
135 Degrees, the Side AB is 77, and the Side AD 36,
you may find the other Angles by Case the fourth of
Oblique Plane Triangles, thus,

Co. Ar.

As the Sum of the Sides —— —— 113 — 7.94693
To Difference of the Sides —— — 41 — 1.61278
So Tangent half Sum unknown Angles 22.30 9.61722

To Tangent of half Difference —— 8.33 9.17693

Hence the Angle A B D is 13.57 and the
Angle ADB is 31.3, to find the Side BD by *Fig.* 77.
Case the Second of Oblique Plane Triangles.

Co. Ar.

As S. ADB —— —— —— 31.3 —— 0.28753
To Side AB —— —— —— 77 —— 1.88649
So is S. BAD ·—— —— 135.0 —— 9 84948

To Side BD —— —— 105.5 —— 2.02350

Also the Angle ADB being 31.3, the Angle B D E
must be 148.57, and although the Sides BE and DE be
not given, yet their Proportion to each other is given,
viz. as 7, is to 3 ; so is BE, to DE, therefore,

Co. Ar.

As Side BE —— —— —— 7 —— 9.15491
To Sine of Angle opposite BDE 148.57 — 9.71247
So is Side DE —— —— 3 —— 0.47712

To Sine of Angle opposite DBE 12.46 — 9.34450

The Angle B D E being 148.57, and DBE 12.46.
D E B must consequently be 18.17, then you have
given the Angle B E D and B D E, and the Side B D

to find the Side BE, the Diſtance ſailed by the Cruiſer before he met with the Fleet.

			· Co. Ar.
As Sine BED — —	18.17	— —	0.50347
To Side oppoſite DB —	105.5	— —	2.02325
So Sine of BDE — —	148.57	— —	9.71247
To Side oppoſite BE —	173.4	— —	2.23919

The whole Angle ABE 26.43 is the Courſe from the South Eaſtward, or S.2.E. 13m. Eaſterly the Cruiſer muſt ſail to fall in with the Fleet, and his Diſtance 173,4 Miles before he comes in with them, which at 7 Miles an Hour, would require 24 Hours 48 Min. nearly in which Time he will fall in with the Fleet.

Queſtion the Fourth.

Plate 8.   Fig. 73:

Two Ships from one Port ſails between the South and Weſt; the Weſtermoſt Ship's Departure was 47 Leagues more than the Eaſtermoſt, and the Eaſtermoſt Ship's Difference of Latitude was 30 Leagues more than the Weſtermoſt. Their Diſtance was equal, namely 115 Leagues, I demand both their Courſes, Difference of Latitude and Departure, according to Plane Sailing.

Geometrical Conſtruction.

Draw the North and South Line GC, and Perpendicular thereto the Line DH, to cut GC at Right-angles in E (both drawn at Length at Pleaſure) from the Interſection at E upon the Line DH ſet the Difference of the Departure 47 from E to D, (towards the Left hand, becauſe the Ship ſails in the South Weſt Quarter) and the Difference of their Latitudes made good, viz. 30 from E to C, ſo ſhall D repreſent the Weſtermoſt Ship, and C the Place of the Southermoſt. Take the Diſtance

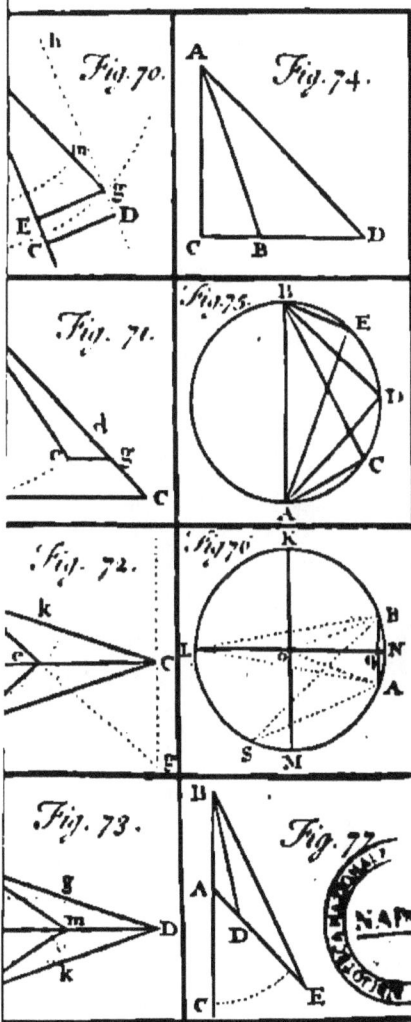

Fig. 70.

Fig. 74.

Fig. 71.

Fig. 75.

Fig. 72.

Fig. 76.

Fig. 73.

Fig. 77.

tance 115 Leagues in your Compaffes, and fetting one Foot in D defcribe the Arch A, and with the fame Extent, and one Foot in the C crofs the fame Arch in A, then is A the Place the Ship failed from. Draw the Line AB parallel to GC to cut DH in H, and from C draw CB parallel to DH. Alfo draw the Perpendicular AL, to divide the Ifcofceles Triangle A D C into two Right-angled Triangles ALC and ALD, and then is the Diagram finifhed.

Arithmetical Calculation.

In the Triangle DEC is given,
The Difference of their Departures DE   — —   47.
The Difference of their N.ing and S.ing EC   —   30.
To find the Angles by Cafe 6, and the Hypothenufe by Cafe 7.

And firft for the Angles by Cafe VI.

As DE	—	—	—	—	47	— — 1.67209
To Radius	—	—	—	—	90	— — 10.00000
So is EC	—	—	—	—	30	— — 1.47712

To Tang. of the Angle CDE   32 33   —   9.80503

For the Hypothenufe DC, by Cafe 7.

As Sine of the Angle EDC	—	32d. 33m.	— 9.73081		
To Side EC	—	—	—	30	— 1.47712
So Radius	—	—	—	90	— 10 00000

To the Hypothenufe DC   —   55.8   —   1.74631

The Ifcofceles Triangle D A C being divided into two Right-angled Triangles by the Perpendicular AL, there is given in the Right-angled Triangle ALC, the Side LC (half of DC) 27.9; the Hypothenufe AC 115, to find the Angle ACL by Cafe 4.

As

As AC ——— ——— —— 115 —— 2.06069
To Radius ——— ——— 90 —— 10.00000
So LC ——— ——— —— 27.9 —— 1.44560

To the Sine of the Angle LAC, 14d. 3m. — 9.38491

The Angle L A C, 14d. 3m. ſubtracted from 90, leaves the Angle LCA, 75.57.

The Angle CDE is 32d. 33m. which ſubtracted from 90, leaves the Angle DCE, 57d. 27m.

From the whole Angle DCA (LCA) 75d. 57m.
Subtract the Angle D C E ——— 57 27

Reſts the Angle ECK ——— —— 18 30

The Angle ECK, is equal to the Angle CAB; becauſe AB and EC are parallel, *Euclid Lib.* 2. *Prop.* 29. Therefore in the Triangle ABC is given AC 115, to find AB the Difference of Latitude, and BC the Departure of the Southermoſt Ship, by the firſt Caſe of Plane Sailing.

As Radius ——— ——— 90 —— 10.00000
To the Diſtance ——— —— 115 —— 2.06069
So is Sine Comp. Courſe —— 71d. 30m. 9.97695

To Diff. Lat. ——— —— 109 —— 2.03764

For the Departure BC.

As Radius —— —— — 90 —— 10.00000
To the Diſtance ——— 115 —— 2.06069
So Sine of the Courſe ——— 18d. 30m. 9 50147

To the Departure BC —— 36½ 1.56216

The firſt Ship's Courſe is South 18d. 30m. Weſt, To which add the whole Angle DAC 28 06
——————
46 : 36

Th

The Sum is the second Ship's Course DAH, which
is South 46 36 West.

The first Ship's Diff. Lat. is AB ——— 109
From which subtract the Excess HB
    equal to EC —— —— —— 30

Rests the second Ship's Diff. Lat. AH — 79
The first Ship's Departure is CB equal
    to EH ·.—— —— —— 36½
To which add DE the Excess —— —— 47

Sum is the second Ship's Depart. DH —— 83½

First Ship's { Course South 18d. 30m. West
    Difference of Latitude —— 109
    Departure —— —— 36½

Sec. Ship's { Course South 46d. 36m. West
    Difference of Latitude —— 79
    Departure —— —— 83½

Question the Fifth.

The Distance and Difference of Latitude in one Sum
given, and the Departure also given, to find the Course,
and also the Difference of Latitude and Distance se-
verally.

Example.

Let the Distance and Difference of Latitude together
be 140, and the Departure 79, to form the Triangle,
and find as above.

Geometrical Construction.

Draw the Right-line AC 140, equal to *Plate* 8.
the Sum of the Distance and Difference of *Fig.* 79.
Latitude, and at the End thereof *c* draw
c d, Perpendicular to A *c*, upon which set off the Departure

ture 79 from *c* to *d*, and draw *Ad*, which divide into two equal Parts in *g*, and from *g*, erect the Perpendicular *g b* to cut *Ac* ſomewhere in *b*, and from *b*, draw *b d*, and then is the Diagram finiſhed, and *b c* is the Difference of Latitude; *b d* the Diſtance, and *c d* the Departure.

Demonſtration.

For the Triangles A *g b* and *d g b* are alike and equal, (*Euclid. Lib* 1. *Prob.* 24.) becauſe A*g* is equal to *g d* by Conſtruction, and *g b* is common to both Triangles, hence *d b* is equal to A*b*, and the Angle *g* A *b* equal to that of *g b*, but the Angle *d b c* is equal to both *g* A *b* and *g d b* (*Euclid. Lib.* 1. *Prob.* 32.) therefore *d* A*c* (*g* A *b* is juſt half the Angle of the Courſe from the Meridian) and from hence we may form this univerſal Proportion for the Solution of Queſtions of this Data.

As Diſtance and Difference of Latitude in one Sum : To Radius : : Departure ·· the Tangent of half the Courſe, which Angle doubled is the Courſe required, and the reſt is found by Caſe the Third of Plane Sailing.

Example.

As Diſt. and Diff. Lat. together ————	140 —	2.14612
To Radius ———— ———— —	90 —	10.00000
So Departure ———— ————	79 —	1.89762
To Tangent of ———— ———	29 26 —	9 75150
Which doubled is ———— ————	58 52	the Courſe
required.		

Then,

As Sine of the Courſe ————	58 52 —	9.93245
To Departure ——— — —	79 —	1.89762
So Radius ———— ———— —	90 ————	10 00000
To the Diſtance . ——— ———	92 3 —	1.96517

And,

And,

As Radius	—— ——	——	90	—— 10.00000
To the Diftance	——	——	92.3	—— 1.96520
So Sine Comp. of Courfe		——	31.8	—— 9.71351

To Diff. Latitude	——	——	47.7	—— 1.67871

The Truth hereof may be further proved by adding
the Difference of Latitude —— 47 : 7
To the Diftance —————— 92 : 3

The Sum is ———— ———— 140 : 0

agreeing with what was firft given, and proving the
whole to be right.

Queftion the Sixtb:

Diftance and Difference of Latitude in one Sum given,
with the Courfe alfo given to conftruct the Diagram to
find each feverally.

We refer in this to the Figure of the laft
Queftion, it being the fame when finifhed. *Plate 8.*
Let it be given as before, *viz.* Diftance and *Fig. 79.*
Difference of Latitude in one Sum 140, and the Courfe
South 58d. 52m. Weft, to find the reft.

Geometrical Conftruction.

Draw A *c* 140, at *c* erect the Perpendicular *c d* at
Pleafure, make the Angle at A equal to 29d. 26m. (*viz.*
half the given Angle 58d. 52m) and draw A *d* to cut
d c in *d*₁ divide A *d* into two equal Parts in *g*, and draw
g b Perpendicular to A *d*, and to cut A *c* in *b*, draw *b d*
and then the Scheme is finifhed, the Demonftration
of which is the fame as *Queftion the Fiftb.*

Arithmetical Calculation.

The Angle *c b d* is 58d. 52m. therefore *c d b* is 31d.
8m. to which add *b d g* (equal to *b* A *g*) 29d. 26m. the
Sum 60d. 34m. is the whole Angle A *d c*, therefore,

A a 2 As

Co. Ar.

As Sine of A *d c* —— 60d. 34m —— 0 06002
To Side oppofite —— 140 —— 2.14612
So Sine of *d* A *c* —— 29 26 —— 9.69144

To Departure *c d* —— 79 —— 1.89758

The Diftance and Difference of Latitude is found as in *Queftion the Fifth*, and need not be here repeated.

Queftion the Seventh.

Given all the three Sides, *viz.* Diftance, Difference of Latitude and Departure in one Sum, the Courfe alfo given to form the Scheme, and find each feverally.

Suppofe the three Sides in one Sum be 157, and the Courfe South 32d. Weft.

Geometrical Conftruction.

Plate 8.   Draw the Line *a b c d* 157, and from *a* draw
Fig. 80.   *a e*, to make with *a d*, an Angle of 16 Degrees, (equal to half the given Courfe) alfo from *d* draw *d e* to make with *a d* an Angle of 45 (always) at *d*, and continue *d e* till it cut *a e* in *e*. From *e* let fall the Perpendicular *e c* upon the Line *a d*, divide *a e* into two equal Parts in *g*, and from *g* draw *g b* perpendicular to *a e*, and to cut *a d* any where, as in *b*, and draw *b e*, and the Diagram is finifhed.

Demonftration.

The Triangle *a g b* and *e g b* are fimilar, (*Euc. Lib.* 1. *Prob.* 24.) becaufe *e g* is equal to *g a* by Conftruction, and *g b* is common to both Triangles, and the Angles *e g b*, and *a g b* are right; therefore *g a b* is equal to *g e b*, but both thefe Angles together, are equal to the Angle *e b c*, (*Euclid Lib.* 1. *Prob.* 32.) and fince the Angle *a e b* (*g a b*) is half the given Courfe by Conftruc-tion ; *e b c* is the true Courfe, and confequently *e b* equal

to

b a the true Diftance, and becaufe *c d* is the Remainder of the whole Line that includes all the three, the Departure muft be equal to it, and therefore the Line *d e* is drawn at the Angle of 45 with *d a* (*d c*) becaufe *e c* is to be equal to *c d*, and then is *a b* proved equal to *b e*, and *b c* is the Difference of Latitude in the Queftion : Alfo *e c* is equal to *c d*, becaufe their oppofite Angles are equal : Therefore the three Sides of the Triangle *b c e* are equal to the whole Line *a d*, and the Angle *e b c* is equal to the given Courfe 32 Degrees, which was to be demonftrated.

Arithmetical Calculation.

We may obferve, that the Angle *g e b* equal to *g a b*, is half the Courfe, and *b e c* is the Complement of the Courfe, and *c e d* is an Angle of 45 Degrees, and thefe three include the whole Angle *a e d* in the Triangle *a e d*, from whence we may deduce this general Proportion, for all Queftions of this Data.

Add half the Courfe, the Complement of the Courfe, and 45 Degrees together, and fay,

As the Sine of thofe three Sums is to the Sum of the three Sides ; fo is the Sine of half the Courfe, to a fourth Term. And as Radius is to that fourth Term ; fo is the Sine of 45 to the Departure.

The Courfe being given, and the Departure thus found, the reft may be obtained by the Common Cafes of Plane Sailing.

Example in the prefent Queftion.

	d. m.
Half the Courfe *g e b* equal to *g a b* —	16 0
Complement of the Courfe *b e c* ——	58 0
Angle *c e d* equal to *e d c* —— ——	45 0
Sum is the whole Angle *a e d* ———	119 0

A a 3 *Then,*

Then,

			Co. Ar.
As Sine of *a e d* —— 119d. om.	——	——	0.05818
To Side opposite *a d* — 157	——	——	2.19590
So Sine of the Angle *a* 16.0	——	——	9.44033
To Side opposite *e d* — 49.5	——	——	1.69441

Then in the Triangle e c d.

As Radius —— 90	——	——	10.00000
To *e d* —— 49.5	——	—	1.69460
So Sine of *e d c* —— 45.0	——	——	9.84948
To Departure *e c* —— 35	——	—	1.54408

In the Triangle *e b c* is given the Course *e b c* 32d. om. and the Departure *e c* 35, to find the Distance *b e*.

As Sine of the Course 32d. om.	——	9.72421
To Departure —— 35	——	1.54406
So Radius —— 90	——	10.00000
To the Distance *b e* — 66	——	1.81985

From the Difference of Latitude.

As Radius —— 90	——	10.00000
To the Distance *b e* — 66	——	1.81954
So Sine Comp. of Course 58d. om.	——	9.92842
To Diff. Latitude *b c* — 56	——	1.74796

Distance ——	66
Diff. Lat. ——	56
Departure —	35
Sum ——	157 as was first given:

From

From what hath been said, we may reduce a general Rule for solving all Questions, whether in Right-angled, or Oblique-angled Triangles, where the three Sides in one Sum are given, and the Angles also are given, *viz.*

Draw a Line in Length equal to the three Sides in one Sum, and from one End thereof, draw a Line to make with the former, an Angle equal to half of any of the given Angles; also from the other End, draw a Line to make with the first an Angle equal to half of another of the given Angles; continue these Lines till they intersect each other, then for the Middle of each, draw a Perpendicular to cut the first Line, then Lines drawn from the Intersection of the Perpendiculars to the first Line, to the Intersection of the two Lines, drawn from the End of the first, shall form the Triangle required, whether Right-angled or Oblique.

Example.

Question the Eighth.

There is an Oblique Triangle, whose three Sides in one Sum is 120, and the three Angles are 40, 60, and 80, the Sides are required separately.

Geometrical Construction.

Draw the Line *a d* 120, the Sum of the Sides, and from *a* draw *a f*, to make with *a d* an Angle of 30 Degrees, *viz.* half the Angle 60, and from *d* draw *d f*, to make with *d a* an Angle of 40 Degrees, *viz.* half the Angle 80, continue *d f* till it cut *a f* in *f*, divide *f d* and *f a* equally in *e* and *g*, and draw the Perpendiculars *e c* and *g b*, to cut *a d* in *b* and *c*, from *b* draw *b f*, and from *c* draw *c f*, then is *b c f* the Triangle required.

Plate 8
Fig. 81.

Demonstration.

For the Angle *f c b* is double to *f d c*, because equal to *c f c* and *c d e* together (*Euclid Lib.* 1. *Prob.* 32.) and
A a 4   the

the Angle *c f e* is equal to *c d e*, becauſe their oppoſite
Sides are equal; but the Angle *f d c* is 40 Degrees by
Conſtruction, therefore *f c b* is 80 Degrees.   By the
ſame Rule *f b c* is 60 Degrees, becauſe *f a b* is 30 by
Conſtruction; therefore *b f c* is 40 (*Euclid Lib.* 1. *Prob.*
32.) and becauſe *a b* is equal to *b f*, and *c d* equal to *c f*;
therefore *b c*, *c f* and *f b* taken together, will be equal to
the Line *a d*; and *b f c* is the Triangle required.

Arithmetical Calculation.

Hence as the Sine of one Angle, added to half the
other two, to the Sum of the three Sides; ſo is half one
of the other Angles to a fourth Term: Then as the
Supplement of the other Angle is to 180, is to that fourth
Term; ſo is the Sine of half the ſaid Angle, to the Side
next it.

Example.

The Angle *a f b* is half *f b c*, becauſe *f a b* is half
f b c by Conſtruction; and *a f b* is equal to *f a b*, becauſe
their oppoſite Sides are equal, and for the ſame Reaſon
c f d is half *f c b*; but *a f b* with *b f c* and *c f d* make the
whole Angle *a f d*; then as the Sine of *a f d*, to Side *a d*
(the three Sides in one Sum) ſo is the Sine of *f a d* 30,
(becauſe half of *f b c* by Conſtruction) to the Side *f d*;
then as Sine of *f c d* (the Supplement of *f c b* to 180) is
to *f d*; ſo is the Sine of *f d c* 40 Degrees to the Side *f c*,
which being found the reſt are alſo eaſily found.

Operations.

As Sine of *a f d* ——	110d: 0m. ——	0.02701
To Side *a d* ——	120 —— ——	2.07918
So Sine of *f a d* ——	30.0 ——	9.69897
To Side *f d* ——	63.8 ——	1.80516

As

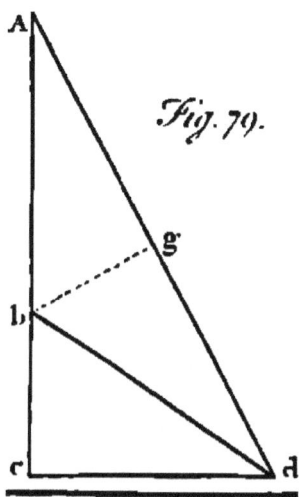

Fig. 79.

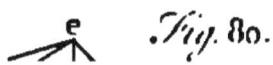

Fig. 80.

                                                   Co. Ar.
As Sine of *f c d* —— 100d. 0m. —— 0.00665
To Side *f d* —— 63.8 —— —— 1.80482
So Sine of *f d c* —— 40.0 —— —— 9.80806
                                                   ――――――
To Side *f c* —— 41.7 —— —— 1.61953

                                                   Co. Ar.
As Sine of *f b c* —— 60d. 0m. —— 0.06247
To Side *f c* —— 41.7 —— —— 1.62013
So Sine of *f c b* —— 80.0 —— —— 9.99335
                                                   ――――――
To Side *f b* —— 47.4 —— —— 1.67595

                                                   Co. Ar.
As Sine of *f b c* —— 60d. 0m: —— 0.06247
To Side *f c* —— 41.7 —— —— 1.62013
So Sine *b c f* —— 40.0 —— —— 9.80806
                                                   ――――――
To Side *b c* —— 30.9 —— —— 1.49066

The Side *f b* —— —— 47.4
The Side *f c* —— —— 41.7
The Side *b c* —— —— 30.9
                                                   ――――――
    Sum —— —— 120.0

The Sum 120, agreeing with what was given, which serves as a Proof of the whole Work.

C H A P.

CHAP. X.

A

New and Exact METHOD,

For finding the Longitude in any Place of the WORLD, any Day at Noon, when the Sun can be feen, without any Regard to, or Dependance upon the Dead Reckoning.

THE often attempted, but never accomplifh'd Work of finding the Longitude by an Obfervation is a Difficulty, which hath hitherto prov'd infuperable, even to the beft Mathematicians in *England,* and elfewhere, although fince the late Act of Parliament, many have exerted their utmoft Induftry for the attaining of that End, fome of whom have been fo far from compleating their intended Defign, that they have on the other Hand, rather rendered themfelves and their Works ridiculous, by publifhing fuch Improper and Improbable Methods, as a certain Artift hath done, (if he deferves that Name) by certain fixed Stars chofen for each Time of the Year, *&c.*

Whereas

Whereas it is evident, that a fixed Star may come upon the Meridian of any Place, at (or near) the same Hour of the Day, that the same Star shall come upon the Meridian of *London*, although not at the same Instant of Time, but sooner or later, 3, 4, 5, or 6 Hours, &c. according as the Place is more or less distant East or West from the Meridian of *London*: But if there be that Difference in Hours as to the Time of a Stars coming upon these two different Meridians, there is also the same Difference in Hours between the Time of the Day or Night at *London*, and that under that Meridian; a South Sun in both Places making 12 o'Clock; and 'tis very evident that at the same Time of the Day or Night, that the Star comes upon the Meridian of *London*, it shall be the same Time of the Day or Night under any other Meridian, when the Star comes upon that Meridian, excepting so much as the Sun's Right Ascension is increased in the Time of the Sun's passing from one Meridian to the other, which will cause the Star to come so much sooner upon a Westerly Meridian than upon an Easterly: But this being so little, being but about 4 Minutes of Time in 24 Hours, or 360 Degrees of Longitude, and but 1 Minute of Time in 90 Degrees, it is imperceptable; and yet (that small Allowance excepted) there is no Difference between the Hour of any fixed Star's coming to the Meridian of *London*, and the Hour of the same Star's coming upon any other Meridian, because the Sun's Motion makes the Hour of the Day, and the Star must needs come upon the Meridian the same Quantity of Time after the Sun in both Places. Indeed, if we could certainly know what Time of the *Day* or *Night* it is at *London*, when the Sun or any known Star is upon the Meridian of any other distant Place, the Longitude might be easily and exactly found; in order to which let the Master provide a good Glass, which may run exactly 24 Hours, or rather a good Watch, that hath been obferv'd ashore, to keep a true and equal Motion, this Watch set to the Time of Day

when

when you depart from any known Meridian, and kept going, will fhew the Difference of Longitude, that you make whether Eaft or Weft; for Example, if you fail Weftwards, obferve juft when the Sun is upon the Meridian, and fee what Hour and Minute it is by your Watch, (which if you fail Weftward it will be paft 12) for thefe Hours and Minutes which the Watch is paft 12 when the Sun is upon the Meridian, reduced to Degrees and Minutes of the Equator, allowing 15 Degrees to one Hour, and one Degree to every 4 Minutes of Time, and 15 Minutes of Longitude to one Minute of Time, fhall fhew the true Difference of Longitude.

Note; If you fail Eaftward, your Watch will want fomething of 12 o'Clock, when the Sun is upon the Meridian, becaufe then you meet the Sun, and have him upon your Meridian, before he comes upon the Meridian of *London*; in this Cafe obferve how much your Watch wants of 12 o'Clock, and that reduced as before, gives your Difference of Longitude Eafterly: This in general is the Method that I fhall recommend to the World for this End:

But I know it will be prefently objected, that this is no new Thing, nor is it practicable at Sea; for fome that have attempted to keep an Account of their Longitude this Way, have found themfelves in an Error, not finding the Difference between the Watch and the Sun, when reduced as above, to give the Difference of Longitude which they were (for fome Reafons) pretty confident they had made, and this Error they impute to the Watch, and hence have inferred, That a Watch will not go fo truly and regularly at Sea as on Shore, by reafon of the falt moift Air that impedes its Motion or makes it uncertain, according to the Variablenefs of the Weather, and thereby renders this Method for finding the Longitude impracticable.

I anfwer, I do not believe at all that this Error is to be imputed to the Watch, if carefully kept, which you may eafily do, if you provide a little fquare or round

Box

Box about four or five Inches broad, and have it filled with fine Cotton, taking Part of the Cotton out, and putting the Watch towards the Middle of the Box upon the Cotton you leave in the Box, and put the rest that you took out upon the Watch, and shut the Box, keeping it in some dry Place, as in your Cabbin, or upon some Shelf near your Bed. I qestion not but a good Watch so kept would go as true at Sea as a-shore.

But I suppose the Error which hath been obferv'd, and which hath caused them to defist from making any Attempts to find the Longitude this Way, is for Want of a right underftanding of the Equation of Time, without which it is impoffible to keep a good Watch right, or even to suppose it so to be either at Sea or a-shore; for a good Watch (or Clock) divides the Time equally, but the Sun, by Reason of some Inequalities in his Motion divides the Time unequally; so that if the Sun and a true Watch be let together at some Times in the Year, yet the Watch will at other Times differ 10, 12, yea, fometimes 16 Minutes from the Time given by the Sun, and yet no Fault in the Watch, and there-fore it is very evident that if for finding the Longitude you only obferve the Time of the Day given by the Watch, without regarding this Equation, efpecially when the Equation is great, as 15 or 16 Minutes, you will be so much wrong in your Account of Longitude, as those 15 or 16 Minutes reduced to the Equator amounts unto, *viz.* about 4 Degrees of Longitude, which is an intolerable Error, and might be prevented by allowing for the Equation of Time.

For further Illuftration hereof, suppose you set your Watch with the Sun the 17th Day of *June*, that the Equation is nothing, being in *London* at the same Time, and continue there 'till the 3d Day of *November*, you will find your Watch to be got 16 Minutes behind the Sun, from hence (if you do not know to the contrary) you might by the foregoing Rule conclude that you had

altered

altered your Longitude 4 Degrees Eaſterly, but if to the Time given by the Watch, you add the Equation 16 Min. the Sun is the true Time of the Day by the Sun, and proves you to be under the ſame Meridian that you were in when you ſet your Watch to the Sun; the Difference in Time, and conſequently the Difference of Longitude being nothing at all.

A TABLE

A TABLE of the *Equation of Time.*

Days	Jan. M.	S.	Feb. M.	S.	March M.	S.	April M.	S.	May M.	S.	June M.	S.
1	4	14	14	15	12	47	3	58	3	14	2	56
2	4	42	14	22	12	35	3	40	3	22	2	48
3	5	9	14	28	12	22	3	22	3	30	2	39
4	5	36	14	34	12	8	3	3	3	37	2	29
5	6	4	14	38	11	54	2	45	3	43	2	19
6	6	30	14	42	11	40	2	26	3	49	2	9
7	6	57	14	45	11	24	2	9	3	53	1	59
8	7	23	14	47	11	9	1	51	3	57	1	48
9	7	48	14	48	10	54	1	34	4	1	1	37
10	8	12	14	49	10	38	1	17	4	4	1	25
11	8	37	14	49	10	21	1	1	4	7	1	14
12	9	2	14	48	10	4	0	45	4	10	1	2
13	9	26	14	46	9	47	0	28	4	11	0	50
14	9	48	14	44	9	30	0	12	4	12	0	37
15	10	10	14	41	9	13	0	3	4	13	0	25
16	10	31	14	37	8	55	0	18	4	12	0	13
17	10	50	14	32	8	37	0	33	4	11	0	0
18	11	9	14	27	8	19	0	48	4	10	0	13
19	11	27	14	21	8	1	1	2	4	8	0	26
20	11	45	14	15	7	43	1	16	4	6	0	39
21	12	2	14	7	7	25	1	28	4	3	0	52
22	12	18	13	59	7	6	1	41	4	0	1	4
23	12	30	13	50	6	47	1	54	3	56	1	17
24	12	47	13	41	6	28	2	6	3	51	1	30
25	13	2	13	31	6	10	2	16	3	46	1	43
26	13	16	13	21	5	51	2	27	3	40	1	56
27	13	28	13	10	5	32	2	38	3	34	2	8
28	13	38	12	59	5	14	2	48	3	28	2	20
29	13	48			4	55	2	57	3	21	2	33
30	13	58			4	36	3	6	3	13	2	45
31	14	7			4	17			3	5		

Jan: Watch too fast
Feb: Too Fast
March: Too Fast
April: Too Fast / Too Slow
May: Too Slow
June: Too Slow / Too Fast

A

A TABLE of the Equation of Time.

Days	July M.	S.	August M.	S.	Sept. M.	S.	October M.	S.	Novem. M.	S.	Decem. M.	S.
1	2	56	5	38	0	18	10	16	15	59	10	20
2	3	7	5	34	0	36	10	34	16	1	9	56
3	3	19	5	30	0	55	10	52	16	0	9	32
4	3	30	5	26	1	14	11	10	15	59	9	8
5	3	41	5	20	1	34	11	28	15	57	8	43
6	3	51	5	14	1	53	11	45	15	54	8	17
7	4	0	5	8	2	12	12	2	15	51	7	51
8	4	8	5	2	2	32	12	18	15	47	7	25
9	4	15	4	52	2	52	12	33	15	41	6	59
10	4	22	4	46	3	13	12	49	15	35	6	30
11	4	29	4	37	3	33	13	4	15	29	6	3
12	4	46	4	28	3	53	13	18	15	21	5	35
13	4	53	4	18	4	14	13	32	15	13	5	6
14	5	0	4	8	4	34	13	46	15	3	4	38
15	5	7	3	57	4	55	13	59	14	53	4	9
16	5	13	3	46	5	15	14	11	14	41	3	40
17	5	18	3	34	5	36	14	23	14	29	3	10
18	5	24	3	21	5	56	14	34	14	17	2	40
19	5	29	3	8	6	17	14	44	14	3	2	10
20	5	33	2	55	6	38	14	54	13	49	1	40
21	5	36	2	42	6	58	15	4	13	34	1	11
22	5	39	2	27	7	19	15	13	13	17	0	41
23	5	42	2	13	7	39	15	21	13	0	0	11
24	5	44	1	58	7	59	15	28	12	43	0	19
25	5	45	1	42	8	19	15	34	12	24	0	49
26	5	46	1	26	8	38	15	40	12	5	1	19
27	5	46	1	9	8	58	15	45	11	45	1	49
28	5	45	0	52	9	18	15	50	11	25	2	18
29	5	44	0	35	9	37	15	53	11	4	2	47
30	5	42	0	17	9	57	15	56	10	42	3	16
31	5	40	0	1			15	58			3	45

Too Fast. Too Fast. Too Fast. Too Fast. Too Slow. Too Slow. Too Fast.

This

This Equation of Time is caufed by an Inequality of the Sun's Motion from Weft to Eaft, according to the Succeffion of Signs; for the fwifter that the Sun is in his Annual Motion from Weft to Eaft, the flower he muft be in his diurnal Motion from Eaft to Weft, as is plain by the Figure.

Suppofe the Wheel A B C D E F to move round upon the Center G, once in 23 Hours, *Plate* 9: according to the Order of the Letters A B C, *Fig.* 78. &c. and in the fame Quantity of Time, *viz.* 23 Hours, a Snail creeps the contrary Way, from A to F; now although the Point A is come to the Place where it was, having got once about, yet the Snail wants the Space F A of a whole Revolution, and will not be got to the Top where A is, 'till A be got fo far as the Point *g*, which will be about another Hour, and if the Snail had moved yet fafter, fo that in 23 Hours fhe had got from A to E, fhe would have been yet fo much longer in arriving at the Top, *viz.* about 25 Hours; from hence it is plain that the fafter the Snail creeps from A towards F, E, &c. the longer fhe is in coming to the Top of the Wheel moving the contrary Way; confequently the fafter the Sun moves from Weft to Eaft in his annual Motion according to the Succeffion of Signs, the longer he is in making one diurnal Revolution; and although the Sun's Revolution from any Meridian to the fame Meridian again always determines the 24 Hours, yet it is plain from hence that every 24 Hours by the Sun, is not exactly the fame equal Space of Time, which in Procefs of Time makes a fmall Difference between the Sun and a good Watch, to adjuft and rectify which, the aforefaid Equation of Time has been inveftigated.

This Inequality of the Sun's Motion proceeds from a Two-fold Caufe.

The firft Caufe is the Obliquity of the Ecliptick, making an Angle with the Equator of 23d.—29m. or thereabouts; now the Ecliptic being properly an Oblique Circle, the Poles of the Equator, and not of the

Ecliptic

Ecliptic, being the Poles of the World, and the Center of all Diurnal Motion, the Right Ascension of the Sun must be accounted upon the Equator: And hence it is manifest, that even tho' the Motion of the Sun in the Ecliptic was always equal, yet this Motion from West to East, or his Right Ascension accounted upon the Equator, could not be so much, or increase so fast in *Aries* or *Libra*, where the Ecliptic makes an Angle of 23 deg. 29 min. with the Equator, as in *Cancer* and *Capricorn*, where the Sun's Way in the Ecliptic is nearly parallel to the Equator.

Plate 9.   For Illustration of what has been said, sup-
Fig. 79. pose the Line ABCD to represent half the E-
quator, and the Arch AFED, to represent half of the Ecliptic, now although the Segments A F, F E, and E D are equal, yet Perpendiculars let fall from the Points F and E upon the Line ABCD, at the Points B and C, do not divide that Line into three equal Parts, from whence it is plain, that the Sun's Right Ascension DC is not so much increased by his running from D to E, at so great an Obliquity to the Equator, as the Sun's Right Ascension C B is, in his running from E to F, where his Motion is almost Parallel to the Equator; and then if his Motion from West to East be swiftest in the Tropics, for the Reason now given, his diurnal Motion must he slower by the first Demonstration.

But a second Cause of this Inequality is occasioned by the Eccentricity of the Sun's Orb (whether we allow the Sun or Earth the Motion it matters not, but in this Case we shall impute the Motion to the Sun) which moves in his Orb sometimes nearer, sometimes farther off from the Earth, by which Means, although the Sun's Motion in his Orb was always equal; yet it would appear to us to be sometimes swifter than at other Times from West to East, and consequently slower in his diurnal Motion, as you see demonstrated in the following Figure.

As

Plate 9. *Fig.* 80.

In this Diagram suppose the Earth at the Center E, a-bout which is described the Circle ABCD, which is e-qually divided into 12 Parts, at the Points A, 1, 2. B, 3, 4, &c. representing the 12 Signs which in themselves are equally divided. Now if the Sun moved in the Circle ABCD, which is equally distant from the Earth its Cen-ter at E, his Motion would be regular and uniform, but, the Earth being not the exact Center of the Sun's Orb, makes the Division of the Signs, although equal in themselves, to appear unequal to us: For the Illustra-tion of which, suppose the Sun to move in the upper-most Circle, marked ♈, ♉, ♊, &c. whose Center is at the Mark ☉, and suppose the Earth at the Center of the other Circle near the Mark E; now although with re-spect to the Center E, the Division or Lines drawn from thence are equally distant in the first Circle, yet these Lines continued to the uppermost Circle in which the Sun is supposed to move, the Divisions upon that Circle are very unequal, and therefore although the Sun's pro-per simple Motion should be always equal, yet it is very plain that the Sun in Capricorn when nearest the Earth, shall appear to us to run more swift, than when in Can-cer, in his Aphelion or greatest Distance from the Earth, by reason of the smaller Division of the Signs, not that they are really so, but appear so to us, because of the Nearness of the Sun's Orb to the Earth, as you see the Points ♈ and ♎ are diametrically opposite to each other supposing the Eye at the Center E, where the Earth is supposed to be, and the far greater Part of the Circle is above the Line ♈ ♎, which yet contains but six Signs, and the far lesser Part is below that Line which contains also six Signs, and yet the Signs equally divided upon that Circle, whereof the Earth is supposed to be the Cen-ter.

Note; I am not here undertaking to determine whe-ther the Sun or Earth be the Center of the World, or whether the Orbs of the Planets be Circular or Ellip-tical; for which Way soever it be, this Demonstration

serves

serves to illustrate what I am now upon, as to the Equation of Time, and to let the Reader see somewhat of the Reason thereof, which is all that is expected from this Diagram.

Now the Occasion of this Equation being Two-fold, as is proved, it is plain, that when both the Inequalities tend one Way, it alters the Equation faster; as about the Middle of *December*, when the Sun is nearest the Earth, he appears to move faster from *West* to *East* by the last Demonstration; and also Running then almost parallel to the Equator, his Motion according to the Succession of Signs, must be swifter, by the second Demonstration; and therefore his Diurnal Motion or Revolution from Noon to Noon, must require more Time, by the first Demonstration, and consequently the Watches must now go faster than the Sun; but in *June*, although that Part of the Ecliptic in which he is then, lies nearly parallel to the Equator, as in *December*, thereby accelerating his *Easterly* Motion, to our Appearance; yet his Distance from the Earth being then in *Aphelion*, helps to retard it, so that the Motion or Alteration of the Equation, is not then so evident.

From this inequality, as grounded by these Two Causes, are the foregoing Tables of Equation of Time calculated, which will serve for many Years, without any sensible Alteration. Their Use is so plain, that every Body may understand it; for find the Month at the Top of the Leaf, and the Day at the Left-hand, and in the common Angle of Meeting, you have the Equation in Minutes and Seconds, whether it be too fast or too slow; as the Title [Watch too fast] or [Watch too slow] directs.

And now for the Application hereof to the finding of the Longitude: When you set Sail, observe in your Table of Equation of Time, how much the Watch is too Fast or too Slow, and set your Watch to it, and not exactly to the Time of the Day, unless it be when the Equation is nothing. As for Example, Suppose I am

bound

bound upon a Voyage any Day, when the Equation found in the Table is Seven Minutes, and the Title [Watch too flow] I conclude from thence, That a good Watch fhould be feven Minutes too flow, or behind the Time given by a true Sun-Dial, therefore I put my Watch feven Minutes behind Time given by the Sun. As, Suppofe I fet my Watch at 12 o'Clock, I put it to 53 Minutes paft 11, or if I fet it at 4, I put it to 53 Minutes paft 3, &c and laying it carefully by, as before directed it is fit for the Voyage.

Suppofe my Watch fet at 53 Min. paft 11, and I have fail'd feveral Days to the *Weftward* of the Meridian departed from, but whether *Northwards* or *Southwards* it matters not, finding the Sun juft upon the Meridian I look at my Watch, and find it 16 Minutes paft 3 o'Clock, and looking in the Equation Table, I find *Watch too faft* 12 Minutes, therefore I fubtract 12 Minutes from the Time given by the Watch, 3 Hours 16 Minutes, the Remainder 3 Hours 4 Minutes, reduced as before directed, gives 46 Deg. o Min. The true Difference of Longitude; but if it had been too flow 12 Min. you muft have added 12 min. to 3 h. 16 m. &c.

N te; If you fet you Watch exactly with the Sun when the Equation is nothing, it will always after that hold the fame Equation found in the Table, whether too faft or too flow, and the fame Quantity, (if your Watch go right) and that is the Reafon, that if you fet your Watch when there is Equation, you muft give it the Equation anfwering to that Day, (whether fwift or flow) and then it will alfo hold the fame Equation.

If any Body will object, that if a Watch proves wrong and erroneous, it may caufe a great Error to be contracted in this Way of finding the Longitude according to the Slownefs or Swiftnefs of the Watch.

I anfwer, That I agree thereto; but neverthelefs when Watches, as well as other Inftruments, are made by a good Workman, and fold to a Gentleman for good and

fubsubstantial, it is commonly expected that they should answer the End for which they are made and bought; and if we will suffer this Objection to prevail yet farther, I answer, that with Respect to the Latitude it may as well be urged, that if our Quadrants, Fore-staves, or other Instruments for that Purpose were made wrong, we should be much deceived in our Observations for the Latitude, and yet how few upon such Suppositions will foolishly desist from the Use of those Instruments for attaining the Latitude; nay, so far are they from that, that when they find any apparent Fault in a Quadrant, they will observe carefully what the Fault is, whether Northerly or Southerly, and how much, thereby to regulate their future Observations; and when all is done, you shall rarely find where there are many Quadrants, or many Observers aboard of one and the same Ship, and observing at one and the same Time, that all their Observations shall be exactly the same, but differ sometimes, 6, 8, 10, or 20 Minutes, or sometimes more; and yet these Differences are thought tolerable, and the Mariners continue to use these Means, and in a great Measure trust to them notwithstanding; for we must not expect to attain to Infallibility in any Respect, whilst we are traversing this Terraqueous Globe——— Not that I would advise any Body to be too credulous, or to take either Watches, or other Instruments, and trust to them as the best, merely upon the Report of another, unless you know you have great Reason to depend upon the Credit of those that so recommend them: But if you think to make Use of this Method for finding the Longitude at Sea, take a Watch along with you, that you have had some Experience of ashore, and if you have found that your Watch has gone well ashore, and yet fear, that for the Reason before mentioned she may not go so well at Sea, take her along with you upon some short Coast Voyage, where you can every now and then observe whether she keeps her true Motion at Sea as well as ashore, and if you find that notwithstand-

ing

ing your keeping her fo carefully as before directed,
the Clamminefs of the Sea-Air, does (as fome fuppofe)
retard her Motion, you may ufe Means to quicken her
Motion as much as to counterballance her Dullnefs occafioned by the Sea Air, if any fuch Thing be, (which
I cannot believe, if fhe be kept warm and dry, as before directed) and by this Means I do not queftion but
that this Method for finding the Longitude might be
rendered as eafy and practicable as the common Methods now in ufe, for finding the Latitude by Obfervation, a Thing very defirable, and therefore deferves to
be encouraged, and put in Practice.

I know it is argued by Sailor's (and with a good Reafon too) that in Places near the Equinoctial, the Degrees of Latitude and Longitude are fo nearly equal, and
that the Mercator's Chart and Plane Chart, are fo much
alike, that the Longitude there need not be much regarded; but the greateft Neceffity for it, and Difficulty in
attaining it, is in all Places nearer the Pole, efpecially
above 60 Degrees of Latitude, where a Degree of Longitude contains not half fo many Miles as a Degree of
Latitude, which makes the Work more difficult.

I anfwer, for the Encouragement of thofe that would
put in Practice what is here delivered, that the nearer
the Poles that you come, and confequently the leffer the
Degrees of Longitude are, the more practicable is this
Method, and the lefs the Errors are that can be fuppofed to be contracted; for in a few Miles Wefting or
Eafting in Latitudes above 60, where the Degrees of
Longitude are not half fo much as a Degree in the Equinoctial, you may much more fenfibly and apparently
difcern your Difference of Longitude, than nearer the
Equinoctial. I have been myfelf running 5 or 6 Knots
due Weft in Latitude 72d. 30m. or thereabouts, (I do
not exactly remember, but it was in coming from
Archangel, about the North Cape of *Finmark*, the Cape
itfelf lies in Latitude 71d. 25m. and we did not make
the Cape coming Home) and in 24 Hours running, I

coulJ

could very fensibly difcern that we had gained about
Half an Hour; fo that when the Glaffes were out for
12 o'Clock, (which we had formerly experienced to be
very right) it wanted about half an Hour of 12 by the
Sun, as near as I could compute, having no Help but
the Ship's Glaffes to compute it by: The Truth of
which is alfo evident; for in Latitude 72d. 30m. there
is but about 18 Miles to a Degree, and if we run about
135 Miles that 24 Hours, which might be done at the
Rate abovefaid, it would anfwer exactly to 7d. —30m.
of Longitude, which is juft half an Hour in Time; and
hence it is plain that the Difference of Longitude by this
Method is more perceptable in great Latitudes than near
the Equinoctial.

And that the Errors here contracted are alfo lefs is
further evident; For fuppofe your Watch to be wrong
by 4 Minutes of Time, this 4 Minutes, I confefs, is a
Degree of Longitude in all Latitudes, and in that Refpect the Error is equal in all Latitudes. But when the
Longitude comes to be reduced to Miles, fuch whereof
60 makes a Degree of a great Circle upon the Earth,
you will find that a Degree at or near the Equinoctial,
is about 60 Miles, and therefore an Error of 4 Minutes
in Time begets an Error of 60 Miles near the Equinoctial; but in Latitude 72d. 30m aforefaid, where there
is but 18 Miles to a Degree, an Error of 4 Minutes of
Time begets an Error of but 18 Miles in Diftance; and
further North where the attaining of Longitude is yet more
difficult, the Errors contracted by this Operation will confequently be lefs, an Error of 6 Leagues being as difcernable in Latitude 72d. 30m. as an Error of 20 Leagues
near the Equinoctial; fo that although Sailing near the
Equinoctial is commonly reckoned the eafieft in all other
Cafes, yet by this Method the moft difficult Cafes are
become the moft practicable and eafy: And I think I
may, without Prefumption, entitle it as ufeful an Help
for finding the Longitude as what has yet been offered,
and may be very affifting in that ufeful Subject, till (if
poffible) the defired Compleatment thereof be accom-
plifhed.   Of

Fig. 93.   Fig. 94.   Fig. 95.

E   F   E   F   H   I

98.   K   Fig. 99.   Fig. 100.

N   N

I   L   P   M

Of ASTRONOMY.

I HAVE in the former Part hereof, treated upon the Projection of the Sphere, and Spherical Trigonometry Right-angled and Oblique, which may serve as a good Introduction to this Chapter; in which I shall lay down all that Part of Astronomy, that is necessary in Navigation, and therewith conclude the Whole.

SECT. I.

Plate 10. *Fig.* 111.

Astronomical Definitions.

I shall bring the Definitions under these five Heads, 1. Points, 2. Great Circles, 3. Lesser Circles, 4. Arches of Circles, 5. Angles.

1. *Points.*

The Zenith is the Point directly over our Heads as Z, and the Nadir is the Point opposite or directly under our Feet, as N.

The Poles of the Equator (commonly called the Poles of the World) are the North Pole P, and the South Pole S, and the Poles of any great Circle are every Way 90 Degrees distant from it, so the Poles of the Ecliptic Æ C are in the Points *c c*, &c.

2. *Great Circles.*

The Azimuth's are great Circles, which intersect one another in the Zenith, and cut the Horizon at right Angles; as Z ☉ R N and Z *d* N.

The Horizon is that great Circle, that divides the upper Hemisphere from the Lower, as H, A, O.

The

The Equator divides the North Hemisphere from the South as E, A, Q. ——the Ecliptic cuts the Equator at an Angle of 23, 29, as Æ, A, C, the Prime Vertical is the East and West Azimuth as Z, A, N, ——The Meridians are great Circles passing from Pole to Pole, as the Primitive PESQP, or the Right Circle PAS, or the Oblique Circle P ☉ S, &c. The Meridians and Azimuths are infinite.

3. *Lesser Circles.*

Lesser Circles are commonly called Parallel Circles, because they are generally Parallel to some great Circle; thus the Parallel of Declination Æ ☉ f is so called, because it is Parallel to the Equinoctial; from whence the Declination is reckoned either North or South. —— The Tropic of Cancer is a Parallel of Declination 23d. 29m. distant from the Equator Northward, and the Tropic of Capricorn at the same Distance Southward from the Equator.

The Parallels of Altitude (by some called Almicanters) are lesser Circles Parallel to the Horizon, as *b* ☉ *e.* The Artic Circle is Parallel to the Equator, distant from it 66d. 31m. Northwards, and the Antartic Circle is the same Distance Southward from the Equator. ——The two Poles of the Ecliptic are in these two Circles.

Note; Arches of lesser Circles are never made Parts of Triangles.

4. *Arches of great Circles.*

The Sun's Altitude is an Arch of an Azimuth Circle comprehended between the Sun and the Horizon, as ☉ R.

Sun's Azimuth is an Arch of the Horizon, contain'd between that Azimuth which the Sun is in, and the Meridian (North or South) thus HR is the Sun's Azimuth from the South, when the Sun is any where in the Azimuth Z ☉ N whatever his Altitude be.

Sun's Amplitude is the Distance that the Sun rises or sets from the East or West, or if you will from the

North

North or South, and is an Arch of the Horizon contained between the Point where the Parallel of Declination, that the Sun moves in, cuts the Horizon, and the
East or West, South or North Points of the Horizon;
thus when the Sun moves in the Parallel Æ ☉ *f,* it rises
at *d,* and the Arch A *d* is the Sun's Amplitude from the
East or West, and the Arch *d* O is the Amplitude from
the North.

Note; The Sun's Amplitude is always of the same
Denomination with the Declination, whether North or
South.

Right Ascension is an Arch of the Equator or Equinoctial contained between the Beginning of *Aries,* and
that Point of the Equator where the Sun's Meridian
intersects it, as AT is the Sun's Right Ascension when
he is any where in the Meridian Ƿ ☉ T S; but in Astronomical Operations, the Distance between the said Intersection; and the nearest Equinoctial Point is the Right
Ascension, and then it can never exceed 90 Degrees.

Ascensional Difference is an Arch of the Equator
contained between that Point of the Equinoctial that
rises with the Sun, and that Point that comes to the
Meridian with him; thus the Point A rises with the
Sun; but the Point *n* comes to the Meridian with him,
therefore A *n* is the Ascensional Difference.

Oblique Ascension is an Arch of the Equator contained between the next Equinoctial Point, and that
Point of the Equator that rises with the Sun, and is
found by adding to the Right Ascension, or subtracting
from it, the Ascensional Difference, the Sum or Remainder is the Oblique Ascension required.

Longitude of the Sun is the Sun's Distance from the
nearest Equinoctial Point, accounted on the Ecliptic
thus, Λ ☉ is the Sun's Longitude.

Declination is an Arch of the Meridian contained
between the Sun and the Equinoctial as ☉ T——

Angles.

Angles.

The Sun's greatest Declination is the Angle that the Ecliptic makes with the Equator, as T A ☉, and is accounted 23d. 29m.

The Hour of the Day is the Angle that the Meridian of the Sun makes with the Meridian of the Place; thus the Angle Æ P' ☉ is the Hour that it wants of, or is past 12, accounting 15d. to an Hour, &c.

The Sun's Azimuth is that Angle at Z, which is made between the Azimuth Circle that the Sun is in, and the North and South Azimuth or Meridian of the Place, as the Angle HZR is reckoned from the South; but RZA is reckoned from East or West.

The Sun's Amplitude is an Angle at Z, made by that Azimuth Circle which intersects the Horizon (where the Sun's Parallel of Declination cuts it also) makes with the Prime Vertical, if reckoned from the East or West, as A Z d, or d Z O, if reckoned from the North.

These Angles may also be reckoned as Arches of great Circles, as the Hour is an Arch of the Equator, the Azimuth and Amplitude are Arches of the Horizon, &c.

SECT. II.

Astronomy Rectangular.

A Lthough most of the necessary Cases in Astronomy may be represented by a Projection upon the Plane of the Meridian, which to a great many Learners seems most natural and intelligible; yet there are some Cases especially in Oblique, that cannot be projected upon that Plane, in which Cases take this general Rule.

When there are given two Sides and an Angle, make a given Side upon the Primitive, and the given Angle

at

at the Primitive; but if two Angles and a Side be given, make both the Angles at the Primitive; and then, if the given Side be contained between the given Angles, it will alfo fall upon the Primitive; but if not it falls within; but when three Sides, or three Angles are given, it matters not which Circle you make Primitive, fo it be generally one of thofe great Circles, an Arch of which is a Part of the Triangle, nor are you always obliged to have the given Triangle at the Primitive; but it may fometimes be drawn within.

There are 12 Signs in the Zodiac, their Names and Characters are as follow.

♈ Aries, ♉ Taurus, ♊ Gemini, ♋ Cancer, ♌ Leo, ♍ Virgo, ♎ Libra, ♏ Scorpia, ♐ Sagittarius, ♑ Capricorn, ♒ Aquarius, ♓ Pifces.

Each Sign contains 30 Degrees, in all 360, the Number of Degrees in every great Circle.

Note, When it is not expreffed what Plane the Projection is upon, it is always fuppofed upon the Plane of the Meridian.

Note alfo, When any Problem is referred to, it means the Problems in Spherical Geometry, annex'd to the Stereographic Projection of the Sphere.

Problem I.

Plate 10. Fig. 112.

Given the Sun's Longitude and greateft Declination, to find the prefent Declination.

Sun's Place ♊ od. om. or Longitude from the Beginning of ♈ 60 greateft Declination 23d. 29m.

Draw the Primitive H Z O N, and crofs it with the Diameters ZN and HO, at Right Angles (which being to be done in all Cafes, I fhall ufe the Contraction PCQ; for the Words, draw the Primitive Circle and crofs it with two Diameters at Right Angles) draw AC the Ecliptic at an Angle of 23, 29, at A by Prob. 3. fet
the

the Longitude 60 from A to ⊙ by Prob. 5. and through
⊙ draw the Circle Z ⊙ B N by Prob. 1. then is A B ⊙
the Triangle required.

Here is given the Hypothenuse A ⊙, the Longitude
60 0, and the Angle A the greateſt Declination 23 29,
to find the Perpendicular B ⊙, the preſent Declination by Caſe 2.

As Radius ——— ——— 90 ——— 10.00000
To Sine of A ⊙ the Longitude 60 0 —— 9.93753
So Sine of A the greateſt Dec. 23 29 —— 9.60041

To Sine of B ☽ the preſent Dec. 20 11 —— 9.53794

Problem 2.

Plate 10. *Fig.* 112:

Given as in Prob. 1. to find the Right Aſcenſion AB;
here is given the Hypothenuſe and an Angle, to find the
Leg adjacent to the given Angle by Caſe 1.

As Tang. Comp. A ⊙ the Longit. 30 0 — 9.76144
To Radius ——— ——— 90 —— 10.00000
So Sine Comp. A the greateſt Dec. 66 31 — 9.96245

To Tang. AB the Right Aſcenſion 57 48 — 10.20101

Problem 3.

Plate 10. *Fig.* 112.

Given the greateſt Declination 23 29, and preſent
Declination 20 11, to find the Longitude.

PCQ, let ÆQ be the Equator, draw EC as in
Prob. 1. for the Ecliptic, and at 20 11 the preſent Declination, Diſtance from the Equator, draw the Pa-
ralel *m* ⊙ *n* by Prob. 4, and through the Point where it
cuts the Ecliptic as in ⊙ draw the Circle Z ⊙ N by
Prob. 1; then is A ⊙ B the Triangle required.

Here is given the Angle A, and Leg oppoſite B ⊙,
to find the Hypothenuſe A ⊙, the Longitude, Caſe 12.

As

As Sine of A the greatest Dec. 23 29 — 9.60040
To Sine of B☉ the present Dec. 20 11 — 9 53785
So Radius ——— ——— ——— — 10.00000

To Sine of A ☉ the Longitude – 60 00 — 9.93745

Problem 4.

Plate 10. Fig. 112.

Given the greatest Declination 23 29, and present Declination 20 11, to find the Right Ascension.

Projection as in Prob. 3, being the same Data, the Angle A and Leg opposite B ☉ are given, to find the other Leg AB by Case 10.

As Radius ——— ——— 90 ——— 10.00000
To Tang. B ☉ the present Dec. 20 11 = 9.56537
So Tang. Comp. A greatest Dec. 66 31 — 10.36204

To Sine AB Right Ascension 57 48 — 9.92741

Problem 5.

Plate 10. Fig. 112.

Given the greatest Declination 23 29, and Right Ascension 57 48, to find the present Declination: PCQ, and let ÆQ be the Equator, and draw EC the Ecliptic to make an Angle of 23 30 the greatest Declination by Prob. 3, with the Equator set the Right Ascension, 57 48, from A upon the Equator to B, by Prob. 5, and thro' the Points ZBN draw the Circle ZBN to cut the Ecliptic in the Point ☉; then is A B ☉ the Triangle required.

Here is the given Angle A, and Leg adjacent **AB**, to find the Leg opposite B ☉; by Case 7.

As Tang. Comp. A greatest Dec. 66 31 — 10.36204
To Radius ——— ——— 90 00 — 10.00000
So Sine AB Right Ascension — 57 48 — 9.92747

To Tang. B ☉, present Dec. — 20 11 — 9.56543

Prob.

Problem. 6.

Plate 10. *Fig.* 112.

Given the greatest Declination 23 29, and the Right Ascension 57 48, to find the Longitude.

Projection as in Problem 5.

Here is given the Angle A, and Leg adjacent AB, to find the Hypothenuse A ⊙, by Case 9.

As Tang. AB Right Ascension   57 48   — 10.20084
To Radius   — — — 90 00   — 10.00000
So Sine Comp. A greatest Dec.   66 31   — 9.96245

To Tang. Com. A ⊙ the Longitude 30 00   — 9.76161

Problem 7.

Plate 10. *Fig.* 113.

Given the Latitude of the Place 50 North, and Sun's Declination 20 North, to find the Ascensional Difference. And also the Sun's Rising and Setting   —   Prob.   8
Length of the Day and Night   ——   ——   9
Sun's Amplitude   ——   ——   ——   10
What Hour the Sun is due East or West   ——   11
Sun's Altitude when East or West   —   ——   12
Sun's Altitude at Six   ——   ——   ——   13
Sun's Azimuth at Six   ——   ——   ——   14
Sun's Decl. and Time of Rising and Setting, to ⎫
    find his Amplitude   ——   ——   ——   ⎭ 15

PCQ, set the Latitude 50 from O to P, and from H to S, and draw P S for the Axis of the World, and at Right Angles thereto draw E Q the Equator, and at an Angle of 23 29 therewith draw ÆC the Ecliptic, draw the Parallel *a* ⊙ *b c d e* distant from the Equator 20d. the given Declination to cut the Ecliptic in ⊙, and through ⊙ draw the Meridian P ⊙ S, and the Azimuth Z ⊙ N ; also through that Point where the Parallel of Declination cuts the Prime Vertical, draw the Meridian P *b* S, and through *c*, the

                          Point

Point where the Parallel of Latitude cuts the Axis P S, draw the Azimuth Z *c* N, and through the Point where the said Parallel cuts the Horizon as at *d*, draw the Meridian P *d* S, and then in the Triangle A *d b* is given the Angle A, the Comp. of Latitude 40; the Side *b d* the Declination, to find A *b* the Ascensional Difference by Case 10.

As Radius —— —— 90 00 — 10,00000
To Tangent of Latitude—— 50 00 — 10,07618
So Tang of *b d* the Declin.— 20 00 — 9,56106

To Sine of Ascensional Differ. 25 42 — 9.63724

Prob. 8.   Plate 10.   Fig. 113.
To find by the Ascensional Difference, what Time the Sun Rises and Sets.

The Ascensional Difference converted into Time (allowing 15 Degrees to an Hour, and 1 Degree to 4 Minutes of Time,) gives the Time that the Sun rises before or sets after 6 in the Summer, or rises after and sets before 6 in the Winter; and at that Rate the Ascensional Difference 25 42 makes 1 Hour 43 Min. which added to 6 makes 7 Hours 43 Min. the Time of the Sun's setting in North Declination, or rising in South Declination; but subtracted from 6 it leaves 4 17 the Time of Sun's Rising in North Declination, or setting in South, &c.

Note; Subtract the Sun's Rising from 12, the Remainder is the Setting.

Prob. 9.   Plate 10.   Fig. 113.
To find the Length of the Day or Night.

Double the Time of Sun's Setting found by *Prob.* 8. it gives the Length of the Night; and double the Time of Sun's Rising it gives the Length of the Day: Then when the Sun rises at 4 17, that doubled makes 8 34, the Length of the Night; but the setting at the same Time 7 43 doubled is 15 26 the Length of the Day.

C c

Prob

Problem 10: *Plate* 10. *Fig.* 113.

To find the Sun's Amplitude.

In the Triangle A *b d* is given the Angle A the Complement of the Latitude, and Side *b d* the Declination, to find A *d* the Amplitude from the East or West, by Case 12.

As Sine A the Comp. Latitude — 40 0 — 9,80807
To Sine of *b d* the Declination — 20 0 — 9,53405
So Radius — — — — — 90 0 — 10,00000
To Sine of A *d* the Amplitude — 32 9 — 9,72598

Or the Amplitude from the North may be found in the Triangle *d* P O, where is given *d* P the Complement of Declination 70 0, and O P the Latitude 50, to find *d* O, by Case 6.

As Sine PO the Comp. of Latitude 40 0 — 9,80807
To Radius — — — — 90 0 — 10,00000
So Sine Comp. P *d* the Comp. Decl. 70 00 — 9,53405
To Sine Comp. *d* O the Amplitude 57 51 — 9,72598

Problem 11. *Plate* 10. *Fig.* 113.

Given the Latitude 50 0, and the Declination 20 00, to find what Time the Sun will be due East or West.

In the Triangle A *m b*, there is given *m b* the Declination, and the Angle A the Latitude, to find *m* A the Hour from 6, by Case 10.

As Radius — — — — — 90 00 — 10,00000
To Tang. *m b* the Declination 20 00 — 9,56106
So Tang. Comp. A the Latitude 50 00 — 9,92381
To Sine A *m* the Hour from 6, 17 47 — 9,48487

Or it may be done by the Quadrantal Triangle A P *b*, the Side AP being a Quadrant, and the Angle P the Hour from 6; and therein is given the Side P *b* the Complement of Declination 70 0; and the Angle *b*AP the Complement of Latitude 40 0 to find *b* P A : And
here

here omitting the Quadrantal Side AP, the Angle P is middle Part, and A and *b* P are Conjuncts; therefore,

As Radius ———— ———— 90 ——— 10,00000
To Tang. A the Comp. Lat. — 40 00 — 9,92381
So Tang. Comp. *b* P — — 20 00 — 9,56106
To Sine of P the Hour from 6, 17 47 — 9,48487

Prob. 12. *Plate* 10. *Fig.* 113.

To find the Sun's Altitude when Eaſt or Weſt.

The Sun is in the Point *b* when Eaſt or Weſt : Then in the Triangle A *m b*, there is given *m b* the Declination, and the Angle *mAb* the Latitude ; to find A*b* the Sun's Altitude, by Caſe 10, which is left for the Learner to perform.

Prob. 13. *Plate* 10. *Fig.* 113.

To find the Sun's Altitud· at 6 o'Clock.

The Sun is in the Point *c* at 6 o'Clock, and in the Triangle A *i c* there is given the Angle A the Latitude, and the Hypothenuſe A *c* the Declination, to find *i c* the Altitude at 6, by Caſe 2.

Prob. 14. *Plate* 10. *Fig.* 113.

To find the Sun's Azimuth at Six o'Clock.

This may be done either by the Right-angled Triangle A *i c*, where the Leg A *i* is the Azimuth at 6, or by the Quadrantal Triangle A *c* Z, where the Angle Z is the ſaid Azimuth, I ſhall only inſtance in the Quadrantal, and leave the other Operation of this and the two laſt Problems for the Reader's Practice.

There is given A *c* the Declination, and the Angle ZA *c* the Complement of the Latitude, to find AZ*c* the Azimuth at 6.

Here A is middle Part, and the other two are Conjuncts ; therefore,

C c 2   A s

As Tang. Comp. of A *c* the Declin. 20 0 = 10,43893
To Radius —— —— —— 90 0 — 10,00000
So Sine of A the Comp. Latitude 40 0 — 9,80807
                                                    ————
To Tang. of Z the Azimuth — 13 10— 9,36914

Prob. 15, *Plate* 10. *Fig.* 113.

Given the Sun's Declination and Time of Rising and Setting, to find the Amplitude.

Suppose the Sun rise at 17m. past Four in the Morning, his Declination 20d. North, I demand the Amplitude.

In the Triangle *d* P O there is given the Angle P, the Hour past Midnight 4 Hours 17m. or reduced to Deg. 64d. 15m. and the Hypothenuse *d* P the Complement of Declination to find the Leg *d* O, by Case 2.

As Radius —— —— —— 90 0 — 10,00000
To Sine of *d* P the Comp. Declin. 70 0 — 9,97298
So Sine of P, the Hour from Midn. 64 15 — 9,95458
                                                    ————
To Sine *d* O the Ampl. from North 57 49 — 9,92756

Or it may be found by the Quadrantal Triangle A*d*P, in which is given the Angle P, the Hour of Sun's Rising before 6; and *d* P the Complement of the Declination to find A *d*, the Amplitude from the East or West.

By a right Understanding of the Nature of Spheric Triangles, the Latitude at Sea may be often attained in an Afternoon or Evening, when you cannot have an Observation at Noon; as in the Triangle A *i c*, you may by having the Declination A *c* and Amplitude A *i* given, find the Latitude, *viz.* the Angle *c* A *i*, or, in the Triangle A *m b* you may by having A *b*, the Sun's Altitude, when East or West, and *m b* the Declination given, find the Angle *m* A *b* the Latitude, &c. but shall leave this to the Reader's Industry.

Prob.

Prob. 16.   *Plate* 10.   *Fig.* 114.

Given the Sun's Altitude, Declination and Latitude of the Place to find the Azimuth.

Latitude 54 40N. Altitude 47 48, Declination 20 12N.

PCQ, set the Latitude 54 40 from O to P, and draw P S, and at Right Angles therewith draw E Q for the Equator; draw the Parallel of Altitude *c* ⊙ *d* at the given Altitude 47 48 above the Horizon, and alfo the Parallel of Declination *a*O*b* 20 12 from the Equator to the Northwards, and through the Interfection of thefe Parallels draw the Azimuth Z ⊙ N, and the Meridian P ⊙ S by Prob. 1. and then by the 11th Cafe of Oblique Sherie Triangles.

Comp. Latitude ———— —— Z P	————	35.20
Comp. Altitude ———— —— Z ⊙	——	42.12
Comp. Declination ———— —— ⊙ P	————	69.48
Sum——		147.20
½ Sum——		73.40
Side oppofite to the required Angle P ⊙ ——		69.48
Remains —		3.52

Z P 35 20 Sine Co. Ar. ——— ·—— ——		0,23782
Z ⊙ 42 12 Sine Co. Ar. ——— —— ——		0,17281
½ Sum of the Sides 73 40 Sine —— ——		9,98211
Diff. 3 52 Sine ——— —— —— ——		8,82888
Sum ——		19,22162

½ Sum is Sine Comp. of 65 55 —— —— 9,61081
Doubled is 131 50 the Azimuth required from the North

Prob. 17.   *Plate* 10.   *Fig.* 115.

Given the Altitude, Declination and Hour, to find the Azimuth.

Altitude 47 48, Declination 20 12, and Hour or Angle at P 32 12.

PCQ, and draw P ⊙ S, to make an Angle at 32 12 with the Primitive by Prob. 3, and thereon fet off the

Comp.

Complement of Declination 69 48 from P to ☉ by
Prob. 5; also draw the Parallel Z *a b c* distant from ☉
42 12 the Complement of Altitude, and it will cut the
Primitive in Z; then through Z ☉, and the opposite
Point N draw the Oblique Circle Z ☉ N, and the Tri-
angle ZP☉ is finished; and by Solution 7.

                                                    Co. Ar.

As Sine Comp. of Altitude Z ☉ 42 12 — 0,17281
To Sine of the Hour ZP☉ ————— 32 12 — 9,72662
So Sine Comp. Declination ☉ P 69 48 — 9,97243
To Sine of the Azimuth ☉ Z P 48 7 — 9,87186

   But being Obtuse it must be subtracted from 180, and
the Remainder 131 53 is the Azimuth required from
the North.
   The Truth of all your Calculations may be examined
and proved by measuring the Side or Angle required by
the Directions given by Prob. 6, and 7, of Spherical
Geometry, which would have been needless to repeat
in all the foregoing Operations.

          Prob. 18.   *Plate* 10.   *Fig.* 113.

   Given the Latitude, Azimuth, and Hour to find the
Declination.
Latitude North, ——— 54 40 ⎫ to find the Sun's Decl.
Azimuth, ————— 131 52 ⎬ by Case the Ninth.
Hour, ——— ——— 32 12 ⎭
   Here the Perpendicular must fall upon the Side ☉Z,
continued, which for Illustration I have annexed to the
Scheme: then in the Triangle Z *e* P, is given *e* Z P
48 8, and Z P 35 20; to find Z P *e*, by Case the third
of Right-angled Spherical Triangles:

As Tang. Com. P Z *e* ——— 41 52 —— 9,95240
To Radius ——— ——— 90 ———— 10,00000
So Sine Comp. Z P. ——— 54 40 —— 9,91158

To Tang. Comp. ZP*e* ——— 42 19 ——— 9,95918

                                              Then

Then by Concluſion the Fourth,

As Sine Comp. *e* P ☉ —— 10 07 Co. Ar. 0,75534
To Tang. PZ— — — 35 20 ——— 9,85059
So Sine Comp. *e* PZ— — 42 19 ——— 9,82816

To Tang. P ☉ — — — 69 48 ——— 10,43409

The Complement of the Declination being 69 48, the Declination required is 20 12.

Prob. 19. *Plate* 10. *Fig.* 113.

Given the Azimuth, Hour, and Angle of Poſition, to find the Latitude.

Azimuth, —— —— 131 52 ⎫ {to find the Latit.
Hour, ——— · —— 32 12 ⎬ by Caſe the 11th.
Angle of Poſition — — 27 29 ⎭

Take the Supplement of the greateſt Angle, to 180, add that and the two leſſer Angles together. and from the half Sum ſubtract the Angle oppoſite to the Side required, and proceed as in Prob. 11.

E X A M P L E.

The greateſt Angle — — — — 131 52

Its Supplement — — — — — — 48 8
Angle Z P ☉, — — — — — — 32 12
Angle Z ☉ P, — — — — — — 27 19

Sum — — — — — — — —107 39

Half Sum — — — — — — — 53 49½
Angle oppoſite to Side required Z ☉ P —. 27 19

Remainder ——— · — 26 30½

CZP

⊙ Z P, its Suppl. —— 48 : 08 S. Co. Ar. 0,12801
Z P ⊙ S. Co. Ar. —— 32 : 12 ———— 0,27337
Half Sum ———— 53 : 49½ — *Sine Log.* 9,906 9
Remainder ———— 26 : 30½ — *Sine Log.* 9,64965

                                                19,95802

Sine Comp. 17 : 40 —— —— —— 9,97901

   ·Doubled is 35 20 the Complement of the Latitude,
or Side Z P required.

    Thus much for the Application of Oblique Spheri-
cal Triangles in Queſtions of Aſtronomy.

    N.B. *The Number* 57.3 *in Chap.* 6. *Page* 141, *has its*
Authority from its being the Semidiameter of a Circle whoſe
Circumference is 360 *Degrees in a great Circle (and conſe-*
quently may be applied otherways to finding Angles than in
Degrees only : This I thought fit to diſcover, leſt ſome Pam-
phletering Navigator ſhould aſſert it to be a fictitious and
groundleſs Number, but ſhall defer any further Account of
the Reaſon of the Steps in the Operation, leſt they ſhould catch
it up and call it their own, as an Attempt was made upon
my Method of finding the Longitude by a Right-angled Plane
Triangle Chap. 6. *Sect* 4. *only I produced it in the firſt Im-*
preſſion of this Book, which was printed above ſeven Years
before this upſtart Anther's Copy was wrote.

                                               A

Franz.

A TA BLE of the *Englifh*, *Latin*, *Italian*, and *Spanifh* Names of the Two and Thirty Points of the COMPASS.

Englifh	Latin.	Italian.	Spanifh.
North	Septentrio	Tramontana	Norte
N by E.	Hypaquilo	4 di Tramontana Greco	Norte 4 a Nord' efte
N N E.	Aquilo	Tramontana Greco	Nord Nord' efte
NE by N.	Mefaquilo	4 di Greco Tramontana	Nord efte, a Norte
N E.	Borrapeliotes	Greco	Nord' efte
NE by E.	Hypocaefias	4 di Greco Levante	Nord 4 al efte
E N E.	Cefias	Levante Greco	Los Nord' a efte
E by N.	Mefocaefias	4 di Levante Greco	Lef 4 a Nord efte
Eaft.	Subfolanus	Levante	L' efte
E by S.	Yfurus	4 di Levante Sirocco	L' efte 4 a Suefte
E S E.	Eurus	Levante Sirocco	L' Suefte
SE by E.	Mefurus	4 di Sirocco Levante	Suefte 4 al efte
S E.	Notapeliotes	Sirocco	Suefte
SE by S.	Hypophaenix	4 di Sirocco Oftro	Suefte 4 al Sur
S S E.	Phaenix	Oftro Sirocco	Sur Suefte
S by E.	Mefophaenix	4 di Oftro Sirocco	Sur 4 a Suefte
South.	Notus	Oftro	Sur
S by W.	Mefolibanotus	4 di Oftro Garbino	Sur 4 al Sudefte
S S W.	Libabotus	Oftro Garbino	Sud Suefte
SW by S.	Hypolibanotus	4 di Garbino Oftro	Sud Veft 4 al Sud
S W.	Nothybicus	Garbino e Libeccio	Sud Veft
SW by W.	Mefafricus	4 di Garbino Ponente	Sud Veft 4 al Oeft
W S W.	Africus	Ponente Garbino	Oeft Sud Oeft
W by S.	Hypafricus	4 di Ponente Garbino	Oeft 4 al Sudueft
Weft.	Zephyrus	Ponente	Oeft
W by N.	Mefocorus	4 di Ponente Maeftro	Oeft 4 al NordOeft
W N W.	Corus	Ponente Maeftro	Oeft Nord Oeft
NW by W.	Hypocorus	4 di Maeftro Ponente	NordOeft 4 al Nort
N W.	Barrohybicus	Maeftro	Nord Oeft
NW by N.	Hypocircius	4 di Maeftro Tramont	Nord Oefte 4 a Nor
N N W.	Circius	Tramontana Maeftro	Nor Nor Ofte
N by W.	Mefocircius	4 di Tramon Maeftro	Nor 4 a Nor Oef

The Variation of the Compaſs, as I obſerved it in the Year 1718, aboard His Majeſty's Ship the *Barfleur*, under the Command of the Right Hon. Sir *George Byng*, Admiral of the Fleet.

Variation Weſt		Bearings of Headlands, &c.
D.	M.	
10	53	USHANT Eaſterly.
8	52	Cape Finiſter, ENE. 12 Leagues,
8	7	Off of Liſbon, within ſight of Land.
8	35	St. Jago, NE. 14 Leagues
8	45	Cape St. Mary, NNE. half E. 5 Leagues.
8	2	Cape St. Mary, NNW. 17 Leagues.
7	0	Before Straits - Mouth, ſometimes without ſight of Land, we had by ſeveral good Obſervations 7 deg. or within half a deg. more or leſs.
9	38	Cape Palos, 10 Leagues WSW.
14	31	Cabrera, [near Majorca) NW. 8 Leagues.
14	52	Eaſt Point of Majorca, WNW. 6 Leagues.
14	8	At Anchor before Port Mahon.
10	54	Cape Tolar, E. half S. 25 Leagues by Eſtimation.
17	16	Cape Carbonera, NW. half N. 7 Leagues.
13	44	Cape Carbonera, WNW. 9 or 10 Leagues.
11	48	Cape Carbonera, W. by S. 11 Leagues.
11	35	Serpentaria, W. by S. Cape Gomera, NW.
11	46	Entry of Naples Bay.
10	50	Off of Palinura.
12	21	Strombolo, W. by S. 1 League.
11	30	Before Syracuſe.
11	45	Cape Paſſero, N. half W. 4 Leagues.
11	38	Before Malta.

Upon the S. Coaſt of Sicily it continued between 12 and 13 deg. but ſailing from Palermo to Naples, it decreaſes to 11.46 as obſerved in Naples Bay.

N.B. It may be thought ſurpriſing, that to the Weſtward of Sardinia, there ſhould be leſs than 11 deg. and South Eaſt from it, it is above 17. There is ſome Reaſon to believe, that there are ſome magnetical Rocks, in or near that Iſland which may affect the Compaſs; but we muſt leave that to Time and further Experience to determine.

A TABLE of LOGARITHMS

For Numbers increasing in their Natural Order, from an Unit to 10,000.

With a Table of Artificial Sines, Tangents, and Secants, the Radius 10,000000; and to every Degree and Minute of the Quadrant.

N.	Logar.	N.	Logar.	N.	Logar.
1	0.000000	34	1.531479	67	1.826075
2	0.301030	35	1.544068	68	1.832509
3	0.477121	36	1.556303	69	1.838849
4	0.602060	37	1.568201	70	1.845098
5	0.698970	38	1.579784	71	1.851258
6	0.778151	39	1.591065	72	1.857332
7	0.845098	40	1.602060	73	1.863323
8	0.903090	41	1.612784	74	1.869232
9	0.954242	42	1.623249	75	1.875061
10	1.000000	43	1.633468	76	1.880814
11	1.041393	44	1.643453	77	1.886491
12	1.079181	45	1.653212	78	1.892095
13	1.113943	46	1.662758	79	1.897627
14	1.146128	47	1.672098	80	1.903090
15	1.176091	48	1.681241	81	1.908485
16	1.204120	49	1.690196	82	1.913814
17	1.230449	50	1.698970	83	1.919078
18	1.255272	51	1.707570	84	1.924279
19	1.278754	52	1.716003	85	1.929419
20	1.301030	53	1.724276	86	1.934498
21	1.322219	54	1.732394	87	1.939519
22	1.342423	55	1.740363	88	1.944483
23	1.361728	56	1.748188	89	1.949390
24	1.380211	57	1.755875	90	1.954243
25	1.397940	58	1.763428	91	1.959041
26	1.414973	59	1.770852	92	1.963788
27	1.431364	60	1.778151	93	1.968483
28	1.447158	61	1.785330	94	1.973128
29	1.462398	62	1.792392	95	1.977724
30	1.477121	63	1.799340	96	1.982271
31	1.491362	64	1.800180	97	1.986772
32	1.505150	65	1.812913	98	1.991226
33	1.518514	66	1.819544	99	1.995635

N°	0	1	2	3	4	Dif
100	2.000000	2.000434	2.000868	2.001301	2.001734	432
101	2.004321	2.004751	2.005180	2.005609	2.006038	428
102	2.008600	2.009026	2.009451	2.009876	2.010300	424
103	2.012837	2.013256	2.013680	2.014100	2.014520	419
104	2.017033	2.017451	2.017868	2.018284	2.018700	416
105	2.021189	2.021603	2.022016	2.022428	2.022841	412
106	2.025306	2.025715	2.026124	2.026533	2.026942	408
107	2.029384	2.029789	2.030195	2.030600	2.031004	404
108	2.033424	2.033826	2.034227	2.034628	2.035029	400
109	2.037426	2.037825	2.038223	2.038620	2.039017	396
110	2.041393	2.041787	2.042182	2.042575	2.042969	393
111	2.045323	2.045714	2.046105	2.046495	2.046885	389
112	2.049218	2.049506	2.049993	2.050380	2.050766	386
113	2.053078	2.053463	2.053846	2.054230	2.054613	383
114	2.056905	2.057286	2.057666	2.058046	2.058426	379
115	2.060698	2.061075	2.061451	2.061829	2.062206	376
116	2.064458	2.064832	2.065206	2.065580	2.065953	373
117	2.068186	2.068557	2.068928	2.069298	2.069668	369
118	2.071882	2.072250	2.072617	2.072985	2.073352	366
119	2.075547	2.075912	2.076276	2.076640	2.077004	363
120	2.079181	2.079543	2.079904	2.080266	2.080626	360
121	2.082785	2.083144	2.083503	2.083861	2.084219	357
122	2.086360	2.086716	2.087071	2.087426	2.087781	355
123	2.089905	2.090258	2.090611	2.090963	2.091315	351
124	2.093422	2.093772	2.094122	2.094471	2.094820	349
125	2.096910	2.097257	2.097604	2.097951	2.098297	346
126	2.100370	2.100715	2.101059	2.101403	2.101747	343
127	2.103804	2.104145	2.104487	2.104828	2.105169	340
128	2.107210	2.107549	2.107888	2.108227	2.108565	338
129	2.110590	2.110926	2.111262	2.111598	2.111934	335
130	2.113943	2.114277	2.114611	2.114944	2.115278	333
131	2.117271	2.117603	2.117934	2.118265	2.118595	330
132	2.120574	2.120903	2.121231	2.121560	2.121888	328
133	2.123852	2.124178	2.124504	2.124830	2.125156	325
134	2.127105	2.127429	2.127752	2.128076	2.128399	323
135	2.130334	2.130655	2.130977	2.131299	2.131619	321
136	2.133539	2.133858	2.134177	2.134496	2.134814	318
137	2.136721	2.137037	2.137354	2.137670	2.137987	315
138	2.139879	2.140194	2.140508	2.140822	2.141136	314
139	2.143015	2.143327	2.143639	2.143951	2.144263	311
140	2.146128	2.146438	2.146748	2.147058	2.147367	309
141	2.149219	2.149527	2.149835	2.150142	2.150449	307
142	2.152288	2.152504	2.152500	2.153205	2.153510	305

Nº	5	6	7	8	9	Dif
100	2.002166	2.002598	2.003029	2.003467	2.003891	433
101	2.006466	2.006894	2.007321	2.007748	2.008174	427
102	2.010724	2.011147	2.011570	2.011993	2.012415	424
103	2.014940	2.015360	2.015779	2.016198	2.016615	419
104	2.019116	2.019532	2.019947	2.020361	2.020775	416
105	2.023252	2.023664	2.024075	2.024486	2.024896	412
106	2.027350	2.027757	2.028164	2.028571	2.028978	408
107	2.031408	2.031812	2.032216	2.032619	2.033021	404
108	2.035430	2.035830	2.036229	2.036629	2.037028	400
109	2.039414	2.039811	2.040207	2.040603	2.040998	396
110	2.043302	2.043755	2.044148	2.044540	2.044931	393
111	2.047275	2.047664	2.048053	2.048442	2.048830	389
112	2.051152	2.051538	2.051924	2.052309	2.052694	385
113	2.054996	2.055378	2.055760	2.056142	2.056524	382
114	2.058805	2.059185	2.059563	2.059942	2.060320	378
115	2.062582	2.062958	2.063333	2.063709	2.064083	376
116	2.066326	2.066699	2.067071	2.067443	2.067814	372
117	2.070038	2.070407	2.070776	2.071145	2.071514	369
118	2.073718	2.074085	2.074451	2.074816	2.075182	366
119	2.077368	2.077731	2.078094	2.078457	2.078819	363
120	2.080987	2.081347	2.081707	2.082067	2.082426	360
121	2.084576	2.084934	2.085291	2.085647	2.086004	357
122	2.088136	2.088490	2.088845	2.089198	2.089552	355
123	2.091667	2.092018	2.092370	2.092721	2.093071	351
124	2.095169	2.095518	2.095866	2.096215	2.096562	347
125	2.098644	2.098990	2.099335	2.099681	2.100026	346
126	2.102090	2.102434	2.102777	2.103119	2.103462	343
127	2.105510	2.105851	2.106191	2.106531	2.106870	340
128	2.108903	2.109241	2.109578	2.109916	2.110253	338
129	2.112270	2.112605	2.112940	2.113275	2.113609	335
130	2.115610	2.115943	2.116276	2.116608	2.116940	333
131	2.118926	2.119256	2.119586	2.119915	2.120245	330
132	2.122216	2.122543	2.122871	2.123198	2.123525	328
133	2.125481	2.125806	2.126131	2.126456	2.126781	325
134	2.128722	2.129045	2.129368	2.129690	2.130012	323
135	2.131939	2.132260	2.132580	2.132900	2.133219	320
136	2.135133	2.135451	2.135768	2.136086	2.136403	318
137	2.138303	2.138618	2.138934	2.139249	2.139564	315
138	2.141450	2.141763	2.142076	2.142389	2.142702	314
139	2.144574	2.144885	2.145196	2.145507	2.145818	311
140	2.147676	2.147985	2.148294	2.148603	2.148911	309
141	2.150756	2.151063	2.151370	2.151676	2.151982	307
142	2.153815	2.154119	2.154424	2.154728	2.155032	305

N°	0	1	2	3	4	Dif
143	2.155336	2.155640	2.155943	2.156246	2.156549	303
144	2.158361	2.158664	2.158965	2.159266	2.159567	301
145	2.161368	2.161667	2.161967	2.162266	2.162564	299
146	2.164353	2.164650	2.164947	2.165244	2.165541	297
147	2.167317	2.167613	2.167908	2.168203	2.168497	295
148	2.170262	2.170555	2.170848	2.171141	2.171434	293
149	2.173186	2.173478	2.173768	2.174060	2.174351	291
150	2.176091	2.176381	2.176670	2.176959	2.177248	289
151	2.178977	2.179264	2.179552	2.179839	2.180126	287
152	2.181844	2.182129	2.182415	2.182700	2.182985	285
153	2.184691	2.184975	2.185259	2.185542	2.185825	283
154	2.187521	2.187803	2.188084	2.188366	2.188647	281
155	2.190332	2.190612	2.190892	2.191171	2.191451	279
156	2.193125	2.193403	2.193681	2.193959	2.194237	278
157	2.195900	2.196176	2.196452	2.196729	2.197005	276
158	2.198657	2.198932	2.199206	2.199481	2.199755	274
159	2.201397	2.201670	2.201943	2.202216	2.202488	272
160	2.204120	2.204391	2.204662	2.204933	2.205204	271
161	2.206826	2.207095	2.207365	2.207634	2.207903	269
162	2.209515	2.209783	2.210051	2.210318	2.210586	267
163	2.212188	2.212454	2.212720	2.212986	2.213252	266
164	2.214844	2.215109	2.215373	2.215638	2.215902	264
165	2.217484	2.217747	2.218010	2.218273	2.218535	262
166	2.220108	2.220370	2.220631	2.220892	2.221153	261
167	2.222716	2.222976	2.223236	2.223496	2.223755	259
168	2.225309	2.225568	2.225826	2.226084	2.226342	258
169	2.227887	2.228144	2.228400	2.228657	2.228913	256
170	2.230449	2.230704	2.230960	2.231215	2.231470	254
171	2.232996	2.233250	2.233504	2.233757	2.234011	253
172	2.235528	2.235781	2.236033	2.236285	2.236537	252
173	2.238046	2.238297	2.238548	2.238799	2.239049	250
174	2.240549	2.240799	2.241048	2.241297	2.241546	249
175	2.243038	2.243286	2.243534	2.243782	2.244030	248
176	2.245513	2.245759	2.246006	2.246252	2.246499	246
177	2.247973	2.248219	2.248464	2.248709	2.248954	245
178	2.250420	2.250664	2.250908	2.251151	2.251395	243
179	2.252853	2.253096	2.253338	2.253580	2.253822	242
180	2.255272	2.255514	2.255755	2.255996	2.256237	241
181	2.257679	2.257918	2.258158	2.258398	2.258637	239
182	2.260071	2.260310	2.260548	2.260787	2.261025	238
183	2.262451	2.262688	2.262925	2.263162	2.263399	237
184	2.264818	2.265054	2.265290	2.265525	2.265761	235
185	2.267172	2.267406	2.267641	2.267875	2.268110	234

N°	5	6	7	8	9	Dif
143	2.156852	2.157154	2.157457	2.157759	2.158061	303
144	2.159868	2.160168	2.160468	2.160769	2.161068	301
145	2.162863	2.163161	2.163460	2.163757	2.164055	299
146	2.165833	2.166134	2.166430	2.166726	2.167022	297
147	2.168792	2.169086	2.169380	2.169674	2.169968	295
148	2.171726	2.172019	2.172311	2.172603	2.172895	292
149	2.174541	2.174932	2.175222	2.175512	2.175802	291
150	2.177536	2.177825	2.178113	2.178401	2.178689	289
151	2.180413	2.180699	2.180986	2.181272	2.181558	287
152	2.183270	2.183554	2.183839	2.184123	2.184407	285
153	2.186108	2.186391	2.186674	2.186956	2.187239	283
154	2.188928	2.189209	2.189490	2.189771	2.190051	281
155	2.191730	2.192010	2.192289	2.192567	2.192846	279
156	2.194514	2.194792	2.195069	2.195346	2.195623	278
157	2.197281	2.197556	2.197832	2.198107	2.198382	276
158	2.200029	2.200303	2.200577	2.200850	2.201124	274
159	2.202761	2.203033	2.203305	2.203577	2.203848	272
160	2.205475	2.205745	2.206016	2.206286	2.206556	271
161	2.208172	2.208441	2.208710	2.208978	2.209247	269
162	2.210853	2.211120	2.211388	2.211654	2.211921	267
163	2.213518	2.213783	2.214049	2.214314	2.214579	266
164	2.216160	2.216430	2.216694	2.216957	2.217221	264
165	2.218798	2.219060	2.219322	2.219584	2.219845	262
166	2.221414	2.221675	2.221936	2.222196	2.222456	261
167	2.224015	2.224274	2.224533	2.224792	2.225051	259
168	2.226600	2.226858	2.227115	2.227372	2.227630	258
169	2.229170	2.229426	2.229682	2.229938	2.230193	256
170	2.231724	2.231979	2.232233	2.232488	2.232742	254
171	2.234264	2.234517	2.234770	2.235023	2.235276	253
172	2.236789	2.237041	2.237292	2.237544	2.237795	251
173	2.239299	2.239550	2.239800	2.240050	2.240300	250
174	2.241795	2.242044	2.242293	2.242541	2.242790	249
175	2.244277	2.244524	2.244772	2.245018	2.245261	248
176	2.246745	2.246991	2.247236	2.247482	2.247728	246
177	2.249198	2.249443	2.249687	2.249932	2.250170	245
178	2.251638	2.251881	2.252125	2.252367	2.252610	243
179	2.254064	2.254306	2.254548	2.254790	2.255031	242
180	2.256477	2.256718	2.256958	2.257198	2.257439	241
181	2.258877	2.259116	2.259355	2.259594	2.259833	239
182	2.261263	2.261501	2.261738	2.261976	2.262214	238
183	2.263635	2.263873	2.264109	2.264345	2.264582	237
184	2.265996	2.266232	2.266467	2.266702	2.266937	235
185	2.268344	2.268578	2.268812	2.269046	2.269279	234

Nº	0	1	2	3	4	Dif
186	2.269513	2.269746	2.269980	2.270213	2.270446	233
187	2.271842	2.272074	2.272306	2.272538	2.272770	232
188	2.274158	2.274389	2.274620	2.274850	2.275081	230
189	2.276462	2.276691	2.276921	2.277151	2.277380	229
190	2.278754	2.278982	2.279210	2.279439	2.279667	228
191	2.281033	2.281261	2.281488	2.281715	2.281942	227
192	2.283301	2.283527	2.283753	2.283979	2.284205	226
193	2.285557	2.285782	2.286007	2.286232	2.286456	225
194	2.287802	2.288025	2.288249	2.288473	2.288696	223
195	2.290035	2.290257	2.290480	2.290702	2.290925	222
196	2.292256	2.292478	2.292699	2.292920	2.293141	221
197	2.294466	2.294687	2.294907	2.295127	2.295347	220
198	2.296665	2.296884	2.297104	2.297323	2.297542	219
199	2.298853	2.299071	2.299289	2.299507	2.299725	218
200	2.301030	2.301247	2.301464	2.301681	2.301898	217
201	2.303196	2.303412	2.303628	2.303844	2.304059	216
202	2.305351	2.305566	2.305781	2.305996	2.306210	215
203	2.307496	2.307710	2.307924	2.308137	2.308351	213
204	2.309630	2.309843	2.310056	2.310268	2.310481	212
205	2.311754	2.311966	2.312177	2.312389	2.312600	211
206	2.313867	2.314078	2.314289	2.314499	2.314710	210
207	2.315970	2.316180	2.316390	2.316599	2.316809	209
208	2.318063	2.318272	2.318481	2.318689	2.318898	208
209	2.320146	2.320354	2.320562	2.320769	2.320977	207
210	2.322219	2.322426	2.322633	2.322839	2.323046	206
211	2.324282	2.324488	2.324694	2.324899	2.325105	205
212	2.326336	2.326541	2.326745	2.326950	2.327154	204
213	2.328380	2.328583	2.328787	2.328991	2.329194	203
214	2.330414	2.330617	2.330819	2.331022	2.331225	202
215	2.332438	2.332640	2.332842	2.333044	2.333246	202
216	2.334454	2.334655	2.334856	2.335056	2.335257	201
217	2.336460	2.336660	2.336860	2.337060	2.337260	200
218	2.338456	2.338656	2.338855	2.339054	2.339253	199
219	2.340444	2.340642	2.340840	2.341039	2.341237	198
220	2.342423	2.342620	2.342817	2.343014	2.343212	197
221	2.344392	2.344589	2.344785	2.344981	2.345178	196
222	2.346353	2.346549	2.346744	2.346939	2.347135	195
223	2.348305	2.348500	2.348694	2.348889	2.349083	194
224	2.350248	2.350442	2.350636	2.350829	2.351023	193
225	2.352182	2.352375	2.352568	2.352761	2.352954	193
226	2.354108	2.354301	2.354493	2.354684	2.354876	192
227	2.356026	2.356217	2.356408	2.356599	2.356790	191
228	2.357935	2.358125	2.358316	2.358506	2.358696	190

N°	5	6	7	8	9	Dif
186	2.270079	2.270912	2.271144	2.271377	2.271609	233
187	2.273001	2.273233	2.273464	2.273696	2.273927	231
188	2.275311	2.275541	2.275771	2.276001	2.276232	230
189	2.277609	2.277838	2.278067	2.278296	2.278525	229
190	2.279895	2.280123	2.280351	2.280578	2.280806	228
191	2.282169	2.282395	2.282622	2.282849	2.283075	227
192	2.284431	2.284656	2.284882	2.285107	2.285332	226
193	2.286691	2.286905	2.287130	2.287354	2.287579	225
194	2.288920	2.289143	2.289361	2.289589	2.289812	223
195	2.291147	2.291369	2.291591	2.291813	2.292034	222
196	2.293362	2.293583	2.293804	2.294025	2.294246	221
197	2.295567	2.295787	2.296007	2.296226	2.296446	220
198	2.297760	2.297979	2.298198	2.298416	2.298635	219
199	2.297943	3.300160	3.300378	2.300595	3.300813	218
200	2.302114	2.302331	2.302547	2.302764	2.302980	217
201	2.304275	2.304490	2.304706	2.304921	2.305136	216
202	2.306425	2.306639	2.306854	2.307068	2.307282	215
203	2.308564	2.308778	2.308991	2.309204	2.309417	213
204	2.310693	2.310906	2.311118	2.311330	2.311542	212
205	2.312812	2.313023	2.313234	2.313445	2.313656	211
206	2.314920	2.315130	2.315340	2.315550	2.315760	210
207	2.317018	2.317227	2.317436	2.317645	2.317854	209
208	2.319106	2.319314	2.319522	2.319730	2.319938	208
209	2.321184	2.321391	2.321598	2.321805	2.322012	207
210	2.323252	2.323458	2.323664	2.323871	2.324077	206
211	2.325310	2.325516	2.325721	2.325926	2.326131	205
212	2.327359	2.327563	2.327767	2.327972	2.328176	204
213	2.329398	2.329601	2.329804	2.330008	2.330211	203
214	2.331427	2.331630	2.331832	2.332034	2.332236	202
215	2.333447	2.333649	2.333850	2.334051	2.334253	202
216	2.335458	2.335658	2.335859	2.336059	2.336259	201
217	2.337459	2.337659	2.337858	2.338058	2.338257	200
218	2.339451	2.339650	2.339849	2.340047	2.340246	199
219	2.341434	2.341632	2.341830	2.342028	2.342225	198
220	2.343409	2.343605	2.343802	2.343999	2.344196	197
221	2.345374	2.345570	2.345766	2.345961	2.346157	196
222	2.347330	2.347525	2.347720	2.347915	2.348110	195
223	2.349277	2.349472	2.349666	2.349860	2.350054	194
224	2.351216	2.351410	2.351603	2.351796	2.351989	193
225	2.353146	2.353339	2.353532	2.353721	2.353916	193
226	2.355068	2.355260	2.355451	2.355613	2.355834	192
227	2.356981	2.357172	2.357363	2.357554	2.357744	191
228	2.358886	2.359076	2.359266	2.359456	2.359646	190

D d

N°	0	1	2	3	4	Dif
229	2.359835	2.360025	2.360215	2.360404	2.360593	189
230	2.361728	2.361917	2.362105	2.362294	2.362482	188
231	2.363612	2.363800	2.363988	2.364176	2.364363	188
232	2.365488	2.365675	2.365862	2.366049	2.366236	187
233	2.367356	2.367542	2.367728	2.367915	2.368101	186
234	2.369216	2.369401	2.369587	2.369772	2.369958	185
235	2.371068	2.371253	2.371437	2.371622	2.371806	184
236	2.372912	2.373096	2.373280	2.373464	2.373647	184
237	2.374748	2.374932	2.375115	2.375298	2.375481	183
238	2.376577	2.376759	2.376942	2.377124	2.377306	182
239	2.378398	2.378580	2.378761	2.378943	2.379124	181
240	2.380211	2.380392	2.380573	2.380754	2.380934	181
241	2.382017	2.382197	2.382377	2.382557	2.382737	180
242	2.383815	2.383995	2.384174	2.384353	2.384533	179
243	2.385606	2.385785	2.385964	2.386142	2.386321	178
244	2.387390	2.387568	2.387746	2.387923	2.388101	178
245	2.389166	2.389343	2.389520	2.389697	2.389874	177
246	2.390935	2.391112	2.391288	2.391464	2.391641	176
247	2.392697	2.392873	2.393048	2.393224	2.393400	176
248	2.394452	2.394627	2.394802	2.394977	2.395152	175
249	2.396199	2.396374	2.396548	2.396722	2.396896	174
250	2.397940	2.398114	2.398287	2.398401	2.398634	173
251	2.399674	2.399847	2.400020	2.400192	2.400365	173
252	2.401400	2.401573	2.401745	2.401917	2.402089	172
253	2.403120	2.403292	2.403464	2.403635	2.403807	171
254	2.404834	2.405005	2.405175	2.405346	2.405517	171
255	2.406540	2.406710	2.406881	2.407051	2.407221	170
256	2.408840	2.408409	2.468579	2.408749	2.408918	169
257	2.409933	2.410103	2.410271	2.410440	2.410608	169
258	2.411620	2.411788	2.411956	2.412124	2.412292	168
259	2.413300	2.413467	2.413635	2.413802	2.413970	167
260	2.414973	2.415140	2.415307	2.415474	2.415641	167
261	2.416640	2.416807	2.416973	2.417139	2.417306	166
262	2.418301	2.418467	2.418633	2.418798	2.418964	165
263	2.419956	2.420121	2.420286	2.420451	2.420516	165
264	2.421604	2.421768	2.421933	2.422097	2.422261	164
265	2.423246	2.423410	2.423573	2.423737	2.423901	164
266	2.424882	2.425045	2.425208	2.425371	2.425534	163
267	2.426511	2.426674	2.426836	2.426999	2.427161	162
268	2.428135	2.428297	2.428459	2.428621	2.428782	162
269	2.429752	2.429914	2.430075	2.430236	2.430398	161
270	2.431364	2.431525	2.431685	2.431846	2.432007	161
271	2.432969	2.433129	2.433290	2.433450	2.433610	160

N°	5	6	7	8	9	Dif
229	2.360783	2.360972	2.361161	2.361350	2.361539	189
230	2.362671	2.362859	2.363048	2.363236	2.363424	188
231	2.364551	2.364739	2.364926	2.365113	2.365301	188
232	2.366423	2.366610	2.366796	2.366983	2.367169	187
233	2.368287	2.368473	2.368659	2.368844	2.369030	186
234	2.370143	2.370328	2.370513	2.370598	2.370883	185
235	2.371991	2.372175	2.372360	2.372544	2.372728	184
236	2.373831	2.374015	2.374198	2.374182	2.374565	184
237	2.375664	2.375846	2.376029	2.376212	2.376394	183
238	2.377488	2.377670	2.377852	2.378034	2.378216	182
239	2.379305	2.379487	2.379668	2.379849	2.380030	181
240	2.381115	2.381296	2.381476	2.381656	2.381837	181
241	2.382917	2.383097	2.383277	2.383456	2.383636	180
242	2.384712	2.384891	2.385070	2.385249	2.385427	179
243	2.386499	2.386677	2.386855	2.387034	2.387212	178
244	2.388279	2.388456	2.388634	2.388811	2.388989	178
245	2.390051	2.390228	2.390405	2.390582	2.390758	177
246	2.391817	2.391993	2.392169	2.392345	2.392521	176
247	2.393575	2.393751	2.393926	2.394101	2.394276	176
248	2.395326	2.395501	2.395676	2.395850	2.396025	175
249	2.397070	2.397245	2.397418	2.397592	2.397766	174
250	2.398808	2.398981	2.399154	2.399327	2.399501	173
251	2.400538	2.400711	2.400883	2.401056	2.401228	173
252	2.402261	2.402433	2.402605	2.402777	2.402949	172
253	2.403978	2.404149	2.404320	2.404492	2.404663	171
254	2.405688	2.405858	2.406029	2.406199	2.406370	171
255	2.407391	2.407561	2.407731	2.407900	2.408070	170
256	2.409087	2.409257	2.409426	2.409595	2.409764	169
257	2.410777	2.410946	2.411114	2.411283	2.411451	169
258	2.412460	2.412628	2.412796	2.412964	2.413132	168
259	2.414137	2.414305	2.414472	2.414639	2.414806	167
260	2.415808	2.415974	2.416141	2.416308	2.416474	167
261	2.417472	2.417638	2.417804	2.417970	2.418135	166
262	2.419129	2.419295	2.419460	2.419625	2.419791	165
263	2.420781	2.420945	2.421110	2.421275	2.421439	165
264	2.422426	2.422590	2.422754	2.422918	2.423082	164
265	2.424064	2.424228	2.424392	2.424555	2.424718	164
266	2.425697	2.425860	2.426023	2.426186	2.426349	163
267	2.427324	2.427486	2.427648	2.427811	2.427973	162
268	2.428944	2.429106	2.429268	2.429429	2.429591	162
269	2.430559	2.430720	2.430881	2.431042	2.431203	161
270	2.432167	2.432328	2.432488	2.432649	2.432809	160
271	2.433770	2.433930	2.434090	2.434249	2.434409	160

A Table of Logarithms,

N°	0	1	2	3	4	Dif
272	2.434569	2.434728	2.434888	2.435048	2.435207	159
273	2.436163	2.436322	2.436481	2.436640	2.436798	159
274	2.437751	2.437909	2.438067	2.438226	2.438384	158
275	2.439333	2.439491	2.439648	2.439806	2.439964	158
276	2.440909	2.441066	2.441224	2.441381	2.441538	157
277	2.442480	2.442536	2.442793	2.442950	2.443106	157
278	2.444045	2.444201	2.444357	2.444513	2.444669	156
279	2.445604	2.445760	2.445915	2.446071	2.446126	155
280	2.447158	2.447313	2.447468	2.447623	2.447778	155
281	2.448706	2.448861	2.449015	2.449170	2.449324	154
282	2.450249	2.450403	2.450557	2.450711	2.450865	154
283	2.451786	2.451940	2.452093	2.452247	2.452400	153
284	2.453318	2.453471	2.453624	2.453777	2.453930	153
285	2.454845	2.454997	2.455149	2.455302	2.455454	152
286	2.456366	2.456518	2.456670	2.456821	2.456973	152
287	2.457882	2.458033	2.458184	2.458336	2.458487	151
288	2.459392	2.459543	2.459694	2.459845	2.459995	151
289	2.460898	2.461048	2.461198	2.461348	2.461498	150
290	2.462398	2.462548	2.462697	2.462847	2.462997	150
291	2.463893	2.464042	2.464191	2.464340	2.464489	149
292	2.465383	2.465532	2.465680	2.465829	2.465977	149
293	2.466868	2.467016	2.467164	2.467312	2.467460	148
294	2.468347	2.468495	2.468643	2.468790	2.468938	147
295	2.469822	2.469969	2.470116	2.470263	2.470410	147
296	2.471292	2.471438	2.471585	2.471732	2.471878	146
297	2.472756	2.472903	2.473049	2.473195	2.473341	146
298	2.474216	2.474362	2.474508	2.474653	2.474799	145
299	2.475671	2.475816	2.475962	2.476107	2.476252	145
300	2.477111	2.477266	2.477411	2.477555	2.477700	145
301	2.478566	2.478711	2.478855	2.478999	2.479143	144
302	2.480007	2.480151	2.480294	2.480438	2.480582	144
303	2.481443	2.481586	2.481729	2.481872	2.482016	143
304	2.482874	2.483016	2.483159	2.483302	2.483445	143
305	2.484300	2.484442	2.484584	2.484727	2.484869	142
306	2.485721	2.485863	2.486005	2.486147	2.486289	142
307	2.487138	2.487280	2.487421	2.487563	2.487704	141
308	2.488551	2.488692	2.488833	2.488973	2.489114	141
309	2.489958	2.490099	2.490239	2.490380	2.490520	140
310	2.491362	2.491502	2.491642	2.491782	2.491922	140
311	2.492760	2.492900	2.493040	2.493179	2.493319	139
312	2.494155	2.494294	2.494433	2.494572	2.494711	139
313	2.495544	2.495683	2.495822	2.495960	2.496099	139
314	2.496930	2.497068	2.497206	2.497344	2.497482	138

Nᵒ	5	6	7	8	9	Dif
172	2.435306	2.435527	2.435685	2.435844	2.436003	159
173	2.436957	2.437116	2.437275	2.437433	2.437592	159
174	2.438542	2.438700	2.438859	2.439017	2.439175	158
175	2.440122	2.440279	2.440137	2.440594	2.440752	158
176	2.441695	2.441852	2.442009	2.442166	2.442323	157
177	2.443213	2.443419	2.443576	2.443732	2.443888	157
178	2.444825	2.444981	2.445137	2.445293	2.445448	156
179	2.446382	2.446537	2.446692	2.446848	2.447003	155
180	2.447933	2.448088	2.448242	2.448397	2.448552	155
281	2.449478	2.449633	2.449787	2.449941	2.450095	154
282	2.451018	2.451172	2.451326	2.451479	2.451633	154
283	2.452553	2.452706	2.452859	2.453012	2.453165	153
284	2.454082	2.454235	2.454387	2.454540	2.454692	153
285	2.455606	2.455758	2.455910	2.456062	2.456214	152
286	2.457125	2.457276	2.457428	2.457579	2.457730	152
287	2.458638	2.458789	2.458940	2.459091	2.459242	151
288	2.460146	2.460296	2.460447	2.460597	2.460747	151
289	2.461649	2.461799	2.461948	2.462098	2.462248	150
290	2.463146	2.463296	2.463445	2.463594	2.463744	150
291	2.464639	2.464787	2.464936	2.465085	2.465234	149
292	2.466126	2.466274	2.466423	2.466571	2.466719	149
293	2.467608	2.467756	2.467904	2.468052	2.468200	148
294	2.469085	2.469233	2.469380	2.469527	2.469675	147
295	2.470557	2.470704	2.470851	2.470998	2.471145	147
296	2.472025	2.472171	2.472317	2.472464	2.472610	146
297	2.473487	2.473633	2.473779	2.473925	2.474070	146
298	2.474944	2.475090	2.475235	2.475381	2.475526	146
299	2.476397	2.476542	2.476687	2.476832	2.476976	145
300	2.477844	2.477989	2.478133	2.478278	2.478422	145
301	2.479287	2.479431	2.479575	2.479719	2.479863	144
302	2.480725	2.480869	2.481012	2.481156	2.481299	144
303	2.482159	2.482302	2.482445	2.482588	2.482731	143
304	2.483587	2.483730	2.483872	2.484015	2.484157	143
305	2.485011	2.485153	2.485295	2.485437	2.485579	142
306	2.486430	2.486572	2.486714	2.486855	2.486997	142
307	2.487845	2.487986	2.488127	2.488269	2.488410	141
308	2.489255	2.489396	2.489537	2.489677	2.489818	141
309	2.490661	2.490801	2.490941	2.491081	2.491222	140
310	2.492062	2.492201	2.492341	2.492481	2.492621	140
311	2.493458	2.493597	2.493737	2.493876	2.494015	139
312	2.494850	2.494989	2.495128	2.495267	2.495406	139
313	2.496237	2.496376	2.496514	2.496653	2.496791	139
314	2.497621	2.497759	2.497897	2.498035	2.498173	138

Nº	0	1	2	3	4	Dif
315	2.498311	2.498448	2.498586	2.498724	2.498862	138
316	2.499687	2.499824	2.499962	2.500100	2.500236	137
317	2.501059	2.501196	2.501333	2.501470	2.501607	137
318	2.502427	2.502564	2.502700	2.502837	2.502973	136
319	2.503791	2.503927	2.504063	2.504199	2.504335	136
320	2.505156	2.505286	2.505421	2.505557	2.505693	136
321	2.506505	2.506640	2.506775	2.506911	2.507046	135
322	2.507856	2.507991	2.508125	2.508260	2.508395	135
323	2.509202	2.509337	2.509471	2.509606	2.509740	134
324	2.510545	2.510679	2.510813	2.510947	2.511081	134
325	2.511882	2.512017	2.512150	2.512284	2.512417	133
326	2.513218	2.513351	2.513484	2.513617	2.513750	133
327	2.514548	2.514680	2.514813	2.514946	2.515079	133
328	2.515874	2.516006	2.516139	2.516271	2.516403	132
329	2.517196	2.517328	2.517460	2.517592	2.517724	132
330	2.518514	2.518645	2.518777	2.518909	2.519040	131
331	2.519828	2.519959	2.520090	2.520221	2.520351	131
332	2.521138	2.521269	2.521400	2.521530	2.521661	131
333	2.522444	2.522575	2.522705	2.522835	2.522966	130
334	2.523746	2.523876	2.524006	2.524136	2.524266	130
335	2.525045	2.525174	2.525304	2.525433	2.525563	129
336	2.526339	2.526468	2.526598	2.526727	2.526856	129
337	2.527630	2.527759	2.527888	2.528016	2.528145	129
338	2.528917	2.529045	2.529174	2.529302	2.529430	128
339	2.530200	2.530328	2.530456	2.530584	2.530712	128
340	2.531479	2.531607	2.531734	2.531862	2.531989	128
341	2.532754	2.532882	2.533009	2.533136	2.533263	127
342	2.534026	2.534153	2.534280	2.534407	2.534534	127
343	2.535294	2.535421	2.535547	2.535674	2.535800	126
344	2.536558	2.536685	2.536811	2.536937	2.537063	126
345	2.537819	2.537945	2.538071	2.538197	2.538322	126
346	2.539076	2.539202	2.539327	2.539452	2.539578	125
347	2.540329	2.540455	2.540580	2.540705	2.540830	125
348	2.541579	2.541704	2.541829	2.541953	2.542078	125
349	2.542825	2.542950	2.543074	2.543199	2.543323	124
350	2.544068	2.544192	2.544316	2.544440	2.544564	124
351	2.545307	2.545431	2.545554	2.545678	2.545802	124
352	2.546543	2.546666	2.546789	2.546913	2.547036	123
353	2.547775	2.547898	2.548021	2.548144	2.548266	123
354	2.549003	2.549126	2.549243	2.549371	2.549494	123
355	2.550228	2.550351	2.550473	2.550595	2.550717	123
356	2.551450	2.551572	2.551694	2.551816	2.551938	122
357	2.552668	2.552790	2.552911	2.553033	2.553154	121

N°	5	6	7	8	9	Dif
315	2.498999	2.499137	2.499275	2.499412	2.499550	138
316	2.500374	2.500511	2.500648	2.500785	2.500922	137
317	2.501744	2.501880	2.502017	2.502154	2.502290	137
318	2.503109	2.503246	2.503382	2.503518	2.503654	136
319	2.504471	2.504607	2.504743	2.504878	2.505014	136
320	2.505828	2.505963	2.506099	2.506234	2.506370	136
321	2.507181	2.507316	2.507451	2.507586	2.507721	135
322	2.508530	2.508664	2.508799	2.508933	2.509068	135
323	2.509874	2.510008	2.510143	2.510277	2.510411	134
324	2.511215	2.511348	2.511482	2.511616	2.511750	134
325	2.512551	2.512684	2.512818	2.512951	2.513084	133
326	2.513883	2.514016	2.514149	2.514282	2.514415	133
327	2.515211	2.515344	2.515476	2.515609	2.515741	133
328	2.516535	2.516668	2.516800	2.516932	2.517064	132
329	2.517855	2.517987	2.518119	2.518251	2.518382	132
330	2.519171	2.519303	2.519434	2.519565	2.519697	132
331	2.520483	2.520614	2.520745	2.520876	2.521007	131
332	2.521792	2.521922	2.522053	2.522183	2.522314	131
333	2.523096	2.523226	2.523356	2.523486	2.523616	130
334	2.524396	2.524526	2.524656	2.524785	2.524915	130
335	2.525692	2.525822	2.525951	2.526081	2.526210	129
336	2.526985	2.527114	2.527243	2.527372	2.527501	129
337	2.528274	2.528402	2.528531	2.528660	2.528788	129
338	2.529559	2.529687	2.529815	2.529943	2.530072	128
339	2.530840	2.530968	2.531095	2.531223	2.531351	128
340	2.532117	2.532245	2.532372	2.532500	2.532627	128
341	2.533391	2.533518	2.533645	2.533772	2.533899	127
342	2.534661	2.534787	2.534914	2.535041	2.535167	127
343	2.535927	2.536053	2.536179	2.536306	2.536432	126
344	2.537189	2.537315	2.537441	2.537567	2.537693	126
345	2.538448	2.538574	2.538699	2.538825	2.538951	126
346	2.539703	2.539829	2.539954	2.540079	2.540204	125
347	2.540955	2.541080	2.541205	2.541330	2.541454	125
348	2.542203	2.542327	2.542452	2.542576	2.542701	125
349	2.543447	2.543571	2.543696	2.543820	2.543944	124
350	2.544688	2.544812	2.544936	2.545060	2.545183	124
351	2.545925	2.546049	2.546172	2.546296	2.546419	124
352	2.547159	2.547282	2.547405	2.547529	2.547652	123
353	2.548389	2.548512	2.548635	2.548758	2.548881	123
354	2.549616	2.549739	2.549861	2.549984	2.550106	123
355	2.550840	2.550962	2.551084	2.551206	2.551328	122
356	2.552059	2.552181	2.552303	2.552425	2.552546	122
357	2.553276	2.553397	2.553519	2.553640	2.553762	122

N°	0	1	2	3	4	Dif
358	2.553883	2.554004	2.554126	2.554247	2.554368	121
359	2.555094	2.555215	2.555336	2.555457	2.555578	121
360	2.556302	2.556423	2.556544	2.556664	2.556785	121
361	2.557507	2.557627	2.557748	2.557868	2.557988	120
362	2.558709	2.558828	2.558948	2.559068	2.559188	120
363	2.559907	2.560026	2.560146	2.560265	2.560385	120
364	2.561101	2.561221	2.561340	2.561459	2.561578	119
365	2.562293	2.562412	2.562531	2.562650	2.562768	119
366	2.563481	2.563600	2.563718	2.563837	2.563955	119
367	2.564666	2.564784	2.564903	2.565021	2.565139	118
368	2.565848	2.565966	2.566084	2.566202	2.566320	118
369	2.567026	2.567144	2.567262	2.567379	2.567497	118
370	2.568202	2.568319	2.568436	2.568554	2.568671	117
371	2.569374	2.569491	2.569608	2.569725	2.569842	117
372	2.570543	2.570660	2.570776	2.570893	2.571010	117
373	2.571709	2.571825	2.571942	2.572058	2.572174	116
374	2.572872	2.572988	2.573104	2.573220	2.573336	116
375	2.574031	2.574147	2.574263	2.574379	2.574494	116
376	2.575188	2.575303	2.575419	2.575534	2.575650	115
377	2.576341	2.576456	2.576572	2.576687	2.576802	115
378	2.577492	2.577607	2.577721	2.577836	2.577951	115
379	2.578639	2.578754	2.578868	2.578983	2.579097	114
380	2.579784	2.579898	2.580012	2.580126	2.580240	114
381	2.580925	2.581039	2.581153	2.581267	2.581381	114
382	2.582063	2.582177	2.582291	2.582404	2.582518	114
383	2.583199	2.583312	2.583425	2.583539	2.583652	113
384	2.584331	2.584444	2.584557	2.584670	2.584783	113
385	2.585461	2.585573	2.585686	2.585799	2.585912	113
386	2.586587	2.586700	2.586812	2.586925	2.587037	112
387	2.587711	2.587823	2.587935	2.588047	2.588160	112
388	2.588832	2.588944	2.589055	2.589167	2.589279	112
389	2.589950	2.590061	2.590173	2.590284	2.590396	112
390	2.591065	2.591176	2.591287	2.591398	2.591510	111
391	2.592177	2.592288	2.592399	2.592510	2.592621	111
392	2.593286	2.593397	2.593508	2.593618	2.593729	111
393	2.594392	2.594503	2.594613	2.594724	2.594834	110
394	2.595496	2.595606	2.595717	2.595827	2.595937	110
395	2.596597	2.596707	2.596817	2.596927	2.597037	110
396	2.597695	2.597805	2.597914	2.598024	2.598134	110
397	2.598790	2.598900	2.599009	2.599119	2.599228	109
398	2.599883	2.599992	2.600101	2.600210	2.600319	109
399	2.600973	2.601082	2.601190	2.601299	2.601408	109
400	2.602060	2.602168	2.602277	2.602386	2.602494	108

N°	5	6	7	8	9	Dif
358	2.554489	2.554610	2.554731	2.554852	2.554973	121
359	2.555699	2.555820	2.555940	2.556061	2.556182	121
360	2.556905	2.557026	2.557146	2.557266	2.557387	121
361	2.558108	2.558228	2.558348	2.558469	2.558589	120
362	2.559308	2.559428	2.559548	2.559667	2.559787	120
363	2.560504	2.560624	2.560743	2.560863	2.560982	119
364	2.561697	2.561817	2.561936	2.562055	2.562174	119
365	2.562887	2.563006	2.563125	2.563244	2.563362	119
366	2.564074	2.564192	2.564311	2.564429	2.564548	119
367	2.565257	2.565375	2.565494	2.565612	2.565730	118
368	2.566437	2.566555	2.566673	2.566791	2.566909	118
369	2.567614	2.567732	2.567849	2.567967	2.568084	118
370	2.568788	2.568905	2.569023	2.569140	2.569257	117
371	2.569959	2.570076	2.570193	2.570309	2.570426	117
372	2.571126	2.571243	2.571359	2.571476	2.571592	117
373	2.572291	2.572407	2.572523	2.572639	2.572755	116
374	2.573452	2.573568	2.573684	2.573800	2.573915	116
375	2.574610	2.574726	2.574841	2.574957	2.575072	116
376	2.575765	2.575880	2.575996	2.576111	2.576226	115
377	2.576917	2.577032	2.577147	2.577262	2.577377	115
378	2.578066	2.578181	2.578295	2.578410	2.578525	125
379	2.579211	2.579326	2.579441	2.579555	2.579669	114
380	2.580355	2.580469	2.580583	2.580697	2.580811	114
381	2.581495	2.581608	2.581722	2.581836	2.581950	114
382	2.582631	2.582745	2.582858	2.582972	2.583085	114
383	2.583765	2.583879	2.583992	2.584105	2.584218	113
384	2.584896	2.585009	2.585122	2.585235	2.585348	113
385	2.586024	2.586137	2.586250	2.586362	2.586475	113
386	2.587149	2.587262	2.587374	2.587486	2.587599	112
387	2.588272	2.588384	2.588496	2.588608	2.588720	112
388	2.589391	2.589503	2.589614	2.589726	2.589838	112
389	2.590507	2.590619	2.590730	2.590842	2.590953	112
390	2.591621	2.591732	2.591843	2.591955	2.592066	111
391	2.592733	2.592843	2.592954	2.593064	2.593175	111
392	2.593840	2.593950	2.594061	2.594171	2.594282	111
393	2.594945	2.595055	2.595165	2.595276	2.595386	110
394	2.596047	2.596157	2.596267	2.596377	2.596487	110
395	2.597146	2.597256	2.597366	2.597476	2.597585	110
396	2.598243	2.598353	2.598462	2.598572	2.598681	110
397	2.599337	2.599446	2.599556	2.599665	2.599774	109
398	2.600428	2.600537	2.600646	2.600755	2.600864	109
399	2.601517	2.601625	2.601734	2.601843	2.601951	109
400	2.602602	2.602711	2.602819	2.602928	2.603036	108

N°	0	1	2	3	4	Dif
401	2.603144	2.603253	2.603301	2.603469	2.603577	108
402	2.604226	2.604334	2.604442	2.604550	2.604658	108
403	2.605305	2.605413	2.605520	2.605628	2.605736	108
404	2.606381	2.606489	2.606596	2.606704	2.606811	107
405	2.607455	2.607562	2.607669	2.607777	2.607884	107
406	2.608526	2.608633	2.608740	2.608847	2.608954	107
407	2.609594	2.609701	2.609808	2.609914	2.610021	107
498	2.610660	2.610767	2.610873	2.610979	2.611086	106
409	2.611723	2.611829	2.611936	2.612042	2.612148	106
410	2.612784	2.612890	2.612996	2.613101	2.613208	106
411	2.613842	2.613947	2.614053	2.614159	2.614264	106
412	2.614897	2.615003	2.615108	2.615213	2.615319	105
413	2.615950	2.616055	2.616160	2.616265	2.616370	105
414	2.617000	2.617105	2.617210	2.617315	2.617420	105
415	2.618048	2.618153	2.618257	2.618362	2.618466	105
416	2.619093	2.619198	2.619302	2.619406	2.619511	504
417	2.620136	2.620240	2.620344	2.620448	2.620552	104
418	2.621176	2.621280	2.621384	2.621488	2.621592	104
419	2.622214	2.622318	2.622421	2.622525	2.622628	104
420	2.623249	2.623353	2.623456	2.623559	2.623663	103
421	2.624282	2.624385	2.624488	2.624591	2.624694	103
422	2.625313	2.625415	2.625518	2.625621	2.625724	103
423	2.626340	2.626443	2.626546	2.626648	2.626751	103
424	2.627366	2.627468	2.627571	2.627673	2.627775	102
425	2.628389	2.628491	2.628593	2.628695	2.628797	102
426	2.629410	2.629511	2.629613	2.629715	2.629817	102
427	2.630428	2.630529	2.630631	2.630733	2.630834	101
428	2.631444	2.631545	2.631647	2.631748	2.631849	101
429	2.632457	2.632558	2.632660	2.632761	2.632862	101
430	2.633468	2.633569	2.633670	2.633771	2.633872	101
431	2.634477	2.634578	2.634679	2.634779	2.634880	100
432	2.635484	2.635584	2.635685	2.635785	2.635886	100
433	2.636488	2.636588	2.636688	2.636789	2.636889	100
434	2.637490	2.637590	2.637690	2.637790	2.637890	100
435	2.638489	2.638589	2.638589	2.638789	2.638888	99
436	2.639486	2.639586	2.639686	2.639785	2.639885	99
437	2.640481	2.640580	2.640680	2.640779	2.640879	99
438	2.641474	2.641573	2.641672	2.641771	2.641870	99
439	2.642464	2.642563	2.642662	2.642761	2.642860	99
440	2.643453	2.643551	2.643650	2.643749	2.643847	98
441	2.644439	2.644537	2.644635	2.644734	2.644832	98
442	2.645422	2.645520	2.645619	2.645717	2.645815	98
443	2.646404	2.646502	2.646600	2.646698	2.646796	98

N°	5	6	7	8	9	Dif
401	2.603085	2.603794	2.603902	2.604010	2.604118	108
402	2.604766	2.604874	2.604982	2.605089	2.605197	108
403	2.605843	2.605951	2.606059	2.606166	2.606274	108
404	2.606918	2.607026	2.607133	2.607240	2.607348	107
405	2.607991	2.608098	2.608205	2.608312	2.608419	107
406	2.609060	2.609167	2.609274	2.609381	2.609488	107
407	2.610128	2.610244	2.610341	2.610447	2.610554	107
408	2.611192	2.611298	2.611405	2.611511	2.611617	106
409	2.612254	2.612360	2.612466	2.612572	2.612678	106
410	2.613313	2.613419	2.613525	2.613630	2.613736	106
411	2.614370	2.614475	2.614581	2.614686	2.614792	106
412	2.615424	2.615529	2.615634	2.615740	2.615845	105
413	2.616475	2.616580	2.616685	2.616790	2.616895	105
414	2.617524	2.617629	2.617734	2.617839	2.617943	105
415	2.618571	2.618675	2.618780	2.618884	2.618989	105
416	2.619615	2.619719	2.619823	2.619928	2.620032	104
417	2.620656	2.620760	2.620864	2.620968	2.621072	104
418	2.621695	2.621799	2.621903	2.622007	2.622110	104
419	2.622732	2.622835	2.622939	2.623042	2.623146	104
420	2.623766	2.623869	2.623972	2.624076	2.624179	103
421	2.624798	2.624901	2.625004	2.625107	2.625209	103
422	2.625827	2.625929	2.626032	2.626135	2.626238	103
423	2.626853	2.626956	2.627058	2.627161	2.627263	103
424	2.627878	2.627980	2.628082	2.628184	2.628287	102
425	2.628900	2.629002	2.629104	2.629206	2.629308	102
426	2.629919	2.630021	2.630123	2.630224	2.630326	102
427	2.630936	2.631038	2.631139	2.631241	2.631342	102
428	2.631951	2.632052	2.632153	2.632255	2.632356	101
429	2.632963	2.633064	2.633165	2.633266	2.633367	101
430	2.633973	2.634074	2.634175	2.634276	2.634376	100
431	2.634981	2.635081	2.635182	2.635283	2.635383	100
432	2.635986	2.636086	2.636187	2.636287	2.636388	100
433	2.636989	2.637089	2.637189	2.637289	2.637390	100
434	2.637990	2.638090	2.638190	2.638289	2.638389	99
435	2.638988	2.639088	2.639188	2.639287	2.639387	99
436	2.639984	2.640084	2.640183	2.640283	2.640382	99
437	2.640978	2.641077	2.641176	2.641276	2.641375	99
438	2.641970	2.642069	2.642168	2.642267	2.642366	99
439	2.642959	2.643058	2.643156	2.643255	2.643354	99
440	2.643946	2.644044	2.644143	2.644242	2.644340	98
441	2.644931	2.645029	2.645127	2.645226	2.645324	98
442	2.645913	2.646011	2.646109	2.646208	2.646306	98
443	2.646894	2.646991	2.647089	2.647187	2.647285	98

N°	0	1	2	3	4	Dif
414	2.647383	2.647481	2.647578	2.647676	2.647774	98
415	2.648360	2.648458	2.648555	2.648653	2.648750	97
416	2.649335	2.649432	2.649530	2.649627	2.649724	97
447	2.650307	2.650405	2.650502	2.650599	2.650696	97
448	2.651278	2.651375	2.651472	2.651569	2.651666	97
449	2.652246	2.652343	2.652440	2.652536	2.652633	97
450	2.653212	2.653309	2.653405	2 653502	2.653598	96
451	2.654176	2.654273	2.654369	2.654465	2.654562	96
452	2.655138	2.655234	2.655331	2.655427	2.655523	96
453	2.656098	2.656194	2.656290	2 656386	2.656481	96
454	2.657056	2.657151	2.657247	2.657343	2.657438	96
455	2.658011	2.658107	2.658202	2.658298	2.658393	95
456	2.658965	2.659060	2.659155	2.65925-	2.659346	95
457	2.659916	2.660011	2.660106	2.660201	2.660296	95
458	2.660865	2.660960	2.661055	2.661150	2.661245	95
459	2.661813	2.661907	2.662002	2.662096	2.662191	95
460	2.662758	2.662852	2.662947	2.663041	2.663135	94
461	2.663701	2.663795	2.663889	2.663983	2.664078	94
462	2.664642	2.664736	2.664830	2.664924	2.665018	94
463	2.665581	2.665675	2.665768	2.665862	2.665956	94
464	2.666518	2.666612	2.666705	2.666759	2.666892	94
465	2.667453	2.667546	2.667640	2.667733	2.667826	93
466	2.668386	2.668479	2.668572	2.668665	2.668758	93
467	2.669317	2.669410	2.669503	2.669596	2.669689	93
468	2.670246	2.670339	2.670431	2.670524	2.670617	93
469	2.671173	2.671265	2.671358	2.671451	2.671543	93
470	2.672098	2.672190	2.672283	2.672375	2.672467	92
471	2.673021	2.673113	2.673205	2.673297	2.673350	92
472	2.673942	2.674034	2.674126	2.674218	2.674310	92
473	2.674861	2.674953	2.675045	2.675136	2.675228	92
474	2.675778	2.675870	2.675961	2.676053	2.676145	92
475	2.676694	2.676785	2.676876	2.676968	2.677059	91
476	2.677607	2.677698	2.677789	2.677881	2.677972	91
477	2.678518	2.678609	2.678700	2.678791	2.678882	91
478	2.679428	2.679519	2.679610	2.679700	2.679791	91
479	2.680335	2.680426	2.680517	2.680607	2.680698	91
480	2.681241	2.681332	2.681422	2.681513	2.681603	90
481	2.682145	2.682235	2.682326	2.682416	2.682506	90
482	2.683047	2.683137	2.683227	2.683317	2.683407	90
483	2.683947	2.684037	2.684127	2.684217	2.684307	90
484	2.684845	2.684935	2.685025	2.685114	2.685204	90
485	2.685742	2.685831	2.685921	2.686010	2.686100	89
486	2.686636	2.686726	2.686815	2.686904	2.686994	89

N°	5	6	7	8	9	Dif
444	2.617872	2.647729	2.648007	2.648165	2.618262	93
445	2.648418	2.648915	2.649043	2.649140	2.649237	97
446	2.649821	2.649919	2.650015	2.650113	2.650210	97
447	2.650293	2.650890	2.650987	2.651084	2.651181	97
448	2.651762	2.651859	2.651956	2.652053	2.652150	97
449	2.652730	2.652826	2.652923	2.653019	2.653116	97
450	2.653609	2.653791	2.653888	2.653984	2.654080	96
451	2.654618	2.654754	2.654850	2.654946	2.655042	96
452	2.655610	2.655714	2.655810	2.655906	2.656002	96
453	2.656577	2.656673	2.656769	2.656864	2.656960	96
454	2.657531	2.657629	2.657725	2.657820	2.657916	96
455	2.658483	2.658584	2.658679	2.658774	2.658870	95
456	2.659441	2.659536	2.659631	2.659726	2.659821	95
457	2.660391	2.660485	2.660581	2.660676	2.660771	95
458	2.661339	2.661434	2.661529	2.661623	2.661718	95
459	2.662285	2.662380	2.662474	2.662569	2.662663	95
460	2.663230	2.663324	2.663418	2.663512	2.663607	94
461	2.664172	2.664266	2.664360	2.664454	2.664548	94
462	2.665112	2.665205	2.665299	2.665393	2.665487	94
463	2.666050	2.666143	2.666237	2.666331	2.666424	94
464	2.666986	2.667079	2.667173	2.667266	2.667359	94
465	2.667920	2.668013	2.668106	2.668199	2.668293	93
466	2.668852	2.668945	2.669038	2.669131	2.669224	93
467	2.669782	2.669874	2.669967	2.670060	2.670153	93
468	2.670710	2.670802	2.670895	2.670988	2.671080	93
469	2.671636	2.671728	2.671821	2.671913	2.672005	93
470	2.672560	2.672652	2.672744	2.672836	2.672929	92
471	2.673482	2.673574	2.673666	2.673758	2.673850	92
472	2.674402	2.674494	2.674586	2.674677	2.674769	92
473	2.675320	2.675412	2.675503	2.675595	2.675687	92
474	2.676236	2.676328	2.676419	2.676511	2.676602	92
475	2.677150	2.677242	2.677333	2.677424	2.677516	91
476	2.678063	2.678154	2.678245	2.678336	2.678427	91
477	2.678973	2.679064	2.679155	2.679246	2.679337	91
478	2.679882	2.679973	2.680063	2.680154	2.680245	91
479	2.680789	2.680879	2.680970	2.681060	2.681151	91
480	2.681693	2.681784	2.681874	2.681964	2.682055	90
481	2.682596	2.682686	2.682777	2.682867	2.682957	90
482	2.683497	2.683587	2.683677	2.683767	2.683857	90
483	2.684396	2.684486	2.684576	2.684666	2.684756	90
484	2.685294	2.685383	2.685473	2.685563	2.685652	90
485	2.686189	2.686279	2.686368	2.686457	2.685547	89
486	2.687083	2.687172	2.687261	2.687351	2.687440	89

A Table of Logarithms,

N°	0	1	2	3	4	Dif
487	2.687529	2.687618	2.687707	2.687796	2.687885	89
488	2.688420	2.688509	2.688598	2.688687	2.688776	89
489	2.689309	2.689398	2.689486	2.689575	2.689664	89
490	2.690196	2.690285	2.690373	2.690462	2.690550	89
491	2.691081	2.691170	2.691258	2.691347	2.691435	88
492	2.691965	2.692053	2.692142	2.692230	2.692318	88
493	2.692847	2.692935	2.693023	2.693111	2.693199	88
494	2.693727	2.693815	2.693903	2.693991	2.694078	88
495	2.694605	2.694693	2.694781	2.694868	2.694956	88
496	2.695482	2.695569	2.695657	2.695744	2.695832	87
497	2.696356	2.696444	2.696531	2.696618	2.696706	87
498	2.697229	2.697316	2.697404	2.697491	2.697578	87
499	2.698100	2.698188	2.698275	2.698362	2.698448	87
500	2.698970	2.699057	2.699144	2.699230	2.699317	87
501	2.699838	2.699924	2.700011	2.700098	2.700184	87
502	2.700704	2.700790	2.700877	2.700963	2.701050	86
503	2.701568	2.701654	2.701741	2.701827	2.701913	86
504	2.702430	2.702517	2.702603	2.702689	2.702775	86
505	2.703291	2.703377	2.703463	2.703549	2.703635	86
506	2.704150	2.704236	2.704322	2.704468	2.704494	86
507	2.705008	2.705094	2.705179	2.705265	2.705350	86
508	2.705864	2.705949	2.706035	2.706120	2.706205	85
509	2.706718	2.706803	2.706888	2.706974	2.707059	85
510	2.707570	2.707655	2.707740	2.707826	2.707911	85
511	2.708421	2.708506	2.708591	2.708676	2.708761	85
512	2.709270	2.709355	2.709440	2.709524	2.709609	85
513	2.710117	2.710202	2.910287	2.710371	2.710456	85
514	2.710963	2.711048	2.711132	2.711216	2.711301	84
515	2.711807	2.711891	2.711976	2.712060	2.712144	84
516	2.712650	2.712734	2.712818	2.712902	2.712986	84
517	2.713490	2.713574	2.713658	2.713742	2.713826	84
518	2.714330	2.714414	2.714497	2.714581	2.714665	84
519	2.715167	2.715251	2.715335	2.715418	2.715502	84
520	2.716003	2.716087	2.716170	2.716254	2.716337	83
521	2.716838	2.716921	2.717004	2.717088	2.717171	83
522	2.717670	2.717754	2.717837	2.717920	2.718003	83
523	2.718502	2.718585	2.718668	2.718751	2.718834	83
524	2.719331	2.719414	2.719497	2.719580	2.719663	83
525	2.720159	2.720242	2.720325	2.720407	2.720490	83
526	2.720986	2.721068	2.721151	2.721233	2.721316	82
527	2.721811	2.721893	2.721975	2.722058	2.722140	82
528	2.722634	2.722716	2.722798	2.722881	2.722963	82
529	2.723456	2.723538	2.723620	2.723702	2.723784	82

Nº	5	6	7	8	9	Dif
487	2.687975	2.688064	2.688153	2.688242	2.688331	89
488	2.688865	2.688953	2.689042	2.689131	2.689220	89
489	2.689753	2.689811	2.689930	2.690019	2.690107	89
490	2.690639	2.690727	2.690816	2.690905	2.690993	89
491	2.691523	2.691611	2.691700	2.691788	2.691877	88
492	2.692406	2.692494	2.692583	2.692671	2.692759	88
493	2.693287	2.693375	2.693463	2.693551	2.693639	88
494	2.694166	2.694254	2.694342	2.694430	2.694517	89
495	2.695044	2.695131	2.695219	2.695306	2.695394	88
496	2.695910	2.696007	2.696094	2.696181	2.696269	87
497	2.696793	2.696880	2.696968	2.697055	2.697142	87
498	2.697665	2.697752	2.697839	2.697926	2.698013	87
499	2.698535	2.698622	2.698709	2.698796	2.698883	87
500	2.699404	2.699491	2.699578	2.699664	2.699751	87
501	2.700271	2.700357	2.700444	2.700531	2.700617	87
502	2.701136	2.701222	2.701309	2.701395	2.701481	86
503	2.701999	2.702086	2.702172	2.702258	2.702344	86
504	2.702861	2.702947	2.703033	2.703119	2.703205	85
505	2.703721	2.703807	2.703893	2.703979	2.704065	86
506	2.704579	2.704665	2.704751	2.704837	2.704922	86
507	2.705436	2.705522	2.705607	2.705693	2.705778	85
508	2.706291	2.706376	2.706462	2.706547	2.706632	85
509	2.707144	2.707229	2.707315	2.707400	2.707485	85
510	2.707996	2.708081	2.708166	2.708251	2.708336	85
511	2.708846	2.708931	2.709015	2.709100	2.709185	85
512	2.709694	2.709779	2.709863	2.709948	2.710033	85
513	2.710540	2.710625	2.710710	2.710794	2.710879	85
514	2.711385	2.711470	2.711554	2.711618	2.711723	84
515	2.712229	2.712313	2.712397	2.712481	2.712565	84
516	2.713070	2.713154	2.713238	2.713322	2.713405	84
517	2.713910	2.713994	2.714078	2.714162	2.714246	84
518	2.714749	2.714832	2.714916	2.715000	2.715084	84
519	2.715586	2.715669	2.715753	2.715836	2.715920	84
520	2.716411	2.716504	2.716588	2.716671	2.716754	83
521	2.717254	2.717338	2.717421	2.717504	2.717587	83
522	2.718086	2.718169	2.718253	2.718336	2.718419	83
523	2.718917	2.719000	2.719083	2.719165	2.719248	83
524	2.719745	2.719828	2.719911	2.719994	2.720077	83
525	2.720572	2.720655	2.720738	2.720821	2.720903	83
526	2.721398	2.721481	2.721563	2.721646	2.721728	82
527	2.722221	2.722305	2.722387	2.722469	2.722552	82
528	2.723045	2.723127	2.723209	2.723291	2.723374	82
529	2.723866	2.723948	2.724030	2.724112	2.724194	82

A Table of Logarithms,

Nº	0	1	2	3	4	Dif
530	2.724276	2.724358	2.724440	2.724522	2.724603	82
531	2.725094	2.725176	2.725258	2.725340	2.725421	82
532	2.725912	2.725993	2.726075	2.726156	2.726238	82
533	2.726727	2.726809	2.726890	2.726972	2.727053	81
534	2.727541	2.727623	2.727704	2.727785	2.727866	81
535	2.728354	2.728435	2.728516	2.728597	2.728678	81
536	2.729165	2.729246	2.729327	2.729408	2.729489	81
537	2.729974	2.730055	2.730136	2.730217	2.730298	81
538	2.730782	2.730863	2.730944	2.731024	2.731105	81
539	2.731589	2.731669	2.731750	2.731830	2.731911	81
540	2.732394	2.732474	2.732555	2.732635	2.732715	80
541	2.733197	2.733277	2.733358	2.733438	2.733518	80
542	2.733999	2.734079	2.734159	2.734240	2.734320	80
543	2.734800	2.734880	2.734960	2.735040	2.735120	80
544	2.735599	2.735679	2.735758	2.735838	2.735918	80
545	2.736396	2.736476	2.736556	2.736635	2.736715	80
546	2.737193	2.737272	2.737352	2.737431	2.737511	79
547	2.737987	2.738067	2.738146	2.738225	2.738305	79
548	2.738781	2.738860	2.738939	2.739018	2.739097	79
549	2.739572	2.739651	2.739730	2.739810	2.739889	79
550	2.740363	2.740442	2.740521	2.740599	2.740678	79
551	2.741152	2.741230	2.741309	2.741388	2.741467	79
552	2.741939	2.742018	2.742096	2.742175	2.742254	79
553	2.742725	2.742804	2.742882	2.742961	2.743039	78
554	2.743510	2.743588	2.743666	2.743745	2.743823	78
555	2.744293	2.744371	2.744449	2.744528	2.744606	78
556	2.745075	2.745153	2.745231	2.745309	2.745387	78
557	2.745855	2.745933	2.746011	2.746089	2.746167	78
558	2.746634	2.746712	2.746790	2.746868	2.746945	78
559	2.747411	2.747489	2.747567	2.747645	2.747722	78
560	2.748188	2.748266	2.748343	2.748421	2.748498	77
561	2.748963	2.749040	2.749118	2.749195	2.749272	77
562	2.749736	2.749814	2.749891	2.749968	2.750045	77
563	2.750508	2.750586	2.750663	2.750740	2.750817	77
564	2.751279	2.751356	2.751433	2.751510	2.751587	77
565	2.752048	2.752125	2.752202	2.752279	2.752356	77
566	2.752816	2.752893	2.752970	2.753047	2.753123	77
567	2.753583	2.753660	2.753736	2.753813	2.753889	77
568	2.754348	2.754425	2.754501	2.754578	2.754654	77
569	2.755112	2.755189	2.755265	2.755341	2.755417	76
570	2.755875	2.755951	2.756027	2.756103	2.756180	76
571	2.756636	2.756712	2.756788	2.756864	2.756940	76
572	2.757396	2.757472	2.757548	2.757624	2.757700	76

N°	5	6	7	8	9	Dif
530	2.724685	2.724767	2.724849	2.724931	2.725013	82
531	2.725503	2.725585	2.725667	2.725748	2.725830	82
532	2.726320	2.726401	2.726443	2.726564	2.726646	62
533	2.727134	2.727216	2.727297	2.727379	2.727460	81
534	2.727948	2.728029	2.728110	2.728191	2.728273	81
535	2.728759	2.728841	2.728922	2.729003	2.729084	81
536	2.729570	2.729651	2.729732	2.729813	2.729893	81
537	2.730378	2.730459	2.730540	2.730621	2.730702	81
538	2.731186	2.731266	2.731347	2.731428	2.731508	81
539	2.731991	2.732072	2.732152	2.732233	2.732313	81
540	2.732796	2.732876	2.732956	2.733037	2.733117	80
541	2.733598	2.733679	2.733759	2.733839	2.733919	80
542	2.734400	2.734480	2.734560	2.734640	2.734720	80
543	2.735200	2.735279	2.735359	2.735439	2.735519	80
544	2.735998	2.736078	2.736157	2.736237	2.736317	80
545	2.736795	2.736874	2.736954	2.737034	2.737113	80
546	2.737590	2.737670	2.737749	2.737829	2.737908	79
547	2.738384	2.738463	2.738543	2.738622	2.738701	79
548	2.739177	2.739256	2.739335	2.739414	2.739493	79
549	2.739968	2.740047	2.740126	2.740205	2.740284	79
550	2.740757	2.740836	2.740915	2.740994	2.741073	79
551	2.741540	2.741624	2.741703	2.741782	2.741860	79
552	2.742332	2.742411	2.742489	2.742568	2.742647	79
553	2.743118	2.743196	2.743275	2.743353	2.743431	78
554	2.743902	2.743980	2.744058	2.744136	2.744215	78
555	2.744684	2.744762	2.744840	2.744919	2.744997	78
556	2.745465	2.745543	2.745621	2.745699	2.745777	78
557	2.746245	2.746323	2.746401	2.746479	2.746556	78
558	2.747023	2.747101	2.747179	2.747256	2.747334	78
559	2.747800	2.747878	2.747955	2.748033	2.748110	78
560	2.748576	2.748653	2.748731	2.748808	2.748885	77
561	2.749350	2.749427	2.749504	2.749582	2.749659	77
562	2.750123	2.750200	2.750277	2.750354	2.750431	77
563	2.750894	2.750971	2.751048	2.751125	2.751202	77
564	2.751664	2.751741	2.751818	2.751895	2.751972	77
565	2.752433	2.752509	2.752586	2.752663	2.752740	77
566	2.753200	2.753277	2.753353	2.753430	2.753506	77
567	2.753966	2.754042	2.754119	2.754195	2.754272	77
568	2.754730	2.754807	2.754883	2.754960	2.755036	76
569	2.755494	2.755570	2.755646	2.755722	2.755799	76
570	2.756256	2.756332	2.756408	2.756484	2.756560	76
571	2.757016	2.757092	2.757168	2.757244	2.757320	76
572	2.757775	2.757851	2.757927	2.758003	2.758079	76

E e

A Table of Logarithms,

Nº	0	1	2	3	4	Dif
573	2.758155	2.758230	2.758306	2.758382	2.758458	76
574	2.758912	2.758988	2.759063	2.759139	2.759214	76
575	2.759668	2.759743	2.759819	2.759894	2.759970	75
576	2.760422	2.760498	2.760573	2.760649	2.760724	75
577	2.761176	2.761251	2.761326	2.761402	2.761477	75
578	2.761928	2.762003	2.762078	2.762153	2.762228	75
579	2.762679	2.762754	2.762829	2.762904	2.762978	75
580	2.763428	2.763503	2.763578	2.763652	2.763727	75
581	2.764176	2.764251	2.764326	2.764400	2.764475	75
582	2.764923	2.764998	2.765072	2.765147	2.765221	75
583	2.765669	2.765743	2.765818	2.765892	2.765966	74
584	2.766413	2.766487	2.766562	2.766636	2.766710	74
585	2.767156	2.767230	2.767304	2.767379	2.767453	74
586	2.767898	2.767972	2.768046	2.768120	2.768194	74
587	2.768638	2.768712	2.768786	2.768860	2.768934	74
588	2.769377	2.769451	2.769525	2.769599	2.769673	74
589	2.770115	2.770189	2.770263	2.770336	2.770410	74
590	2.770852	2.770926	2.770999	2.771073	2.771146	74
591	2.771587	2.771661	2.771734	2.771808	2.771881	73
592	2.772322	2.772395	2.772468	2.772542	2.772615	73
593	2.773055	2.773128	2.773201	2.773274	2.773348	73
594	2.773786	2.773860	2.773933	2.774005	2.774079	73
595	2.774517	2.774590	2.774663	2.774736	2.774809	73
596	2.775246	2.775319	2.775392	2.775465	2.775538	73
597	2.775974	2.776047	2.776120	2.776193	2.776265	73
598	2.776701	2.776774	2.776846	2.776919	2.776992	73
599	2.777427	2.777499	2.777572	2.777644	2.777717	72
600	2.778151	2.778224	2.778296	2.778368	2.778441	72
601	2.778874	2.778947	2.779019	2.779091	2.779163	72
602	2.779596	2.779669	2.779741	2.779811	2.779885	72
603	2.780317	2.780389	2.780461	2.780533	2.780605	72
604	2.781037	2.781109	2.781181	2.781253	2.781324	72
605	2.781755	2.781827	2.781899	2.781971	2.782042	72
606	2.782473	2.782544	2.782616	2.782688	2.782759	72
607	2.783189	2.783260	2.783332	2.783403	2.783475	71
608	2.783904	2.783975	2.784046	2.784118	2.784189	71
609	2.784617	2.784689	2.784760	2.784831	2.784902	71
610	2.785330	2.785401	2.785472	2.785543	2.785615	71
611	2.786041	2.786112	2.786183	2.786254	2.786325	71
612	2.786751	2.786822	2.786893	2.786964	2.787035	71
613	2.787460	2.787531	2.787602	2.787673	2.787744	71
614	2.788168	2.788239	2.788310	2.788381	2.788451	71
615	2.788875	2.788946	2.789016	2.789087	2.789157	71

N°	5	6	7	8	9	Dif
573	2.758533	2.758609	2.758685	2.758761	2.758836	76
574	2.759290	2.759366	2.759441	2.759517	2.759592	76
575	2.760045	2.760121	2.760196	2.760272	2.760347	75
576	2.760799	2.760875	2.760950	2.761025	2.761101	75
577	2.761552	2.761627	2.751702	2.761778	2.761853	75
578	2.762303	2.762378	2.762453	2.762529	2.762604	75
579	2.763053	2.763128	2.763203	2.763278	2.763353	75
580	2.763802	2.763877	2.763952	2.764027	2.764101	75
581	2.764550	2.764624	2.764699	2.764774	2.764848	75
582	2.765296	2.765370	2.765445	2.765520	2.765594	75
583	2.766041	2.766115	2.766190	2.766264	2.766338	74
584	2.766785	2.766859	2.766933	2.767007	2.767082	74
585	2.767527	2.767601	2.767675	2.767749	2.767823	74
586	2.768268	2.768342	2.768416	2.768490	2.768564	74
587	2.769008	2.769082	2.769156	2.769230	2.769303	74
588	2.769746	2.769820	2.769894	2.769968	2.770041	74
589	2.770484	2.770557	2.770631	2.770705	2.770778	74
590	2.771220	2.771293	2.771367	2.771440	2.771514	74
591	2.771955	2.772028	2.772102	2.772175	2.772248	73
592	2.772688	2.772762	2.772835	2.772908	2.772981	73
593	2.773421	2.773494	2.773567	2.773640	2.773713	73
594	2.774152	2.774225	2.774298	2.774371	2.774444	73
595	2.774882	2.774955	2.775028	2.775100	2.775173	73
596	2.775610	2.775583	2.775756	2.775829	2.775902	73
597	2.776338	2.776411	2.776483	2.776556	2.776629	73
598	2.777064	2.777137	2.777209	2.777282	2.777354	73
599	2.777789	2.777862	2.777934	2.778006	2.778079	72
600	2.778513	2.778585	2.778658	2.778730	2.778802	72
601	2.779236	2.779308	2.779380	2.779452	2.779524	72
602	2.779957	2.780029	2.780101	2.780173	2.780245	72
603	2.780677	2.780749	2.780821	2.780893	2.780965	72
604	2.781396	2.781468	2.781540	2.781612	2.781684	72
605	2.782114	2.782186	2.782258	2.782329	2.782401	72
606	2.782831	2.782902	2.782974	2.783046	2.783117	72
607	2.783546	2.783618	2.783689	2.783761	2.783832	71
608	2.784261	2.784332	2.784403	2.784475	2.784546	71
609	2.784974	2.785045	2.785116	2.785187	2.785259	71
610	2.785686	2.785757	2.785828	2.785899	2.785970	71
611	2.786396	2.786467	2.786538	2.786609	2.786680	71
612	2.787106	2.787177	2.787248	2.787319	2.787390	71
613	2.787815	2.787885	2.787956	2.788027	2.788098	71
614	2.788522	2.788593	2.788663	2.788734	2.788804	71
615	2.789228	2.789299	2.789369	2.789440	2.789510	71

N°	0	1	2	3	4	Dif
616	2.789581	2.789651	2.789722	2.789792	2.789863	70
617	2.790285	2.790356	2.790426	2.790496	2.790567	70
618	2.790988	2.791059	2.791129	2.791199	2.791269	70
619	2.791691	2.791761	2.791831	2.791901	2.791971	70
620	2.792392	2.792462	2.792532	2.792602	2.792672	70
621	2.793092	2.793162	2.793231	2.793301	2.793371	70
622	2.793790	2.793860	2.793930	2.794000	2.794070	70
623	2.794488	2.794558	2.794627	2.794697	2.794767	70
624	2.795185	2.795254	2.795324	2.795393	2.795463	69
625	2.795880	2.795949	2.796019	2.796088	2.796158	69
626	2.796574	2.796644	2.796713	2.796782	2.796852	69
627	2.797268	2.797337	2.797406	2.797475	2.797545	69
628	2.797960	2.798029	2.798098	2.798167	2.798236	69
629	2.798651	2.798720	2.798789	2.798856	2.798927	69
630	2.799341	2.799409	2.799478	2.799547	2.799616	69
631	2.800029	2.800098	2.800166	2.800236	2.800305	69
632	2.800717	2.800786	2.800854	2.800923	2.800992	69
633	2.801404	2.801472	2.801541	2.801609	2.801678	69
634	2.802089	2.802158	2.802226	2.802295	2.802363	68
635	2.802774	2.802842	2.802910	2.802979	2.803047	68
636	2.803457	2.803525	2.803594	2.803662	2.803730	68
637	2.804139	2.804208	2.804276	2.804344	2.804412	68
638	2.804821	2.804889	2.804957	2.805025	2.805093	68
639	2.805501	2.805569	2.805637	2.805705	2.805773	68
640	2.806180	2.806248	2.806316	2.806384	2.806451	68
641	2.806858	2.806926	2.806994	2.807061	2.807129	68
642	2.807535	2.807603	2.807670	2.807738	2.807806	68
643	2.808211	2.808279	2.808346	2.808414	2.808481	67
644	2.808886	2.808953	2.809021	2.809088	2.809155	67
645	2.809560	2.809627	2.809694	2.809762	2.809829	67
646	2.810232	2.810300	2.810367	2.810434	2.810501	67
647	2.810904	2.810971	2.811038	2.811106	2.811173	67
648	2.811575	2.811642	2.811709	2.811776	2.811843	67
649	2.812245	2.812312	2.812378	2.812445	2.812512	67
650	2.812913	2.812980	2.813047	2.813114	2.813180	67
651	2.813581	2.813648	2.813714	2.813781	2.813848	67
652	2.814248	2.814314	2.814381	2.814447	2.814514	67
653	2.814913	2.814980	2.815046	2.815113	2.815179	66
654	2.815578	2.815644	2.815710	2.815777	2.815843	66
655	2.816241	2.816308	2.816374	2.816440	2.816506	66
656	2.816904	2.816970	2.817036	2.817102	2.817169	66
657	2.817565	2.817631	2.817698	2.817764	2.817830	66
658	2.818226	2.818292	2.818358	2.818424	2.818490	66

N°	5	6	7	8	9	Dif
616	2.789933	2.790003	2.790074	2.790144	2.790215	70
617	2.790637	2.790707	2.790778	2.790848	2.790918	70
618	2.791340	2.791410	2.791480	2.791550	2.791620	70
619	2.792041	2.792111	2.792181	2.792252	2.792322	70
620	2.792742	2.792812	2.792882	2.792952	2.793022	70
621	2.793441	2.793511	2.793581	2.793651	2.793721	70
622	2.794139	2.794209	2.794279	2.794349	2.794418	70
623	2.794836	2.794906	2.794976	2.795045	2.795115	70
624	2.795532	2.795602	2.795671	2.795741	2.795810	69
625	2.796227	2.796297	2.796366	2.796436	2.796505	69
626	2.796921	2.796990	2.797060	2.797129	2.797198	69
627	2.797614	2.797683	2.797752	2.797821	2.797890	69
628	2.798305	2.798374	2.798443	2.798512	2.798582	69
629	2.798996	2.799065	2.799134	2.799203	2.799272	69
630	2.799685	2.799754	2.799823	2.799892	2.799960	69
631	2.800373	2.800442	2.800511	2.800580	2.800648	69
632	2.801060	2.801129	2.801198	2.801267	2.801335	69
633	2.801747	2.801815	2.801884	2.801952	2.802021	69
634	2.802432	2.802500	2.802568	2.802637	2.802705	68
635	2.803116	2.803184	2.803252	2.803321	2.803389	68
636	2.803798	2.803867	2.803953	2.804003	2.804071	68
637	2.804480	2.804548	2.804616	2.804685	2.804753	68
638	2.805161	2.805229	2.805297	2.805365	2.805433	68
639	2.805840	2.805908	2.805976	2.806044	2.806112	68
640	2.806519	2.806587	2.806655	2.806723	2.806790	68
641	2.807196	2.807264	2.807332	2.807400	2.807467	68
642	2.807873	2.807941	2.808008	2.808076	2.808143	68
643	2.808548	2.808616	2.808683	2.808751	2.808818	67
644	2.809223	2.809290	2.809358	2.809425	2.809492	67
645	2.809896	2.809963	2.810031	2.810098	2.810165	67
646	2.810568	2.810636	2.810703	2.810770	2.810837	67
647	2.811240	2.811307	2.811374	2.811441	2.811508	67
648	2.811910	2.811977	2.812044	2.812111	2.812178	67
649	2.812579	2.812646	2.812713	2.812780	2.812846	67
650	2.813247	2.813314	2.813381	2.813447	2.813514	67
651	2.813914	2.813981	2.814048	2.814114	2.814181	67
652	2.814580	2.814647	2.814714	2.814780	2.814847	67
653	2.815246	2.815312	2.815378	2.815445	2.815511	66
654	2.815910	2.815976	2.816042	2.816109	2.816175	66
655	2.816573	2.816639	2.816705	2.816771	2.816838	66
656	2.817235	2.817301	2.817367	2.817433	2.817499	66
657	2.817896	2.817962	2.818028	2.818094	2.818160	66
658	2.818556	2.818622	2.818688	2.818754	2.818819	66

E 3

A Table of Logarithms,

N°	0	1	2	3	4	Dif
659	2.818885	7.818951	2.819017	2.819083	2.819149	66
660	2.819544	2.819610	2.819675	2.819741	2.819807	66
661	2.820201	2.820267	2.820333	2.820398	2.820464	66
662	2.820858	2.820924	2.820989	2.821055	2.521120	66.
663	2.821513	2.821579	2.821644	2.821710	2.821775	65
664	2.822168	2.822233	2.822299	2.822364	2.822430	65
665	2.822822	2.822887	2.822952	2.823017	2.823083	65
666	2.823474	2.823539	2.823605	2.823670	2.823735	65
667	2.824126	2.824191	2.824256	2.824321	2.824386	65
668	2.824776	2.824841	2.824906	2.824971	2.825036	65
669	2.825426	2.825491	2.825556	2.825621	2.825686	65
670	2.826075	2.826140	2.826204	2.826269	2.826334	65
671	2.826722	2.826787	2.826852	2.826917	2.826981	65
672	2.827369	2.827434	2.827498	2.827563	2.827628	65
673	2.828015	2.828080	2.828144	2.828209	2.828273	64
674	2.828660	2.828724	2.828789	2.828853	2.828918	64
675	2.829304	2.829368	2.829432	2.829497	2.829561	64
676	2.829947	2.830011	2.830075	2.830139	2.830204	64
677	2.830589	2.830653	2.830717	2.830781	2.830845	64
678	2.831230	2.831294	2.831358	2.831422	2.831486	64
679	2.831870	2.831934	2.831998	2.832062	2.832125	64
680	2.832509	2.832573	2.832637	2.832700	2.832764	64
681	2.833147	2.833211	2.833275	2.833338	2.833402	64
682	2.833784	2.833848	2.833912	2.833975	2.834039	64
683	2.834421	2.834484	2.834548	2.834611	2.834675	64
684	2.835056	2.835120	2.835183	2.835246	2.835310	63
685	2.835691	2.835754	2.835817	2.835881	2.835944	63
686	2.836324	2.836387	2.836451	2.836514	2.836577	63
687	2.836957	2.837020	2.837083	2.837146	2.837209	63
688	2.837588	2.837652	2.837715	2.837778	2.837841	63
689	2.838219	2.838282	2.838345	2.838408	2.838471	63
690	2.838849	2.838912	2.838975	2.839038	2.839101	63
691	2.839478	2.839541	2.839604	2.839667	2.839729	63
692	2.840106	2.840169	2.840232	2.840294	2.840357	63
693	2.840733	2.840796	2.840859	2.840921	2.840984	63
694	2.841359	2.841422	2.841485	2.841547	2.841610	63
695	2.841985	2.842047	2.842110	2.842172	2.842235	62
696	2.842609	2.842672	2.842734	2.842796	2.842859	62
697	2.843233	2.843295	2.843357	2.843420	2.843482	62
698	2.843855	2.843918	2.843980	2.844042	2.844104	62
699	2.844477	2.844539	2.844601	2.844663	2.844726	62
700	2.845098	2.845160	2.845222	2.845284	2.845346	62
701	2.845718	2.845780	2.845842	2.845904	2.845966	62

N°	5	6	7	8	9	Dif
659	2.819215	2.819281	2.819346	2.819412	2.819478	66
660	2.819873	2.819939	2.820004	2.820070	2.820136	66
661	2.820530	2.820595	2.820661	2.820727	2.820792	66
662	2.821186	2.821251	2.821317	2.821382	2.821448	66
663	2.821841	2.821906	2.821972	2.822037	2.822103	65
664	2.822495	2.822560	2.822626	2.822691	2.822756	65
665	2.823148	2.823213	2.823279	2.823344	2.823409	65
666	2.823800	2.823865	2.823930	2.823996	2.824061	65
667	2.824451	2.824516	2.824581	2.824646	2.824711	65
668	2.825101	2.825166	2.825231	2.825296	2.825361	65
669	2.825751	2.825815	2.825880	2.825945	2.826010	65
670	2.826399	2.826464	2.826528	2.826593	2.826658	65
671	2.827046	2.827111	2.827175	2.827240	2.827305	65
672	2.827692	2.827757	2.827821	2.827886	2.827950	65
673	2.828338	2.828402	2.828467	2.828531	2.828595	64
674	2.828982	2.829046	2.829111	2.829175	2.829239	64
675	2.829625	2.829690	2.829754	2.829818	2.829882	64
676	2.830268	2.830332	2.830396	2.830460	2.830524	64
677	2.830909	2.830973	2.831037	2.831102	2.831166	64
678	2.831550	2.831614	2.831678	2.831742	2.831805	64
679	2.832189	2.832253	2.832317	2.832381	2.832445	64
680	2.832828	2.832892	2.832956	2.833019	2.833083	64
681	2.833466	2.833530	2.833593	2.833657	2.833721	64
682	2.834103	2.834166	2.834230	2.834293	2.834357	64
683	2.834738	2.834802	2.834866	2.834929	2.834993	64
684	2.835373	2.835437	2.835500	2.835564	2.835627	63
685	2.836007	2.836071	2.836134	2.836197	2.836261	63
686	2.836640	2.836704	2.836767	2.836830	2.836893	63
687	2.837273	2.837336	2.837399	2.837462	2.837525	63
688	2.837904	2.837967	2.838030	2.838093	2.838156	63
689	2.838534	2.838597	2.838660	2.838723	2.838785	63
690	2.839164	2.839227	2.839289	2.839352	2.839415	63
691	2.839792	2.839855	2.839918	2.839981	2.840043	63
692	2.840420	2.840482	2.840545	2.840608	2.840671	63
693	2.841046	2.841109	2.841172	2.841234	2.841297	63
694	2.841672	2.841735	2.841797	2.841860	2.841922	63
695	2.842297	2.842360	2.842422	2.842484	2.842547	62
696	2.842921	2.842983	2.843046	2.843108	2.843170	62
697	2.843544	2.843606	2.843669	2.843731	2.843793	62
698	2.844166	2.844229	2.844291	2.844353	2.844415	62
699	2.844788	2.844850	2.844912	2.844974	2.845036	62
700	2.845408	2.845470	2.845532	2.845594	2.845656	62
701	2.846028	2.846090	2.846151	2.846213	2.846275	62

A Table of Logarithms,

N°	0	1	2	3	4	Dif
702	2.846337	7.846399	2.846461	2.846523	2.846584	62
703	2.846955	2.847017	2.847079	2.847141	2.847202	62
704	2.847573	2.847634	2.847696	2.847758	2.847819	62
705	2.848189	2.848251	2.848312	2.848374	2.848435	62
706	2.848805	2.848866	2.848928	2.848989	2.849051	61
707	2.849419	2.849481	2.849542	2.849604	2.849665	61
708	2.850033	2.850095	2.850156	2.850217	2.850279	61
709	2.850646	2.850707	2.850769	2.850830	2.850891	61
710	2.851258	2.851319	2.851381	2.851442	2.851503	61
711	2.851870	2.851931	2.851952	2.852053	2.852114	61
712	2.852480	2.852541	2.852602	2.852663	2.852724	61
713	2.853089	2.853150	2.853211	2.853272	2.853333	61
714	2.853698	2.853759	2.853820	2.853881	2.853941	61
715	2.854306	2.854367	2.854427	2.854488	2.854549	61
716	2.854913	2.854974	2.855034	2.855095	2.855156	61
717	2.855519	2.855580	2.855640	2.855701	2.855761	60
718	2.856124	2.856185	2.856245	2.856306	2.856366	60
719	2.856729	2.856789	2.856850	2.856910	2.856970	60
720	2.857332	2.857393	2.857453	2.657513	2.857574	60
721	2.857935	2.857995	2.858056	2.858116	2.858176	60
722	2.858537	2.858597	2.858657	2.858718	2.858778	60
723	2.859138	2.859198	2.859258	2.859318	2.859378	60
724	2.859739	2.859798	2.859858	2.859918	2.859978	60
725	2.860338	2.860398	2.860458	2.860518	2.860578	60
726	2.860937	2.860996	2.861056	2.861116	2.861176	60
727	2.861534	2.861594	2.861654	2.861714	2.861773	60
728	2.862131	2.862191	2.862251	2.862310	2.862370	60
729	2.862727	2.862787	2.862847	2.862906	2.862966	60
730	2.863323	2.863382	2.863442	2.863501	2.863561	59
731	2.863917	2.863977	2.864036	2.864096	2.864155	59
732	2.864511	2.864570	2.864630	2.864689	2.864748	59
733	2.865104	2.865163	2.865222	2.865282	2.865341	59
734	2.865696	2.865755	2.865814	2.865873	2.865933	59
735	2.866287	2.866346	2.866405	2.866465	2.866524	59
736	2.866878	2.866937	2.866996	2.867055	2.867114	59
737	2.867467	2.867526	2.867585	2.867644	2.867703	59
738	2.868056	2.868115	2.868174	2.868233	2.868292	59
739	2.868644	2.868703	2.868762	2.868821	2.868879	59
740	2.869232	2.869290	2.869349	2.869408	2.869466	59
741	2.869818	2.869877	2.869935	2.869994	2.870053	59
742	2.870404	2.870462	2.870521	2.870579	2.870638	58
743	2.870989	2.871047	2.871106	2.871164	2.871223	58
744	2.871573	2.871631	2.871690	2.871748	2.871806	58

N°	5	6	7	8	9	Dif
702	2.846046	2.846708	2.846770	2.846832	2.846893	62
703	2.847264	2.847326	2.847388	2.847449	2.847511	62
704	2.847881	2.847943	2.848004	2.848066	2.848127	62
705	2.848497	2.848559	2.848620	2.848682	2.848743	62
706	2.849112	2.849174	2.849235	2.849296	2.849356	61
707	2.849726	2.849786	2.849847	2.849911	2.849972	61
708	2.850340	2.850401	2.850462	2.850524	2.850585	61
709	2.850952	2.851014	2.851075	2.851136	2.851197	61
710	2.851564	2.851625	2.851686	2.851747	2.851808	61
711	2.852175	2.852236	2.852297	2.852358	2.852419	61
712	2.852785	2.852846	2.852907	2.852968	2.853029	61
713	2.853394	2.853455	2.853516	2.853576	2.853637	61
714	2.854002	2.854063	2.854124	2.854184	2.854245	61
715	2.854610	2.854670	2.854731	2.854792	2.854852	61
716	2.855216	2.855277	2.855337	2.855398	2.855459	61
717	2.855822	2.855882	2.855943	2.856003	2.856064	60
718	2.856427	2.856487	2.856548	2.856628	2.856668	60
719	2.857031	2.857091	2.857151	2.857212	2.857272	60
720	2.857634	2.857694	2.857754	2.857815	2.857875	60
721	2.858136	2.858296	2.858357	2.858417	2.858477	60
722	2.858838	2.858898	2.858958	2.859018	2.859078	60
723	2.859438	2.859498	2.859559	2.859619	2.859679	60
724	2.860038	2.860098	2.860158	2.860218	2.860278	60
725	2.860637	2.860697	2.860757	2.860817	2.860877	60
726	2.861236	2.861295	2.861355	2.861415	2.861475	60
727	2.861833	2.861893	2.861952	2.862012	2.862072	60
728	2.862430	2.862489	2.862549	2.862608	2.862668	60
729	2.863025	2.863085	2.863144	2.863204	2.863263	60
730	2.863620	2.863680	2.863739	2.863798	2.863858	59
731	2.864214	2.864274	2.864333	2.864392	2.864452	59
732	2.864808	2.864867	2.864926	2.864985	2.865045	59
733	2.865400	2.865459	2.865518	2.865578	2.865637	59
734	2.865992	2.866051	2.866110	2.866169	2.866228	59
735	2.866583	2.866642	2.866701	2.866760	2.866819	59
736	2.867173	2.867232	2.867291	2.867350	2.867409	59
737	2.867762	2.867821	2.867890	2.867939	2.867997	59
738	2.868350	2.868409	2.868488	2.868527	2.868586	59
739	2.868938	2.868997	2.869056	2.869114	2.869173	59
740	2.869525	2.869584	2.869642	2.869701	2.869760	59
741	2.870111	2.870170	2.870228	2.870287	2.870345	59
742	2.870696	2.870755	2.870813	2.870872	2.870930	58
743	2.871281	2.871339	2.871398	2.871456	2.871515	58
744	2.871865	2.871923	2.871981	2.872040	2.872098	58

N°	0	1	2	3	4	Dif
745	2.872156	2.872215	2.872273	2.872331	2.872389	58
746	2.872719	2.872797	2.872855	2.872913	2.872972	58
747	2.873321	2.873379	2.873437	2.873495	2.873553	58
748	2.873902	2.873960	2.874018	2.874076	2.874134	58
749	2.874482	2.874540	2.874598	2.874656	2.874714	58
750	2.875061	2.875119	2.875179	2.875235	2.875293	58
751	2.875640	2.875698	2.875756	2.875813	2.875871	58
752	2.876218	2.876276	2.876333	2.876391	2.876449	58
753	2.876795	2.876853	2.876910	2.876968	2.877026	58
754	2.877371	2.877429	2.877486	2.877544	2.877602	58
755	2.877947	2.878004	2.878062	2.878119	2.878177	57
756	2.878522	2.878579	2.878637	2.878694	2.878751	57
757	2.879096	2.879153	2.879211	2.879268	2.879325	57
758	2.879669	2.879726	2.879784	2.879841	2.879898	57
759	2.880242	2.880299	2.880356	2.880413	2.880471	57
760	2.880814	2.880871	2.880928	2.880985	2.881042	57
761	2.881385	2.881442	2.881499	2.881556	2.881613	57
762	2.881955	2.882012	2.882069	2.882126	2.882183	57
763	2.882524	2.882581	2.882638	2.882695	2.882752	57
764	2.883093	2.883150	2.883207	2.883264	2.883321	57
765	2.883661	2.883718	2.883775	2.883832	2.883889	57
766	2.884229	2.884285	2.884342	2.884399	2.884455	57
767	2.884795	2.884852	2.884909	2.884965	2.885022	57
768	2.885361	2.885418	2.885474	2.885531	2.885587	57
769	2.885926	2.885983	2.886039	2.886096	2.886152	56
770	2.886491	2.886547	2.886603	2.886660	2.886716	56
771	2.887054	2.887111	2.887167	2.887223	2.887280	56
772	2.887617	2.887673	2.887730	2.887786	2.887842	56
773	2.888179	2.888236	2.888292	2.888348	2.888404	56
774	2.888741	2.888797	2.888853	2.888909	2.888965	56
775	2.889302	2.889358	2.889414	2.889470	2.889526	56
776	2.889862	2.889918	2.889974	2.890030	2.890086	56
777	2.890421	2.890477	2.890533	2.890589	2.890646	56
778	2.890980	2.891035	2.891091	2.891147	2.891203	56
779	2.891537	2.891593	2.891649	2.891705	2.891760	56
780	2.892095	2.892150	2.892206	2.892262	2.892317	56
781	2.892651	2.892707	2.892762	2.892818	2.892873	56
782	2.893207	2.893262	2.893318	2.893373	2.893429	56
783	2.893762	2.893817	2.893673	2.893928	2.893984	55
784	2.894316	2.894371	2.894427	2.894482	2.894538	55
785	2.894870	2.894925	2.894980	2.895036	2.895091	55
786	2.895422	2.895478	2.895533	2.895588	2.895643	55
787	2.895975	2.896030	2.896085	2.896140	2.896195	55

N°	5	6	7	8	9	Dif
745	2.872448	2.872506	2.872564	2.872622	2.872681	58
746	2.873030	2.873088	2.873146	2.873204	2.873262	58
747	2.873611	2.873669	2.873727	2.873785	2.873843	58
748	2.874192	2.874250	2.874308	2.874366	2.874424	58
749	2.874772	2.874830	2.874887	2.874945	2.875003	58
750	2.875351	2.875409	2.875466	2.875524	2.875582	58
751	2.875929	2.875987	2.876044	2.876102	2.876160	58
752	2.876506	2.876564	2.876622	2.876680	2.876737	58
753	2.877083	2.877141	2.877198	2.877256	2.877314	58
754	2.877659	2.877717	2.877774	2.877832	2.877889	58
755	2.878234	2.878292	2.878349	2.878407	2.878464	57
756	2.878809	2.878866	2.878924	2.878981	2.879038	57
757	2.879383	2.879440	2.879497	2.879555	2.879612	57
758	2.879956	2.880013	2.880070	2.880127	2.880185	57
759	2.880528	2.880585	2.880642	2.880699	2.880756	57
760	2.881099	2.881156	2.881213	2.881270	2.881328	57
761	2.881670	2.881727	2.881784	2.881841	2.881898	57
762	2.882240	2.882297	2.882354	2.882411	2.882468	57
763	2.882809	2.882866	2.882923	2.882980	2.883036	57
764	2.883377	2.883434	2.883491	2.883548	2.883605	57
765	2.883945	2.884002	2.884059	2.884115	2.884172	57
766	2.884512	2.884569	2.884625	2.884682	2.884739	57
767	2.885078	2.885135	2.885191	2.885248	2.885305	57
768	2.885644	2.885700	2.885757	2.885813	2.885870	57
769	2.886209	2.886265	2.886321	2.886378	2.886434	56
770	2.886773	2.886829	2.886885	2.886942	2.886998	56
771	2.887336	2.887392	2.887448	2.887505	2.887561	56
772	2.887898	2.887955	2.888011	2.888068	2.888123	56
773	2.888460	2.888516	2.888573	2.888629	2.888685	56
774	2.889021	2.889077	2.889134	2.889190	2.889246	56
775	2.889582	2.889638	2.889694	2.889750	2.889806	56
776	2.890141	2.890197	2.890253	2.890309	2.890365	56
777	2.890700	2.890756	2.890812	2.890868	2.890924	56
778	2.891259	2.891314	2.891370	2.891426	2.891482	56
779	2.891816	2.891872	2.891927	2.891983	2.892039	56
780	2.892373	2.892428	2.892484	2.892540	2.892595	56
781	2.892929	2.892985	2.893040	2.893096	2.893151	56
782	2.893484	2.893540	2.893595	2.893651	2.893706	56
783	2.894039	2.894094	2.894150	2.894205	2.894261	55
784	2.894593	2.894648	2.894704	2.894759	2.894814	55
785	2.895146	2.895201	2.895257	2.895312	2.895367	55
786	2.895699	2.895754	2.895809	2.895864	2.895919	55
787	2.896231	2.896306	2.896361	2.896416	2.896471	55

N°	0	1	2	3	4	Dif
788	2.896520	2.896581	2.896636	2.896691	2.896747	55
789	2.897077	2.897132	2.897187	2.897242	2.897297	55
790	2.897627	2.897682	2.897737	2.897792	2.897847	55
791	2.898176	2.898231	2.898286	2.898341	2.898396	55
792	2.898725	2.898780	2.898835	2.898890	2.898944	55
793	2.899273	2.899328	2.899383	2.899437	2.899492	55
794	2.899820	2.899875	2.899930	2.899985	2.900039	55
795	2.900367	2.900422	2.900476	2.900531	2.900586	55
796	2.900913	2.900968	2.901022	2.901077	2.901131	55
797	2.901458	2.901513	2.901567	2.901622	2.901676	54
798	2.902003	2.902057	2.902112	2.902166	2.902220	54
799	2.902547	2.902601	2.902655	2.902710	2.902764	54
800	2.903090	2.903144	2.903198	2.903253	2.903307	54
801	2.903632	2.903687	2.903741	2.903795	2.903849	54
802	2.904174	2.904228	2.904283	2.904337	2.904391	54
803	2.904715	2.904770	2.904824	2.904878	2.904932	54
804	2.905256	2.905310	2.905364	2.905418	2.905472	54
805	2.905796	2.905850	2.905904	2.905958	2.906012	54
806	2.906335	2.906389	2.906443	2.906497	2.906550	54
807	2.906873	2.906927	2.906981	2.907035	2.907089	54
808	2.907411	2.907465	2.907519	2.907573	2.907626	54
809	2.907948	2.908002	2.908056	2.908109	2.908163	54
810	2.908485	2.908539	2.908592	2.908646	2.908699	54
811	2.909021	2.909074	2.909128	2.909181	2.909235	53
812	2.909556	2.909609	2.909663	2.909716	2.909770	53
813	2.910090	2.910144	2.910197	2.910251	2.910304	53
814	2.910624	2.910678	2.910731	2.910784	2.910838	53
815	2.911158	2.911211	2.911264	2.911317	2.911371	53
816	2.911690	2.911743	2.911797	2.911850	2.911903	53
817	2.912222	2.912275	2.912328	2.912381	2.912435	53
818	2.912753	2.912806	2.912859	2.912912	2.912966	53
819	2.913284	2.913337	2.913390	2.913443	2.913496	53
820	2.913814	2.913867	2.913920	2.913973	2.914026	53
821	2.914343	2.914396	2.914449	2.914502	2.914555	53
822	2.914872	2.914925	2.914977	2.915030	2.915083	53
823	2.915400	2.915453	2.915505	2.915558	2.915611	53
824	2.915927	2.915980	2.916033	2.916085	2.916138	53
825	2.916454	2.916507	2.916559	2.916612	2.916664	53
826	2.916980	2.917033	2.917085	2.917138	2.917190	53
827	2.917505	2.917558	2.917610	2.917663	2.917715	52
828	2.918030	2.918083	2.918135	2.918188	2.918240	52
829	2.918555	2.918607	2.918659	2.918712	2.918765	52
830	2.919078	2.919130	2.919183	2.919236	2.919287	52

Nᵒ	5	6	7	8	9	Dif
788	2.896802	2.896857	2.896911	2.896967	2.897022	55
789	2.897352	2.897407	2.897462	2.897517	2.897572	55
790	2.897902	2.897957	2.898012	2.898067	2.898122	55
791	2.898451	2.898506	2.898561	2.898615	2.898670	55
792	2.898999	2.899054	2.899109	2.899164	2.899218	55
793	2.899547	2.899602	2.899656	2.899711	2.899765	55
794	2.900094	2.900149	2.900205	2.900258	2.900312	55
795	2.900640	2.900695	2.900749	2.900804	2.900858	55
796	2.901186	2.901240	2.901295	2.901349	2.901404	55
797	2.901731	2.901785	2.901840	2.901894	2.901948	54
798	2.902275	2.902329	2.902384	2.902438	2.902492	54
799	2.902818	2.902873	2.902927	2.902981	2.903035	54
800	2.903361	2.903416	2.903470	2.903524	2.903578	54
801	2.903903	2.903958	2.904012	2.904066	2.904120	54
802	2.904445	2.904499	2.904553	2.904607	2.904661	54
803	2.904986	2.905040	2.905094	2.905148	2.905202	54
804	2.905526	2.905580	2.905634	2.905688	2.905742	54
805	2.906065	2.906119	2.906173	2.906227	2.906281	54
806	2.906604	2.906658	2.906712	2.906766	2.906820	54
807	2.907142	2.907196	2.907250	2.907304	2.907358	54
808	2.907680	2.907734	2.907787	2.907841	2.907895	54
809	2.908217	2.908270	2.908324	2.908378	2.908431	54
810	2.908753	2.908807	2.908860	2.908914	2.908967	54
811	2.909288	2.909342	2.909395	2.909449	2.909502	54
812	2.909823	2.909877	2.909930	2.909984	2.910037	53
813	2.910358	2.910411	2.910464	2.910518	2.910571	53
814	2.910891	2.910944	2.910998	2.911051	2.911104	53
815	2.911424	2.911477	2.911530	2.911584	2.911637	53
816	2.911956	2.912009	2.912063	2.912116	2.912169	53
817	2.912488	2.912541	2.912594	2.912647	2.912700	53
818	2.913019	2.913072	2.913125	2.913178	2.913241	53
819	2.913549	2.913602	2.913655	2.913708	2.913761	53
820	2.914079	2.914131	2.914184	2.914237	2.914290	53
821	2.914608	2.914660	2.914713	2.914766	2.914819	53
822	2.915136	2.915189	2.915241	2.915294	2.915347	53
823	2.915664	2.915716	2.915769	2.915822	2.915874	53
824	2.916191	2.916243	2.916296	2.916349	2.916401	53
825	2.916717	2.916770	2.916822	2.916875	2.916927	53
826	2.917243	2.917295	2.917348	2.917400	2.917453	53
827	2.917768	2.917820	2.917873	2.917925	2.917978	53
828	2.918292	2.918345	2.918397	2.918450	2.918502	52
829	2.918816	2.918869	2.918921	2.918973	2.919026	52
830	2.919340	2.919392	2.919444	2.919496	2.919549	52

N°	0	1	2	3	4	Dif
831	2.919601	2.919653	2.919705	2.919758	2.919810	52
832	2.920123	2.920175	2.920228	2.920280	2.920332	52
833	2.920645	2.920697	2.920749	2.920801	2.920853	52
834	2.921165	2.921218	2.921270	2.921322	2.921374	52
835	2.921686	2.921738	2.921790	2.921842	2.921894	52
836	2.922206	2.922258	2.922310	2.922362	2.922414	52
837	2.922725	2.922777	2.922829	2.922881	2.922933	52
838	2.923244	2.923296	2.923348	2.923399	2.923451	52
839	2.923762	2.923814	2.923865	2.923917	2.923969	52
840	2.924279	2.924331	2.924383	2.924434	2.924486	52
841	2.924796	2.924848	2.924899	2.924951	2.925002	52
842	2.925312	2.925364	2.925415	2.925467	2.925518	52
843	2.925828	2.925879	2.925931	2.925982	2.926034	52
844	2.926342	2.926394	2.926445	2.926497	2.926548	51
845	2.926857	2.926908	2.926959	2.927011	2.927062	51
846	2.927370	2.927422	2.927473	2.927524	2.927576	51
847	2.927883	2.927935	2.927986	2.928037	2.928088	51
848	2.928396	2.928447	2.928498	2.928549	2.928601	51
849	2.928908	2.928959	2.929010	2.929061	2.929112	51
850	2.929419	2.929470	2.929521	2.929572	2.929623	51
851	2.929930	2.929981	2.930032	2.930083	2.930134	51
852	2.930440	2.930491	2.930541	2.930592	2.930643	51
853	2.930949	2.931000	2.931051	2.931102	2.931153	51
854	2.931458	2.931509	2.931560	2.931610	2.931661	51
855	2.931966	2.932017	2.932068	2.932118	2.932169	51
856	2.932474	2.932524	2.932575	2.932626	2.932677	51
857	2.932981	2.933031	2.933082	2.933133	2.933183	51
858	2.933487	2.933538	2.933588	2.933639	2.933690	51
859	2.933993	2.934044	2.934094	2.934145	2.934195	51
860	2.934498	2.934549	2.934599	2.934650	2.934700	50
861	2.935003	2.935054	2.935104	2.935154	2.935205	50
862	2.935507	2.935558	2.935608	2.935658	2.935709	50
863	2.936011	2.936061	2.936111	2.936162	2.936212	50
864	2.936514	2.936564	2.936614	2.936664	2.936715	50
865	2.937016	2.937066	2.937116	2.937167	2.937217	50
866	2.937518	2.937568	2.937618	2.937668	2.937718	50
867	2.938019	2.938069	2.938119	2.938169	2.938219	50
868	2.938520	2.938570	2.938620	2.938670	2.938720	50
869	2.939020	2.939070	2.939120	2.939170	2.939220	50
870	2.939519	2.939569	2.939619	2.939669	2.939719	50
871	2.940018	2.940068	2.940118	2.940168	2.940218	50
872	2.940516	2.940566	2.940616	2.940666	2.940716	50
873	2.941014	2.941064	2.941114	2.941163	2.941213	50

N°	5	6	7	8	9	Dif
831	2.919862	2.919914	2.919967	2.920019	2.920071	52
832	2.920384	2.920436	2.920489	2.920541	2.920593	52
833	2.920906	2.920958	2.921010	2.921062	2.921114	52
834	2.921426	2.921478	2.921530	2.921582	2.921634	52
835	2.921946	2.921998	2.922050	2.922102	2.922154	52
836	2.922466	2.922518	2.922570	2.922622	2.922674	52
837	2.922985	2.923037	2.923088	2.923140	2.923192	52
838	2.923503	2.923555	2.923607	2.923658	2.923710	52
839	2.924021	2.924073	2.924124	2.924176	2.924228	52
840	2.924518	2.924589	2.924641	2.924693	2.924744	52
841	2.925054	2.925106	2.925157	2.925209	2.925260	52
842	2.925570	2.925621	2.925673	2.925724	2.925776	52
843	2.926085	2.926137	2.926188	2.926239	2.926291	52
844	2.926600	2.926651	2.926702	2.926754	2.926805	51
845	2.927114	2.927165	2.927216	2.927268	2.927319	51
846	2.927627	2.927678	2.927730	2.927781	2.927832	51
847	2.928140	2.928191	2.928242	2.928293	2.928345	51
848	2.928652	2.928703	2.928754	2.928805	2.928856	51
849	2.929163	2.929214	2.929266	2.929317	2.929368	51
850	2.929674	2.929725	2.929776	2.929827	2.929878	51
851	2.930185	2.930236	2.930287	2.930338	2.930389	51
852	2.930694	2.930745	2.930796	2.930847	2.930898	51
853	2.931203	2.931254	2.931305	2.931356	2.931407	51
854	2.931712	2.931763	2.931814	2.931864	2.931915	51
855	2.932220	2.932271	2.932321	2.932372	2.932423	51
856	2.932727	2.932778	2.932829	2.932879	2.932930	51
857	2.933234	2.933285	2.933335	2.933386	2.933437	51
858	2.933740	2.933791	2.933841	2.933892	2.933943	51
859	2.934246	2.934296	2.934347	2.934397	2.934448	51
860	2.934751	2.934801	2.934852	2.934902	2.934953	50
861	2.935255	2.935306	2.935356	2.935406	2.935457	50
862	2.935759	2.935809	2.935860	2.935910	2.935960	50
863	2.936262	2.936313	2.936363	2.936413	2.936463	50
864	2.936765	2.936815	2.936865	2.936916	2.936966	50
865	2.937267	2.937317	2.937367	2.937418	2.937468	50
866	2.937769	2.937819	2.937869	2.937919	2.937969	50
867	2.938269	2.938319	2.938370	2.938420	2.938470	50
868	2.938770	2.938820	2.938870	2.938920	2.938970	50
869	2.939270	2.939319	2.939369	2.939419	2.939469	50
870	2.939769	2.939819	2.939868	2.939918	2.939968	50
871	2.940267	2.940317	2.940367	2.940417	2.940467	50
872	2.940765	2.940815	2.940865	2.940915	2.940965	50
873	2.941263	2.941313	2.941362	2.941412	2.941462	50

N°	0	1	2	3	4	Dif
874	2.941511	2.941561	2.941611	2.941660	2.941710	50
875	2.942008	2.942058	2.942107	2.942157	2.942206	50
876	2.942504	2.942554	2.942604	2.942653	2.942702	50
877	2.943000	2.943049	2.943099	2.943148	2.943198	49
878	2.943494	2.943544	2.943593	2.943643	2.943692	49
879	2.943989	2.944038	2.944088	2.944137	2.944186	49
880	2.944483	2.944532	2.944581	2.944631	2.944680	49
881	2.944976	2.945025	2.945074	2.945124	2.945173	49
882	2.945469	2.945518	2.945567	2.945616	2.945665	49
883	2.945961	2.946010	2.946059	2.946108	2.946157	49
884	2.946452	2.946501	2.946550	2.946600	2.946649	49
885	2.946943	2.946992	2.947041	2.947090	2.947139	49
886	2.947434	2.947483	2.947532	2.947581	2.947630	49
887	2.947924	2.947973	2.948021	2.948070	2.948119	49
888	2.948413	2.948462	2.948511	2.948560	2.948608	49
889	2.948902	2.948951	2.948999	2.949048	2.949097	49
890	2.949390	2.949439	2.949488	2.949536	2.949585	49
891	2.949878	2.949926	2.949975	2.950024	2.950073	49
892	2.950365	2.950413	2.950462	2.950511	2.950560	49
893	2.950851	2.950900	2.950949	2.950997	2.951046	49
894	2.951337	2.951386	2.951435	2.951483	2.951532	49
895	2.951823	2.951872	2.951920	2.951969	2.952017	49
896	2.952308	2.952356	2.952405	2.952453	2.952502	49
897	2.952792	2.952841	2.952889	2.952938	2.952986	48
898	2.953276	2.953325	2.953373	2.953421	2.953470	48
899	2.953760	2.953808	2.953856	2.953905	2.953953	48
900	2.954242	2.954291	2.954339	2.954387	2.954435	48
901	2.954725	2.954773	2.954821	2.954869	2.954918	48
902	2.955206	2.955255	2.955303	2.955351	2.955399	48
903	2.955688	2.955736	2.955784	2.955832	2.955880	48
904	2.956168	2.956216	2.956264	2.956312	2.956360	48
905	2.956649	2.956697	2.956745	2.956792	2.956840	48
906	2.957128	2.957176	2.957224	2.957272	2.957320	48
907	2.957607	2.957655	2.957703	2.957751	2.957799	48
908	2.958086	2.958134	2.958181	2.958229	2.958277	48
909	2.958564	2.958612	2.958659	2.958707	2.958755	48
910	2.959041	2.959089	2.959137	2.959184	2.959232	48
911	2.959518	2.959566	2.959614	2.959661	2.959709	48
912	2.959995	2.960042	2.960090	2.960138	2.960185	48
913	2.960471	2.960518	2.960566	2.960613	2.960661	48
914	2.960946	2.960994	2.961041	2.961089	2.961136	47
915	2.961421	2.961468	2.961516	2.961563	2.961611	47
916	2.961895	2.961943	2.961990	2.962038	2.962085	47

N°	5	6	7	8	9	Dif
874	2.941760	2.941809	2.941859	2.941909	2.941958	49
875	2.942256	2.942306	2.942355	2.942405	2.942455	49
876	2.942752	2.942801	2.942851	2.942900	2.942950	49
877	2.943247	2.943297	2.943346	2.943396	2.943445	49
878	2.943742	2.943791	2.943841	2.943890	2.943939	49
879	2.944236	2.944285	2.944335	2.944384	2.944433	49
880	2.944729	2.944779	2.944828	2.944877	2.944927	49
881	2.945222	2.945272	2.945321	2.945370	2.945419	49
882	2.945715	2.945764	2.945813	2.945862	2.945911	49
883	2.946207	2.946256	2.946305	2.946354	2.946403	49
884	2.946694	2.946747	2.946796	2.946845	2.946894	49
885	2.947189	2.947238	2.947287	2.947336	2.947385	49
886	2.947679	2.947728	2.947777	2.947826	2.947875	49
887	2.948168	2.948217	2.948266	2.948315	2.948364	49
888	2.948657	2.948706	2.948755	2.948804	2.948853	49
889	2.949146	2.949195	2.949244	2.949292	2.949341	49
890	2.949633	2.949683	2.949731	2.949780	2.949829	49
891	2.950121	2.950170	2.950219	2.950267	2.950316	49
892	2.950608	2.950657	2.950705	2.950754	2.950803	49
893	2.951095	2.951143	2.951192	2.951241	2.951289	49
894	2.951580	2.951629	2.951677	2.951726	2.951774	49
895	2.952066	2.952114	2.952163	2.952211	2.952259	48
896	2.952550	2.952599	2.952647	2.952696	2.952744	48
897	2.953034	2.953083	2.953131	2.953180	2.953228	48
898	2.953518	2.953566	2.953615	2.953663	2.953711	48
899	2.954001	2.954049	2.954098	2.954146	2.954194	48
900	2.954484	2.954532	2.954580	2.954628	2.954677	48
901	2.954966	2.955014	2.955062	2.955110	2.955158	48
902	2.955447	2.955495	2.955543	2.955591	2.955640	48
903	2.955928	2.955976	2.956024	2.956072	2.956120	48
904	2.956409	2.956457	2.956505	2.956553	2.956601	48
905	2.956888	2.956936	2.956984	2.957032	2.957080	48
906	2.957368	2.957416	2.957464	2.957511	2.957559	48
907	2.957847	2.957894	2.957942	2.957990	2.958038	48
908	2.958325	2.958373	2.958420	2.958468	2.958516	48
909	2.958803	2.958850	2.958898	2.958946	2.958994	48
910	2.959280	2.959328	2.959375	2.959423	2.959471	48
911	2.959757	2.959804	2.959852	2.959900	2.959947	48
912	2.960233	2.960281	2.960328	2.960376	2.960423	48
913	2.960709	2.960756	2.960804	2.960851	2.960899	48
914	2.961184	2.961231	2.961279	2.961326	2.961374	47
915	2.961658	2.961706	2.961753	2.961801	2.961848	47
916	2.962132	2.962180	2.962227	2.962275	2.962322	47

A Table of Logarithms,

N°	0	1	2	3	4	Dif
917	2.962307	2.962417	2.962464	2.962511	2.962559	47
918	2.962843	2.962890	2.961937	2.961985	2.963032	47
919	2.963315	2.963363	2.963410	2.963457	2.963504	47
920	2.963788	2.963835	2.963882	2.963929	2.963977	47
921	2.964260	2.964307	2.964354	2.964401	2.964448	47
922	2.964731	2.964778	2.964825	2.964872	2.964919	47
923	2.965202	2.965219	2.965296	2.965343	2.965390	47
924	2.965672	2.965719	2.965766	2.965813	2.965860	47
925	2.966142	2.966189	2.966236	2.966283	2.966329	47
926	2.966611	2.966658	2.966705	2.966752	2.966798	47
927	2.967080	2.967127	2.967173	2.967220	2.967267	47
928	2.967548	2.967595	2.967642	2.967688	2.967735	47
929	2.968016	2.968062	2.968109	2.968156	2.968203	47
930	2.968483	2.968530	2.968576	2.968623	2.968670	47
931	2.968950	2.968996	2.969043	2.969090	2.969136	47
932	2.969416	2.969462	2.969509	2.969556	2.969502	47
933	2.969882	2.969928	2.969975	2.970021	2.970068	47
934	2.970347	2.970393	2.970440	2.970486	2.970533	46
935	2.970812	2.970858	2.970904	2.970951	2.970997	46
936	2.971276	2.971322	2.971369	2.971415	2.971461	46
937	2.971740	2.971786	2.971832	2.971879	2.971925	46
938	2.972203	2.972249	2.972295	2.972342	2.972388	46
939	2.972666	2.972712	2.972758	2.972804	2.972851	46
940	2.973128	2.973174	2.973220	2.973266	2.973313	46
941	2.973590	2.973636	2.973682	2.973728	2.973774	46
942	2.974051	2.974097	2.974143	2.974189	2.974235	46
943	2.974512	2.974558	2.974604	2.974650	2.974696	46
944	2.974972	2.975018	2.975064	2.975110	2.975156	46
945	2.975432	2.975478	2.975524	2.975570	2.975616	45
946	2.975891	2.975937	2.975983	2.976023	2.976075	46
947	2.976350	2.976396	2.976442	2.976487	2.976533	46
948	2.976808	2.976854	2.976900	2.976946	2.976991	46
949	2.977266	2.977312	2.977358	2.977403	2.977449	46
950	2.977724	2.977769	2.977815	2.977861	2.977906	46
951	2.978180	2.978226	2.978272	2.978317	2.978363	46
952	2.978637	2.978683	2.978728	2.978774	2.978819	46
953	2.979093	2.979138	2.979184	2.979230	2.979275	46
954	2.979548	2.979594	2.979639	2.979685	2.979730	46
955	2.980003	2.980049	2.980094	2.980140	2.980185	45
956	2.980458	2.980503	2.980549	2.980594	2.980640	45
957	2.980912	2.980957	2.981003	2.981048	2.981093	45
958	2.981365	2.981411	2.981456	2.981501	2.981547	45
959	2.981819	2.981864	2.981909	2.981954	2.982000	45

N°	5	6	7	8	9	Dif
917	2.962606	2.962653	2.962701	2.962748	2.962795	47
918	2.963079	2.963126	2.963174	2.963221	2.963268	47
919	2.963552	2.963599	2.963646	2.963693	2.963741	47
920	2.964024	2.964071	2.964118	2.964165	2.964212	47
921	2.964475	2.964542	2.964590	2.964637	2.964684	47
922	2.964966	2.965013	2.965060	2.965108	2.965155	47
923	2.965437	2.965484	2.965531	2.965578	2.965625	47
924	2.965907	2.965954	2.966001	2.966048	2.966095	47
925	2.966376	2.966423	2.966470	2.966517	2.966564	47
926	2.966845	2.966892	2.966939	2.966986	2.967011	47
927	2.967314	2.967361	2.967408	2.967454	2.967501	47
928	2.967782	2.967829	2.967875	2.967922	2.967969	47
929	2.968249	2.968296	2.968343	2.968389	2.968436	47
930	2.968716	2.968763	2.968810	2.968856	2.968903	47
931	2.969183	2.969229	2.969276	2.969323	2.969369	47
932	2.969649	2.969695	2.969742	2.969788	2.969835	47
933	2.970114	2.970161	2.970207	2.970254	2.970300	47
934	2.970579	2.970626	2.970672	2.970719	2.970765	47
935	2.971044	2.971090	2.971137	2.971183	2.971229	46
936	2.971508	2.971554	2.971600	2.971647	2.971693	46
937	2.971971	2.972018	2.972064	2.972110	2.972156	45
938	2.972434	2.972480	2.972527	2.972573	2.972619	46
939	2.972897	2.972943	2.972989	2.973035	2.973082	46
940	2.973359	2.973405	2.973451	2.973497	2.973543	46
941	2.973820	2.973856	2.973913	2.973959	2.974005	46
942	2.974281	2.974327	2.974373	2.974420	2.974466	46
943	2.974742	2.974788	2.974834	2.974880	2.974926	46
944	2.975202	2.975248	2.975294	2.975340	2.975386	46
945	2.975661	2.975707	2.975753	2.975799	2.975845	46
946	2.976121	2.976166	2.976212	2.976258	2.976304	46
947	2.976579	2.976625	2.976671	2.976717	2.976762	46
948	2.977037	2.977083	2.977129	2.977175	2.977220	46
949	2.977495	2.977541	2.977586	2.977632	2.977678	46
950	2.977952	2.977998	2.978043	2.978089	2.978135	46
951	2.978409	2.978454	2.978500	2.978546	2.978591	46
952	2.978865	2.978911	2.978956	2.979002	2.979047	45
953	2.979321	2.979366	2.979412	2.979457	2.979503	46
954	2.979776	2.979821	2.979867	2.979912	2.979958	45
955	2.980231	2.980276	2.980322	2.980367	2.980412	45
956	2.980685	2.980730	2.980776	2.980821	2.980867	45
957	2.981139	2.981184	2.981229	2.981275	2.981320	45
958	2.981592	2.981637	2.981683	2.981728	2.981773	45
959	2.982045	2.982066	2.982135	2.982181	2.982226	45

N°	0	1	2	3	4	Dif
960	2.982271	2.982316	2.982362	2.982407	2.982452	45
961	2.982723	2.982769	2.982814	2.982859	2.982904	45
962	2.983175	2.983220	2.983265	2.983310	2.983356	45
963	2.983626	2.983671	2.983716	2.983762	2.983807	45
964	2.984077	2.984122	2.984167	2.984212	2.984257	45
965	2.984527	2.984572	2.984617	2.984662	2.984707	45
966	2.984977	2.985022	2.985067	2.985112	2.985157	45
967	2.985426	2.985471	2.985516	2.985561	2.985606	45
968	2.985875	2.985920	2.985965	2.986010	2.986055	45
969	2.986324	2.986369	2.986413	2.986458	2.986503	45
970	2.986772	2.986816	2.986861	2.986906	2.986951	45
971	2.987219	2.987264	2.987309	2.987353	2.987398	45
972	2.987666	2.987711	2.987756	2.987800	2.987845	45
973	2.988113	2.988157	2.988202	2.988247	2.988291	45
974	2.988559	2.988603	2.988648	2.988693	2.988737	45
975	2.989005	2.989049	2.989094	2.989138	2.989183	45
976	2.989450	2.989494	2.989539	2.989583	2.989628	44
977	2.989895	2.989939	2.989983	2.990029	2.990072	44
978	2.990339	2.990383	2.990428	2.990472	2.990516	44
979	2.990783	2.990827	2.990871	2.990916	2.990960	44
980	2.991226	2.991270	2.991315	2.991359	2.991403	44
981	2.991669	2.991713	2.991757	2.991802	2.991846	44
982	2.992111	2.992156	2.992200	2.992244	2.992289	44
983	2.992553	2.992598	2.992642	2.992686	2.992730	44
984	2.992995	2.993039	2.993083	2.993127	2.993172	44
985	2.993436	2.993480	2.993524	2.993568	2.993613	44
986	2.993877	2.993921	2.993965	2.994009	2.994053	44
987	2.994317	2.994361	2.994405	2.994449	2.994493	44
988	2.994757	2.994801	2.994845	2.994889	2.994933	44
989	2.995196	2.995240	2.995284	2.995328	2.995372	44
990	2.995635	2.995679	2.995723	2.995767	2.995811	44
991	2.996074	2.996117	2.996161	2.996205	2.996249	44
992	2.996512	2.996555	2.996599	2.996643	2.996687	44
993	2.996949	2.996993	2.997037	2.997080	2.997124	44
994	2.997386	2.997430	2.997474	2.997517	2.997561	44
995	2.997823	2.997867	2.997910	2.997954	2.997998	44
996	2.998259	2.998303	2.998346	2.998390	2.998434	44
997	2.998695	2.998739	2.998782	2.998826	2.998869	44
998	2.999130	2.999174	2.999218	2.999261	2.999305	44
999	2.999565	2.999609	2.999652	2.999696	2.999739	43

Nᵉ	5	6	7	8	9	Dif
960	2.982497	2.982543	2.982588	2.982633	2.982678	45
961	2.982919	2.982994	2.983040	2.983085	2.983130	45
962	2.983401	2.983446	2.983491	2.983536	2.983581	45
963	2.983852	2.983897	2.983942	2.983987	2.984032	45
964	2.984302	2.984347	2.984392	2.984437	2.984482	45
965	2.984752	2.984797	2.984842	2.984887	2.984932	45
966	2.985202	2.985247	2.985292	2.985337	2.985382	45
967	2.985651	2.985696	2.985741	2.985786	2.985830	45
968	2.986100	2.986144	2.986189	2.986234	2.986279	45
969	2.986548	2.986593	2.986637	2.986682	2.986727	45
970	2.986995	2.987040	2.987085	2.987130	2.987174	45
971	2.987443	2.987487	2.987532	2.987577	2.987621	45
972	2.987890	2.987934	2.987979	2.988024	2.988068	45
973	2.988336	2.988381	2.988425	2.988470	2.988514	45
974	2.988782	2.988826	2.988871	2.988915	2.988960	45
975	2.989227	2.989272	2.989316	2.989361	2.989405	45
976	2.989672	2.989717	2.989761	2.989806	2.989850	44
977	2.990117	2.990161	2.990206	2.990250	2.990294	44
978	2.990561	2.990605	2.990650	2.990694	2.990738	44
979	2.991004	2.991049	2.991093	2.991137	2.991182	44
980	2.991448	2.991492	2.991536	2.991580	2.991625	44
981	2.991890	2.991934	2.991979	2.992023	2.992067	44
982	2.992333	2.992377	2.992421	2.992465	2.992509	44
983	2.992774	2.992818	2.992863	2.992907	2.992951	44
984	2.993216	2.993260	2.993304	2.993348	2.993392	44
985	2.993657	2.993701	2.993745	2.993789	2.993833	44
986	2.994097	2.994141	2.994185	2.994229	2.994273	44
987	2.994537	2.994581	2.994625	2.994669	2.994713	44
988	2.994977	2.995021	2.995064	2.995108	2.995152	44
989	2.995416	2.995460	2.995504	2.995547	2.995591	44
990	2.995854	2.995898	2.995942	2.995985	2.996030	44
991	2.996291	2.996336	2.996380	2.996424	2.996468	44
992	2.996730	2.996774	2.996818	2.996862	2.996905	44
993	2.997168	2.997212	2.997255	2.997299	2.997343	44
994	2.997605	2.997648	2.997692	2.997736	2.997779	44
995	2.998041	2.998085	2.998128	2.998172	2.998216	44
996	2.998477	2.998521	2.998564	2.998608	2.998652	44
997	2.998913	2.998956	2.999000	2.999043	2.999087	44
998	2.999348	2.999392	2.999435	2.999478	2.999522	44
999	2.999783	2.999826	2.999870	2.999913	2.999957	43

0 Degree.

Minutes	Sine		Tangent	Secant			
0	0.000000	10.000000	0.000000	Infinite	10.000000	Infinite	60
1	6.463726	9.999999	6.463726	13.536274	10.000000	13.536274	59
2	6.764756	9.999999	6.764756	13.235244	10.000000	13.235244	58
3	6.940847	9.999999	6.940847	13.059153	10.000000	13.059153	57
4	7.065786	9.999999	7.065786	12.934214	10.000000	13.934214	56
5	7.162696	9.999999	7.162696	12.837304	10.000000	12.837304	55
6	7.241877	9.999999	7.241878	12.758122	10.000001	12.758123	54
7	7.308824	9.999999	7.308825	12.691175	10.000001	12.691176	53
8	7.366816	9.999999	7.366817	12.633183	10.000001	12.633184	52
9	7.417968	9.999999	7.417970	12.582010	10.000001	12.582032	51
10	7.563725	9.999998	7.463727	12.536273	10.000001	12.536275	50
11	7.505118	9.999998	7.505120	12.494880	10.000002	12.494882	49
12	7.542907	9.999997	7.542909	12.457091	10.000003	12.457093	48
13	7.577668	9.999997	7.577671	12.422328	10.000003	12.422332	47
14	7.609853	9.999896	7.609857	12.390143	10.000004	12.390147	46
15	7.639816	9.999996	7.639820	12.360180	10.000004	12.360184	45
16	7.667844	9.999995	7.667849	12.332151	10.000005	12.332156	44
17	7.694173	9.999995	7.694179	12.305821	10.000005	12.305827	43
18	7.718997	9.999994	7.719003	12.280997	10.000006	12.281003	42
19	7.742478	9.999993	7.742484	12.257516	10.000007	12.257521	41
20	7.764754	9.999993	7.764761	12.235239	10.000007	12.235246	40
21	7.785943	9.999992	7.785951	12.214049	10.000008	12.214057	39
22	7.806146	9.999991	7.806155	12.193845	10.000009	12.193854	38
23	7.825451	9.999990	7.825460	12.174540	10.000010	12.174549	37
24	7.843914	9.999989	7.843944	12.156066	10.000011	12.156066	36
25	7.861662	9.999988	7.861674	12.138326	10.000012	12.138338	35
26	7.878695	9.999988	7.878708	12.121292	10.000012	12.121305	34
27	7.895085	9.999987	7.895099	12.104901	10.000013	12.104915	33
28	7.910879	9.999986	7.910894	12.089106	10.000014	12.089121	32
29	7.926119	9.999984	7.926134	12.073866	10.000016	12.073881	31
30	7.940842	9.999983	7.940858	12.059142	10.000017	12.059158	30
	Sin		Tangent	Secant			Minutes

89 Degrees.

o Degrees.

Minutes	Sine		Tangent		Secant		Minutes
30	7.940843	9.999984	7.940858	12.059142	10.000016	12.059158	30
31	7.955048	9.999982	7.955100	12.044900	10.000018	12.044918	29
32	7.968870	9.999981	7.968889	12.031111	10.000019	12.031130	28
33	7.982233	9.999980	7.982253	12.017747	10.000020	12.017767	27
34	7.995198	9.999979	7.995219	12.004781	10.000021	12.004802	26
35	8.007787	9.999977	8.007809	11.992191	10.000023	11.992213	25
36	8.020021	9.999976	8.020045	11.979955	10.000024	11.979979	24
37	8.031919	9.999975	8.031945	11.968055	10.000025	11.968081	23
38	8.043501	9.999974	8.043527	11.956473	10.000026	11.956499	22
39	8.054781	9.999972	8.054809	11.945191	10.000028	11.945219	21
40	8.065776	9.999971	8.065806	11.934194	10.000029	11.934224	20
41	8.076500	9.999969	8.076531	11.923469	10.000031	11.923500	19
42	8.086965	9.999968	8.086997	11.913003	10.000032	11.913035	18
43	8.097183	9.999966	8.097217	11.902783	10.000034	11.902817	17
44	8.107167	9.999964	8.107202	11.892798	10.000036	11.892833	16
45	8.116926	9.999963	8.116963	11.883037	10.000037	11.883074	15
46	8.126471	9.999961	8.126510	11.873490	10.000039	11.873529	14
47	8.135810	9.999959	8.135851	11.864149	10.000041	11.864190	13
48	8.144953	9.999958	8.144995	11.855004	10.000042	11.855047	12
49	8.153925	9.999956	8.153952	11.846048	10.000044	11.846092	11
50	8.162681	9.999954	8.162727	11.837273	10.000046	11.837319	10
51	8.171280	9.999952	8.171328	11.828672	10.000048	11.828720	9
52	8.179713	9.999950	8.179763	11.820237	10.000050	11.820287	8
53	8.187985	9.999948	8.188036	11.811964	10.000052	11.812016	7
54	8.196102	9.999946	8.196156	11.803844	10.000054	11.803898	6
55	8.204070	9.999944	8.204126	11.795874	10.000056	11.795930	5
56	8.211895	9.999942	8.211953	11.788047	10.000058	11.788105	4
57	8.219581	9.999940	8.219641	11.780359	10.000060	11.780419	3
58	8.227134	9.999938	8.227195	11.772805	10.000062	11.772866	2
59	8.234557	9.999936	8.234621	11.765379	10.000064	11.765443	1
60	8.241855	9.999934	8.241922	11.758078	10.000066	11.758145	0
	Sine		Tangent		Secant		Minutes

89 Degrees.

1 Degree.

Minutes	Sine		Tangent		Secant		Minutes
0	8.241855	9.999934	8.241921	11.758079	10.000066	11.758145	60
1	8.249033	9.999932	8.249102	11.750898	10.000068	11.750967	59
2	8.256094	9.999929	8.256165	11.743835	10.000071	11.743906	58
3	8.263042	9.999927	8.263115	11.736885	10.000073	11.736958	57
4	8.269881	9.999925	8.266956	11.730044	10.000075	11.730119	56
5	8.276614	9.999922	8.276691	11.723309	10.000078	11.723386	55
6	8.283243	9.999920	8.283323	11.716677	10.000080	11.716757	54
7	8.289773	9.999917	8.289856	11.710144	10.000082	11.710227	53
8	8.296207	9.999915	8.296292	11.703708	10.000085	11.703793	52
9	8.302546	9.999912	8.302633	11.697367	10.000087	11.697454	51
10	8.308794	9.999910	8.308884	11.691116	10.000090	11.691200	50
11	8.314954	9.999907	8.315045	11.684954	10.000093	11.685046	49
12	8.321027	9.999905	8.321122	11.678878	10.000095	11.678973	48
13	8.327016	9.999902	8.327114	11.672886	10.000098	11.672984	47
14	8.332924	9.999900	8.333025	11.666975	10.000101	11.667076	46
15	8.338753	9.999897	8.338856	11.661144	10.000103	11.661247	45
16	8.344505	9.999894	8.344610	11.655389	10.000106	11.655496	44
17	8.350184	9.999891	8.350289	11.649710	10.000109	11.649814	43
18	8.355783	9.999888	8.355895	11.644105	10.000112	11.644216	42
19	8.361315	9.999885	8.361430	11.638570	10.000115	11.638685	41
20	8.366777	9.999882	8.366894	11.633105	10.000118	11.633223	40
21	8.372171	9.999879	8.372291	11.627708	10.000121	11.627829	39
22	8.377496	9.999876	8.377622	11.622378	10.000124	11.622501	38
23	8.382763	9.999873	8.382889	11.617111	10.000127	11.617238	37
24	8.387962	9.999870	8.388092	11.611908	10.000130	11.612038	36
25	8.393101	9.999867	8.393234	11.606766	10.000133	11.606899	35
26	8.398179	9.999864	8.398315	11.601685	10.000136	11.601821	34
27	8.403199	9.999861	8.403338	11.596662	10.000139	11.596801	33
28	8.408161	9.999858	8.408304	11.591696	10.000142	11.591839	32
29	8.413068	9.999854	8.413213	11.586787	10.000146	11.586932	31
30	8.417919	9.999851	8.418068	11.581932	10.000149	11.582081	30
	Sin		Tangent		Secant		Minutes

88 Degrees.

1 Degree.

Minutes	Sine		Tangent		Secant		Minutes
30	8.417919	9.999851	8.418006	11.581932	10.000149	11.582081	30
31	8.422717	9.999848	8.422869	11.577131	10.000152	11.577283	29
32	8.427462	9.999844	8.427618	11.572382	10.000156	11.572538	28
33	8.432156	9.999841	8.432315	11.567685	10.000159	11.567844	27
34	8.436800	9.999838	8.436962	11.563038	10.000162	11.563200	26
35	8.441394	9.999834	8.441560	11.558440	10.000166	11.558606	25
36	8.445941	9.999831	8.446110	11.553890	10.000169	11.554059	24
37	8.450440	9.999827	8.450613	11.549387	10.000173	11.549560	23
38	8.454893	9.999823	8.455070	11.544930	10.000177	11.545107	22
39	8.459301	9.999820	8.459481	11.540519	10.000180	11.540699	21
40	8.463665	9.999816	8.463849	11.536151	10.000184	11.536335	20
41	8.467985	9.999812	8.468172	11.531827	10.000188	11.532015	19
42	8.472263	9.999809	8.472454	11.527546	10.000191	11.527737	18
43	8.476498	9.999805	8.476691	11.523307	10.000195	11.523502	17
44	8.480693	9.999801	8.480892	11.519108	10.000199	11.519307	16
45	8.484848	9.999797	8.485050	11.514949	10.000203	11.515152	15
46	8.488963	9.999793	8.489170	11.510830	10.000207	11.511037	14
47	8.493040	9.999790	8.493250	11.506750	10.000210	11.506960	13
48	8.497078	9.999786	8.497293	11.502707	10.000214	11.501922	12
49	8.501080	9.999782	8.501298	11.498702	10.000218	11.498920	11
50	8.505045	9.999778	8.505267	11.494733	10.000222	11.494955	10
51	8.508974	9.999774	8.509200	11.490800	10.000226	11.491026	9
52	8.512867	9.999769	8.513098	11.486902	10.000231	11.487133	8
53	8.516726	9.999765	8.516961	11.483039	10.000235	11.483274	7
54	8.520551	9.999761	8.520790	11.479210	10.000239	11.479449	6
55	8.524343	9.999757	8.524586	11.475414	10.000243	11.475657	5
56	8.528102	9.999753	8.528349	11.471651	10.000247	11.471898	4
57	8.531828	9.999748	8.532080	11.467920	10.000252	11.468172	3
58	8.535523	9.999744	8.535779	11.464221	10.000256	11.464477	2
59	8.539186	9.999740	8.539447	11.460553	10.000260	11.460814	1
60	8.542819	9.999735	8.543084	11.456916	10.000265	11.457181	0
	Sine		Tangent		Secant		Minutes

88 Degrees.

2 Degrees.

Minutes	Sine		Tangent		Secant		Minutes
0	8.542819	9.999735	8.543084	11.456916	10.000265	11.457181	60
1	8.546422	9.999731	8.546691	11.453309	10.000269	11.453578	59
2	8.549995	9.999727	8.550268	11.449732	10.000273	11.450005	58
3	8.553539	9.999722	8.553817	11.446183	10.000278	11.446461	57
4	8.557054	9.999717	8.557336	11.442664	10.000283	11.442946	56
5	8.560541	9.999713	8.560828	11.439172	10.000287	11.439460	55
6	8.563999	9.999708	8.564291	11.435709	10.000292	11.436001	54
7	8.567431	9.999704	8.567727	11.432273	10.000296	11.432569	53
8	8.570836	9.999699	8.571137	11.428863	10.000301	11.429164	52
9	8.574214	9.999694	8.574520	11.425480	10.000306	11.425786	51
10	8.577566	9.999689	8.577877	11.422123	10.000311	11.422434	50
11	8.580892	9.999685	8.581208	11.418792	10.000315	11.419108	49
12	8.584193	9.999680	8.584514	11.415486	10.000320	11.415807	48
13	8.587469	9.999675	8.587795	11.412205	10.000325	11.412531	47
14	8.590721	9.999670	8.591051	11.408949	10.000330	11.409279	46
15	8.593948	9.999665	8.594283	11.405717	10.000335	11.406052	45
16	8.597152	9.999660	8.597492	11.402508	10.000340	11.402848	44
17	8.600332	9.999655	8.600677	11.399323	10.000345	11.399668	43
18	8.603489	9.999650	8.603839	11.396161	10.000350	11.396511	42
19	8.606623	9.999645	8.606978	11.393022	10.000355	11.393377	41
20	8.609734	9.999640	8.610094	11.389906	10.000360	11.390266	40
21	8.612823	9.999635	8.613189	11.386811	10.000365	11.387177	39
22	8.615891	9.999629	8.616262	11.383738	10.000371	11.384109	38
23	8.618937	9.999624	8.619313	11.380687	10.000376	11.381063	37
24	8.621962	9.999619	8.622343	11.377657	10.000381	11.378038	36
25	8.624965	9.999614	8.625352	11.374648	10.000386	11.375035	35
26	8.627948	9.999608	8.628340	11.371660	10.000392	11.372052	34
27	8.630911	9.999603	8.631308	11.368692	10.000397	11.369089	33
28	8.633854	9.999597	8.634256	11.365744	10.000304	11.366146	32
29	8.636776	9.999592	8.637185	11.362815	10.000308	11.363224	31
30	8.639680	9.999587	8.640093	11.359907	10.000313	11.360310	30

| | | Secant | | Tangent | | Secant | Minutes |

87 Degrees.

2 Degrees.

Minutes	Secant		Tangent		Secant		Minutes
30	8.639680	9.999580	8.640093	11.359907	10.000414	11.300310	30
31	8.641563	9.999581	8.641981	11.357018	10.000419	11.357417	29
32	8.645428	9.999575	8.645853	11.354147	10.000425	11.354572	28
33	8.648274	9.999570	8.648704	11.351296	10.000430	11.351726	27
34	8.651102	9.999564	8.651538	11.348462	10.000436	11.348898	26
35	8.653911	9.999558	8.654352	11.345648	10.000442	11.340089	25
36	8.656702	9.999553	8.657149	11.342851	10.000447	11.343298	24
37	8.659475	9.999547	8.659928	11.340072	10.000453	11.340525	23
38	8.662230	9.999541	8.662689	11.337311	10.000459	11.337770	22
39	8.664968	9.999535	8.665433	11.334567	10.000465	11.335032	21
40	8.667689	9.999529	8.668160	11.331840	10.000471	11.332311	20
41	8.670393	9.999523	8.670870	11.329130	10.000476	11.329607	19
42	8.673080	9.999518	8.673563	11.326437	10.000482	11.326920	18
43	8.675751	9.999512	8.676239	11.323761	10.000488	11.324249	17
44	8.678405	9.999506	8.678900	11.321100	10.000494	11.321595	16
45	8.681043	9.999500	8.681544	11.318456	10.000500	11.318957	15
46	8.683665	9.999494	8.684172	11.315828	10.000506	11.316335	14
47	8.686272	9.999487	8.686784	11.313216	10.000513	11.313728	13
48	8.688863	9.999481	8.689381	11.310618	10.000519	11.311137	12
49	8.691438	9.999475	8.691963	11.308037	10.000525	11.308562	11
50	8.693998	9.999469	8.694529	11.405471	10.000531	11.306003	10
51	8.696543	9.999462	8.697081	11.302919	10.000538	11.303457	9
52	8.699073	9.999456	8.699617	11.300383	10.000544	11.300927	8
53	8.701589	9.999450	8.702139	11.297861	10.000550	11.298411	7
54	8.704090	9.999444	8.704646	11.295354	10.000556	11.295910	6
55	8.706577	9.999437	8.707139	11.292861	10.000563	11.293423	5
56	8.709049	9.999431	8.709619	11.290381	10.000569	11.290951	4
57	8.711508	9.999424	8.712083	11.287917	10.000576	11.288492	3
58	8.713952	9.999418	8.714535	11.285465	10.000582	11.286048	2
59	8.716383	9.999411	8.716972	11.283028	10.000589	11.283617	1
60	8.718800	9.999404	8.719396	11.280604	10.000596	11.281200	0
		Sine		Tangent		Secant	Minutes

87 Degrees.

A Table of Artificial Sines,

3 Degrees.

Minutes	Sine		Tangent		Secant		
0	8.718800	9.999404	8.719396	11.280604	10.000596	11.281400	60
1	8.721204	9.999398	8.721806	11.278194	10.000602	11.278796	59
2	8.723595	9.999391	8.724203	11.275797	10.000609	11.276405	58
3	8.725972	9.999384	8.726588	11.273412	10.000616	11.274028	57
4	8.728337	9.999378	8.728959	11.271041	10.000622	11.271663	56
5	8.730688	9.999371	8.731317	11.268683	10.000629	11.269312	55
6	8.733027	9.999364	8.733663	11.266337	10.000636	11.266973	54
7	8.735353	9.999357	8.735996	11.264004	10.000643	11.264647	53
8	8.737668	9.999350	8.738317	11.261683	10.000650	11.262332	52
9	8.739969	9.999343	8.740626	11.259374	10.000657	11.260031	51
10	8.742259	9.999336	8.742922	11.257078	10.000663	11.257741	50
11	8.744536	9.999329	8.745207	11.254793	10.000671	11.255464	49
12	8.746802	9.999322	8.747479	11.252521	10.000678	11.253198	48
13	8.749055	9.999315	8.749740	11.250260	10.000685	11.250945	47
14	8.751297	9.999308	8.751989	11.248011	10.000692	11.248703	46
15	8.753528	9.999301	8.754227	11.245773	10.000699	11.246472	45
16	8.755747	9.999294	8.756453	11.243547	10.000706	11.244253	44
17	8.757955	9.999286	8.758668	11.241332	10.000714	11.242045	43
18	8.760151	9.999279	8.760872	11.239128	10.000721	11.239849	42
19	8.762337	9.999272	8.763066	11.236935	10.000728	11.237663	41
20	8.764511	9.999265	8.765246	11.234754	10.000735	11.235489	40
21	8.766675	9.999257	8.767418	11.232582	10.000743	11.233325	39
22	8.768827	9.999250	8.769578	11.230422	10.000750	11.231173	38
23	8.770970	9.999242	8.771727	11.228273	10.000758	11.229030	37
24	8.773101	9.999235	8.773866	11.226134	10.000765	11.226899	36
25	8.775223	9.999227	8.775995	11.224005	10.000773	11.224777	35
26	8.777333	9.999220	8.778114	11.221886	10.000780	11.222667	34
27	8.779434	9.999212	8.780222	11.219778	10.000788	11.220566	33
28	8.781524	9.999205	8.782320	11.217680	10.000795	11.218476	32
29	8.783605	9.999197	8.784408	11.215592	10.000803	11.216395	31
30	8.785675	9.999189	8.786486	11.213514	10.000811	11.214325	30
	Secant		Tangent		Secant		Minutes

83 Degrees.

3 Degrees.

Minutes	Sine		Tangent		Secant		Minutes
30	8.785675	9.999184	8.786486	11.213514	10.000811	11.214325	30
31	8.787736	9.999181	8.788554	11.211446	10.000819	11.212264	29
32	8.789787	9.999174	8.790613	11.209187	10.000826	11.210213	28
33	8.791828	9.999166	8.792662	11.207338	10.000834	11.208172	27
34	8.793859	9.999158	8.794701	11.205299	10.000842	11.206141	26
35	8.795881	9.999150	8.796731	11.203269	10.000850	11.204119	25
36	8.797894	9.999142	8.798752	11.201248	10.000858	11.202106	24
37	8.799897	9.999134	8.800763	11.199237	10.000866	11.200103	23
38	8.801892	9.999126	8.802765	11.197235	10.000874	11.198108	22
39	8.803876	9.999118	8.804758	11.195242	10.000882	11.196124	21
40	8.805852	9.999110	8.806742	11.193258	10.000890	11.194148	20
41	8.807819	9.999102	8.808717	11.191283	10.000898	11.192181	19
42	8.809777	9.999094	8.810683	11.189317	10.000906	11.190223	18
43	8.811726	9.999086	8.812641	11.187359	10.000914	11.188274	17
44	8.813667	9.999077	8.814589	11.185411	10.000923	11.186333	16
45	8.815598	9.999069	8.816529	11.183471	10.000931	11.184402	15
46	8.817522	9.999061	8.818461	11.181539	10.000939	11.182478	14
47	8.819436	9.999052	8.820384	11.179616	10.000948	11.180564	13
48	8.821343	9.999044	8.822298	11.177702	10.000956	11.178657	12
49	8.823240	9.999036	8.824205	11.175795	10.000964	11.176760	11
50	8.825130	9.999027	8.826103	11.173897	10.000973	11.174870	10
51	8.827011	9.999019	8.827992	11.172008	10.000981	11.172989	9
52	8.828884	9.999010	8.829874	11.170126	10.000990	11.171116	8
53	8.830749	9.999002	8.831748	11.168252	10.000998	11.169251	7
54	8.832607	9.998993	8.833613	11.166387	10.001007	11.167393	6
55	8.834456	9.998984	8.835471	11.164529	10.001016	11.165544	5
56	8.836297	9.998976	8.837321	11.162679	10.001024	11.163703	4
57	8.838130	9.998967	8.839163	11.160837	10.001033	11.161870	3
58	8.839956	9.998958	8.840998	11.159002	10.001042	11.160044	2
59	8.841774	9.998950	8.842825	11.157175	10.001050	11.158226	1
60	8.843584	9.998941	8.844644	11.155356	10.001059	11.156416	0
	Sine		Tangent		Secant		Minutes

86 Degrees.

Tangents and Secants.

4 Degrees.

Minutes	Sine		Tangent		Secant		
0	8.843584	9.998941	8.844644	11.155356	10.001059	11.150415	60
1	8.845387	9.998932	8.846455	11.153545	10.001068	11.154613	59
2	8.847183	9.998923	8.848260	11.151740	10.001077	11.152817	58
3	8.848971	9.998914	8.850057	11.149943	10.001086	11.151029	57
4	8.850751	9.998905	8.851846	11.148154	10.001095	11.149249	56
5	8.852524	9.998896	8.853628	11.146372	10.001104	11.147476	55
6	8.854291	9.998887	8.855403	11.144597	10.001113	11.145709	54
7	8.856049	9.998878	8.857171	11.142829	10.001121	11.143951	53
8	8.857801	9.998869	8.858932	11.141068	10.001131	11.142199	52
9	8.859546	9.998860	8.860686	11.139314	10.001140	11.140454	51
10	8.861283	9.998851	8.862433	11.137567	10.001149	11.138717	50
11	8.863014	9.998841	8.864173	11.135827	10.001159	11.136986	49
12	8.864738	9.998832	8.865905	11.134095	10.001168	11.135262	48
13	8.866455	9.998823	8.867632	11.132368	10.001177	11.133545	47
14	8.868165	9.998813	8.869351	11.130649	10.001187	11.131835	46
15	8.869868	9.998804	8.871064	11.128936	10.001196	11.130132	45
16	8.871565	9.998795	8.872770	11.127230	10.001205	11.128435	44
17	8.873255	9.998785	8.874469	11.125531	10.001215	11.126745	43
18	8.874938	9.998776	8.876162	11.123838	10.001224	11.125062	42
19	8.876615	9.998766	8.877849	11.122151	10.001234	11.123385	41
20	8.878285	9.998757	8.879529	11.120471	10.001243	11.121715	40
21	8.879945	9.998747	8.881202	11.118798	10.001253	11.120055	39
22	8.881607	9.998737	8.882869	11.117131	10.001263	11.118393	38
23	8.883258	9.998728	8.884530	11.115470	10.001272	11.116741	37
24	8.884903	9.998718	8.886185	11.113815	10.001282	11.115097	36
25	8.886542	9.998708	8.887833	11.112167	10.001292	11.113458	35
26	8.888174	9.998699	8.889476	11.110524	10.001301	11.111816	34
27	8.889801	9.998689	8.891112	11.108888	10.001311	11.110199	33
28	8.891421	9.998679	8.892742	11.107258	10.001321	11.108579	32
29	8.893035	9.998669	8.894366	11.105634	10.001331	11.106965	31
30	8.894643	9.998659	8.895984	11.104016	10.001341	11.105357	30
		Sine		Tangent		Secant	Minutes

85 Degrees.

4 Degrees.

Minutes	Sine		Tangent		Secant		Minutes
30	8.894643	9.998659	8.895984	11.104016	10.001341	11.105357	30
31	8.896245	9.998649	8.897596	11.102404	10.001351	11.103754	29
32	8.897842	9.998639	8.899203	11.100797	10.001361	11.102158	28
33	8.899432	9.998629	8.900803	11.099197	10.001371	11.100568	27
34	8.901017	9.998619	8.902398	11.097602	10.001381	11.098983	26
35	8.902596	9.998609	8.903987	11.096013	10.001391	11.097404	25
36	8.904169	9.998599	8.905570	11.094430	10.001401	11.095831	24
37	8.905736	9.998589	8.907147	11.092853	10.001411	11.094264	23
38	8.907297	9.998578	8.908719	11.091281	10.001422	11.092703	22
39	8.908854	9.998568	8.910285	11.089715	10.001432	11.091146	21
40	8.910404	9.998558	8.911846	11.088154	10.001442	11.089596	20
41	8.911949	9.998548	8.913401	11.086599	10.001452	11.088051	19
42	8.913488	9.998537	8.914951	11.085049	10.001463	11.086512	18
43	8.915022	9.998527	8.916495	11.083505	10.001473	11.084978	17
44	8.916550	9.998516	8.918034	11.081966	10.001484	11.083450	16
45	8.918073	9.998506	8.919567	11.080433	10.001494	11.081927	15
46	8.919591	9.998495	8.921096	11.078904	10.001505	11.080409	14
47	8.911103	9.998485	8.922619	11.077181	10.001515	11.078897	13
48	8.922611	9.998474	8.924136	11.075864	10.001526	11.077389	12
49	8.924113	9.998464	8.925649	11.074351	10.001536	12.075888	11
50	8.925609	9.998453	8.927156	11.072844	10.001547	11.074391	10
51	8.927100	9.998442	8.928658	11.071342	10.001558	11.072900	9
52	8.928587	9.998431	8.930155	11.069845	10.001569	11.071413	8
53	8.930068	9.998421	8.931647	11.068353	10.001579	11.069932	7
54	8.931544	9.998410	8.933134	11.066866	10.001590	11.068426	6
55	8.933015	9.998399	8.934616	11.065384	10.001601	11.066985	5
56	8.934481	9.998388	8.936093	11.063907	10.001612	11.065519	4
57	8.935942	9.998377	8.937565	11.062435	10.001623	11.064058	3
58	8.937398	9.998366	8.939032	11.060968	10.001634	11.062602	2
59	8.938850	9.998355	8.940494	11.059506	10.001645	11.061150	1
60	8.940296	9.998344	8.941952	11.058048	10.001656	11.059704	0
	Sine		Tangent		Secant		Minutes

85 Degrees.

5 Degrees.

Minutes	Sine		Tangent		Secant		
0	8.940295	9.998344	8.941952	11.058048	10.001656	11.059704	60
1	8.941738	9.998333	8.943404	11.056596	10.001667	11.058262	59
2	8.943174	9.998322	8.944852	11.055148	10.001678	11.056826	58
3	8.944600	9.998311	8.946295	11.053705	10.001689	11.055394	57
4	8.946034	9.998300	8.947734	11.052266	10.001700	11.053966	56
5	8.947456	9.698289	8.949168	11.050832	10.001711	11.052544	55
6	8.948874	9.998277	8.950597	11.049403	10.001723	11.051120	54
7	8.950287	9.998266	8.952021	11.047979	10.001734	11.049713	53
8	8.951696	9.998255	8.953441	11.046559	10.001745	11.048304	52
9	8.953100	9.998243	8.954856	11.045144	10.001757	11.046900	51
10	8.954499	9.998232	8.956267	11.043733	10.001768	11.045501	50
11	8.955894	9.998220	8.957673	11.042327	10.001780	11.044106	49
12	8.957284	9.998209	8.959075	11.040925	10.001791	11.042716	48
13	8.958670	9.998197	8.960473	11.039527	10.001803	11.041330	47
14	8.960052	9.998186	8.961866	11.038134	10.001814	11.039948	46
15	9.901429	9.598174	8.963254	11.036746	10.001826	11.038571	45
16	8.962801	9.998163	8.964639	11.035361	10.001837	11.037199	44
17	8.964170	9.998151	8.966019	11.033981	10.001849	11.035830	43
18	8.965534	9.998139	8.967394	11.032606	10.001861	11.034406	42
19	8.966893	9.998127	8.968766	11.031234	10.001873	11.033107	41
20	8.968249	9.998116	8.970133	11.029867	10.001884	11.031751	40
21	8.969600	9.998104	8.971496	11.028504	10.001896	11.030400	39
22	8.970947	9.998092	8.972855	11.027145	10.001908	11.029053	38
23	8.972289	9.998080	8.974209	11.025791	10.001920	11.027711	37
24	8.973628	9.998068	8.975560	11.024440	10.001932	11.026372	36
25	8.974902	9.998056	8.976906	11.023094	10.001944	11.025038	35
26	8.976293	9.998044	8.978248	11.021752	10.001956	11.023707	34
27	8.977619	9.998032	8.979586	11.020414	10.001968	11.022381	33
28	8.978941	9.998020	8.980921	11.019079	10.001980	11.021059	32
29	8.980259	9.998008	8.982251	11.017749	10.001992	11.019741	31
30	8.981573	9.997996	8.983577	11.016423	10.002004	11.018427	30
	Sine		Tangent		Secant		Minutes

84 Degrees.

5 Degrees.

Minutes	Sine	Tangent	Secant				Minutes
30	8.981573	9.997990	8.983577	11.016423	10.002004	11.018427	30
31	8.982883	9.997984	8.984899	11.015101	10.002016	11.017117	29
32	8.984189	9.997972	8.986217	11.013783	10.002028	11.015811	28
33	8.985491	9.997959	8.987532	11.012468	10.002041	11.014509	27
34	8.986789	9.997947	8.988842	11.011158	10.002053	11.013211	26
35	8.988083	9.997935	8.990149	11.009851	10.002065	11.011917	25
36	8.989374	9.997922	8.991451	11.008549	10.002078	11.010626	24
37	8.990660	9.997910	8.992750	11.007250	10.002090	11.009340	23
38	8.991943	9.997897	8.994045	11.005955	10.002103	11.008057	22
39	8.993222	9.997885	8.995337	11.004663	10.002115	11.006778	21
40	8.994497	9.997873	8.996624	11.003376	10.002127	11.005503	20
41	8.995768	9.997859	8.997908	11.002092	10.002141	11.004232	19
42	8.997036	9.997847	8.999188	11.000812	10.002153	11.002964	18
43	8.998299	9.997835	9.000465	11.999535	10.002165	11.001701	17
44	8.999559	9.997822	9.001738	11.998262	10.002178	11.000441	16
45	9.000816	9.997829	9.003007	11.996963	10.002191	10.999184	15
46	9.002069	9.997797	9.004272	11.995728	10.002203	10.997931	14
47	9.003318	9.997784	9.005534	11.994466	10.002216	10.996682	13
48	9.004563	9.997771	9.006792	11.993208	10.002229	10.995437	12
49	9.005805	9.997758	9.008046	11.991953	10.002242	10.994195	11
50	9.007044	9.997745	9.009298	11.990702	10.002255	10.992956	10
51	9.008278	9.997732	9.010546	11.989454	10.002268	10.991722	9
52	9.009510	9.997719	9.011790	11.988210	10.002281	10.990490	8
53	9.010737	9.997706	9.013031	11.986969	10.002294	10.989163	7
54	9.011962	9.997693	9.014268	11.985732	10.002307	10.988038	6
55	9.013181	9.997680	9.015502	11.984498	10.002320	10.986818	5
56	9.014400	9.997667	9.016733	11.983267	10.002333	10.985600	4
57	9.015613	9.997654	9.017959	11.982041	10.002346	10.984387	3
58	9.016824	9.997641	9.019183	11.980817	10.002359	10.983176	2
59	9.018031	9.997628	9.020403	11.979597	10.002372	10.981959	1
60	9.019215	9.997614	9.021620	11.978380	10.002386	10.980076	0
	Sine	Tangent	Secant				Minutes

84 Degrees.

G g

8 Degrees.

Minutes	Sine		Tangent		Secant		
0	9.019235	9.997614	9.021620	10.978380	10.002386	10.980765	60
1	9.020435	9.997601	9.022834	10.977166	10.002399	10.979565	59
2	9.021632	9.997588	9.024044	10.975956	10.002412	10.978368	58
3	9.022825	9.997574	9.025251	10.974749	10.002426	10.977175	57
4	9.024014	9.997561	9.026365	10.973545	10.002439	10.975984	56
5	9.025203	9.997547	9.027655	10.972345	10.002453	10.974797	55
6	9.026386	9.997534	9.028852	10.971148	10.002466	10.973614	54
7	9.027567	9.997521	9.030046	10.969954	10.002479	10.972433	53
8	9.028744	9.997507	9.031237	10.968763	10.002493	10.971256	52
9	9.029918	9.997493	9.032425	10.967575	10.002507	10.970082	51
10	9.031089	9.997480	9.033609	10.966391	10.002520	10.968911	50
11	9.032257	9.997466	9.034791	10.965209	10.002534	10.967743	49
12	9.033421	9.997452	9.035969	10.964031	10.002548	10.966579	48
13	9.034582	9.997439	9.037144	10.962856	10.002561	10.965418	47
14	9.035741	9.997425	9.038316	10.961684	10.002575	10.964259	46
15	9.036896	9.997411	9.039485	10.960515	10.002589	10.963104	45
16	9.038049	9.997397	9.040651	10.959349	10.002602	10.961951	44
17	9.039197	9.997383	9.041813	10.958187	10.002617	10.960803	43
18	9.040342	9.997369	9.042973	10.957027	10.002631	10.959658	42
19	9.041485	9.997355	9.044130	10.955870	10.002645	10.958515	41
20	9.042625	9.997341	9.045284	10.954716	10.002659	10.957375	40
21	9.043762	9.997327	9.046434	10.953566	10.002673	10.956238	39
22	9.044895	9.997313	9.047581	10.952418	10.002687	10.955105	38
23	9.046026	9.997299	9.048727	10.951273	10.002701	10.953974	37
24	9.047154	9.997285	9.049859	10.950131	10.002715	10.952846	36
25	9.048279	9.997271	9.051008	10.948992	10.002729	10.951721	35
26	9.049400	9.997257	9.052144	10.947856	10.002743	10.950600	34
27	9.050515	9.997242	9.053277	10.946723	10.002758	10.949461	33
28	9.051635	9.997228	9.054407	10.945593	10.002772	10.948365	32
29	9.052748	9.997214	9.055535	10.944465	10.002786	10.947252	31
30	9.053850	9.997200	9.056659	10.943341	10.002801	10.946141	30
		Secant		Tangent		Secant	Minutes

83 Degrees.

6 Degrees.

Minutes	Sine	Tangent		Secant			Minutes
30	9.053859	9.977199	9.056659	10.943341	10.002801	10.946141	30
31	9.054950	9.997185	7.057781	10.942219	10.002815	10.945034	29
32	9.056071	9.977170	9.058900	10.941100	10.002830	10.943929	28
33	9.057172	9.977156	9.060016	10.939984	10.002844	10.942828	27
34	9.058271	9.997141	9.051130	10.938870	10.002859	10.941729	26
35	9.059367	9.997127	9.062240	10.937760	10.002873	10.940633	25
36	9.060460	9.997112	9.063348	10.936652	10.002888	10.939540	24
37	9.061551	9.997098	9.064453	10.935547	10.002902	10.938449	23
38	9.062639	9.997083	9.065556	10.934444	10.002917	10.937361	22
39	9.061721	9.997068	9.066655	10.933345	10.002932	10.936279	21
40	9.064806	9.997053	9.057752	10.932248	10.002947	10.935194	20
41	9.065885	9.997019	9.068846	10.931154	10.002961	10.934115	19
42	9.066961	9.977024	9.069938	10.930062	10.002976	10.933038	18
43	9.068036	9.997009	9.071027	10.928973	10.002991	10.931964	17
44	9.069107	9.996994	9.072113	10.927887	10.003006	10.930893	16
45	9.070176	9.996979	9.073197	10.926803	10.003021	10.929824	15
46	9.071242	9.996964	9.074278	10.925722	10.003036	10.918758	14
47	9.072305	9.996949	9.075356	10.914644	10.003058	10.927695	13
48	9.073366	9.996934	9.076432	10.923468	10.003066	10.926634	12
49	9.074424	9.996919	9.077505	10.922495	10.003081	10.925576	11
50	9.075480	9.996904	9.078576	10.921424	10.003096	10.924520	10
51	9.076533	9.996889	9.079644	10.920356	10.003111	10.923467	9
52	9.077583	9.996874	9.080710	10.919290	10.003126	10.922417	8
53	9.078631	9.996858	9.081773	10.918227	10.003142	10.921369	7
54	9.079676	9.996843	9.082833	10.917167	10.003157	10.920324	6
55	9.080719	9.996828	9.083891	10.916109	10.003172	10.919281	5
56	9.081759	9.996812	9.084947	10.915053	10.003188	10.918241	4
57	9.082797	9.996797	9.086000	10.914000	10.003203	10.917203	3
58	9.083832	9.996782	9.087050	10.912950	10.003218	10.916168	2
59	9.084864	9.996765	9.088098	10.911902	10.003234	10.915136	1
60	9.085894	9.996751	9.089144	10.910856	10.003249	10.914106	0
	Sine	Tangent		Secant			Minutes

83 Degrees.

A Table of Artificial Sines,

7 Degrees.

Minutes	Sine		Tangent		Secant		Minutes
0	9.085895	9.996751	9.089144	10.910856	10.003249	10.914105	60
1	9.086922	9.996735	9.090187	10.909813	10.003265	10.913078	59
2	9.087947	9.996720	9.091228	10.908772	10.003280	10.912053	58
3	9.088970	9.996704	9.092266	10.907734	10.003296	10.911030	57
4	9.089990	9.996688	9.093302	10.906698	10.003312	10.910010	56
5	9.091008	9.996673	9.094315	10.905665	10.003327	10.908992	55
6	9.092024	9.996657	9.095367	10.904633	10.003343	10.907976	54
7	9.093037	9.996641	9.096395	10.903605	10.003359	10.906963	53
8	9.094047	9.996625	9.097422	10.902578	10.003375	10.905953	52
9	9.095056	9.996610	9.098446	10.901554	10.003390	10.904944	51
10	9.096062	9.996594	9.099468	10.900532	10.003406	10.903938	50
11	9.097065	9.996578	9.100487	10.899513	10.003422	10.902935	49
12	9.098066	9.996562	9.101504	10.898496	10.003438	10.901934	48
13	9.099065	9.996546	9.102519	10.897481	10.003454	10.900935	47
14	9.100062	9.996530	9.103532	10.896468	10.003470	10.899938	46
15	9.101056	9.996514	9.104542	10.895458	10.003486	10.898944	45
16	9.102048	9.996498	9.105550	10.894450	10.003502	10.897952	44
17	9.103037	9.996482	9.106556	10.893444	10.003518	10.896963	43
18	9.104025	9.996465	9.107559	10.892441	10.003535	10.895975	42
19	9.105010	9.996449	9.108560	10.891440	10.003551	10.894990	41
20	9.105992	9.996433	9.109559	10.890441	10.003567	10.894008	40
21	9.106973	9.996417	9.110556	10.889444	10.003583	10.893027	39
22	9.107951	9.996400	9.111551	10.888449	10.003600	10.892049	38
23	9.108927	9.996384	9.112543	10.887457	10.003616	10.891073	37
24	9.109901	9.996368	9.113533	10.886467	10.003632	10.890099	36
25	9.110873	9.996351	9.114521	10.885479	10.003649	10.889127	35
26	9.111842	9.996335	9.115507	10.884493	10.003665	10.888158	34
27	9.112809	9.996318	9.116491	10.883509	10.003682	10.887191	33
28	9.113774	9.996302	9.117472	10.882528	10.003698	10.886226	32
29	9.114737	9.996285	9.118452	10.881548	10.003715	10.885263	31
30	9.115698	9.996269	9.119429	10.880571	10.003731	10.884302	30
	Secant		Tangent		Secant		Minutes

82 Degrees.

7 Degrees.

Minutes	Sine		Tangent		Secant		
30	9.115698	9.996269	9.119429	10.880571	10.003731	10.884302	30
31	9.116656	9.996252	9.120404	10.879596	10.003749	10.883344	29
32	9.117612	9.996235	9.121377	10.878623	10.003765	10.882386	28
33	9.118567	9.996218	9.122348	10.877652	10.003782	10.881433	27
34	9.119519	9.996202	9.123317	10.876683	10.003798	10.880481	26
35	9.120469	9.996185	9.124284	10.875716	10.003815	10.879531	25
36	9.121417	9.996168	9.125249	10.874751	10.003832	10.878581	24
37	9.122362	9.996151	9.126211	10.873789	10.003849	10.877638	23
38	9.123306	9.996134	9.127172	10.872828	10.003866	10.876694	22
39	9.124248	9.996117	9.128130	10.871870	10.003883	10.875752	21
40	9.125187	9.996100	9.129087	10.870913	10.003900	10.874813	20
41	9.126125	9.996083	9.130041	10.869959	10.003917	10.873875	19
42	9.127060	9.996066	9.130994	10.869006	10.003934	10.872940	18
43	9.127993	9.996049	9.131944	10.868056	10.003951	10.872007	17
44	9.128925	9.996032	9.132893	10.867107	10.003968	10.871075	16
45	9.129854	9.996015	9.133839	10.866161	10.003985	10.870146	15
46	9.130781	9.995998	9.134783	10.865217	10.004002	10.869219	14
47	9.131706	9.995980	9.135726	10.864274	10.004020	10.868294	13
48	9.132630	9.995963	9.136666	10.863334	10.004037	10.867370	12
49	9.133551	9.995946	9.137605	10.862395	10.004054	10.866449	11
50	9.134470	9.995928	9.138542	10.861458	10.004072	10.865530	10
51	9.135387	9.995911	9.139476	10.860524	10.004089	10.864613	9
52	9.136303	9.995894	9.140409	10.859591	10.004106	10.863697	8
53	9.137216	9.995876	9.141340	10.858660	10.004124	10.862784	7
54	9.138127	9.995859	9.142269	10.857731	10.004141	10.861873	6
55	9.139037	9.995841	9.143196	10.856804	10.004159	10.860963	5
56	9.139944	9.995823	9.144121	10.855879	10.004177	10.860056	4
57	9.140850	9.995806	9.145044	10.854956	10.004194	10.859150	3
58	9.141754	9.995788	9.145965	10.854035	10.004212	10.858245	2
59	9.142655	9.995770	9.146885	10.853115	10.004230	10.857345	1
60	9.143555	9.995751	9.147802	10.852198	10.004247	10.856445	0
	Sine		Tangent		Secant		Minutes

82 Degrees.

8 Degrees.

Minutes	Sine	Tangent		Secant			
0	9.143555	9.995753	9.147802	10.652198	10.004247	10.856445	60
1	9.144453	9.995735	9.148718	10.851282	10.004265	10.855547	59
2	9.145349	9.995717	9.149632	10.850368	10.004283	10.854651	58
3	9.146243	9.995699	9.150544	10.849456	10.004301	10.853757	57
4	9.147136	9.995681	9.151454	10.848546	10.004319	10.852864	56
5	9.148026	9.995663	9.152363	10.847637	10.004337	10.851974	55
6	9.148915	9.995646	9.153269	10.846731	10.004354	10.851085	54
7	9.149801	9.995628	9.154174	10.845826	10.004372	10.850199	53
8	9.150686	9.995609	9.155077	10.844923	10.004391	10.849714	52
9	9.151560	9.995591	9.155978	10.844022	10.004409	10.848431	51
10	9.152451	9.995573	9.156877	10.843123	10.004427	10.847549	50
11	9.153330	9.995555	9.157775	10.842225	10.004445	10.846670	49
12	9.154208	9.995537	9.158671	10.841329	10.004463	10.845792	48
13	9.155083	9.995519	9.159565	10.840435	10.004481	10.844917	47
14	9.155957	9.995500	9.160457	10.839543	10.004500	10.844043	46
15	9.156830	9.995482	9.161347	10.838653	10.004518	10.843170	45
16	9.157700	9.995464	9.162236	10.837764	10.004536	10.842300	44
17	9.158569	9.995445	9.163123	10.836877	10.004555	10.841431	43
18	9.159435	9.995427	9.164008	10.835992	10.004573	10.840565	42
19	9.160300	9.995409	9.164892	10.835108	10.004591	10.839700	41
20	9.161164	9.995390	9.165774	10.834226	10.004610	10.838836	40
21	9.162025	9.995372	9.166654	10.833346	10.004628	10.837975	39
22	9.162885	9.995353	9.167532	10.832468	10.004647	10.837115	38
23	9.163743	9.995334	9.168409	10.831591	10.004666	10.836257	37
24	9.164600	9.995316	9.169284	10.830716	10.004684	10.835400	36
25	9.165454	9.995297	9.170157	10.829843	10.004703	10.834546	35
26	9.166307	9.995278	9.171029	10.828971	10.004722	10.833693	34
27	9.167159	9.995260	9.171899	10.828101	10.004740	10.832841	33
28	9.168008	9.995241	9.172767	10.827233	10.004759	10.831992	32
29	9.168856	9.995222	9.173634	10.826366	10.004778	10.831144	31
30	9.169702	9.995203	9.174500	10.825501	10.004797	10.830298	30
	Sine	Tangent		Secant			Minutes

81 Degrees.

8 Degrees.

Minutes	Sine	Tangent		Secant		
30	9.169702	9.995203	9.174499	10.825501	10.004797	10.830198	30
31	9.170546	9.995184	9.175362	10.824638	10.004816	10.829454	29
32	9.171389	9.995165	9.176224	10.823776	10.004835	10.828611	28
33	9.172230	9.995146	9.177084	10.822916	10.004854	10.827770	27
34	9.173070	9.995127	9.177942	10.822058	10.004873	10.826930	26
35	9.173908	9.995108	9.178799	10.821201	10.004892	10.826092	25
36	9.174744	9.995089	9.179655	10.820345	10.004911	10.825256	24
37	9.175578	9.995070	9.180508	10.819491	10.004930	10.824422	23
38	9.176411	9.995051	9.181360	10.818640	10.004949	10.823589	22
39	9.177242	9.995032	9.182211	10.817789	10.004968	10.822758	21
40	9.178072	9.995013	9.183059	10.816941	10.004987	10.821928	20
41	9.178900	9.994993	9.183907	10.816093	10.005007	10.821100	19
42	9.179726	9.994974	9.184752	10.815248	10.005026	10.820274	18
43	9.180551	9.994955	9.185597	10.814403	10.005045	10.819449	17
44	9.181374	9.994935	9.186440	10.813560	10.005065	10.818626	16
45	9.182196	9.994916	9.187280	10.812720	10.005084	10.817804	15
46	9.183016	9.994896	9.188120	10.811880	10.005104	10.816984	14
47	9.183834	9.994877	9.188957	10.811043	10.005123	10.816166	13
48	9.184651	9.994857	9.189794	10.810206	10.005143	10.815349	12
49	9.185466	9.994838	9.190629	10.809371	10.005162	10.814534	11
50	9.186280	9.994818	9.191462	10.808538	10.005182	10.813720	10
51	9.187092	9.994798	9.192294	10.807706	10.005202	10.812908	9
52	9.187903	9.994779	9.193124	10.806876	10.005221	10.812097	8
53	9.188712	9.994759	9.193953	10.806047	10.005241	10.811288	7
54	9.189519	9.994739	9.194780	10.805220	10.005261	10.810481	6
55	9.190325	9.994719	9.195606	10.804394	10.005281	10.809675	5
56	9.191132	9.994700	9.196430	10.803570	10.005300	10.808871	4
57	9.191933	9.994680	9.197253	10.802747	10.005320	10.808067	3
58	9.192734	9.994660	9.198074	10.801926	10.005340	10.807266	2
59	9.193534	9.994640	9.198894	10.801106	10.005360	10.806466	1
60	9.194332	9.994620	9.199712	10.800288	10.005380	10.805668	0

| | Sine | | Tangent | | Secant | | Minutes |

81 Degrees.

9 Degrees.

Minutes	Sine		Tangent		Secant		Minutes
0	9.191332	9.994623	9.199712	10.700188	10.005380	10.805668	60
1	9.191329	9.994600	9.200529	10.799471	10.005400	10.804871	59
2	9.195925	9.994580	9.201345	10.798655	10.005420	10.804075	58
3	9.196719	9.994560	9.202159	10.797841	10.005440	10.803281	57
4	9.197511	9.994540	9.202971	10.797029	10.005460	10.802469	56
5	9.198302	9.994519	9.203782	10.796218	10.005481	10.801698	55
6	9.199091	9.994199	9.204592	10.795408	10.005501	10.800909	54
7	9.199879	9.994475	9.205400	10.794600	10.005521	10.800121	53
8	9.200666	9.994459	9.206207	10.793793	10.005541	10.799334	52
9	9.201451	9.994438	9.207013	10.792987	10.005562	10.798549	51
10	9.202234	9.994418	9.207817	10.792183	10.005582	10.797766	50
11	9.203017	9.994397	9.208619	10.791381	10.005603	10.796983	49
12	9.203797	9.994377	9.209420	10.790580	10.005623	10.796203	48
13	9.204577	9.994357	9.210220	10.789780	10.005643	10.795423	47
14	9.205354	9.994336	9.211018	10.788982	10.005664	10.794646	46
15	9.206131	9.994316	9.211815	10.788185	10.005684	10.793869	45
16	9.206906	9.994295	9.212611	10.787389	10.005705	10.793094	44
17	9.207679	9.994274	9.213405	10.786595	10.005726	10.792371	43
18	9.208452	9.994254	9.214198	10.785802	10.005746	10.791548	42
19	9.209223	9.994233	9.214989	10.785011	10.005767	10.790777	41
20	9.209992	9.994212	9.215779	10.784221	10.005788	10.790008	40
21	9.210760	9.994191	9.216568	10.783432	10.005809	10.789240	39
22	9.211526	9.994171	9.217356	10.782644	10.005829	10.788474	38
23	9.212291	9.994150	9.218142	10.781858	10.005850	10.787709	37
24	9.213056	9.994129	9.218926	10.781074	10.005871	10.786945	36
25	9.213818	9.994108	9.219710	10.780290	10.005892	10.786182	35
26	9.214579	9.994087	9.220492	10.779508	10.005913	10.785421	34
27	9.215338	9.994066	9.221272	10.778728	10.005934	10.784662	33
28	9.216097	9.994045	9.222052	10.777948	10.005955	10.783903	32
29	9.216854	9.994024	9.222830	10.777170	10.005976	10.783146	31
30	9.217609	9.994003	9.223606	10.776394	10.005997	10.782391	30

	Sine		Tangent		Secant		Minutes

80 Degrees.

9 Degrees.

Minutes	Sine		Tangent		Secant		Minutes
30	9.217609	9.994003	9.223606	10.776894	10.005997	10.782391	30
31	9.218363	9.993981	9.224382	10.775610	10.006919	10.781637	29
32	9.219116	9.993960	9.225156	10.774844	10.006040	10.780884	28
33	9.219868	9.993939	9.225929	10.774071	10.006061	10.780132	27
34	9.220618	9.993918	9.225700	10.773300	10.006082	10.779382	26
35	9.221367	9.993896	9.227471	10.772529	10.006104	10.778633	25
36	9.222115	9.993875	9.228239	10.771761	10.006125	10.777885	24
37	9.222861	9.993854	9.229007	10.770993	10.006146	10.777139	23
38	9.223606	9.993832	9.229774	10.770226	10.006168	10.776394	22
39	9.224349	9.993811	9.230539	10.769461	10.006189	10.775651	21
40	9.225092	9.993789	9.231302	10.768698	10.006211	10.774908	20
41	9.225833	9.993768	9.232065	10.767935	10.006232	10.774167	19
42	9.226573	9.993746	9.232826	10.767174	10.006254	10.773427	18
43	9.227311	9.993725	9.233586	10.766414	10.006275	10.772689	17
44	9.228048	9.993703	9.234345	10.765655	10.006297	10.771952	16
45	9.228784	9.993681	9.235103	10.764897	10.006319	10.771216	15
46	9.229519	9.993660	9.235859	10.764141	10.006340	10.770481	14
47	9.230252	9.993638	9.236614	10.763386	10.006362	10.769748	13
48	9.230984	9.993616	9.237368	10.762632	10.006384	10.769016	12
49	9.231715	9.993594	9.238120	10.761880	10.006406	10.768285	11
50	9.232444	9.993572	9.238872	10.761128	10.006428	10.767556	10
51	9.233172	9.993550	9.239622	10.760378	10.006450	10.766828	9
52	9.233899	9.993528	9.240371	10.759629	10.006472	10.766101	6
53	9.234625	9.993507	9.241118	10.758882	10.006493	10.765375	7
54	9.235349	9.993484	9.241865	10.758135	10.006516	10.764651	6
55	9.236073	9.993462	9.242610	10.757390	10.006538	10.763927	5
56	9.236795	9.993440	9.243354	10.756646	10.006560	10.763205	4
57	9.237515	9.993418	9.244097	10.755903	10.006582	10.762485	3
58	9.238235	9.993396	9.244839	10.755161	10.006604	10.761765	2
59	9.238953	9.993374	9.245579	10.754421	10.006626	10.761047	1
60	9.239670	9.99 3352	9.246319	10.753681	10.006648	10.760330	0
	Sine		Tangent		Secant		Minutes

80 Degrees.

10 Degrees.

Minutes	Sine		Tangent		Secant		Minutes
0	9.239670	9.993351	9.246319	10.753681	10.006649	10.700330	60
1	9.240386	9.993329	9.247057	10.752943	10.006671	10.759614	59
2	9.241101	9.993307	9.247794	10.752206	10.006693	10.758899	58
3	9.241814	9.993284	9.248530	10.751470	10.006716	10.758186	57
4	9.242526	9.993262	9.249264	10.750736	10.006738	10.757474	56
5	9.243237	9.993240	9.249998	10.750002	10.006760	10.756763	55
6	9.243947	9.993217	9.250730	10.749270	10.006783	10.756053	54
7	9.244656	9.993195	9.251461	10.748539	10.006805	10.755344	53
8	9.245363	9.993172	9.252191	10.747809	10.006828	10.754637	52
9	9.246069	9.993149	9.252920	10.747080	10.006851	10.753931	51
10	9.246775	9.993127	9.253648	10.746352	10.006873	10.753225	50
11	9.247478	9.993104	9.254374	10.745626	10.006896	10.752522	49
12	9.248181	9.993081	9.255100	10.744900	10.006919	10.751819	48
13	9.248883	9.993059	9.255824	10.744176	10.006941	10.751117	47
14	9.249583	9.993036	9.256547	10.743453	10.006964	10.750417	46
15	9.250282	9.993013	9.257269	10.742731	10.006987	10.749718	45
16	9.250980	9.992990	9.257990	10.742010	10.007010	10.749020	44
17	9.251677	9.992967	9.258710	10.741290	10.007033	10.748323	43
18	9.252373	9.992944	9.259428	10.740572	10.007056	10.747627	42
19	9.253067	9.992921	9.260146	10.739854	10.007079	10.746933	41
20	9.253761	9.992898	9.260862	10.739138	10.007102	10.746239	40
21	9.254453	9.992875	9.261578	10.738422	10.007125	10.745547	39
22	9.255144	9.992852	9.262292	10.737708	10.007148	10.744856	38
23	9.255833	9.992829	9.263005	10.736995	10.007171	10.744166	37
24	9.256522	9.992806	9.263717	10.736283	10.007194	10.743477	36
25	9.257211	9.992783	9.264428	10.735572	10.007217	10.742789	35
26	9.257898	9.992759	9.265138	10.734862	10.007241	10.742102	34
27	9.258583	9.992736	9.265847	10.734153	10.007264	10.741417	33
28	9.259268	9.992713	9.266555	10.733445	10.007287	10.740732	32
29	9.259951	9.992689	9.267261	10.732739	10.007311	10.740049	31
30	9.260633	9.992666	9.267967	10.732033	10.007334	10.739367	30
		Sine		Tangent		Secant	Minutes

79 Degrees.

10 Degrees.

Minutes	Sine		Tangent		Secant		
30	9.260633	9.992566	9.26796:	10.732033	10.007334	10.739367	30
31	9.261314	9.992643	9.263671	10.731329	10.007357	10.738686	29
32	9.261994	9.992619	9.269375	10.730625	10.007381	10.738006	28
33	9.262673	9.992591	9.27007:	10.729923	10.007464	10.737327	27
34	9.263351	9.992572	9.270779	10.729221	10.007428	10.736649	26
35	9.264016	9.992549	9.271470	10.728521	10.007451	10.735973	25
36	9.264703	9.992525	9.272178	10.727822	10.007475	10.735297	24
37	9.265377	9.992501	9.272876	10.727124	10.007499	10.734623	23
38	9.266051	9.992578	9.273573	10.726427	10.007522	10.733949	22
39	9.266723	9.992454	9.274269	10.725731	10.007546	10.733277	21
40	9.267395	9.992430	9.274904	10.725036	10.007570	10.732605	20
41	9.268065	9.992406	9.275658	10.724342	10.007594	10.731935	19
42	9.268734	9.992382	9.276351	10.723649	10.007618	10.731266	18
43	9.269402	9.992358	9.277043	10.722957	10.007642	10.730598	17
44	9.270069	9.992335	9.277734	10.722266	10.007665	10.729931	16
45	9.270735	9.992311	9.278424	10.721576	10.007689	10.729265	15
46	9.271400	9.992287	9.279113	10.720887	10.007713	10.728600	14
47	9.272063	9.992263	9.279801	10.720199	10.007737	10.727937	13
48	9.272726	9.992238	9.280483	10.719513	10.007762	10.727274	12
49	9.273388	9.992214	9.281174	10.718826	10.007786	10.726612	11
50	9.274049	9.992190	9.281858	10.718142	10.007810	10.725951	10
51	9.274708	9.992166	9.282542	10.717458	10.007834	10.725292	9
52	9.275367	9.992142	9.283225	10.716775	10.007858	10.724633	8
53	9.276024	9.992117	9.283907	10.716093	10.007883	10.723976	7
54	9.276681	9.992093	9.284588	10.715412	10.007907	10.723319	6
55	9.277337	9.992069	9.285268	10.714732	10.007931	10.722663	5
56	9.277991	9.992044	9.285947	10.714053	10.007956	10.722009	4
57	9.278644	9.992020	9.286624	10.713376	10.007980	10.721356	3
58	9.279297	9.991996	9.287301	10.712699	10.008004	10.720703	2
59	9.279948	9.991971	9.287997	10.712023	10.008029	10.720052	1
60	9.280599	9.991947	9.288652	10.711348	10.008053	10.719401	0

| | Sine | | Tangent | | Secant | | Minutes |

11 Degrees.

Minutes	Sine		Tangent		Secant		
0	9.280599	9.991947	9.288652	10.711348	10.008053	10.719401	60
1	9.281248	9.991922	9.289326	10.710674	10.008078	10.718752	59
2	9.281897	9.991897	9.289999	10.710001	10.008103	10.718103	58
3	9.282544	9.991873	9.290671	10.709329	10.008127	10.717456	57
4	9.283190	9.991848	9.291342	10.708658	10.008152	10.716810	56
5	9.283836	9.991823	9.292013	10.707987	10.008177	10.716164	55
6	9.284480	9.991799	9.292682	10.707318	10.008201	10.715520	54
7	9.285124	9.991774	9.293350	10.706650	10.008226	10.714876	53
8	9.285766	9.991749	9.294017	10.705983	10.008251	10.714234	52
9	9.286408	9.991724	9.294684	10.705316	10.008276	10.713592	51
10	9.287048	9.991699	9.295349	10.704651	10.008301	10.712952	50
11	9.287687	9.991674	9.296013	10.703987	10.008326	10.712313	49
12	9.288326	9.991649	9.296677	10.703323	10.008351	10.711674	48
13	9.288964	9.991624	9.297339	10.702661	10.008376	10.711036	47
14	9.289600	9.991599	9.298001	10.701999	10.008401	10.710400	46
15	9.290236	9.991574	9.298662	10.701338	10.008426	10.709764	45
16	9.290870	9.991549	9.299322	10.700678	10.008451	10.709130	44
17	9.291504	9.991524	9.299980	10.700020	10.008476	10.708496	43
18	9.292137	9.991498	9.300638	10.699362	10.008502	10.707863	42
19	9.292768	9.991473	9.301295	10.698705	10.008527	10.707232	41
20	9.293399	9.991448	9.301951	10.698049	10.008552	10.706601	40
21	9.294029	9.991423	9.302607	10.697393	10.008577	10.705971	39
22	9.294658	9.991397	9.303261	10.696739	10.008603	10.705342	38
23	9.295286	9.991372	9.303914	10.696086	10.008628	10.704714	37
24	9.295913	9.991346	9.304567	10.695433	10.008654	10.704087	36
25	9.296539	9.991321	9.305218	10.694782	10.008679	10.703461	35
26	9.297164	9.991295	9.305869	10.694131	10.008705	10.702836	34
27	9.297788	9.991270	9.306519	10.693481	10.008730	10.702212	33
28	9.298412	9.991244	9.307167	10.692833	10.008756	10.701588	32
29	9.299034	9.991218	9.307815	10.692185	10.008782	10.700966	31
30	9.299655	9.991193	9.308463	10.691537	10.008807	10.700345	30
	Sine		Tangent		Secant		Minutes

78 Degrees.

11 Degrees.

Minutes	Secant		Tangent		Secant		Minutes
30	9.399655	9.991195	9.308463	10.691537	10.008607	10.700345	30
31	9.300276	9.991167	9.309109	10.690891	10.008833	10.699724	29
32	9.300895	9.991141	9.309754	10.690246	10.008859	10.699105	28
33	9.301514	9.991115	9.310398	10.689602	10.008885	10.698486	27
34	9.302132	9.991090	9.311042	10.688958	10.008910	10.697868	26
35	9.302749	9.991064	9.311685	10.688315	10.008936	10.697251	25
36	9.303364	9.991038	9.322327	10.687673	10.008962	10.696636	24
37	9.303979	9.991012	9.322967	10.687063	10.008988	10.696021	23
38	9.304593	9.990986	9.333608	10.686392	10.009014	10.695407	22
39	9.305207	9.990960	9.344247	10.685753	10.009040	10.694793	21
40	9.305819	9.990934	9.344885	10.685115	10.009066	10.694181	20
41	9.306430	9.990908	9.355523	10.684477	10.009092	10.693570	19
42	9.307041	9.990882	9.366159	10.683841	10.009118	10.692959	18
43	9.307650	9.990855	9.366795	10.683205	10.009145	10.692350	17
44	9.308259	9.990829	9.377430	10.682570	10.009171	10.691741	16
45	9.308867	9.990803	9.388064	10.681936	10.009197	10.691133	15
46	9.309474	9.990777	9.388697	10.681303	10.009223	10.690520	14
47	9.310080	9.990750	9.399320	10.680571	10.009250	10.689920	13
48	9.310685	9.990724	9.399961	10.680039	10.009276	10.689315	12
49	9.311289	9.990697	9.300592	10.679408	10.009303	10.688711	11
50	9.311893	9.990671	9.311222	10.678778	10.009329	10.688107	10
51	9.312495	9.990644	9.311851	10.678149	10.009356	10.687505	9
52	9.313097	9.990615	9.322479	10.677521	10.009381	10.686903	8
53	9.313698	9.990591	9.333106	10.676894	10.009409	10.686302	7
54	9.314297	9.990565	9.333733	10.676267	10.009435	10.685703	6
55	9.314896	9.990538	9.344358	10.675642	10.009462	10.685104	5
56	9.315495	9.990511	9.344983	10.675017	10.009489	10.684505	4
57	9.316092	9.990485	9.355607	10.674393	10.009515	10.683908	3
58	9.316688	9.990458	9.366230	10.673770	10.009542	10.683312	2
59	9.317284	9.990431	9.366853	10.673147	10.009569	10.682716	1
60	9.317879	9.990404	9.377474	10.672526	10.009596	10.682121	0
		Sine		Tangent		Secant	Minutes

1 2 Degrees.

M	Sine		Tangent		Secant		M
0	9.317879	9.990404	9.327474	10.672526	10.009596	10.682121	60
1	9.318473	9.990377	9.328095	10.671905	10.009623	10.681527	59
2	9.319066	9.990351	9.328715	10.671285	10.009649	10.680934	58
3	9.319658	9.990324	9.329334	10.670666	10.009676	10.680342	57
4	9.320249	9.990297	9.329953	10.670047	10.009704	10.679751	56
5	9.320840	9.990270	9.330570	10.669430	10.009730	10.679160	55
6	9.321430	9.990243	9.331187	10.668813	10.009757	10.678570	54
7	9.322019	9.990215	9.331803	10.668197	10.009785	10.677981	53
8	9.322607	9.990188	9.332418	10.667582	10.009812	10.677393	52
9	9.323194	9.990161	9.333033	10.666967	10.009839	10.676806	51
10	9.323780	9.990134	9.333646	10.666354	10.009866	10.676220	50
11	9.324366	9.990107	9.334259	10.665741	10.009893	10.675634	49
12	9.324951	9.990079	9.334871	10.665129	10.009921	10.675049	48
13	9.325534	9.990052	9.335482	10.664518	10.009948	10.674466	47
14	9.326117	9.990025	9.336093	10.663907	1.009975	10.673883	46
15	9.326700	9.989997	9.336702	10.663298	10.010003	10.673300	45
16	9.327281	9.989970	9.337311	10.662689	10.010030	10.672719	44
17	9.327861	9.989942	9.337919	10.662081	10.010058	10.672138	43
18	9.328442	9.989915	9.338527	10.661473	10.010085	10.671568	42
19	9.329021	9.989887	9.339133	10.660867	10.010113	10.670979	41
20	9.329599	9.989860	9.339739	10.660261	10.010140	10.660401	40
21	9.330176	9.989832	9.340344	10.659656	10.010168	10.669824	39
22	9.330753	9.989804	9.340948	10.659052	10.010196	10.669247	38
23	9.331329	9.989777	9.341552	10.658448	10.010203	10.668672	37
24	9.331903	9.989749	9.342155	10.657845	10.010251	10.668097	36
25	9.332478	9.989721	9.342757	12.657243	10.010279	10.667522	35
26	9.333051	9.989693	9.343358	12.656642	10.010307	10.666949	34
27	9.333624	9.989665	9.343958	12.656042	10.010335	10.666376	33
28	9.334195	9.989637	9.344558	12.655442	10.010363	10.665805	32
29	9.334766	9.989609	9.345157	12.654843	10.010391	10.665234	31
30	9.335337	9.989581	9.345755	12.655245	10.010419	10.664663	30
	Sine		Tangent		Secant		M

77 Degrees.

12 Degrees.

Minutes	Sine		Tangent	Secant			
30	9.335337	9.989581	9.345755	10.654245	10.010419	10.664063	30
31	9.335906	9.989553	9.346353	10.653647	10.010447	10.664094	29
32	9.336475	9.989525	9.346949	10.653051	10.010475	10.663525	28
33	9.337043	9.989497	9.347545	10.652455	10.010503	10.662957	27
34	9.337610	9.989469	9.348141	10.651859	10.010531	10.662390	26
35	9.338176	9.989441	9.348735	10.651265	10.010559	10.661824	25
36	9.338742	9.989413	9.349329	10.650671	10.010587	10.661258	24
37	9.339306	9.989384	9.349922	10.650078	10.010616	10.660694	23
38	9.339871	9.989356	9.350514	10.649486	10.010644	10.660129	22
39	9.340434	9.989328	9.351106	10.648894	10.010672	10.659566	21
40	9.340996	9.989299	9.351697	10.648303	10.010701	10.659004	20
41	9.341558	9.989271	9.352287	10.647713	10.010729	10.658442	19
42	9.342119	9.989243	9.352876	10.647124	10.010757	10.657881	18
43	9.342679	9.989214	9.353465	10.646535	10.010786	10.657321	17
44	9.343239	9.989186	9.354053	10.645947	10.010814	10.656761	16
45	9.343797	9.989157	9.354640	10.645360	10.010843	10.656203	15
46	9.344355	9.989128	9.355227	10.644773	10.010872	10.655645	14
47	9.344912	9.989100	9.355813	10.644187	10.010900	10.655088	13
48	9.345469	9.989071	9.356398	10.643602	10.010929	10.654531	12
49	9.346024	9.989042	9.356982	10.643018	10.010958	10.653976	11
50	9.346579	9.989014	9.357566	10.642434	10.010986	10.653421	10
51	9.347134	9.988985	9.358149	10.641851	10.011015	10.652866	9
52	9.347687	9.988956	9.358731	10.641269	10.011044	10.652313	8
53	9.348240	9.988927	9.359313	10.640687	10.011073	10.651760	7
54	9.348792	9.988898	9.359893	10.640107	10.011102	10.651208	6
55	9.349343	9.988869	9.360474	10.639526	10.011131	10.650657	5
56	9.349893	9.988840	9.361053	10.638947	10.011160	10.650107	4
57	9.350443	9.988811	9.361632	10.638368	10.011189	10.649557	3
58	9.350992	9.988782	9.362210	10.637790	10.011218	10.649008	2
59	9.351540	9.988753	9.362787	10.637213	10.011247	10.648460	1
60	9.352088	9.988724	9.363364	10.636636	10.011276	10.647912	0
	Sine		Tangent	Secant			Minutes

77 Degrees.

13 Degree.

Minutes	Sine	Tangent		Secant		Minutes
0	9.352088	9.988724	9.363364 10.636636	10.011276	10.647912	60
1	9.352635	9.988695	9.363940 10.636060	10.011305	10.647365	59
2	9.353181	9.988666	9.364516 10.635484	10.011334	10.646819	58
3	9.353726	9.988636	9.365090 10.634910	10.011364	10.646274	57
4	9.354271	9.988607	9.365664 10.634336	10.011393	10.645729	56
5	9.354815	9.988578	9.366237 10.633763	10.011422	10.645185	55
6	9.355358	9.988548	9.366810 10.633190	10.011452	10.644642	54
7	9.355901	9.988519	9.367382 10.632618	10.011481	10.644099	53
8	9.356443	9.988489	9.367953 10.632047	10.011511	10.643557	52
9	9.356984	9.988460	9.368524 10.631476	10.011540	10.643016	51
10	9.357524	9.988430	9.369094 10.630906	10.011570	10.642476	50
11	9.358064	9.988401	9.369663 10.630337	10.011599	10.641936	49
12	9.358603	9.988371	9.370231 10.629769	10.011629	10.641397	48
13	9.359141	9.988342	9.370799 10.629201	10.011658	10.640859	47
14	9.359679	9.988312	9.371367 10.628633	10.011688	10.640321	46
15	9.360215	9.988283	9.371933 10.628067	10.011718	10.639785	45
16	9.360751	9.988253	9.372499 10.627501	10.011748	10.639249	44
17	9.361287	9.988223	9.373064 10.626936	10.011777	10.638713	43
18	9.361822	9.988193	9.373629 10.626371	10.011807	10.638178	42
19	9.362356	9.988163	9.374193 10.625807	10.011837	10.637644	41
20	9.362889	9.988133	9.374756 10.625244	10.011867	10.637111	40
21	9.363422	9.988103	9.375319 10.624681	10.011897	10.636578	39
22	9.363954	9.988073	9.375881 10.624119	10.011927	10.636046	38
23	9.364485	9.988043	9.376442 10.623558	10.011957	10.635515	37
24	9.365016	9.988013	9.377003 10.622997	10.011987	10.634984	36
25	9.365546	9.987983	9.377563 12.622437	10.011017	10.634454	35
26	9.366075	9.987952	9.378123 12.621877	10.012048	10.633925	34
27	9.366604	9.987922	9.378681 12.621319	10.012078	10.633396	33
28	9.367132	9.987892	9.379239 12.620761	10.012108	10.632868	32
29	9.367659	9.987862	9.379797 12.620203	10.012138	10.632341	31
30	9.368186	9.987831	9.380354 12.619646	10.012169	10.631815	30
	Sine	Tangent		Secant		Minutes

76 Degrees.

13 Degrees.

Minutes	Sine		Tangent		Secant		Minutes
30	9.368185	9.987831	9.380355	10.619046	10.013169	10.031815	30
31	9.368711	9.987801	9.380910	10.619090	10.012119	10.631289	29
32	9.369236	9.987771	9.381465	10.618535	10.012229	10.630764	28
33	9.369761	9.987740	9.382020	10.617980	10.012260	10.630239	27
34	9.370285	9.987710	9.382575	10.617425	10.012290	10.629715	26
35	9.370808	9.987679	9.383129	10.616871	10.012321	10.629192	25
36	9.371330	9.987649	9.383682	10.616318	10.012351	10.628670	24
37	9.371852	9.987618	9.384234	10.615766	10.012382	10.628148	23
38	9.372373	9.987588	9.384786	10.615214	10.012412	10.627627	22
39	9.372894	9.987557	9.385337	10.614663	10.012443	10.627106	21
40	9.373414	9.987526	9.385888	10.614112	10.012474	10.626586	20
41	9.373933	9.987495	9.386438	10.613562	10.012505	10.626067	19
42	9.374452	9.987465	9.386987	10.613013	10.012535	10.625548	18
43	9.374970	9.987434	9.387535	10.612464	10.012566	10.625030	17
44	9.375487	9.987403	9.388084	10.611916	10.012597	10.624513	16
45	9.376003	9.987372	9.388631	10.611369	10.012628	10.623997	15
46	9.376519	9.987341	9.389178	10.610822	10.012659	10.623481	14
47	9.377035	9.987310	9.389724	10.610276	10.012690	10.622965	13
48	9.377549	9.987279	9.390270	10.609730	10.012721	10.622451	12
49	9.378063	9.987248	9.390815	10.609185	10.012752	10.621937	11
50	9.378577	9.987217	9.391359	10.608641	10.012783	10.621423	10
51	9.379089	9.987186	9.391903	10.608097	10.012814	10.620911	9
52	9.379601	9.987155	9.392447	10.607553	10.012845	10.620399	8
53	9.380113	9.987124	9.392989	10.607011	10.012876	10.619887	7
54	9.380624	9.987092	9.393511	10.606469	10.012908	10.619376	6
55	9.381134	9.987061	9.394073	10.605927	10.012939	10.618866	5
56	9.381643	9.987030	9.394614	10.605386	10.012970	10.618357	4
57	9.382152	9.987998	9.395154	10.604846	10.013002	10.617848	3
58	9.382660	9.987967	9.395693	10.604307	10.013033	10.617340	2
59	9.383168	9.987936	9.396233	10.603767	10.013064	10.616832	1
60	9.383675	9.987904	9.396771	10.603229	10.013096	10.616326	0
	Sine		Tangent		Secant		Minutes

76 Degrees.

H h

A Table of Artificial Sines,

14 Degrees.

Minutes	Sine		Tangent		Secant		
0	9.383075	9.986904	9.396771	10.603229	10.013096	10.616325	60
1	9.384181	9.986873	9.397309	10.602691	10.013127	10.615819	59
2	9.385287	9.986841	9.397846	10.602154	10.013159	10.615313	58
3	9.385192	9.986809	9.398383	10.601617	10.013191	10.614808	57
4	9.385697	9.986778	9.398919	10.601081	10.013222	10.614303	56
5	9.386201	9.986746	9.399455	10.600545	10.013254	10.613799	55
6	9.386704	9.986714	9.399990	10.600010	10.013286	10.613296	54
7	9.387207	9.986683	9.400524	10.599476	10.013317	10.612793	53
8	9.387709	9.986651	9.401058	10.598942	10.013349	10.612291	52
9	9.388211	9.986619	9.401591	10.598409	10.013381	10.611790	51
10	9.388711	9.986587	9.402124	10.597876	10.013413	10.611289	50
11	9.389211	9.986555	9.402656	10.597344	10.013445	10.610789	49
12	9.389711	9.986523	9.403187	10.596813	10.013477	10.610289	48
13	9.390210	9.986491	9.403718	10.596282	10.013509	10.609790	47
14	9.390708	9.986459	9.404249	10.595751	10.013541	10.609292	46
15	9.391206	9.986427	9.404778	10.595222	10.013573	10.608794	45
16	9.391703	9.986395	9.405308	10.594692	10.013605	10.608297	44
17	9.392199	9.986363	9.405836	10.594164	10.013637	10.607801	43
18	9.392695	9.986331	9.406364	10.593636	10.013669	10.607305	42
19	9.393190	9.986299	9.406892	10.593108	10.013701	10.606810	41
20	9.393685	9.986266	9.407419	10.592581	10.013734	10.606315	40
21	9.394179	9.986234	9.407945	10.592055	10.013766	10.605821	39
22	9.394673	9.986202	9.408471	10.591529	10.013798	10.605327	38
23	9.395166	9.986169	9.408997	10.591003	10.013831	10.604834	37
24	9.395659	9.986137	9.409521	10.590479	10.013863	10.604342	36
25	9.396150	9.986104	9.410045	10.589955	10.013896	10.603850	35
26	9.396641	9.986072	9.410569	10.589431	10.013928	10.603359	34
27	9.397132	9.986039	9.411092	10.588908	10.013961	10.602868	33
28	9.397621	9.986007	9.411615	10.588385	10.013993	10.602379	32
29	9.398111	9.985974	9.412137	10.587863	10.014026	10.601889	31
30	9.398600	9.985942	9.412658	10.587342	10.014058	10.601400	30
	Sine		Tangent		Secant		Minutes

75 Degrees.

14 Degrees.

Minutes	Sine		Tangent		Secant		
30	9.397002	9.985913	9.413056	10.587142	10.014058	10.601400	30
31	9.397083	9.985990	9.413179	10.586821	10.014091	10.600912	29
32	9.399575	9.985876	9.413633	10.586301	10.014121	10.600425	28
33	9.400002	9.985843	9.414219	10.585781	10.014157	10.599938	27
34	9.400519	9.985811	9.414731	10.585262	10.014180	10.599451	26
35	9.401015	9.985773	9.415237	10.584743	10.014222	10.598905	25
36	9.401520	9.985743	9.415773	10.584225	10.014255	10.598480	24
37	9.402005	9.985712	9.416293	10.583707	10.014288	10.597995	23
38	9.402489	9.985679	9.416810	10.583190	10.014321	10.597511	22
39	9.402972	9.985644	9.417326	10.581674	10.014354	10.597029	21
40	9.403155	9.985613	9.417842	10.58218	10.014387	10.596547	20
41	9.403935	9.985580	9.418358	10.581642	10.014410	10.596062	19
42	9.404442	9.985547	9.418873	10.581127	10.014453	10.595580	18
43	9.404901	9.985515	9.419387	10.580613	10.014487	10.595099	17
44	9.405382	9.985483	9.419901	10.580099	10.014520	10.594618	16
45	9.405604	9.985449	9.420415	10.579585	10.014553	10.594138	15
46	9.406341	9.985414	9.420927	10.579073	10.014580	10.593659	14
47	9.406820	9.985386	9.421440	10.578560	10.014620	10.593180	13
48	9.407297	9.985347	9.421951	10.578049	10.014653	10.592701	12
49	9.407777	9.985114	9.422463	10.577537	10.014685	10.592223	11
50	9.408254	9.985280	9.422973	10.577027	10.014720	10.591746	10
51	9.408731	9.985247	9.423484	10.576516	10.014753	10.591269	9
52	9.409207	9.985211	9.423991	10.576007	10.014787	10.590793	8
53	9.409682	9.985180	9.424503	10.575497	10.014820	10.590318	7
54	9.410156	9.985146	9.425011	10.574989	10.014853	10.589841	6
55	9.410632	9.985112	9.425519	10.574481	10.014888	10.589368	5
56	9.411106	9.985079	9.426027	10.573973	10.014921	10.588894	4
57	9.411579	9.985045	9.426534	10.573466	10.014955	10.588421	3
58	9.412052	9.985011	9.427041	10.572959	10.014989	10.587948	2
59	9.412524	9.984978	9.427547	10.572453	10.015022	10.587476	1
60	9.412996	9.984944	9.428052	10.571948	10.015056	10.587004	0

| | Sine | | Tangent | | Secant | | Minutes |

75 Degrees.

A Table of Artificial Sines,

15 Degrees.

Minutes	Sine		Tangent		Secant		
0	9.412990	9.984944	9.428052	10.571948	10.015056	10.587004	60
1	9.413467	9.984910	9.428557	10.571443	10.015090	10.586533	59
2	9.413938	9.984876	9.429062	10.570938	10.015124	10.586062	58
3	9.414408	9.984842	9.429566	10.570434	10.015158	10.585592	57
4	9.414878	9.984808	9.430070	10.569930	10.015192	10.585122	56
5	9.415347	9.984774	9.430573	10.569427	10.015226	10.584653	55
6	9.415815	9.984740	9.431075	10.568925	10.015260	10.584185	54
7	9.416283	9.984706	9.431577	10.568423	10.015294	10.583717	53
8	9.416751	9.984672	9.432079	10.567921	10.015328	10.583249	52
9	9.417217	9.984637	9.432580	10.567420	10.015363	10.582783	51
10	9.417684	9.984603	9.433080	10.566920	10.015397	10.582316	50
11	9.418149	9.984569	9.433580	10.566420	10.015431	10.581851	49
12	9.418615	9.984535	9.434080	10.565920	10.015465	10.581385	48
13	9.419079	9.984500	9.434579	10.565421	10.015500	10.580921	47
14	9.419544	9.984466	9.435078	10.564922	10.015534	10.580456	46
15	9.420007	9.984432	9.435576	10.564424	10.015568	10.579993	45
16	9.420470	9.984397	9.436073	10.563927	10.015603	10.579530	44
17	9.420933	9.984363	9.436570	10.563430	10.015637	10.579067	43
18	9.421395	9.984328	9.437067	10.562933	10.015672	10.578605	42
19	9.421857	9.984293	9.437563	10.562437	10.015707	10.578143	41
20	9.422318	9.984259	9.438059	10.561941	10.015741	10.577682	40
21	9.422778	9.984224	9.438554	10.561446	10.015776	10.577222	39
22	9.423238	9.984189	9.439048	10.560952	10.015811	10.576762	38
23	9.423697	9.984155	9.439543	10.560457	10.015845	10.576303	37
24	9.424156	9.984110	9.440036	10.559964	10.015880	10.575844	36
25	9.424615	9.984085	9.440529	10.559471	10.015915	10.575385	35
26	9.425073	9.984050	9.441022	10.558978	10.015950	10.574927	34
27	9.425530	9.984015	9.441514	10.558486	10.015985	10.574470	33
28	9.425987	9.983981	9.442006	10.557994	10.016019	10.574013	32
29	9.426443	9.983946	9.442497	10.557503	10.016054	10.573557	31
30	9.426899	9.983910	9.442988	10.557012	10.016090	10.573101	30
	Sine		Tangent		Secant		Minutes

74 Degrees.

15 Degrees.

Minutes	Sine		Tangent		Secant		
30	9.426899	9.983910	9.442958	10.557018	10.016090	10.573101	30
31	9.427354	9.983875	9.443479	10.556521	10.016125	10.572646	29
32	9.427809	9.983840	9.443968	10.556032	10.016160	10.572191	28
33	9.428263	9.983805	9.444455	10.555544	10.016195	10.571737	27
34	9.428717	9.983770	9.444947	10.555053	10.016230	10.571283	26
35	9.429170	9.983735	9.445435	10.554505	10.016205	10.570830	25
36	9.429623	9.983700	9.445923	10.554077	10.016300	10.570377	24
37	9.430075	9.983664	9.446411	10.553589	10.016336	10.569925	23
38	9.430527	9.983629	9.446898	10.553102	10.016371	10.569473	22
39	9.430978	9.983594	9.447384	10.552616	10.016406	10.569022	21
40	9.431429	9.983558	9.447870	10.552130	10.016441	10.568571	20
41	9.431879	9.983523	9.448356	10.551644	10.016477	10.568121	19
42	9.432328	9.983487	9.448841	10.551159	10.016513	10.567672	18
43	9.432778	9.983452	9.449326	10.550674	10.016548	10.567222	17
44	9.433226	9.983416	9.449810	10.550190	10.016584	10.566774	16
45	9.433675	9.983380	9.450294	10.549700	10.016620	10.566325	15
46	9.434122	9.983345	9.450777	10.549223	10.016655	10.565878	14
47	9.434569	9.983309	9.451260	10.548740	10.016691	10.565431	13
48	9.435016	9.983273	9.451742	10.548257	10.016727	10.564984	12
49	9.435462	9.983238	9.452225	10.547775	10.016762	10.564538	11
50	9.435908	9.983202	9.452706	10.547294	10.016798	10.564092	10
51	9.436353	9.983166	9.453187	10.546813	10.016834	10.563647	9
52	9.436798	9.983130	9.453668	10.546332	10.016870	10.563202	8
53	9.437242	9.983094	9.454148	10.545852	10.016906	10.562758	7
54	9.437685	9.983058	9.454628	10.545372	10.016942	10.562314	6
55	9.438129	9.983022	9.455107	10.544893	10.016978	10.561871	5
56	9.438572	9.982986	9.455586	10.544414	10.017014	10.561428	4
57	9.439014	9.982950	9.456064	10.543936	10.017050	10.560986	3
58	9.439456	9.982914	9.456542	10.543458	10.017086	10.560544	2
59	9.439897	9.982878	9.457019	10.542981	10.017122	10.560103	1
60	9.440338	9.982842	9.457496	10.542504	10.017158	10.559662	0

| | Sine | | Tangent | | Secant | | Minutes |

74 Degrees.

16 Degrees.

Minutes	Sine	Tangent	Secant			
0	9.440338	9.982842	9.457496	10.542504	10.017158	10.559662	60
1	9.440778	9.982805	9.457973	10.542027	10.017195	10.559222	59
2	9.441218	9.982769	9.458449	10.541551	10.017231	10.558782	58
3	9.441658	9.982733	9.458925	10.541075	10.017267	10.558342	57
4	9.442096	9.982696	9.459100	10.540900	10.017304	10.557904	56
5	9.442535	9.982660	9.459875	10.540125	10.017340	10.557465	55
6	9.442973	9.982624	9.460349	10.539651	10.017376	10.557027	54
7	9.443410	9.982587	9.460823	10.539177	10.017413	10.556590	53
8	9.443847	9.982551	9.461297	10.538703	10.017449	10.556153	52
9	9.444284	9.982514	9.461770	10.538230	10.017486	10.555716	51
10	9.444720	9.982477	9.462242	10.537758	10.017523	10.555280	50
11	9.445155	9.982441	9.462714	10.537286	10.017559	10.554845	49
12	9.445590	9.982404	9.463186	10.536814	10.017596	10.554410	48
13	9.446025	9.982367	9.463658	10.536342	10.017633	10.553975	47
14	9.446459	9.982331	9.464128	10.535872	10.017669	10.553541	46
15	9.446893	9.982294	9.464599	10.535401	10.017706	10.553107	45
16	9.447326	9.982257	9.465069	10.534931	10.017743	10.552674	44
17	9.447759	9.982220	9.465539	10.534461	10.017780	10.552241	43
18	9.448191	9.982183	9.466008	10.533992	10.017817	10.551809	42
19	9.448623	9.982146	9.466476	10.533524	10.017854	10.551377	41
20	9.449054	9.982109	9.466945	10.533055	10.017891	10.550946	40
21	9.449485	9.982072	9.467413	10.532587	10.017928	10.550515	39
22	9.449915	9.982035	9.467880	10.532120	10.017965	10.550085	38
23	9.450345	9.981998	9.468347	10.531653	10.018002	10.549655	37
24	9.450775	9.981961	9.468814	10.531186	10.018039	10.549225	36
25	9.451204	9.981924	9.469280	10.530720	10.018076	10.548796	35
26	9.451632	9.981886	9.469746	10.530254	10.018114	10.548368	34
27	9.452060	9.981849	9.470211	10.529789	10.018151	10.547940	33
28	9.452488	9.981812	9.470676	10.529324	10.018188	10.547512	32
29	9.452915	9.981774	9.471141	10.528859	10.018226	10.547085	31
30	9.453342	9.981717	9.471605	10.528395	10.018263	10.546658	30

	Sine	Tangent	Secant	Minutes

16 Degrees.

Minutes	Sine	Tangent	,	Secant	
30	9.453342	9.981737	9.471605	10.528395	10.018163	10.54 656	30
31	9.453768	9.981699	9.472068	10.527932	10.018301	10.545232	41
32	9.454194	9.981662	9.472533	10.527468	10.018338	10.545806	2
33	9.454619	9.981625	9.472995	10.527005	10.018375	10.545381	4
34	9.455044	9.981587	9.473457	10.526543	10.018413	10.54495	26
35	9.455469	9.981549	9.473919	10.526081	10.018451	10.544531	25
36	9.455893	9.981512	9.174381	10.525619	10.018488	10.544107	24
37	9.456316	9.981474	9.474842	10.525158	10.018526	10.543684	23
38	9.456739	9.981436	9.475303	10.524697	10.018564	10.543261	22
39	9.457162	9.981399	9.475763	10.524237	10.018601	10.542838	21
40	9.457584	9.981361	9.476222	10.523777	10.018639	10.542416	20
41	9.458006	9.981323	9.476681	10.523317	10.018677	10.541994	19
42	9.458427	9.981285	9.477142	10.522858	10.018715	10.541573	18
43	9.458848	9.981247	9.477601	10.522399	10.018753	10.541152	17
44	9.459268	9.981209	9.478059	10.521941	10.018791	10.540732	16
45	9.459688	9.981171	9.478517	10.521483	10.018829	10.540312	15
46	9.460108	9.981133	9.478975	10.521025	10.018867	10.539892	14
47	9.460527	9.981095	9.479432	10.520568	10.018905	10.539473	13
48	9.460946	9.981057	9.479889	10.520111	10.018943	10.539054	12
49	9.461364	9.981019	9.480345	10.519655	10.018981	10.538636	11
50	9.461782	9.980980	9.480801	10.519199	10.019020	10.538218	10
51	9.462199	9.980942	9.481257	10.518743	10.019058	10.537801	9
52	9.462616	9.980904	9.481712	10.518288	10.019096	10.537384	8
53	9.463032	9.980866	9.482167	10.517833	10.019134	10.536968	7
54	9.463448	9.980827	9.482621	10.517379	10.019173	10.536552	6
55	9.463864	9.980789	9.483075	10.516925	10.019211	10.536136	5
56	9.464279	9.980751	9.483528	10.516471	10.019249	10.535721	4
57	9.464694	9.980712	9.483982	10.516018	10.019288	10.535306	3
58	9.465108	9.980673	9.484435	10.515565	10.019327	10.534892	2
59	9.465522	9.980635	9.484887	10.515113	10.019365	10.534478	1
60	9.465915	9.980596	9.485339	10.514661	10.019404	10.534065	0
	Sine		Tangen		Secant	Minutes

73 Degrees.

17 Degrees.

Minutes	Sine		Tangent		Secant		Minutes
0	9.465935	9.980536	9.485339	10.514661	10.019404	10.534065	60
1	9.466348	9.980558	9.485791	10.514209	10.019442	10.533652	59
2	9.466761	9.980519	9.486242	10.513758	10.019481	10.533239	58
3	9.467173	9.980480	9.486693	10.513307	10.019520	10.532827	57
4	9.467585	9.980441	9.487143	10.512857	10.019559	10.532415	56
5	9.467996	9.980403	9.487593	10.512407	10.019597	10.532004	55
6	9.468407	9.980364	9.488043	10.511957	10.019636	10.531593	54
7	9.468817	9.980325	9.488492	10.511508	10.019675	10.531183	53
8	9.469227	9.980286	9.488941	10.511059	10.019714	10.530773	52
9	9.469637	9.980247	9.489390	10.510610	10.019753	10.530363	51
10	9.470046	9.980208	9.489838	10.510162	10.019792	10.529953	50
11	9.470455	9.980169	9.490286	10.509714	10.019831	10.529545	49
12	9.470863	9.980130	9.490733	10.509267	10.019870	10.529137	48
13	9.471271	9.980091	9.491180	10.508820	10.019909	10.528729	47
14	9.471678	9.980052	9.491627	10.508373	10.019948	10.528322	46
15	9.472086	9.980012	9.492073	10.507927	10.019988	10.527914	45
16	9.472492	9.979973	9.492519	10.507481	10.020027	10.527508	44
17	9.472898	9.979934	9.492965	10.507035	10.020066	10.527102	43
18	9.473304	9.979895	9.493410	10.506590	10.020105	10.526696	42
19	9.473710	9.979855	9.493854	10.506146	10.020145	10.526290	41
20	9.474115	9.979816	9.494299	10.505701	10.020184	10.525885	40
21	9.474519	9.979776	9.494743	10.505257	10.020224	10.525481	39
22	9.474923	9.979737	9.495186	10.504814	10.020263	10.525077	38
23	9.475327	9.979697	9.495630	10.504370	10.020303	10.524673	37
24	9.475730	9.979658	9.496073	10.503927	10.020342	10.524270	36
25	9.476133	9.979618	9.496515	10.503485	10.020382	10.523867	35
26	9.476536	9.979578	9.496957	10.503043	10.020422	10.523464	34
27	9.476938	9.979539	9.497399	10.502601	10.020461	10.523062	33
28	9.477340	9.979499	9.497841	10.502159	10.020501	10.522660	32
29	9.477741	9.979459	9.498282	10.501718	10.020541	10.522259	31
30	9.478142	9.979419	9.498722	10.501278	10.020581	10.521858	30
	Sine		Tangent		Secant		Minutes

72 Degrees.

17 Degrees.

Minutes	Sine	Tangent		Secant		Minutes
30	9.478142	9.979419	9.498722	10.501278	10.020581	10.521858	30
31	9.478542	9.979380	9.499163	10.500837	10.020620	10.521458	29
32	9.478942	9.979340	9.499603	10.500397	10.020660	10.521058	28
33	9.479342	9.979300	9.500042	10.499958	10.020700	10.520658	27
34	9.479711	9.979260	9.500481	10.499519	10.020740	10.520259	26
35	9.480140	9.979220	9.500920	10.499080	10.020780	10.519860	25
36	9.480538	9.979180	9.501359	10.498641	10.020820	10.519461	24
37	9.480937	9.979140	9.501797	10.498203	10.020860	10.519063	23
38	9.481334	9.979100	9.502235	10.497765	10.020900	10.518666	22
39	9.481711	9.979059	9.502672	10.497328	10.020941	10.518269	21
40	9.482128	9.979019	9.503109	10.496891	10.020981	10.517872	20
41	9.482525	9.978979	9.503546	10.496454	10.021021	10.517475	19
42	9.482921	9.978939	9.503982	10.496018	10.021061	10.517079	18
43	9.483316	9.978898	9.504418	10.495582	10.021102	10.516684	17
44	9.483712	9.978858	9.504854	10.495146	10.021142	10.516288	16
45	9.484107	9.978817	9.505289	10.494711	10.021183	10.515893	15
46	9.484501	9.978777	9.505724	10.494276	10.021223	10.515499	14
47	9.484895	9.978736	9.506159	10.493841	10.021264	10.515105	13
48	9.485289	9.978696	9.506593	10.493407	10.021304	10.514717	12
49	9.485681	9.978655	9.507027	10.492973	10.021345	10.514311	11
50	9.486075	9.978615	9.507460	10.492540	10.021385	10.513925	10
51	9.486467	9.978574	9.507893	10.492107	10.021426	10.513533	9
52	9.486859	9.978533	9.508326	10.491674	10.021467	10.513141	8
53	9.487251	9.978493	9.508759	10.491241	10.021507	10.512749	7
54	9.487643	9.978452	9.509191	10.490809	10.021548	10.512357	6
55	9.488033	9.978411	9.509622	10.490378	10.021589	10.511967	5
56	9.488424	9.978370	9.510054	10.489946	10.021630	10.511576	4
57	9.488814	9.978329	9.510485	10.489515	10.021671	10.511186	3
58	9.489204	9.978288	9.510916	10.489084	10.021712	10.510796	2
59	9.489593	9.978247	9.511346	10.488654	10.021753	10.510407	1
60	9.489982	9.978206	9.511776	10.488224	10.021794	10.510018	0

| | Sine | | Tangent | | Secant | | Minutes |

72 Degrees.

A Table of Artificial Sines,

18 Degrees.

Minutes	Sine		Tangent		Secant		
0	9.480982	9.978206	9.511776	10.488224	10.021794	10.500018	60
1	9.490371	9.978165	9.512166	10.487794	10.021835	10.509629	59
2	9.490759	9.978124	9.512635	10.487365	10.021876	10.509241	58
3	9.491147	9.978083	9.513064	10.486936	10.021917	10.508853	57
4	9.491534	9.978042	9.513493	10.486507	10.021958	10.508466	56
5	9.491922	9.978001	9.513921	10.486079	10.021999	10.508078	55
6	9.492308	9.977959	9.514349	10.485651	10.022041	10.507692	54
7	9.492695	9.977918	9.514777	10.485223	10.022082	10.507305	53
8	9.493081	9.977878	9.515204	10.484796	10.022122	10.506919	52
9	9.493466	9.977835	9.515631	10.484369	10.022165	10.506534	51
10	9.493851	9.977794	9.516057	10.483943	10.022206	10.506149	50
11	9.494236	9.977752	9.516484	10.483516	10.022248	10.505764	49
12	9.494620	9.977711	9.516910	10.483090	10.022289	10.505380	48
13	9.495005	9.977669	9.517335	10.482665	10.022331	10.504995	47
14	9.495388	9.977628	9.517761	10.482239	10.022372	10.504612	46
15	9.495772	9.977586	9.518185	10.481815	10.022414	10.504228	45
16	9.496154	9.977544	9.518610	10.481390	10.022456	10.503846	44
17	9.496537	9.977503	9.519034	10.480966	10.022497	10.503463	43
18	9.496919	9.977461	9.519458	10.480542	10.022539	10.503081	42
19	9.497301	9.977419	9.519882	10.480118	10.022581	10.502699	41
20	9.497682	9.977377	9.520305	10.479695	10.022623	10.502318	40
21	9.498063	9.977335	9.520728	10.479272	10.022665	10.501937	39
22	9.498444	9.977293	9.521151	10.478849	10.022707	10.501556	38
23	9.498824	9.977251	9.521573	10.478427	10.022749	10.501176	37
24	9.499204	9.977209	9.521995	10.478005	10.022791	10.500796	36
25	9.499584	9.977167	9.522417	10.477583	10.022833	10.500416	35
26	9.499963	9.977125	9.522838	10.477162	10.022875	10.500037	34
27	9.500342	9.977083	9.523259	10.476741	10.022917	10.499658	33
28	9.500721	9.977041	9.523679	10.476321	10.022959	10.499279	32
29	9.501099	9.976999	9.524100	10.475900	10.023001	10.498901	31
30	9.501476	9.976957	9.524520	10.475480	10.023043	10.498424	30
	Sine		Tangent		Secant		Minutes

71 Degrees.

18 Degrees.

Minutes	Sine	Tangent		Secant			
30	9.501476	9.976957	9.524529	10.475480	10.023043	10.498524	30
31	9.501854	9.976914	9.524939	10.475061	10.023086	10.498146	29
32	9.502231	9.976872	9.525359	10.474641	10.023128	10.497769	28
33	9.502607	9.976830	9.525778	10.474222	10.023170	10.497393	27
34	9.502984	9.976787	9.526197	10.473803	10.023213	10.497016	26
35	9.503360	9.976745	9.526615	10.473385	10.023255	10.496640	25
36	9.503735	9.976702	9.527033	10.472967	10.023298	10.496265	24
37	9.504110	9.976660	9.527451	10.472549	10.023340	10.495889	23
38	9.504485	9.976617	9.527868	10.472132	10.023383	10.495515	22
39	9.504860	9.976574	9.528286	10.471715	10.023426	10.495140	21
40	9.505234	9.976532	9.528702	10.471298	10.023468	10.494766	20
41	9.505608	9.976489	9.529119	10.470881	10.023511	10.494392	19
42	9.505981	9.976446	9.529535	10.470465	10.023554	10.494019	18
43	9.506354	9.976404	9.529950	10.470050	10.023596	10.493646	17
44	9.506727	9.976361	9.530366	10.469634	10.023639	10.493273	16
45	9.507099	9.976318	9.530781	10.469219	10.023682	10.492901	15
46	9.507471	9.976275	9.531196	10.468804	10.023725	10.492529	14
47	9.507843	9.976232	9.531611	10.468389	10.023768	10.492157	13
48	9.508214	9.976189	9.532025	10.467975	10.023811	10.491786	12
49	9.508585	9.976146	9.532439	10.467561	10.023854	10.491415	11
50	9.508955	9.976103	9.532853	10.467147	10.023897	10.491044	10
51	9.509326	9.976060	9.533266	10.466734	10.023940	10.490574	9
52	9.509696	9.976017	9.533679	10.466321	10.023983	10.490304	8
53	9.510065	9.975974	9.534092	10.465908	10.024026	10.489915	7
54	9.510434	9.975930	9.534504	10.465496	10.024070	10.489566	6
55	9.510803	9.975887	9.534916	10.465084	10.024113	10.489197	5
56	9.511172	9.975844	9.535328	10.464672	10.024156	10.488828	4
57	9.511540	9.975800	9.535739	10.464261	10.024200	10.488460	3
58	9.511907	9.975757	9.536150	10.463850	10.024243	10.488093	2
59	9.512275	9.975713	9.536561	10.463439	10.024287	10.487725	1
60	9.512642	9.975670	9.536972	10.463028	10.024330	10.487358	0

	Sine	Tangent	Secant	Minutes

71 Degrees.

A Table of Artificial Sines,

19 Degrees.

Minutes	Sine		Tangent		Secant		Minutes
0	9.512642	9.975670	9.536972	10.463028	10.024330	10.487358	60
1	9.513009	9.975626	9.537382	10.462618	10.024374	10.486991	59
2	9.513375	9.975583	9.537792	10.462208	10.024417	10.486625	58
3	9.513741	9.975539	9.538202	10.461798	10.024461	10.486259	57
4	9.514107	9.975496	9.538611	10.461389	10.024504	10.485893	56
5	9.514472	9.975452	9.539020	10.460980	10.024548	10.485528	55
6	9.514837	9.975408	9.539429	10.460571	10.024592	10.485163	54
7	9.515202	9.975365	9.539837	10.460163	10.024635	10.484798	53
8	9.515566	9.975321	9.540245	10.459755	10.024679	10.484434	52
9	9.515930	9.975277	9.540653	10.459347	10.024723	10.484070	51
10	9.516294	9.975233	9.541061	10.458939	10.024767	10.483706	50
11	9.516657	9.975189	9.541468	10.458532	10.024811	10.483343	49
12	9.517020	9.975145	9.541875	10.458125	10.024855	10.482980	48
13	9.517382	9.975101	9.542281	10.457719	10.024899	10.482618	47
14	9.517745	9.975057	9.542688	10.457312	10.024943	10.482255	46
15	9.518107	9.975013	9.543094	10.456906	10.024987	10.481893	45
16	9.518468	9.974969	9.543499	10.456501	10.025031	10.481532	44
17	9.518829	9.974925	9.543905	10.456095	10.025075	10.481171	43
18	9.519190	9.974880	9.544310	10.455690	10.025120	10.480810	42
19	9.519551	9.974836	9.544715	10.455285	10.025164	10.480449	41
20	9.519911	9.974792	9.545119	10.454881	10.025208	10.480089	40
21	9.520271	9.974747	9.545524	10.454476	10.025253	10.479729	39
22	9.520631	9.974703	9.545928	10.454072	10.025297	10.479369	38
23	9.520990	9.974659	9.546331	10.453669	10.025341	10.479010	37
24	9.521349	9.974614	9.546735	10.453265	10.025386	10.478651	36
25	9.521707	9.974570	9.547138	10.452862	10.025430	10.478293	35
26	9.522066	9.974525	9.547540	10.452460	10.025475	10.477934	34
27	9.522423	9.974481	9.547943	10.452057	10.025519	10.477577	33
28	9.522781	9.974436	9.548345	10.451655	10.025564	10.477219	32
29	9.523138	9.974391	9.548747	10.451253	10.025609	10.476862	31
30	9.523495	9.974347	9.549149	10.450851	10.025653	10.476505	30
		Secant		Tangent		Secant	Minutes

70 Degrees.

19 Degrees.

Minutes	Sine	Tangent	Secant				
30	9.523495	9.974347	9.549149	10.450851	10.025653	10.476505	30
31	9.523862	9.974302	9.549550	10.450450	10.025698	10.476148	29
32	9.524208	9.974257	9.549951	10.450049	10.025743	10.475792	28
33	9.524564	9.974212	9.550352	10.449648	10.025788	10.475436	27
34	9.524920	9.974167	9.550752	10.449248	10.025813	10.475080	26
35	9.525275	9.974122	9.551152	10.448848	10.025878	10.474725	25
36	9.525630	9.974077	9.551552	10.448448	10.025923	10.474370	24
37	9.525984	9.974032	9.551952	10.448048	10.025968	10.474016	23
38	9.526339	9.973987	9.552351	10.447649	10.026013	10.473661	22
39	9.526693	9.973942	9.552750	10.447250	10.026058	10.473307	21
40	9.527040	9.973897	9.553149	10.446851	10.026103	10.472954	20
41	9.527400	9.973852	9.553548	10.446452	10.026148	10.472600	19
42	9.527753	9.973807	9.553946	10.446054	10.026193	10.472247	18
43	9.528105	9.973761	9.554344	10.445656	10.026239	10.471895	17
44	9.528458	9.973716	9.554741	10.445256	10.026284	10.471542	16
45	9.528810	9.973671	9.555139	10.444861	10.026329	10.471190	15
46	9.529161	9.973625	9.555536	10.444464	10.026375	10.470839	14
47	9.529513	9.973580	9.555933	10.444067	10.026420	10.470487	13
48	9.529864	9.973535	9.556329	10.443671	10.026465	10.470136	12
49	9.530215	9.973489	9.556725	10.443275	10.026511	10.469785	11
50	9.530565	9.973443	9.557121	10.442879	10.026557	10.469435	10
51	9.530915	9.973398	9.557517	10.442483	10.026602	10.469085	9
52	9.531265	9.973352	9.557912	10.442088	10.026648	10.468735	8
53	9.531614	9.973307	9.558308	10.441692	10.026693	10.468386	7
54	9.531963	9.973261	9.558702	10.441298	10.026739	10.468037	6
55	9.532312	9.973215	9.559097	10.440903	10.026785	10.467688	5
56	9.532661	9.973169	9.559491	10.440509	10.026831	10.467339	4
57	9.533009	9.973124	9.559885	10.440115	10.026876	10.466991	3
58	9.533357	9.973078	9.560279	10.439720	10.026922	10.466643	2
59	9.533704	9.973032	9.560673	10.439327	10.026968	10.466296	1
60	9.534052	9.972986	9.561066	10.438934	10.027014	10.465948	0
	Sine	Tangent	Secant				Minutes

70 Degrees.

A Table of Artificial Sines,

20 Degrees.

Minutes	Sine		Tangent		Secant		
0	9.534052	9.972986	9.561066	10.438934	10.027014	10.465948	60
1	9.534399	9.972940	9.561459	10.438541	10.027060	10.465601	59
2	9.534745	9.972894	9.561851	10.438149	10.027106	10.465255	58
3	9.535092	9.972848	9.562244	10.437756	10.027152	10.464908	57
4	9.535437	9.972802	9.562636	10.437364	10.027198	10.464563	56
5	9.535783	9.972755	9.563028	10.436972	10.027245	10.464217	55
6	9.536129	9.972709	9.563419	10.436581	10.027291	10.463871	54
7	9.536474	9.972663	9.563811	10.436189	10.027337	10.463526	53
8	9.536818	9.972617	9.564202	10.435798	10.027383	10.463182	52
9	9.537163	9.972570	9.564592	10.435408	10.027430	10.462837	51
10	9.537507	9.972524	9.564983	10.435017	10.027476	10.462493	50
11	9.537851	9.972477	9.565373	10.434627	10.027523	10.462149	49
12	9.538194	9.972431	9.565763	10.434237	10.027569	10.461806	48
13	9.538538	9.972385	9.566153	10.433847	10.027615	10.461462	47
14	9.538880	9.972338	9.566542	10.433458	10.027662	10.461120	46
15	9.539223	9.972291	9.566932	10.433068	10.027709	10.460777	45
16	9.539565	9.972245	9.567321	10.432679	10.027755	10.460435	44
17	9.539907	9.972198	9.567709	10.432291	10.027802	10.460093	43
18	9.540249	9.972151	9.568098	10.431902	10.027849	10.459751	42
19	9.540590	9.972105	9.568486	10.431514	10.027895	10.459410	41
20	9.540931	9.972058	9.568873	10.431127	10.027942	10.459069	40
21	9.541272	9.972011	9.569261	10.430739	10.027989	10.458728	39
22	9.541613	9.971964	9.569648	10.430352	10.028036	10.458387	38
23	9.541953	9.971917	9.570035	10.429965	10.028083	10.458047	37
24	9.542293	9.971870	9.570422	10.429578	10.028130	10.457707	36
25	9.542632	9.971823	9.570809	10.429191	10.028177	10.457368	35
26	9.542971	9.971776	9.571195	10.428805	10.028224	10.457029	34
27	9.543310	9.971729	9.571581	10.428419	10.028271	10.456690	33
28	9.543649	9.971682	9.571967	10.428033	10.028318	10.456351	32
29	9.543987	9.971635	9.572352	10.427648	10.028365	10.456013	31
30	9.544325	9.971588	9.572738	10.427262	10.028412	10.455675	30
		Secant		Tangent		Secant	Minutes

69 Degrees.

20 Degrees.

Minutes	Sine	Tangent		Secant		Minutes
30	9.544325	9.971588	9.572738	10.427262	10.028412	10.455675	30
31	9.544665	9.971540	9.573123	10.426877	10.028460	10.455337	29
32	9.545001	9.971493	9.573507	10.426493	10.028507	10.454999	28
33	9.545338	9.971446	9.573892	10.426108	10.028554	10.454662	27
34	9.545674	9.971398	9.574276	10.425724	10.028602	10.454326	26
35	9.546011	9.971351	9.574660	10.425340	10.028649	10.453989	25
36	9.546347	9.971303	9.575044	10.424956	10.028697	10.453653	24
37	9.546683	9.971256	9.575427	10.424573	10.028744	10.453317	23
38	9.547019	9.971208	9.575810	10.424190	10.028792	10.452981	22
39	9.547354	9.971161	9.576193	10.423807	10.028839	10.452646	21
40	9.547689	9.971113	9.576570	10.423424	10.028887	10.452311	20
41	9.548024	9.971065	9.576958	10.423042	10.028935	10.451976	19
42	9.548358	9.971018	9.577341	10.422659	10.028982	10.451642	18
43	9.548693	9.970970	9.577723	10.422277	10.029030	10.451307	17
44	9.549027	9.970922	9.578104	10.421896	10.029078	10.450973	16
45	9.549360	9.970874	9.578486	10.421514	10.029126	10.450640	15
46	9.549693	9.970826	9.578867	10.421133	10.029174	10.450307	14
47	9.550026	9.970779	9.579248	10.420752	10.029221	10.449974	13
48	9.550359	9.970731	9.579629	10.420371	10.029269	10.449641	12
49	9.550692	9.970683	9.580000	10.419991	10.029317	10.449308	11
50	9.551024	9.970635	9.580389	10.419611	10.029365	10.448976	10
51	9.551356	9.970586	9.580765	10.419231	10.029414	10.448644	9
52	9.551687	9.970538	9.581149	10.418851	10.029462	10.448313	8
53	9.552018	9.970490	9.581528	10.418472	10.029510	10.447982	7
54	9.552349	9.970442	9.581907	10.418093	10.029558	10.447651	6
55	9.552680	9.970394	9.582286	12.417714	10.029606	10.447320	5
56	9.553010	9.970345	9.582665	12.417335	10.029655	10.446990	4
57	9.553341	9.970297	9.583043	12.416957	10.029703	10.446659	3
58	9.553670	9.970249	9.583421	12.416578	10.029751	10.446330	2
59	9.554000	9.970200	9.583800	12.416200	10.029800	10.446000	1
60	9.554329	9.970152	9.584177	12.415823	10.029848	10.445671	0

| | Sine | Tangent | | Secant | | Minutes |

69 Degrees.

21 Degrees.

Minutes	Sine	Tangent		Secant		Minutes
0	9.554329	9.970152	9.584177	10.415823	10.029848	10.445671	60
1	9.554658	9.970103	9.584555	10.415445	10.029897	10.445342	59
2	9.554987	9.970055	9.584932	10.415068	10.029945	10.445013	58
3	9.555315	9.970006	9.585309	10.414691	10.029994	10.444685	57
4	9.555643	9.969957	9.585686	10.414314	10.030043	10.444357	56
5	9.555971	9.969909	9.586062	10.413938	10.030091	10.444029	55
6	9.556299	9.969860	9.586439	10.413561	10.030140	10.443701	54
7	9.556626	9.969811	9.586815	10.413185	10.030189	10.443374	53
8	9.556953	9.969762	9.587190	10.412810	10.030238	10.443047	52
9	9.557280	9.969714	9.587566	10.412434	10.030286	10.442720	51
10	9.557606	9.969665	9.587941	10.412059	10.030335	10.442394	50
11	9.557932	9.969616	9.588316	10.411684	10.030384	10.442068	49
12	9.558258	9.969567	9.588691	10.411309	10.030433	10.441742	48
13	9.558583	9.969518	9.589066	10.410934	10.030482	10.441417	47
14	9.558909	9.969469	9.589440	10.410560	10.030531	10.441091	46
15	9.559234	9.969420	9.589814	10.410186	10.030580	10.440766	45
16	9.559558	9.969370	9.590188	10.409812	10.030630	10.440442	44
17	9.559883	9.969321	9.590562	10.409438	10.030679	10.440117	43
18	9.560207	9.969272	9.590935	10.409065	10.030728	10.439793	42
19	9.560531	9.969223	9.591308	10.408692	10.030777	10.439469	41
20	9.560855	9.969173	9.591681	10.408319	10.030827	10.439145	40
21	9.561178	9.969124	9.592054	10.407946	10.030876	10.438822	39
22	9.561501	9.969075	9.592426	10.407574	10.030925	10.438499	38
23	9.561824	9.969025	9.592798	10.407202	10.030975	10.438176	37
24	9.562146	9.968976	9.593170	10.406830	10.031024	10.437854	36
25	9.562468	9.968926	9.593542	10.406458	10.031074	10.437532	35
26	9.562790	9.968877	9.593914	10.406086	10.031123	10.437210	34
27	9.563112	9.968827	9.594285	10.405715	10.031173	10.436988	33
28	9.563433	9.968777	9.594656	10.405344	10.031223	10.436566	32
29	9.563755	9.968728	9.595027	10.404973	10.031272	10.436245	31
30	9.564075	9.968678	9.595397	10.404603	10.031322	10.435925	30
	Sine	Tangent		Secant		Minutes

68 Degrees.

21 Degrees.

Minutes	Sine	Tangent		Secant		Minutes
30	9.564075	9.968678	9.595397	10.404603	10.031322	10.435925	30
31	9.564396	9.968628	9.595768	10.404232	10.031372	10.435604	29
32	9.564716	9.968578	9.596138	10.403862	10.031422	10.435284	28
33	9.565036	9.968528	9.596508	10.403492	10.031472	10.434964	27
34	9.565356	9.968479	9.596878	10.403122	10.031521	10.434644	26
35	9.565676	9.968429	9.597247	10.402753	10.031571	10.434324	25
36	9.565995	9.968379	9.597616	10.402384	10.031621	10.434005	24
37	9.566314	9.968329	9.597985	10.402015	10.031672	10.433686	23
38	9.566633	9.968278	9.598354	10.401646	10.031722	10.433368	22
39	9.566951	9.968228	9.598722	10.401278	10.031772	10.433049	21
40	9.567269	9.968178	9.599091	10.400929	10.031822	10.432731	20
41	9.567587	9.968128	9.599459	10.400541	10.031872	10.432413	19
42	9.567904	9.968078	9.599827	10.400173	10.031922	10.432096	18
43	9.568222	9.968027	9.600194	10.399806	10.031973	10.431778	17
44	9.568539	9.967977	9.600562	10.399438	10.032023	10.431461	16
45	9.568855	9.967927	9.600929	10.399071	10.032073	10.431145	15
46	9.569172	9.967876	9.601296	10.398704	10.032124	10.430828	14
47	9.569488	9.967826	9.601662	10.398338	10.032174	10.430512	13
48	9.569804	9.967775	9.602029	10.397971	10.032225	10.430196	12
49	9.570120	9.967725	9.602395	10.397605	10.032275	10.429880	11
50	9.570435	9.967674	9.602761	10.397239	10.032326	10.429565	10
51	9.570751	9.967623	9.603127	10.396873	10.032377	10.429249	9
52	9.571066	9.967573	9.603493	10.396507	10.032427	10.428934	8
53	9.571380	9.967522	9.603858	10.396142	10.032478	10.428620	7
54	9.571695	9.967471	9.604223	10.395777	10.032529	10.428305	6
55	9.572009	9.967420	9.604588	10.395412	10.032580	10.427991	5
56	9.572323	9.967370	9.604953	10.395047	10.032630	10.427677	4
57	9.572636	9.967319	9.605317	10.394683	10.032681	10.427364	3
58	9.572949	9.967268	9.605682	10.394318	10.032732	10.427051	2
59	9.573263	9.967217	9.606046	10.393954	10.032783	10.426737	1
60	9.573575	9.967166	9.606410	10.393590	10.032834	10.426425	0

	Sine	Tangent		Secant		Minutes

A Table of Artificial Sines,

22 Degrees.

Minutes	Sine	Tangent		Secant		Minutes
0	9.573575	9.967166	9.606410	10.393590	10.032834	10.420425	60
1	9.573888	9.967115	9.606773	10.393227	10.032885	10.426112	59
2	9.574200	9.967064	9.607137	10.392863	10.032936	10.425800	58
3	9.574512	9.967013	9.607500	10.392500	10.032987	10.425488	57
4	9.574824	9.966961	9.607863	10.392137	10.033039	10.425176	56
5	9.575136	9.966910	9.608225	10.391775	10.033090	10.424864	55
6	9.575447	9.966859	9.608588	10.391412	10.033141	10.424553	54
7	9.575759	9.966807	9.608950	10.391050	10.033193	10.424242	53
8	9.576068	9.966756	9.609312	10.390688	10.033244	10.423932	52
9	9.576379	9.966705	9.609674	10.390326	10.033295	10.423621	51
10	9.576689	9.966653	9.610036	10.389964	10.033347	10.423311	50
11	9.576999	9.966602	9.610397	10.389603	10.033398	10.423001	49
12	9.577309	9.966550	9.610759	10.389241	10.033450	10.422691	48
13	9.577618	9.966499	9.611120	10.388880	10.033501	10.422382	47
14	9.577927	9.966447	9.611480	10.388520	10.033553	10.422073	46
15	9.578236	9.966395	9.611841	10.388159	10.033605	10.421764	45
16	9.578545	9.966344	9.612201	10.387799	10.033656	10.421455	44
17	9.578854	9.966292	9.612561	10.387439	10.033708	10.421146	43
18	9.579162	9.966240	9.612921	10.387079	10.033760	10.420838	42
19	9.579469	9.966188	9.613281	10.386719	10.033812	10.420531	41
20	9.579777	9.966137	9.613641	10.386359	10.033863	10.420223	40
21	9.580084	9.966085	9.614000	10.386000	10.033915	10.419915	39
22	9.580392	9.966033	9.614359	10.385641	10.033967	10.419608	38
23	9.580699	9.965981	9.614718	10.385282	10.034019	10.419301	37
24	9.581005	9.965929	9.615077	10.384923	10.034072	10.418995	36
25	9.581312	9.965876	9.615435	10.384565	10.034124	10.418688	35
26	9.581618	9.965824	9.615793	10.384207	10.034176	10.418382	34
27	9.581924	9.965772	9.616151	10.383849	10.034228	10.418076	33
28	9.582229	9.965720	9.616509	10.383491	10.034280	10.417771	32
29	9.582535	9.965668	9.616867	10.383133	10.034332	10.417465	31
30	9.582847	9.965615	9.617224	10.382770	10.034385	10.417160	30

	Sine	Tangent		Secant		Minutes

67 Degrees.

22 Degrees.

Minutes	Sine	Tangent	Secant		Minutes
30	9.582840	9.965015	9.617224	10.382770 10.034985	10.417100	30
31	9.583145	9.965563	9.617582	10.382418 10.034437	10.416855	29
32	9.583449	9.965511	9.617939	10.382061 10.034489	10.416551	28
33	9.583754	9.965458	9.618295	10.381705 10.034542	10.416246	27
34	9.584058	9.965406	9.618652	10.381348 10.034594	10.415942	26
35	9.584361	9.965353	9.619008	10.380992 10.034647	10.415639	25
36	9.584665	9.965301	9.619365	10.380635 10.034699	10.415335	24
37	9.584968	9.965248	9.619721	10.380279 10.034752	10.415032	23
38	9.585272	9.965195	9.620076	10.379924 10.034805	10.414728	22
39	9.585574	9.965143	9.620432	10.379568 10.034857	10.414426	21
40	9.585877	9.965090	9.620787	10.379213 10.034910	10.414123	20
41	9.586180	9.965037	9.621142	10.378858 10.034963	10.413820	19
42	9.586482	9.964984	9.621497	10.378503 10.035016	10.413518	18
43	9.586784	9.964931	9.621852	10.378148 10.035069	10.413216	17
44	9.587085	9.964879	9.622207	10.377793 10.035121	10.412915	16
45	9.587386	9.964825	9.622561	10.377439 10.035175	10.412614	15
46	9.587688	9.964773	9.622915	10.377085 10.035227	10.412312	14
47	9.587988	9.964719	9.623269	10.376731 10.035281	10.412012	13
48	9.588289	9.964666	9.623623	10.376377 10.035334	10.411711	12
49	9.588590	9.964613	9.623976	10.376024 10.035387	10.411410	11
50	9.588890	9.964560	9.624330	10.375670 10.035440	10.411110	10
51	9.589190	9.964507	9.624683	10.375317 10.035493	10.410810	9
52	9.589489	9.964454	9.625036	10.374964 10.035546	10.410511	8
53	9.589789	9.964400	9.625388	10.374612 10.035600	10.410211	7
54	9.590088	9.964347	9.625741	10.374259 10.035653	10.409912	6
55	9.590387	9.964294	9.626093	10.373907 10.035706	10.409613	5
56	9.590686	9.964240	9.626445	10.373555 10.035760	10.409314	4
57	9.590984	9.964187	9.626797	10.373203 10.035813	10.409016	3
58	9.591282	9.964133	9.627149	10.372851 10.035867	10.408718	2
59	9.591580	9.964080	9.627501	10.372499 10.035920	10.408420	1
60	9.591878	9.964026	9.627852	10.372148 10.035974	10.408122	0
	Sine		Tangent		Secant	Minutes

67 Degrees.

A Table of Artificial Sines,

23 Degrees.

Minutes	Sine		Tangent		Secant		
0	9.591876	9.964026	9.627852	10.372148	10.035974	10.408122	60
1	9.592175	9.963972	9.628203	10.371797	10.036028	10.407825	59
2	9.592473	9.963919	9.628554	10.371446	10.036081	10.407527	58
3	9.592770	9.963865	9.628905	10.371095	10.036135	10.407230	57
4	9.593067	9.963811	9.629255	10.370745	10.036189	10.406933	56
5	9.593363	9.963757	9.629606	10.370394	10.036243	10.406637	55
6	9.593659	9.963704	9.629955	10.370044	10.036296	10.406341	54
7	9.593955	9.963650	9.630306	10.369694	10.036350	10.406045	53
8	9.594251	9.963596	9.630656	10.369344	10.036404	10.405749	52
9	9.594547	9.963542	9.631005	10.368995	10.036458	10.405453	51
10	9.594842	9.963488	9.631354	10.368646	10.036512	10.405158	50
11	9.595137	9.963434	9.631704	10.368296	10.036566	10.404863	49
12	9.595432	9.963379	9.632053	10.367947	10.036621	10.404568	48
13	9.595727	9.963325	9.632401	10.367599	10.036675	10.404273	47
14	9.596021	9.963271	9.632750	10.367250	10.036729	10.403979	46
15	9.596315	9.963217	9.633098	10.366902	10.036783	10.403685	45
16	9.596609	9.963162	9.633447	10.366553	10.036838	10.403391	44
17	9.596903	9.963108	9.633795	10.366205	10.036892	10.403097	43
18	9.597196	9.963054	9.634143	10.365857	10.036946	10.402804	42
19	9.597490	9.962999	9.634490	10.365510	10.037001	10.402510	41
20	9.597783	9.962915	9.634838	10.365162	10.037055	10.402217	40
21	9.598075	9.962840	9.635185	10.364815	10.037110	10.401925	39
22	9.598368	9.962836	9.635532	10.364468	10.037164	10.401632	38
23	9.598660	9.962781	9.635879	10.364121	10.037219	10.401340	37
24	9.598052	9.962727	9.636226	10.363774	10.037273	10.401048	36
25	9.599244	9.962672	9.636572	10.363428	10.037328	10.400756	35
26	9.599536	9.962617	9.636918	10.363082	10.037383	10.400464	34
27	9.599827	9.962562	9.637265	10.362735	10.037438	10.400173	33
28	9.600118	9.962508	9.637611	10.362389	10.037492	10.399882	32
29	9.600409	9.962453	9.637956	10.362044	10.037547	10.399591	31
30	9.600700	9.962398	9.638302	10.361698	10.037602	10.399300	30
	Secant		Tangent		Secant		Minutes

66 Degrees.

23 Degrees.

Minutes	Sine	Tangent	Secant			
30	9.600700	9.962398	9.638302	10 361698	10.037602	10.399303	30
31	9 600990	9.962343	9.638647	10 361353	10.037657	10.399010	29
32	9.601280	9.962288	9.638992	10.361008	10.037712	10.398720	28
33	9.601570	9.962233	9.639337	10 360663	10.037767	10.398430	27
34	9.601860	9.962178	9.639682	10.360318	10.037822	10.398142	26
35	9.602149	9.952123	9.640027	10.359973	10.037877	10.397851	25
36	9.602439	9.962067	9.640371	10.359629	10.037933	10.397561	24
37	9.602728	9.962012	9.640716	10.359284	10.037988	10.397272	23
38	9.603017	9.961957	9.641060	10.358940	10.038043	10.396982	22
39	9.603305	9.961902	9.641404	10.358596	10.038098	10.396695	21
40	9.603594	9.961846	9.641747	10.358253	10.038154	10.396406	20
41	9.603882	9.961791	9.642091	10.357909	10.038209	10.396118	19
42	9.604170	9.961735	9.642434	10.357566	10.038265	10.395830	18
43	9.604457	9.961680	9.642777	10.357223	10.038320	10.395543	17
44	9.604745	9.961624	9.643120	10.356880	10.038376	10.395255	16
45	9.605032	9.961569	9.643463	10.356537	10.038431	10.394968	15
46	9.605319	9.961513	9.643806	10.356194	10.038487	10.394681	14
47	9.605606	9.961458	9.644148	10.355852	10.038542	10.394394	13
48	9.605892	9.961402	9.644490	10.355510	10.038598	10.394108	12
49	9.606179	9.961346	9.644832	10.355168	10.038654	10.393821	11
50	9.606465	9.961290	9.645174	10.354826	10.038710	10.393535	10
51	9.606751	9.961235	9.645516	10.354484	10.038765	10.393249	9
52	9.607036	9.961179	9.645857	10.354143	10.038821	10.392964	8
53	9.607322	9.961123	9.646199	10.353801	10.038877	10.392678	7
54	9.607607	9.961067	9.646540	10.353460	10.038933	10.392393	6
55	9.607892	9.961011	9.646881	10.353119	10.038989	10.392108	5
56	9.608176	9.960955	9.647222	10.352778	10.039045	10.391824	4
57	9.608461	9.960899	9.647562	10.352438	10.039101	10.391539	3
58	9.608745	9.960843	9.647903	10.352097	10.039157	10.391255	2
59	9.609029	9.960786	9.648243	10.351757	10.039214	10.390971	1
60	9.609313	9.960730	9.648583	10.351417	10.039270	10.390687	0

	Sine	.	Tangent		Secant		Minutes

66 Degrees.

24 Degrees.

Minutes	Sine		Tangent		Secant		
0	9.609313	9.960730	9.648583	10.351417	10.039270	10.390000	60
1	9.609597	9.960574	9.648923	10.351077	10.039326	10.390940	59
2	9.609880	9.560618	9.649263	10.350737	10.039382	10.390120	58
3	9.610163	9.960561	9.649602	10.350398	10.039439	10.389837	57
4	9.610446	9.960505	9.649942	10.350058	10.039495	10.389554	56
5	9.610729	9.960448	9.650281	10.349719	10.039552	10.389271	55
6	9.611012	9.960392	9.650620	10.349380	10.039608	10.388988	54
7	9.611294	9.960335	9.650959	10.349041	10.039665	10.388706	53
8	9.611576	9.960279	9.651297	10.348703	10.039721	10.388424	52
9	9.611858	9.960222	9.651636	10.348364	10.039778	10.388142	51
10	9.612140	9.960166	9.651974	10.348026	10.039834	10.387860	50
11	9.612421	9.960100	9.652312	10.347688	10.039891	10.387579	49
12	9.612702	9.960052	9.652650	10.347350	10.039948	10.387298	48
13	9.612983	9.959995	9.652988	10.347012	10.040005	10.387017	47
14	9.613264	9.959936	9.653326	10.346674	10.040062	10.386736	46
15	9.613545	9.959882	9.653663	10.346337	10.040118	10.386455	45
16	9.613825	9.959825	9.654000	10.346000	10.040175	10.386175	44
17	9.614105	9.959768	9.654337	10.345663	10.040232	10.385895	43
18	9.614385	9.959711	9.654674	10.345326	10.040289	10.385615	42
19	9.614665	9.959653	9.655011	10.344989	10.040347	10.385335	41
20	9.614944	9.959595	9.655348	10.344652	10.040404	10.385056	40
21	9.615223	9.959539	9.655684	10.344316	10.040461	10.384777	39
22	9.615502	9.959482	9.656020	10.343980	10.040518	10.384498	38
23	9.615781	9.959425	9.656356	10.343644	10.040575	10.384219	37
24	9.616060	9.959368	9.656692	10.343308	10.040632	10.383940	36
25	9.616338	9.959310	9.657028	10.342972	10.040690	10.383662	35
26	9.616616	9.959253	9.657364	10.342636	10.040747	10.383384	34
27	9.616894	9.959195	9.657699	10.342301	10.040805	10.383106	33
28	9.617172	9.959138	9.658034	10.341966	10.040862	10.382828	32
29	9.617450	9.959081	9.658369	10.341631	10.040919	10.382550	31
30	9.617727	9.959023	9.658704	10.341296	10.040977	10.382273	30
		Secant	Tangent		Secant		Minutes

65 Degrees.

24 Degrees.

Minutes	Sine		Tangent		Secant		Minutes
30	9.617727	9.959023	9.658704	10.341296	10.041977	10.382273	30
31	9.618004	9.958965	9.659039	10.340961	10.041035	10.381591	29
32	9.618281	9.958908	9.659373	10.340627	10.041092	10.381719	28
33	9.618558	9.958850	9.659708	10.340292	10.041150	10.381442	27
34	9.618814	9.958792	9.660042	10.339958	10.041208	10.381166	26
35	9.619110	9.958734	9.660376	10.339624	10.041266	10.380890	25
36	9.619386	9.958677	9.660710	10.339290	10.041323	10.380614	24
37	9.619662	9.958619	9.661043	10.338957	10.041381	10.380338	23
38	9.619938	9.958561	9.661377	10.338623	10.041439	10.380062	22
39	9.620213	9.958503	9.661710	10.338290	10.041497	10.379787	21
40	9.620488	9.958445	9.662043	10.337957	10.041555	10.379512	20
41	9.620763	9.958387	9.662376	10.337624	10.041613	10.379237	19
42	9.621038	9.958329	9.662709	10.337291	10.041671	10.378962	18
43	9.621313	9.958271	9.663042	10.336958	10.041729	10.378687	17
44	9.621587	9.958212	9.663375	10.336625	10.041788	10.378413	16
45	9.621861	9.958154	9.663707	10.336293	10.041846	10.378139	15
46	9.622135	9.958096	9.664039	10.335961	10.041904	10.377865	14
47	9.622409	9.958038	9.664371	10.335629	10.041962	10.377591	13
48	9.622682	9.957979	9.664703	10.335297	10.042021	10.377318	12
49	9.622956	9.957921	9.665035	10.334965	10.042079	10.377044	11
50	9.623229	9.957863	9.665366	10.334634	10.042137	10.376771	10
51	9.623502	9.957804	9.665697	10.334303	10.042196	10.376498	9
52	9.623774	9.957746	9.666029	10.333971	10.042254	10.376226	8
53	9.624047	9.957687	9.666360	10.333640	10.042313	10.375953	7
54	9.624319	9.957628	9.666691	10.333309	10.042372	10.375681	6
55	9.624591	9.957570	9.667021	10.332979	10.042430	10.375409	5
56	9.624863	9.957511	9.667352	10.332648	10.042489	10.375137	4
57	9.625135	9.957452	9.667682	10.332318	10.042548	10.374865	3
58	9.625405	9.957393	9.668013	10.331987	10.042607	10.374594	2
59	9.625677	9.957335	9.668343	10.331657	10.042665	10.374323	1
60	9.625948	9.957276	9.668672	10.331328	10.042724	10.374052	0
	Sine		Tangen		Secant		Minutes

65 Degrees.

25 Degrees.

Minutes	Sine		Tangent		Secant		
0	9.625948	9.957276	9.668672	10.331328	10.042724	10.374052	60
1	9.626219	9.957217	9.669002	10.330998	10.042783	10.373781	59
2	9.626490	9.957158	9.669332	10.330668	10.042842	10.373510	58
3	9.626760	9.957099	9.669661	10.330339	10.042901	10.373240	57
4	9.627030	9.957040	9.669991	10.330009	10.042960	10.372970	56
5	9.627300	9.956981	9.670320	10.329680	10.043019	10.372700	55
6	9.627570	9.956922	9.670649	10.329351	10.043071	10.372430	54
7	9.627840	9.956862	9.670977	10.329023	10.043138	10.372160	53
8	9.628109	9.956803	9.671306	10.328694	10.043197	10.371891	52
9	9.628378	9.956744	9.671634	10.328366	10.043256	10.371622	51
10	9.628647	9.956684	9.671963	10.328037	10.043316	10.371353	50
11	9.628916	9.956625	9.672291	10.327709	10.043375	10.371084	49
12	9.629184	9.956566	9.672619	10.327381	10.043434	10.370816	48
13	9.629453	9.956506	9.672947	10.327053	10.043494	10.370547	47
14	9.629721	9.956447	9.673274	10.326726	10.043553	10.370279	46
15	9.629989	9.956387	9.673602	10.326398	10.043613	10.370011	45
16	9.630257	9.956327	9.673929	10.326071	10.043673	10.369743	44
17	9.630524	9.956268	9.674257	10.325743	10.043732	10.369476	43
18	9.630792	9.956208	9.674584	10.325416	10.043792	10.369208	42
19	9.631059	9.956148	9.674910	10.325090	10.043852	10.368941	41
20	9.631326	9.956089	9.675237	10.324763	10.043911	10.368674	40
21	9.631593	9.956029	9.675564	10.324436	10.043971	10.368407	39
22	9.631859	9.955969	9.675890	10.324110	10.044031	10.368141	38
23	9.632126	9.955909	9.676216	10.323784	10.044091	10.367874	37
24	9.632392	9.955849	9.676543	10.323457	10.044151	10.367608	36
25	9.632658	9.955789	9.676869	10.323131	10.044211	10.367342	35
26	9.632923	9.955729	9.677194	10.322806	10.044271	10.367077	34
27	9.633189	9.955669	9.677520	10.322480	10.044331	10.366811	33
28	9.633454	9.955609	9.677846	10.322154	10.044391	10.366546	32
29	9.633719	9.955548	9.678171	10.321829	10.044452	10.366281	31
30	9.633984	9.955488	9.678496	10.321504	10.044512	10.366016	30

	Sine		Tangent		Secant		Minutes

64 Degrees.

25 Degrees.

Minutes	Sine		Tangen		Secant		Minutes
30	9.63398	9.955488	9.678496	10.321504	10.044512	10.360016	30
31	9.634249	9.955428	9.678821	10.321179	10.044572	10.365751	29
32	9.634514	9.955368	9.679146	10.320854	10.044632	10.365486	28
33	9.634778	9.955307	9.679471	10.320529	10.044693	10.365222	27
34	9.635042	9.955247	9.679795	10.320205	10.044753	10.364956	26
35	9.635306	9.955180	9.680120	10.319880	10.044814	10.364694	25
36	9.635570	9.955126	9.680444	10.319556	10.044874	10.364430	24
37	9.635833	9.955065	9.680768	10.319232	10.044935	10.364167	23
38	9.636097	9.955005	9.681092	10.318908	10.044995	10.363903	22
39	9.636360	9.954944	9.681416	10.318584	10.045056	10.363640	21
40	9.636623	9.954883	9.681740	10.318260	10.045117	10.363377	20
41	9.636886	9.954823	9.682063	10.317937	10.045177	10.363114	19
42	9.637148	9.954762	9.682386	10.317614	10.045238	10.362852	18
43	9.637411	9.954701	9.682710	10.317290	10.045299	10.362589	17
44	9.637673	9.954640	9.683033	10.316967	10.045360	10.362327	16
45	9.637935	9.954579	9.683356	10.316644	10.045421	10.362065	15
46	9.638197	9.954518	9.683678	10.316322	10.045482	10.361803	14
47	9.638458	9.954457	9.684001	10.315999	10.045543	10.361542	13
48	9.638720	9.954396	9.684324	10.315676	10.045604	10.361280	12
49	9.638981	9.954335	9.684646	10.315354	10.045665	10.361019	11
50	9.639243	9.954274	9.684968	10.315032	10.045726	10.360758	10
51	9.639503	9.954213	9.685290	10.314710	10.045787	10.360497	9
52	9.639764	9.954152	9.685612	10.314388	10.045848	10.360236	8
53	9.640024	9.954090	9.685934	10.314066	10.045910	10.359976	7
54	9.640284	9.954029	9.686255	10.313745	10.045971	10.359716	6
55	9.640544	9.953968	9.686577	10.313423	10.046032	10.359456	5
56	9.640804	9.953906	9.686898	10.313102	10.046094	10.359196	4
57	9.641064	9.953845	9.687219	10.312781	10.046155	10.358936	3
58	9.641323	9.953783	9.687540	10.312460	10.046217	10.358677	2
59	9.641583	9.953722	9.687861	10.312139	10.046278	10.358417	1
60	9.641842	9.953660	9.688182	10.311818	10.046340	10.358158	0
	Sine		Tangen		Secant		Minutes

64 Degrees.

A Table of Artificial Sines,

26 Degrees.

Minutes	Sine		Tangent		Secant		Minutes
0	9.641842	9.953660	9.688182	10.311818	10.046340	10.358158	60
1	9.642101	9.953598	9.688502	10.311498	10.046402	10.357899	59
2	9.642360	9.953537	9.688823	10.311177	10.046463	10.357640	58
3	9.642618	9.953475	9.689143	10.310857	10.046525	10.357382	57
4	9.642876	9.953413	9.689463	10.310537	10.046587	10.357124	56
5	9.643135	9.953352	9.689783	10.310217	10.046648	10.356865	55
6	9.643393	9.953290	9.690103	10.309897	10.046710	10.356607	54
7	9.643650	9.953228	9.690423	10.309577	10.046772	10.356350	53
8	9.643908	9.953166	9.690742	10.309258	10.046834	10.356092	52
9	9.644165	9.953104	9.691062	10.308938	10.046896	10.355835	51
10	9.644423	9.953042	9.691381	10.308619	10.046958	10.355577	50
11	9.644680	9.952980	9.691700	10.308300	10.047020	10.355320	49
12	9.644937	9.952917	9.692019	10.307981	10.047083	10.355063	48
13	9.645193	9.952855	9.692338	10.307662	10.047145	10.354807	47
14	9.645450	9.952793	9.692657	10.307343	10.047207	10.354550	46
15	9.645706	9.952731	9.692975	10.307025	10.047269	10.354294	45
16	9.645962	9.952669	9.693293	10.306707	10.047331	10.354038	44
17	9.646218	9.952606	9.693612	10.306388	10.047394	10.353782	43
18	9.646474	9.952544	9.693930	10.306070	10.047456	10.353526	42
19	9.646729	9.952481	9.694248	10.305752	10.047519	10.353271	41
20	9.646984	9.952419	9.694566	10.305434	10.047581	10.353016	40
21	9.647239	9.952356	9.694883	10.305117	10.047644	10.352761	39
22	9.647495	9.952294	9.695201	10.304799	10.047706	10.352505	38
23	9.647749	9.952231	9.695518	10.304482	10.047769	10.352251	37
24	9.648004	9.952168	9.695836	10.304164	10.047832	10.351996	36
25	9.648258	9.952105	9.696153	10.303847	10.047895	10.351742	35
26	9.648512	9.952043	9.696470	10.303530	10.047957	10.351488	34
27	9.648766	9.951980	9.696787	10.303213	10.048020	10.351234	33
28	9.649020	9.951917	9.697103	10.302897	10.048083	10.350980	32
29	9.649274	9.951854	9.697420	10.302580	10.048146	10.350726	31
30	9.649527	9.951791	9.697736	10.302264	10.048209	10.350473	30
	Sine		Tangent		Secant		Minutes

63 Degrees.

26 Degrees.

Minutes	Sine	Tangent		Secant		Minutes
30	9.649527	9.951791	9.697736	10.302264	10.048209	10.350473	30
31	9.649781	9.951728	9.698053	10.301947	10.048272	10.350219	29
32	9.650034	9.951665	9.698369	10.301631	10.048335	10.349966	28
33	9.650287	9.951602	9.698685	10.301315	10.048398	10.349713	27
34	9.650510	9.951539	9.699001	10.300999	10.048461	10.349461	26
35	9.650792	9.951476	9.699311	10.300684	10.048524	10.349208	25
36	9.651044	9.951412	9.699632	10.300368	10.048588	10.348956	24
37	9.651297	9.951349	9.699947	10.300053	10.048651	10.348703	23
38	9.651549	9.951286	9.700263	10.299737	10.048714	10.348451	22
39	9.651800	9.951221	9.700578	10.299422	10.048778	10.348200	21
40	9.652052	9.951159	9.700893	10.299107	10.048841	10.347948	20
41	9.652304	9.951096	9.701208	10.298792	10.048904	10.347696	19
42	9.652555	9.951032	9.701523	10.298477	10.048968	10.347445	18
43	9.652806	9.950968	9.701837	10.298163	10.049032	10.347194	17
44	9.653057	9.950905	9.702152	10.297848	10.049095	10.346943	16
45	9.653307	9.950841	9.702466	10.297534	10.049159	10.346693	15
46	9.653558	9.950778	9.702780	10.297220	10.049222	10.346442	14
47	9.653808	9.950714	9.703075	10.296905	10.049286	10.346192	13
48	9.654059	9.950650	9.703409	10.296591	10.049350	10.345941	12
49	9.654310	9.950686	9.703723	10.296277	10.049414	10.345691	11
50	9.654558	9.950522	9.704036	10.295964	10.049478	10.345442	10
51	9.654808	9.950458	9.704350	10.295650	10.049542	10.345192	9
52	9.655057	9.950394	9.704663	10.295337	10.049506	10.344943	8
53	9.655307	9.950330	9.704977	10.295023	10.049670	10.344693	7
54	9.655556	9.950266	9.705290	10.294710	10.049734	10.344444	6
55	9.655805	9.950202	9.705603	10.294397	10.049798	10.344195	5
56	9.656054	9.950136	9.705916	10.294084	10.049862	10.343946	4
57	9.656302	9.950074	9.706228	10.293772	10.049926	10.343698	3
58	9.656551	9.950009	9.706541	10.293459	10.049991	10.343449	2
59	9.656799	9.949944	9.706854	10.293146	10.050055	10.343201	1
60	9.657047	9.949881	9.707166	10.292834	10.050119	10.342953	0

| | Sine | Tangent | | Secant | | Minutes |

63 Degrees.

27 Degrees.

Minutes	Sine	Tangent	Secant				Minutes
0	9.657047	9.949461	9.707166	10.292834	10.050119	10.342953	60
1	9.657295	9.949816	9.707478	10.292522	10.050184	10.342705	59
2	9.657542	9.949752	9.707790	10.292210	10.050248	10.342458	58
3	9.657790	9.949688	9.708102	10.291898	10.050312	10.342210	57
4	9.658037	9.949623	9.708414	10.291586	10.050377	10.341963	56
5	9.658184	9.949558	9.708726	10.291274	10.050442	10.341716	55
6	9.658531	9.949494	9.709037	10.290963	10.050506	10.341469	54
7	9.658778	9.949429	9.709349	10.290651	10.050571	10.341222	53
8	9.659025	9.949364	9.709660	10.290340	10.050636	10.340975	52
9	9.659271	9.949300	9.709971	10.290029	10.050700	10.340729	51
10	9.659517	9.949235	9.710282	10.289718	10.050765	10.340483	50
11	9.659763	9.949170	9.710593	10.289407	10.050830	10.340237	49
12	9.660009	9.949105	9.710904	10.289096	10.050895	10.339991	48
13	9.660255	9.949040	9.711215	10.288785	10.050960	10.339745	47
14	9.660500	9.948975	9.711525	10.288475	10.051025	10.339500	46
15	9.660746	9.948910	9.711836	10.288164	10.051090	10.339254	45
16	9.660991	9.948845	9.712146	10.287854	10.051155	10.339009	44
17	9.661236	9.948780	9.712456	10.287544	10.051220	10.338764	43
18	9.661481	9.948715	9.712766	10.287234	10.051285	10.338519	42
19	9.661726	9.948649	9.713076	10.286924	10.051351	10.338274	41
20	9.661970	9.948584	9.713386	10.286614	10.051416	10.338030	40
21	9.662214	9.948519	9.713696	10.286304	10.051481	10.337786	39
22	9.662459	9.948453	9.714005	10.285995	10.051547	10.337541	38
23	9.662703	9.948388	9.714314	10.285686	10.051612	10.337297	37
24	9.662946	9.948323	9.714624	10.285376	10.051677	10.337054	36
25	9.663190	9.948257	9.714933	10.285067	10.051743	10.336810	35
26	9.663433	9.948192	9.715242	10.284758	10.051808	10.336567	34
27	9.663677	9.948126	9.715551	10.284449	10.051874	10.336323	33
28	9.663920	9.948060	9.715859	10.284141	10.051940	10.336080	32
29	9.664163	9.947995	9.716168	10.283832	10.052005	10.335837	31
30	9.664406	9.947929	9.716477	10.283523	10.052071	10.335594	30
	Sine	Tangent	Secant				Minutes

62 Degrees.

27 Degrees.

Minutes	Sine		Tangent		Secant		Minutes
30	9.664406	9.947929	9.716477	10.283523	10.052071	10.335594	30
31	9.664648	9.947863	9.716785	10.283215	10.052137	10.335352	29
32	9.664891	9.947797	9.717093	10.282907	10.052203	10.335109	28
33	9.665133	9.947731	9.717401	10.282599	10.052269	10.334867	27
34	9.665375	9.947665	9.717709	10.282291	10.052335	10.334625	26
35	9.665617	9.947599	9.718017	10.281983	10.052401	10.334383	25
36	9.665859	9.947533	9.718325	10.281675	10.052467	10.334141	24
37	9.666100	9.947467	9.718633	10.281367	10.052533	10.333900	23
38	9.666341	9.947401	9.718941	10.281060	10.052599	10.333659	22
39	9.666581	9.947335	9.719248	10.280752	10.052665	10.333417	21
40	9.666822	9.947269	9.719555	10.280445	10.052731	10.333176	20
41	9.667065	9.947203	9.719862	10.280138	10.052797	10.332935	19
42	9.667305	9.947136	9.720169	10.279831	10.052864	10.332695	18
43	9.667546	9.947070	9.720477	10.279524	10.052930	10.332454	17
44	9.667786	9.947004	9.720781	10.279217	10.052996	10.332214	16
45	9.668026	9.946937	9.721088	10.278911	10.053063	10.331974	15
46	9.668266	9.946871	9.721396	10.278604	10.053129	10.331734	14
47	9.668506	9.946804	9.721702	10.278298	10.053196	10.331494	13
48	9.668746	9.946738	9.722008	10.277992	10.053262	10.331254	12
49	9.668986	9.946671	9.722115	10.277685	10.053329	10.331014	11
50	9.669225	9.946604	9.722621	10.277379	10.053396	10.330775	10
51	9.669464	9.946538	9.722927	10.277073	10.053462	10.330536	9
52	9.669703	9.946471	9.723232	10.276768	10.053529	10.330297	8
53	9.669942	9.946404	9.723538	10.276462	10.053595	10.330058	7
54	9.670181	9.946337	9.723844	10.276156	10.053663	10.329819	6
55	9.670419	9.946270	9.724149	10.275851	10.053730	10.329581	5
56	9.670658	9.946203	9.724454	10.275546	10.053797	10.329342	4
57	9.670896	9.946136	9.724759	10.275241	10.053864	10.329104	3
58	9.671134	9.946069	9.725065	10.274935	10.053931	10.328866	2
59	9.671372	9.946002	9.725369	10.274631	10.053998	10.328628	1
60	9.671609	9.945935	9.725674	10.274326	10.054065	10.328391	0
	Sine		Tangent		Secant		Minutes

62 Degrees.

28 Degrees.

Minutes	Sine		Tangent		Secant		
0	9.671609	9.945935	9.725674	10.274326	10.054065	10.328391	60
1	9.671847	9.945868	9.725979	10.274021	10.054132	10.328153	59
2	9.672084	9.945801	9.726284	10.273716	10.054199	10.327916	58
3	9.672321	9.945733	9.726588	10.273412	10.054267	10.327679	57
4	9.672558	9.945666	9.726893	10.273107	10.054334	10.327442	56
5	9.672795	9.945598	9.727197	10.272803	10.054402	10.327205	55
6	9.673032	9.945531	9.727501	10.272499	10.054469	10.326968	54
7	9.673268	9.945464	9.727805	10.272195	10.054536	10.326732	53
8	9.673505	9.945396	9.728109	10.271891	10.054604	10.326495	52
9	9.673741	9.945328	9.728412	10.271588	10.054672	10.326259	51
10	9.673977	9.945261	9.728716	10.271284	10.054739	10.326023	50
11	9.674213	9.945193	9.729020	10.270980	10.054807	10.325787	49
12	9.674448	9.945126	9.729323	10.270677	10.054874	10.325552	48
13	9.674684	9.945058	9.729626	10.270374	10.054942	10.325316	47
14	9.674919	9.944990	9.729929	10.270071	10.055010	10.325081	46
15	9.675155	9.944922	9.730233	10.269767	10.055078	10.324845	45
16	9.675390	9.944854	9.730535	10.269465	10.055146	10.324610	44
17	9.675625	9.944786	9.730838	10.269162	10.055214	10.324375	43
18	9.675859	9.944718	9.731141	10.268859	10.055282	10.324141	42
19	9.676094	9.944650	9.731444	10.268556	10.055350	10.323906	41
20	9.676328	9.944582	9.731746	10.268254	10.055418	10.323672	40
21	9.676562	9.944514	9.732048	10.267952	10.055486	10.323438	39
22	9.676796	9.944446	9.732351	10.267649	10.055554	10.323204	38
23	9.677030	9.944377	9.732653	10.267347	10.055623	10.322970	37
24	9.677261	9.944309	9.732955	10.267045	10.055691	10.322736	36
25	9.677498	9.944241	9.733257	10.266743	10.055759	10.322502	35
26	9.677731	9.944173	9.733558	10.266442	10.055827	10.322269	34
27	9.677964	9.944104	9.733860	10.266140	10.055896	10.322036	33
28	9.678197	9.944036	9.734162	10.265838	10.055964	10.321803	32
29	9.678430	9.943967	9.734463	10.265537	10.056033	10.321570	31
30	9.678663	9.943898	9.734764	10.265236	10.056102	10.321337	30
		Sine		Tangent		Secant	Minutes

61 Degrees.

28 Degrees.

Minutes	Sine	Tangent		Secant		
30	9.678003	9.943898	9.734764	10.265236	10.056102	10.321337	30
31	9.678895	9.943830	9.735066	10.264934	10.056170	10.321105	29
32	9.679128	9.943761	9.735367	10.264633	10.056239	10.320873	28
33	9.679360	9.943693	9.735668	10.264332	10.056307	10.320640	27
34	9.679592	9.943624	9.735958	10.264032	10.056376	10.320408	26
35	9.679824	9.943555	9.736269	10.263731	10.056445	10.320170	25
36	9.680056	9.943486	9.736570	10.263430	10.056514	10.319944	24
37	9.680288	9.943417	9.736871	10.263129	10.056583	10.319712	23
38	9.680519	9.943348	9.737171	10.262829	10.056652	10.319481	22
39	9.680750	9.943279	9.737471	10.262529	10.056721	10.319250	21
40	9.680981	9.943210	9.737771	10.262229	10.056790	10.319018	20
41	9.681212	9.943141	9.738071	10.261929	10.056859	10.318787	19
42	9.681442	9.943072	9.738371	10.261629	10.056928	10.318557	18
43	9.681672	9.943003	9.738671	10.261329	10.056997	10.318326	17
44	9.681902	9.942934	9.738971	10.261029	10.057066	10.318096	16
45	9.682131	9.942864	9.739271	10.260729	10.057136	10.317805	15
46	9.682361	9.942795	9.739570	10.260430	10.057205	10.317635	14
47	9.682590	9.942725	9.739870	10.260130	10.057275	10.317405	13
48	9.682820	9.942656	9.740169	10.259831	10.057344	10.317175	12
49	9.683050	9.942587	9.740468	10.259532	10.057413	10.316945	11
50	9.683280	9.942517	9.740767	10.259233	10.057483	10.316710	10
51	9.683510	9.942448	9.741066	10.258934	10.057552	10.316486	9
52	9.683741	9.942378	9.741365	10.258635	10.057622	10.316257	8
53	9.683972	9.942308	9.741664	10.258336	10.057692	10.316028	7
54	9.684201	9.942239	9.741962	10.258038	10.057761	10.315793	-
55	9.684431	9.942169	9.742261	10.257739	10.057831	10.315570	5
56	9.684656	9.942099	9.742559	10.257441	10.057901	10.315341	4
57	9.684880	9.942029	9.742858	10.257142	10.057971	10.315113	3
58	9.685115	9.941959	9.743156	10.256844	10.058041	10.314885	2
59	9.685343	9.941889	9.743454	10.256546	10.058111	10.314657	1
60	9.685571	9.941819	9.743752	10.256248	10.058181	10.314430	0

| | Sine. | Tangent | | Secant | | Minutes |

61 Degrees.

29 Degrees.

Minutes	Sine	Tangent		Secant		
0	9.685571	9.941819	9.743752	10.256248	10.058181	10.314429	60
1	9.685799	9.941749	9.744050	10.255950	10.058251	10.314201	59
2	9.686027	9.941679	9.744348	10.255652	10.058321	10.313973	58
3	9.686254	9.941609	9.744645	10.255355	10.058391	10.313746	57
4	9.686482	9.941539	9.744943	10.255057	10.058461	10.313518	56
5	9.686709	9.941468	9.745240	10.254760	10.058532	10.313291	55
6	9.686936	9.941398	9.745538	10.254462	10.058602	10.313064	54
7	9.687163	9.941328	9.745835	10.254165	10.058672	10.312837	53
8	9.687189	9.941258	9.746132	10.253868	10.058742	10.312611	52
9	9.687616	9.941187	9.746429	10.253571	10.058813	10.312384	51
10	9.687843	9.941117	9.746726	10.253274	10.058883	10.312157	50
11	9.688069	9.941046	9.747023	10.252977	10.058954	10.311931	49
12	9.688295	9.940975	9.747319	10.252681	10.059025	10.311705	48
13	9.688521	9.940905	9.747616	10.252384	10.059095	10.311479	47
14	9.688747	9.940834	9.747912	10.252088	10.059166	10.311253	46
15	9.688972	9.940763	9.748209	10.251791	10.059237	10.311028	45
16	9.689198	9.940693	9.748505	10.251495	10.059307	10.310802	44
17	9.689423	9.940622	9.748801	10.251199	10.059378	10.310577	43
18	9.689648	9.940551	9.749097	10.250903	10.059449	10.310352	42
19	9.689873	9.940480	9.749393	10.250607	10.059520	10.310127	41
20	9.690098	9.940409	9.749689	10.250311	10.059591	10.309902	40
21	9.690323	9.940338	9.749985	10.250015	10.059662	10.309677	39
22	9.690548	9.940267	9.750281	10.249719	10.059733	10.309452	38
23	9.690772	9.940196	9.750576	10.249424	10.059804	10.309228	37
24	9.690996	9.940125	9.750872	10.249128	10.059875	10.309004	36
25	9.691220	9.940053	9.751167	10.248833	10.059947	10.308780	35
26	9.691444	9.939982	9.751462	10.248538	10.060018	10.308556	34
27	9.691668	9.939911	9.751757	10.248243	10.060089	10.308332	33
28	9.691892	9.939840	9.752052	10.247948	10.060160	10.308108	32
29	9.692115	9.939768	9.752347	10.247653	10.060232	10.307885	31
30	9.692339	9.939697	9.752642	10.247358	10.060303	10.307661	30
	Sine	Tangent		Secant		Minutes

60 Degrees.

29 Degrees.

Minutes	Sine		Tangent		Secant		Minutes
30	9.6923379	9.939697	9.752042	10.247356	10.060303	10.307561	30
31	9.692562	9.939645	9.752937	10.247063	10.060375	10.307436	29
32	9.692785	9.939554	9.753231	10.246769	10.060446	10.307215	28
33	9.693009	9.93:482	9.753526	10.246474	10.060518	10.306991	27
34	9.693211	9.939410	9.753820	10.246180	10.060597	10.306700	26
35	9.693453	9.939339	9.754115	10.245885	10.060661	10.306547	25
36	9.693676	9.939267	9.754409	10.245591	10.060733	10.306324	24
37	9.693898	9.939195	9.754703	10.245297	10.060805	10.306102	23
38	9.694120	9.939123	9.754997	10.245003	10.060877	10.305880	22
39	9.694342	9.939051	9.755291	10.244709	10.060949	10.305658	21
40	9.694564	9.938980	9.755585	10.244415	10.061020	10.305436	20
41	9.694785	9.938908	9.755878	10.244122	10.061092	10.305215	19
42	9.695007	9.938836	9.756172	10.243828	10.061164	10.304993	18
43	9.695228	9.938763	9.756465	10.243535	10.061237	10.304771	17
44	9.695450	9.938691	9.756759	10.243241	10.061309	10.304550	16
45	9.695671	9.938619	9.757052	10.242948	10.061381	10.304329	15
46	9.695892	9.938547	9.757345	10.242655	10.061453	10.304108	14
47	9.696113	9.938475	9.757638	10.242362	10.061525	10.303887	13
48	9.696334	9.938402	9.757931	10.242069	10.061598	10.303666	12
49	9.696554	9.938330	9.758224	10.241776	10.061670	10.303446	11
50	9.696774	9.938258	9.758517	10.241483	10.061742	10.303226	10
51	9.696995	9.938185	9.758810	10.241190	10.061815	10.303005	9
52	9.697215	9.938113	9.759102	10.240898	10.061888	10.302785	8
53	9.697435	9.938040	9.759395	10.240605	10.061960	10.302565	7
54	9.697654	9.937968	9.759688	10.240312	10.062033	10.302346	6
55	9.697874	9.937895	9.759979	10.240021	10.062106	10.302126	5
56	9.698094	9.937822	9.760272	10.239728	10.062178	10.301906	4
57	9.698313	9.937749	9.760564	10.239436	10.062251	10.321687	3
58	9.698532	9.937676	9.760805	10.239144	10.062324	10.301468	2
59	9.698751	9.937603	9.761148	10.238852	10.062397	10.301249	1
60	9.698970	9.937531	9.761430	10.238561	10.062469	10.301030	0

| | Sine | | Tangent | | Secant | | Minutes |

60 Degrees.

K k

30 Degrees.

Minutes	Sine		Tangent		Secant		Minutes
0	9.698970	9.937531	9.761439	10.238501	10.062469	10.301030	60
1	9.699189	9.937458	9.761731	10.238269	10.062541	10.300811	59
2	9.699407	9.937385	9.762023	10.237977	10.062615	10.300593	58
3	9.699626	9.937312	9.762314	10.237686	10.062688	10.300374	57
4	9.699844	9.937238	9.762606	10.237394	10.062762	10.300156	56
5	9.700062	9.937165	9.762897	10.237103	10.062835	10.299938	55
6	9.700280	9.937092	9.763188	10.236812	10.062908	10.299720	54
7	9.700498	9.937019	9.763479	10.236521	10.062981	10.299502	53
8	9.700716	9.936946	9.763770	10.236230	10.063054	10.299284	52
9	9.700933	9.936872	9.764061	10.235939	10.063128	10.299067	51
10	9.701151	9.936799	9.764352	10.235648	10.063201	10.298849	50
11	9.701368	9.936725	9.764643	10.235357	10.063275	10.298632	49
12	9.701585	9.936652	9.764933	10.235067	10.063348	10.298415	48
13	9.701802	9.936578	9.765224	10.234776	10.063422	10.298198	47
14	9.702019	9.936505	9.765514	10.234486	10.063495	10.297981	46
15	9.702236	9.936431	9.765805	10.234195	10.063569	10.297764	45
16	9.702452	9.936357	9.766095	10.233905	10.063643	10.297548	44
17	9.702669	9.936283	9.766385	10.233615	10.063716	10.297331	43
18	9.702885	9.936210	9.766675	10.233325	10.063790	10.297115	42
19	9.703101	9.936136	9.766965	10.233035	10.063864	10.296899	41
20	9.703317	9.936062	9.767255	10.232745	10.063938	10.296683	40
21	9.703533	9.935988	9.767545	10.232455	10.064012	10.296467	39
22	9.703849	9.935914	9.767834	10.232166	10.064086	10.296251	38
23	9.703964	9.935840	9.768124	10.231876	10.064160	10.296036	37
24	9.704170	9.935766	9.768413	10.231587	10.064234	10.295821	36
25	9.704395	9.935692	9.768703	10.231297	10.064308	10.295605	35
26	9.704610	9.935618	9.768992	10.231008	10.064382	10.295390	34
27	9.704825	9.935543	9.769281	10.230719	10.064457	10.295175	33
28	9.705040	9.935469	9.769570	10.230430	10.064531	10.294960	32
29	9.705254	9.935395	9.769860	10.230140	10.064605	10.294746	31
30	9.705469	9.935320	9.770148	10.229852	10.064680	10.294531	30
		Secant		Tangent		Secant	Minutes

59 Degrees.

30 Degrees.

Minutes	Sine	Tangent		Secant		
30	9.703410	9.935343	9.770144	10.229852	10.064686	10.294551	30
31	9.703683	9.935269	9.770437	10.229563	10.064754	10.294317	29
32	9.703895	9.935171	9.770726	10.229274	10.064829	10.294103	28
33	9.706112	9.935509	9.771015	10.228985	10.064703	10.293888	27
34	9.706322	9.935022	9.771303	10.228697	10.064976	10.293674	26
35	9.706535	9.934913	9.771592	10.228408	10.005052	10.293461	25
36	9.706755	9.934683	9.771880	10.228120	10.065127	10.293247	24
37	9.706967	9.934769	9.772168	10.227832	10.065202	10.293033	23
38	9.707178	9.934723	9.772457	10.227543	10.065277	10.292810	22
39	9.707391	9.934649	9.772745	10.227255	10.065351	10.292607	21
40	9.707601	9.934574	9.773033	10.226967	10.065424	10.292394	20
41	9.707810	9.934449	9.773321	10.226679	10.065501	10.292181	19
42	9.708032	9.934424	9.773608	10.226392	10.065576	10.291968	18
43	9.708240	9.934349	9.773896	10.226104	10.065651	10.291755	17
44	9.708455	9.934274	9.774184	10.225816	10.065726	10.291543	16
45	9.708670	9.934199	9.774471	10.225529	10.065801	10.291330	15
46	9.708883	9.934123	9.774759	10.225241	10.065877	10.291118	14
47	9.709094	9.934044	9.775046	10.224954	10.065952	10.290906	13
48	9.709306	9.933973	9.775333	10.224667	10.066027	10.290694	12
49	9.709518	9.933898	9.775621	10.224379	10.066102	10.290482	11
50	9.709730	9.933822	9.775908	10.224092	10.066178	10.290270	10
51	9.709941	9.933747	9.776195	10.223805	10.066253	10.290059	9
52	9.710153	9.933671	9.776482	10.223518	10.066329	10.289847	8
53	9.710364	9.933596	9.776769	10.223231	10.066404	10.289636	7
54	9.710575	9.933520	9.777055	10.222945	10.066480	10.289425	6
55	9.710787	9.933444	9.777342	10.222658	10.066556	10.289214	5
56	9.710997	9.933369	9.777628	10.222372	10.066631	10.289003	4
57	9.711208	9.933293	9.777915	10.222085	10.066707	10.288792	3
58	9.711419	9.933217	9.778201	10.221799	10.066783	10.288581	2
59	9.711629	9.933141	9.778487	10.221513	10.066859	10.288371	1
60	9.711839	9.933066	9.778774	10.221226	10.066934	10.288161	0

| | Sine | Tangent | | Secant | | Minutes |

59 Degrees.

K k 2

31 Degrees.

Minutes	Sine		Tangent		Secant		
0	9.711839	9.932066	9.778774	10.221226	10.066934	10.288161	60
1	9.712049	9.932990	9.779060	10.220940	10.067010	10.287951	59
2	9.712259	9.932914	9.779346	10.220654	10.067086	10.287741	58
3	9.712469	9.932838	9.779632	10.220368	10.067162	10.287531	57
4	9.712679	9.932762	9.779918	10.220082	10.067238	10.287521	56
5	9.712889	9.932685	9.780203	10.219797	10.067315	10.287111	55
6	9.713098	9.932609	9.780489	10.219511	10.067391	10.286902	54
7	9.713308	9.932533	9.780775	10.219225	10.067467	10.286692	53
8	9.713517	9.932457	9.781060	10.218940	10.067543	10.286483	52
9	9.713726	9.932360	9.781346	10.218654	10.067620	10.286274	51
10	9.713935	9.932304	9.781631	10.218369	10.067696	10.286065	50
11	9.714144	9.932228	9.781916	10.218084	10.067772	10.285856	49
12	9.714352	9.932151	9.782201	10.217799	10.067849	10.285648	48
13	9.714561	9.932075	9.782486	10.217514	10.067925	10.285439	47
14	9.714769	9.931998	9.782771	10.217229	10.068002	10.285231	46
15	9.714978	9.931921	9.783056	10.216944	10.068079	10.285022	45
16	9.715186	9.931845	9.783341	10.216659	10.068155	10.284814	44
17	9.715394	9.931768	9.783626	10.216374	10.068232	10.284606	43
18	9.715601	9.931691	9.783910	10.216090	10.068309	10.284399	42
19	9.715809	9.931614	9.784195	10.215805	10.068386	10.284191	41
20	9.716017	9.931537	9.784479	10.215521	10.068463	10.283983	40
21	9.716224	9.931460	9.784764	10.215236	10.068540	10.283776	39
22	9.716432	9.931383	9.785048	10.214952	10.068617	10.283568	38
23	9.716639	9.931306	9.785332	10.214668	10.068694	10.283361	37
24	9.716846	9.931229	9.785616	10.214384	10.068771	10.283154	36
25	9.717053	9.931154	9.785900	10.214100	10.068848	10.282947	35
26	9.717259	9.931075	9.786184	10.213816	10.068925	10.282741	34
27	9.717466	9.930998	9.786468	10.213532	10.069002	10.282534	33
28	9.717672	9.930920	9.786752	10.213248	10.069080	10.282328	32
29	9.717879	9.930843	9.787036	10.212964	10.069157	10.282121	31
30	9.718085	9.930766	9.787319	10.212681	10.069234	10.281915	30

	Sine		Tangent		Secant		Minutes

58 Degrees.

31 Degrees.

Minutes	Sine	Tangent		Secant			Minutes
30	9.718065	9.930765	9.787319	10.211681	10.069234	10.281915	30
31	9.718291	9.930638	9.787603	10.212397	10.069312	10.281709	29
32	9.718497	9.930611	9.787886	10.212114	10.069389	10.281503	28
33	9.718703	9.930533	9.788170	10.211830	10.069467	10.281297	27
34	9.718909	9.930456	9.788453	10.211547	10.069544	10.281091	26
35	9.719114	9.930378	9.788736	10.211264	10.069622	10.280886	25
36	9.719320	9.930300	9.789019	10.210981	10.069700	10.280680	24
37	9.719525	9.930223	9.789302	10.210668	10.069777	10.280475	23
38	9.719730	9.930145	9.789585	10.210415	10.069855	10.280270	22
39	9.719935	9.930007	9.789868	10.210132	10.069913	10.280065	21
40	9.720140	9.929989	9.790151	10.209849	10.070011	10.279860	20
41	9.720345	9.929911	9.790433	10.209567	10.070089	10.279655	19
42	9.720549	9.929833	9.790716	10.209284	10.070167	10.279451	18
43	9.720754	9.929755	9.790999	10.209001	10.070245	10.279246	17
44	9.720958	9.929677	9.791281	10.208719	10.070323	10.279042	16
45	9.721162	9.929599	9.791563	10.208437	10.070401	10.278838	15
46	9.721365	9.929521	9.791846	10.208154	10.070479	10.278634	14
47	9.721570	9.929442	9.792128	10.207872	10.070558	10.278430	13
48	9.721774	9.929364	9.792410	10.207590	10.070636	10.278226	12
49	9.721978	9.929286	9.792692	10.207308	10.070714	10.278022	11
50	9.722181	9.929207	9.792974	10.207026	10.070793	10.277819	10
51	9.722385	9.929129	9.793256	10.206744	10.070871	10.277615	9
52	9.722588	9.929050	9.793538	10.206462	10.070950	10.277412	8
53	9.722791	9.928972	9.793819	10.206181	10.071028	10.277209	7
54	9.722994	9.928893	9.794101	10.205899	10.071107	10.277005	6
55	9.723197	9.928814	9.794383	10.205617	10.071186	10.276803	5
56	9.723400	9.928736	9.794664	10.205336	10.071264	10.276600	4
57	9.723603	9.928657	9.794945	10.205055	10.071343	10.276397	3
58	9.723805	9.928578	9.795227	10.204773	10.071422	10.276195	2
59	9.724007	9.928499	9.795508	10.204492	10.071501	10.275993	1
60	9.724210	9.928420	9.795780	10.204211	10.071580	10.275790	0
		Sine		Tangent		Secant	Minutes

58 Degrees.

32 Degrees.

Minutes	Sine		Tangent		Secant		Minutes
0	9.724210	9.928420	9.795789	10.204211	10.071580	10.275790	60
1	9.724412	9.928341	9.796070	10.203930	10.071659	10.275588	59
2	9.724614	9.928262	9.796351	10.203649	10.071738	10.275386	58
3	9.724816	9.928183	9.796632	10.203368	10.071817	10.275184	57
4	9.725017	9.928104	9.796913	10.203087	10.071896	10.274983	56
5	9.725219	9.928025	9.797194	10.202806	10.071975	10.274781	55
6	9.725420	9.927946	9.797474	10.202526	10.072054	10.274580	54
7	9.725622	9.927867	9.797755	10.202245	10.072133	10.274378	53
8	9.725823	9.927787	9.798030	10.201964	10.072213	10.274177	52
9	9.726024	9.927708	9.798316	10.201684	10.072292	10.273976	51
10	9.726225	9.927628	9.798593	10.201404	10.072372	10.273775	50
11	9.726426	9.927549	9.798877	10.201123	10.072451	10.273574	49
12	9.726624	9.927469	9.799157	10.200843	10.072531	10.273374	48
13	9.726827	9.927390	9.799437	10.200563	10.072610	10.273173	47
14	9.727027	9.927310	9.799717	10.200283	10.072690	10.272973	46
15	9.727278	9.927231	9.799997	10.200003	10.072769	10.272772	45
16	9.727428	9.927151	9.800277	10.199723	10.072849	10.272572	44
17	9.727628	9.927071	9.800557	10.199443	10.072929	10.272372	43
18	9.727828	9.926991	9.800836	10.199164	10.073009	10.272172	42
19	9.728027	9.926911	9.801116	10.198884	10.073089	10.271973	41
20	9.728227	9.926831	9.801396	10.198604	10.073169	10.271773	40
21	9.728427	9.926751	9.801675	10.198325	10.073249	10.271573	39
22	9.728526	9.926671	9.801955	10.198045	10.073329	10.271374	38
23	9.728825	9.926591	9.802234	10.197766	10.073409	10.271175	37
24	9.729024	9.926511	9.802513	10.197487	10.073489	10.270976	36
25	9.729223	9.926431	9.802792	10.197208	10.073569	10.270777	35
26	9.729422	9.926351	9.803072	10.196928	10.073649	10.270578	34
27	9.729621	9.926270	9.803351	10.196649	10.073730	10.270379	33
28	9.729816	9.926190	9.803630	10.196370	10.073810	10.270184	32
29	9.730018	9.926110	9.803908	10.196092	10.073890	10.269982	31
30	9.730216	9.926029	9.804187	10.195813	10.073971	10.269784	30
	Sine		Tangent		Secant		Minutes

57 Degrees.

31 Degrees.

Minutes	Sine	Tangent	Secant			Minutes
30	9.730216	9.926029	9.804187	10.195813	10.073971	10.269764	30
31	9.730415	9.925949	9.804466	10.195534	10.074051	10.269585	29
32	9.730613	9.925868	9.804745	10.195255	10.074132	10.269387	28
33	9.730811	9.925787	9.805023	10.194977	10.074213	10.269189	27
34	9.731009	9.925707	9.805302	10.194698	10.074293	10.268991	26
35	9.731206	9.925626	9.805580	10.194420	10.074374	10.268794	25
36	9.731404	9.925545	9.805859	10.194141	10.074455	10.268596	24
37	9.731601	9.925465	9.806137	10.193863	10.074535	10.268399	23
38	9.731799	9.925384	9.806415	10.193585	10.074616	10.268201	22
39	9.731996	9.925301	9.806693	10.193307	10.074697	10.268004	21
40	9.732193	9.925222	9.806971	10.193029	10.074778	10.267807	20
41	9.732390	9.925141	9.807249	10.192751	10.074859	10.267610	19
42	9.732587	9.925060	9.807527	10.192473	10.074940	10.267413	18
43	9.732784	9.924979	9.807805	10.192195	10.075021	10.267216	17
44	9.732980	9.924897	9.808083	10.191917	10.075103	10.267020	16
45	9.733177	9.924816	9.808361	10.191639	10.075184	10.266823	15
46	9.733373	9.924735	9.808618	10.191382	10.075265	10.266627	14
47	9.733569	9.924653	9.808916	10.191084	10.075347	10.266431	13
48	9.733765	9.924572	9.809193	10.190827	10.075428	10.266235	12
49	9.733961	9.924490	9.809471	10.190529	10.075510	10.266039	11
50	9.734157	9.924409	9.809748	10.190252	10.075591	10.265843	10
51	9.734353	9.924328	9.810025	10.189975	10.075672	10.265647	9
52	9.734548	9.924216	9.810302	10.189658	10.075754	10.265452	8
53	9.734744	9.924164	9.810580	10.189420	10.075836	10.265256	7
54	9.734939	9.924083	9.810857	10.189143	10.075917	10.265061	6
55	9.735134	9.924001	9.811134	10.188866	10.075999	10.264866	5
56	9.735330	9.923919	9.811410	10.188590	10.076081	10.264670	4
57	9.735525	9.923837	9.811687	10.188311	10.076163	10.264475	3
58	9.735719	9.923755	9.811964	10.188036	10.076245	10.264281	2
59	9.735914	9.923673	9.812241	10.187759	10.076327	10.264086	1
60	9.736009	9.923591	9.812517	10.187484	10.076409	10.263891	0

| | Sine | Tangent | Secant | | | Minutes |

33 Degrees.

Minutes	Sine	Tangent		Secant		Minutes
0	9.736109	9.923591	9.812517	10.187483	10.076409	10.263691	60
1	9.736303	9.923569	9.812794	10.187206	10.076493	10.263697	59
2	9.736498	9.923427	9.813070	10.186930	10.076573	10.263502	58
3	9.736692	9.923345	9.813347	10.186653	10.076655	10.263308	57
4	9.736886	9.923263	9.813623	10.186377	10.076737	10.263114	56
5	9.737080	9.923180	9.813899	10.186101	10.076820	10.262920	55
6	9.737274	9.923098	9.814175	10.185825	10.076902	10.262726	54
7	9.737467	9.923016	9.814452	10.185548	10.076984	10.262533	53
8	9.737661	9.922933	9.814728	10.185272	10.077067	10.262339	52
9	9.737855	9.922651	9.815004	10.184996	10.077149	10.262145	51
10	9.738048	9.922768	9.815279	10.184721	10.077232	10.261952	50
11	9.738241	9.922686	9.815555	10.184445	10.077314	10.261759	49
12	9.738434	9.922603	9.815831	10.184169	10.077397	10.261566	48
13	9.738627	9.922520	9.816107	10.183893	10.077480	10.261373	47
14	9.738820	9.922438	9.816382	10.183618	10.077562	10.261180	46
15	9.739013	9.922355	9.816658	10.183342	10.077645	10.260987	45
16	9.739205	9.922272	9.816933	10.183067	10.077728	10.260795	44
17	9.739398	9.922189	9.817209	10.182791	10.077811	10.260602	43
18	9.739590	9.922106	9.817484	10.182516	10.077894	10.260410	42
19	9.739782	9.922023	9.817759	10.182241	10.077977	10.260217	41
20	9.739975	9.921940	9.818035	10.181965	10.078060	10.260025	40
21	9.740167	9.921857	9.818310	10.181690	10.078143	10.259833	39
22	9.740359	9.921774	9.818585	10.181415	10.078226	10.259641	38
23	9.740550	9.921691	9.818860	10.181140	10.078309	10.259450	37
24	9.740742	9.921607	9.819135	10.180865	10.078393	10.259258	36
25	9.740934	9.921524	9.819410	10.180590	10.078476	10.259066	35
26	9.741125	9.921441	9.819684	10.180316	10.078559	10.258875	34
27	9.741316	9.921357	9.819959	10.180041	10.078643	10.258684	33
28	9.741507	9.921274	9.820234	10.179766	10.078726	10.258493	32
29	9.741699	9.921190	9.820508	10.179492	10.078810	10.258301	31
30	9.741889	9.921107	9.820783	10.179217	10.078893	10.258111	30
		Secant		Tangent		Secant	Minutes

56 Degrees.

33 Degrees.

Minute	Sine	Tangent		Secant		
30	9.741889	9.921107	9.620783	10.179217	10.078893	10.258111	30
31	9.742080	9.921023	9.821057	10.178943	10.078977	10.257920	29
32	9.742271	9.920939	9.821332	10.178668	10.079061	10.257729	28
33	9.742462	9.920855	9.821606	10.178394	10.079145	10.257538	27
34	9.742652	9.920772	9.821880	10.178120	10.079228	10.257348	26
35	9.742842	9.920688	9.822154	10.177846	10.079312	10.257158	25
36	9.743033	9.920604	9.822429	10.177571	10.079396	10.256967	24
37	9.743223	9.920520	9.822703	10.177297	10.079480	10.256777	23
38	9.743413	9.920436	9.822977	10.177023	10.079564	10.256587	22
39	9.743602	9.920352	9.823250	10.176750	10.079648	10.256398	21
40	9.743792	9.920268	9.823524	10.176476	10.079732	10.256208	20
41	9.743982	9.920184	9.823798	10.176202	10.079816	10.256018	19
42	9.744171	9.920099	9.824072	10.175928	10.079901	10.255829	18
43	9.744361	9.920015	9.824345	10.175655	10.079985	10.255639	17
44	9.744550	9.919931	9.824619	10.175381	10.080069	10.255450	16
45	9.744739	9.919846	9.824893	10.175107	10.080154	10.255261	15
46	9.744928	9.919762	9.825166	10.174834	10.080238	10.255072	14
47	9.745117	9.919677	9.825439	10.174561	10.080323	10.254883	13
48	9.745306	9.919593	9.825713	10.174287	10.080407	10.254694	12
49	9.745494	9.919508	9.825986	10.174014	10.080492	10.254506	11
50	9.745683	9.919424	9.826259	10.173741	10.080576	10.254317	10
51	9.745871	9.919339	9.826532	10.173468	10.080661	10.254129	9
52	9.746059	9.919254	9.826805	10.173195	10.080746	10.253941	8
53	9.746247	9.919169	9.827078	10.172922	10.080831	10.253752	7
54	9.746436	9.919084	9.827351	10.172649	10.080916	10.253564	6
55	9.746624	9.919000	9.827624	10.172376	10.081000	10.253375	5
56	9.746811	9.918915	9.827897	10.172103	10.081085	10.253189	4
57	9.746999	9.918830	9.828170	10.171830	10.081170	10.253001	3
58	9.747187	9.918744	9.828442	10.171558	10.081256	10.252813	2
59	9.747374	9.918659	9.828715	10.171285	10.081341	10.252626	1
60	9.747562	9.918574	9.828987	10.171013	10.081426	10.252438	0

| | Sine | Tangent | | Secant | | Minute |

56 Degrees.

34 Degrees.

Minutes	Sine		Tangent		Secant		Minutes
0	9.747562	9.918574	9.828987	10.171043	10.081426	10.252438	60
1	9.747749	9.918489	9.829260	10.170740	10.081511	10.252251	59
2	9.747936	9.918404	9.829532	10.170468	10.081596	10.252064	58
3	9.748123	9.918316	9.829805	10.170195	10.081682	10.251877	57
4	9.748310	9.918233	9.830077	10.169923	10.081767	10.251690	56
5	9.748497	9.918147	9.830349	10.169651	10.081853	10.251503	55
6	9.748683	9.918062	9.830621	10.169379	10.081938	10.251317	54
7	9.748870	9.917976	9.830893	10.169107	10.082024	10.251130	53
8	9.749056	9.917891	9.831165	10.168835	10.082109	10.250944	52
9	9.749242	9.917805	9.831437	10.168563	10.082195	10.250758	51
10	9.749429	9.917719	9.831709	10.168291	10.082281	10.250571	50
11	9.749615	9.917634	9.831981	10.168019	10.082366	10.250385	49
12	9.749801	9.917548	9.832253	10.167747	10.082452	10.250199	48
13	9.749987	9.917462	9.832525	10.167475	10.082538	10.250013	47
14	9.750172	9.917376	9.832796	10.167204	10.082624	10.249828	46
15	9.750358	9.917290	9.833008	10.166932	10.082710	10.249642	45
16	9.750543	9.917204	9.833339	10.166661	10.082796	10.249457	44
17	9.750729	9.917118	9.833611	10.166389	10.082882	10.249271	43
18	9.750914	9.917032	9.833882	10.166118	10.082968	10.249086	42
19	9.751099	9.916945	9.834154	10.165846	10.083055	10.248901	41
20	9.751284	9.916859	9.834425	10.165575	10.083141	10.248716	40
21	9.751469	9.916773	9.834696	10.165304	10.083227	10.248531	39
22	9.751654	9.916687	9.834967	10.165033	10.083313	10.248346	38
23	9.751838	9.916600	9.835238	10.164763	10.083400	10.248162	37
24	9.752023	9.916514	9.835509	10.164491	10.083486	10.247977	36
25	9.752207	9.916427	9.835780	10.164220	10.083573	10.247793	35
26	9.752392	9.916341	9.836051	10.163949	10.083659	10.247608	34
27	9.752576	9.916254	9.836322	10.163678	10.083746	10.247424	33
28	9.752760	9.916167	9.836593	10.163407	10.083833	10.247240	32
29	9.752944	9.916080	9.836864	10.163136	10.083920	10.247056	31
30	9.753128	9.915994	9.837134	10.162866	10.084006	10.246872	30
	Sine		Tangent		Secant		Minutes

55 Degrees.

34 Degrees.

Minutes	Sine		Tangent			Secant	Minutes
30	9.753128	9.915994	9.837134	10.162866	10.084006	10.246872	30
31	9.753312	9.915907	9.837405	10.162595	10.084093	10.246688	29
32	9.753495	9.915820	9.837675	10.162325	10.084180	10.246505	28
33	9.753679	9.915733	9.837946	10.162054	10.084267	10.246321	27
34	9.753862	9.915646	9.838216	10.161784	10.084354	10.246138	26
35	9.754046	9.915559	9.838487	10.161513	10.084441	10.245954	25
36	9.754229	9.915472	9.838757	10.161243	10.084528	10.245771	24
37	9.754412	9.915385	9.839027	10.160973	10.084615	10.245588	23
38	9.754595	9.915297	9.839297	10.160703	10.084703	10.245405	22
39	9.754778	9.915210	9.839508	10.160432	10.084790	10.245222	21
40	9.754960	9.915123	9.839838	10.160162	10.084877	10.245040	20
41	9.755141	9.915035	9.840108	10.159892	10.084965	10.244857	19
42	9.755326	9.914948	9.840378	10.159622	10.085052	10.244674	18
43	9.755508	9.914860	9.840647	10.159353	10.085140	10.244492	17
44	9.755690	9.914773	9.840917	10.159083	10.085227	10.244310	16
45	9.755872	9.914685	9.841187	10.158813	10.085315	10.244128	15
46	9.756054	9.914598	9.841457	10.158543	10.085402	10.243946	14
47	9.756235	9.914510	9.841726	10.158274	10.085490	10.243764	13
48	9.756418	9.914422	9.841995	10.158004	10.085578	10.243582	12
49	9.756600	9.914334	9.842265	10.157734	10.085666	10.243400	11
50	9.756781	9.914246	9.842535	10.157465	10.085754	10.243219	10
51	9.756963	9.914158	9.842805	10.157195	10.085842	10.243037	9
52	9.757144	9.914070	9.843074	10.156926	10.085930	10.242856	8
53	9.757326	9.913982	9.843343	10.156657	10.086018	10.242674	7
54	9.757507	9.913894	9.843612	10.156388	10.086106	10.242493	6
55	9.757688	9.913806	9.843882	10.156118	10.086194	10.242312	5
56	9.757869	9.913718	9.844151	10.155849	10.086282	10.242131	4
57	9.758049	9.913630	9.844420	10.155580	10.086370	10.241951	3
58	9.758230	9.913541	9.844689	10.155311	10.086459	10.241770	2
59	9.758411	9.913453	9.844958	10.155042	10.086547	10.241589	1
60	9.758591	9.913364	9.845227	10.154773	10.086636	10.241409	0

	Sine		Tangen			Secant	Minutes

55 Degrees.

35 Degrees.

Minutes	Sine		Tangent		Secant		Minutes
0	9.758591	9.913364	9.845227	10.154773	10.086616	10.241409	60
1	9.758772	9.913276	9.845496	10.154504	10.086724	10.241228	59
2	9.758952	9.913187	9.845764	10.154236	10.086813	10.241048	58
3	9.759132	9.913099	9.846033	10.153967	10.086902	10.240868	57
4	9.759312	9.913010	9.846302	10.153698	10.086990	10.240688	56
5	9.759492	9.912921	9.846570	10.153430	10.087079	10.240508	55
6	9.759672	9.912833	9.846839	10.153161	10.087167	10.240328	54
7	9.759851	9.912744	9.847107	10.152893	10.087256	10.240149	53
8	9.760031	9.912655	9.847375	10.152624	10.087345	10.239969	52
9	9.760211	9.912566	9.847644	10.152356	10.087434	10.239780	51
10	9.760390	9.912477	9.847913	10.152087	10.087523	10.239610	50
11	9.760569	9.912388	9.848181	10.151819	10.087612	10.239431	49
12	9.760748	9.912299	9.848449	10.151551	10.087701	10.239252	48
13	9.760927	9.912210	9.848717	10.151283	10.087790	10.239073	47
14	9.761106	9.912121	9.848985	10.151015	10.087879	10.238894	46
15	9.761285	9.912031	9.849254	10.150746	10.087969	10.238715	45
16	9.761464	9.911942	9.849522	10.150478	10.088058	10.238536	44
17	9.761642	9.911853	9.849790	10.150210	10.088147	10.238358	43
18	9.761831	9.911763	9.850057	10.149943	10.088237	10.238179	42
19	9.761999	9.911674	9.850325	10.149675	10.088326	10.238001	41
20	9.762177	9.911584	9.850593	10.149407	10.088416	10.237823	40
21	9.762356	9.911495	9.850861	10.149139	10.088505	10.237644	39
22	9.762534	9.911405	9.851128	10.148872	10.088595	10.237466	38
23	9.762712	9.911315	9.851396	10.148604	10.088685	10.237288	37
24	9.762889	9.911226	9.851664	10.148336	10.088774	10.237111	36
25	9.763067	9.911136	9.851931	10.148069	10.088864	10.236933	35
26	9.763245	9.911046	9.852199	10.147801	10.088954	10.236755	34
27	9.763422	9.910956	9.852466	10.147534	10.089044	10.236578	33
28	9.763600	9.910866	9.852733	10.147267	10.089134	10.236400	32
29	9.763777	9.910776	9.853001	10.146999	10.089224	10.236223	31
30	9.763954	9.910686	9.853268	10.146732	10.089314	10.236046	30
	Sine		Tangent		Secant		Minutes

54 Degrees.

35 Degrees.

Minutes	Sine		Tangent		Secant		Minutes
30	9.763954	9.910686	9.853268	10.146732	10.089314	10.236046	30
31	9.764131	9.910596	9.853535	10.146465	10.089404	10.235869	29
32	9.764308	9.910506	9.853802	10.146198	10.089494	10.235692	28
33	9.764485	9.910415	9.854069	10.145931	10.089585	10.235515	27
34	9.764662	9.910325	9.854336	10.145664	10.089675	10.235338	26
35	9.764838	9.910235	9.854603	10.145397	10.089765	10.235162	25
36	9.765015	9.910144	9.854870	10.145130	10.089856	10.234985	24
37	9.765191	9.910054	9.855137	10.144863	10.089946	10.234809	23
38	9.765367	9.909963	9.855404	10.144596	10.090037	10.234633	22
39	9.765544	9.909873	9.855671	10.144329	10.090127	10.234456	21
40	9.765720	9.909782	9.855938	10.144062	10.090218	10.234280	20
41	9.765896	9.909691	9.856204	10.143796	10.090309	10.234104	19
42	9.766071	9.909601	9.856471	10.143529	10.090399	10.233929	18
43	9.766247	9.909510	9.856737	10.143263	10.090490	10.233753	17
44	9.766423	9.909419	9.857004	10.142996	10.090581	10.233577	16
45	9.766599	9.909328	9.857270	10.142730	10.090672	10.233402	15
46	9.766774	9.909237	9.857537	10.142463	10.090763	10.233226	14
47	9.766949	9.909146	9.857803	10.142197	10.090854	10.233051	13
48	9.767124	9.909055	9.858069	10.141931	10.090945	10.232876	12
49	9.767300	9.908964	9.858336	10.141664	10.091036	10.232700	11
50	9.767475	9.908873	9.858602	10.141398	10.091127	10.232525	10
51	9.767649	9.908781	9.858868	10.141132	10.091219	10.232351	9
52	9.767824	9.908690	9.859134	10.140866	10.091310	10.232176	8
53	9.767999	9.908599	9.859400	10.140600	10.091401	10.232001	7
54	9.768173	9.908507	9.859666	10.140334	10.091493	10.231827	6
55	9.768348	9.908416	9.859932	10.140068	10.091584	10.231652	5
56	9.768522	9.908324	9.860198	10.139802	10.091676	10.231478	4
57	9.768697	9.908233	9.860464	10.139536	10.091767	10.231303	3
58	9.768871	9.908141	9.860730	10.139270	10.091859	10.231129	2
59	9.769045	9.908049	9.860995	10.139005	10.091951	10.230955	1
60	9.769219	9.907958	9.861261	10.138739	10.092042	10.230781	0
	Sine		Tangent		Secant		Minutes

54 Degrees.

36 Degrees.

Minutes	Sine		Tangent		Secant		Minutes
0	9.769819	9.907958	9.861261	10.138739	10.092042	10.230781	60
1	9.769392	9.907866	9.861527	10.138473	10.092134	10.230608	59
2	9.769566	9.907774	9.861792	10.138208	10.092226	10.230434	58
3	9.769740	9.907682	9.862058	10.137942	10.092318	10.230260	57
4	9.769912	9.907590	9.862323	10.137677	10.092410	10.230088	56
5	9.770087	9.907498	9.862589	10.137411	10.092502	10.229913	55
6	9.770260	9.907406	9.862854	10.137146	10.092594	10.229740	54
7	9.770433	9.907314	9.863119	10.136881	10.092680	10.229567	53
8	9.770606	9.907222	9.863385	10.136615	10.092778	10.229394	52
9	9.770779	9.907129	9.863650	10.136350	10.092871	10.229221	51
10	9.770952	9.907037	9.863915	10.136085	10.092903	10.229048	50
11	9.771125	9.906945	9.864180	10.135820	10.093055	10.228875	49
12	9.771298	9.906852	9.864445	10.135555	10.093148	10.228702	48
13	9.771470	9.906760	9.864711	10.135289	10.093240	10.228530	47
14	9.771643	9.906667	9.864975	10.135025	10.093333	10.228357	46
15	9.771815	9.906574	9.865240	10.134760	10.093426	10.228185	45
16	9.771987	9.906482	9.865505	10.134495	10.093518	10.228013	44
17	9.772159	9.906389	9.865770	10.134230	10.093611	10.227841	43
18	9.772331	9.906296	9.866035	10.133965	10.093704	10.227669	42
19	9.772503	9.906204	9.866300	10.133700	10.093797	10.227497	41
20	9.772675	9.906111	9.866564	10.133436	10.093889	10.227325	40
21	9.772847	9.906018	9.866829	10.133171	10.093982	10.227153	39
22	9.773018	9.905925	9.867094	10.132906	10.094075	10.226982	38
23	9.773190	9.905832	9.867358	10.132642	10.094168	10.226810	37
24	9.773361	9.905739	9.867623	10.132377	10.094261	10.226639	36
25	9.773533	9.905645	9.867887	10.132113	10.094355	10.226467	35
26	9.773704	9.905552	9.868152	10.131848	10.094448	10.226296	34
27	9.773875	9.905459	9.868416	10.131584	10.094541	10.226125	33
28	9.774046	9.905366	9.868680	10.131320	10.094634	10.225954	32
29	9.774217	9.905272	9.868945	10.131055	10.094728	10.225783	31
30	9.774388	9.905179	9.869209	10.130791	10.094821	10.225612	30
	Sine		Tangent		Secant		Minutes

53 Degrees.

36 Degrees.

Minutes	Sine	Tangent	Secant		Minutes
30	9.774388 9.905179	9.869209 10.130791	10.094821 10.225612		30
31	9.774558 9.905085	9.869473 10.130527	10.094915 10.225442		29
32	9.774729 9.904992	9.869737 10.130263	10.095008 10.225271		28
33	9.774899 9.904898	9.870001 10.129999	10.095102 10.225101		27
34	9.775070 9.904804	9.870265 10.129735	10.095195 10.224930		26
35	9.775240 9.904711	9.870529 10.129471	10.095289 10.224760		25
36	9.775410 9.904617	9.870793 10.129207	10.095383 10.224590		24
37	9.775580 9.904523	9.871057 10.128943	10.095477 10.224420		23
38	9.775750 9.904429	9.871321 10.128679	10.095571 10.224250		22
39	9.775920 9.904335	9.871585 10.128415	10.095665 10.224080		21
40	9.776090 9.904241	9.871849 10.128151	10.095759 10.223910		20
41	9.776259 9.904147	9.872112 10.127888	10.095853 10.223741		19
42	9.776429 9.904053	9.872376 10.127624	10.095947 10.223571		18
43	9.776598 9.903959	9.872640 10.127360	10.096011 10.223402		17
44	9.776768 9.903864	9.872903 10.127097	10.096136 10.223232		16
45	9.776937 9.903770	9.873167 10.126833	10.096233 10.223063		15
46	9.777106 9.903676	9.873430 10.126570	10.096324 10.222894		14
47	9.777275 9.903581	9.873694 10.126306	10.096419 10.222725		13
48	9.777444 9.903487	9.873957 10.126043	10.096513 10.222556		12
49	9.777613 9.903392	9.874220 10.125780	10.096608 10.222387		11
50	9.777781 9.903298	9.874484 10.125516	10.096702 10.222219		10
51	9.777950 9.903203	9.874747 10.125253	10.096797 10.222050		9
52	9.778119 9.903108	9.875010 10.124990	10.096892 10.221882		8
53	9.778287 9.903014	9.875273 10.124727	10.096985 10.221713		7
54	9.778455 9.902919	9.875536 10.124464	10.097081 10.221545		6
55	9.778623 9.902824	9.875800 10.124200	10.097176 10.221377		5
56	9.778792 9.902729	9.876063 10.123937	10.097271 10.221208		4
57	9.778960 9.902634	9.876326 10.123674	10.097365 10.221040		3
58	9.779128 9.902539	9.876589 10.123411	10.097461 10.220872		2
59	9.779296 9.902444	9.876851 10.123149	10.097555 10.220705		1
60	9.779464 9.902349	9.877114 10.122886	10.097651 10.220537		0

| | Sine | Tangent | Secant | | Minutes |

53 Degrees.

37 Degrees.

Minute	Sine		Tangent		Secant		Minute
0	9.779463	9.902349	9.877114	10.122886	10.097651	10.220537	60
1	9.779631	9.902253	9.877377	10.122623	10.097747	10.220309	59
2	9.779798	9.902158	9.877940	10.122360	10.097842	10.220202	58
3	9.779965	9.902063	9.877903	10.122097	10.097937	10.220035	57
4	9.780133	9.901967	9.878165	10.121835	10.098033	10.219867	56
5	9.780300	9.501872	9.878428	10.121572	10.098128	10.219700	55
6	9.780467	9.901776	9.878691	10.121309	10.098224	10.219533	54
7	9.780634	9.901681	9.878953	10.121047	10.098319	10.219366	53
8	9.780801	9.901585	9.879216	10.120784	10.098415	10.219199	52
9	9.780968	9.901489	9.879478	10.120522	10.098511	10.219032	51
10	9.781134	9.901394	9.879741	10.120259	10.098606	10.218866	50
11	9.781301	9.901298	9.880003	10.119997	10.098702	10.218699	49
12	9.781467	9.901202	9.880265	10.119735	10.098798	10.218533	48
13	9.781634	9.901106	9.880528	10.119472	10.098894	10.218366	47
14	9.781800	9.901010	9.880790	10.119210	10.098990	10.218200	46
15	9.781966	9.900914	9.881052	10.118948	10.099086	10.217034	45
16	9.782132	9.900818	9.881314	10.118686	10.099182	10.217868	44
17	9.782298	9.900722	9.881576	10.118424	10.099278	10.217702	43
18	9.782464	9.900626	9.881839	10.118161	10.099374	10.217536	42
19	9.782630	9.900529	9.882101	10.117899	10.099471	10.217370	41
20	9.782796	9.900433	9.882363	10.117637	10.099567	10.217204	40
21	9.782961	9.900337	9.882625	10.117375	10.099663	10.217039	39
22	9.783127	9.900240	9.882887	10.117113	10.099760	10.216873	38
23	9.783292	9.900144	9.883148	10.116852	10.099856	10.216708	37
24	9.783457	9.900047	9.283410	10.116590	10.099953	10.216543	36
25	9.783623	9.899951	9.883672	10.116328	10.100049	10.216377	35
26	9.783788	9.899854	9.883934	10.116066	10.100149	10.216212	34
27	9.783953	9.899757	9.884196	10.115804	10.100243	10.216047	33
28	9.784118	9.899660	9.884457	10.115543	10.100340	10.215882	32
29	9.784282	9.899564	9.884719	10.115231	10.100436	10.215718	31
30	9.784447	9.899467	9.884980	10.115020	10.100533	10.215553	30
		Sine		Tangent		Secant	Minutes

52 Degrees.

37 Degrees.

Minutes	Sine	Tangent		Secant		Minutes
30	9.784447	9.899467	9.884980	10.115020	10.100533	10.215553	30
31	9.784612	9.899170	9.885242	10.114758	10.100630	10.215380	29
32	9.784776	9.899273	9.885503	10.114497	10.100727	10.215224	28
33	9.784941	9.899176	9.885765	10.114235	10.100824	10.215059	27
34	9.785105	9.899007	9.886026	10.113974	10.100922	10.214895	26
35	9.785269	9.898981	9.886288	10.113712	10.101019	10.214731	25
36	9.785433	9.898884	9.886549	10.113451	10.101116	10.214567	24
37	9.785597	9.898787	9.886810	10.113190	10.101213	10.214403	23
38	9.785761	9.898689	9.887072	10.112928	10.101311	10.214239	22
39	9.785925	9.898592	9.887333	10.112667	10.101408	10.214075	21
40	9.786089	9.898494	9.887594	10.112405	10.101506	10.213911	20
41	9.786252	9.898397	9.887855	10.112145	10.101603	10.213748	19
42	9.786416	9.898299	9.888116	10.111884	10.101701	10.213584	18
43	9.786579	9.898202	9.888377	10.111623	10.101798	10.213421	17
44	9.786742	9.898104	9.888639	10.111361	10.101896	10.213258	16
45	9.786906	9.898006	9.888900	10.111100	10.101994	10.213094	15
46	9.787069	9.897908	9.889160	10.100840	10.102092	10.212931	14
47	9.787232	9.897810	9.889421	10.100579	10.102190	10.212768	13
48	9.787395	9.897712	9.889682	10.100318	10.102288	10.212605	12
49	9.787557	9.897614	9.889943	10.100057	10.102386	10.212443	11
50	9.787720	9.897516	9.890204	10.109796	10.102484	10.212280	10
51	9.787883	9.897418	9.890465	10.109535	10.102582	10.212117	9
52	9.788045	9.897320	9.890725	10.109275	10.102680	10.211955	8
53	9.788208	9.897222	9.890986	10.109014	10.102778	10.211792	7
54	9.788370	9.897123	9.891247	10.108753	10.102877	10.211630	6
55	9.788532	9.897025	9.891507	10.108493	10.102975	10.211468	5
56	9.788694	9.896926	9.891768	10.108232	10.103074	10.211306	4
57	9.788856	9.896828	9.892028	10.107972	10.103172	10.211144	3
58	9.789018	9.896729	9.892289	10.107711	10.103271	10.210982	2
59	9.789180	9.896631	9.892549	10.107451	10.103369	10.210820	1
60	9.789342	9.896532	9.892810	10.107190	10.103468	10.210658	0

| | Sine | Tangent | | Secant | | Minutes |

52 Degrees.

L l

A Table of Artificial Sines,

38 Degrees.

Minutes	Sine		Tangent		Secant		
0	9.789342	9.806533	9.892810	10.107190	10.103468	10.210658	60
1	9.789504	9.806433	9.893070	10.106930	10.103567	10.210496	59
2	9.789665	9.806335	9.893331	10.106669	10.103665	10.210335	58
3	9.789827	9.806236	9.893591	10.106409	10.103764	10.210173	57
4	9.789988	9.806137	9.893851	10.106149	10.103863	10.210012	56
5	9.790149	9.806038	9.894111	10.105889	10.103962	10.209851	55
6	9.790310	9.805939	9.894371	10.105629	10.104061	10.209690	54
7	9.790471	9.805840	9.894632	10.105368	10.104160	10.209529	53
8	9.790632	9.805741	9.894892	10.105108	10.104259	10.209368	52
9	9.790791	9.805641	9.895152	10.104847	10.104359	10.209207	51
10	9.790954	9.805542	9.895412	10.104588	10.104458	10.209046	50
11	9.791115	9.805443	9.895672	10.104328	10.104557	10.208885	49
12	9.791275	9.805343	9.895932	10.104068	10.104657	10.208725	48
13	9.791436	9.805244	9.896192	10.103808	10.104756	10.208564	47
14	9.791596	9.805144	9.896452	10.103548	10.104856	10.208404	46
15	9.791757	9.805045	9.896712	10.103288	10.104955	10.208243	45
16	9.791917	9.804945	9.896971	10.103029	10.105055	10.208083	44
17	9.792077	9.804846	9.897231	10.102769	10.105154	10.207923	43
18	9.792237	9.804746	9.897491	10.102509	10.105254	10.207763	42
19	9.792397	9.804646	9.897751	10.102249	10.105354	10.207603	41
20	9.792557	9.804546	9.898010	10.101990	10.105454	10.207443	40
21	9.792716	9.804447	9.898270	10.101730	10.105553	10.207284	39
22	9.792876	9.804347	9.898530	10.101470	10.105654	10.207124	38
23	9.793035	9.804247	9.898789	10.101211	10.105754	10.206965	37
24	9.793195	9.804146	9.899049	10.100951	10.105854	10.206805	36
25	9.793354	9.804046	9.899308	10.100692	10.105954	10.206645	35
26	9.793513	9.803946	9.899568	10.100432	10.106054	10.206487	34
27	9.793673	9.803846	9.899827	10.100173	10.106154	10.206327	33
28	9.793832	9.803745	9.900086	10.099914	10.106255	10.206168	32
29	9.793991	9.803645	9.900346	10.099654	10.106355	10.206009	31
30	9.794150	9.803544	9.900605	10.099395	10.106456	10.205850	30
		Sine		Tangent		Secant	Minutes

51 Degrees.

38 Degrees.

M.	Sine	Tangent		Secant		
30	9.794150	9.893544	9.900605	10.0.9395	10.106456	10.205851	30
31	9.794308	9.893144	9.900804	10.099136	10.106556	10.205672	29
32	9.794467	9.893343	9.901124	10.068876	10.106657	10.205533	28
33	9.794626	9.893243	9.901383	10.098617	10.106757	10.205374	27
34	9.794784	9.893142	9.901612	10.098357	10.106858	10.205216	26
35	9.794942	9.893041	9.901901	10.098099	10.106959	10.205056	25
36	9.795101	9.892940	9.902160	10.097840	10.107060	10.204899	24
37	9.795259	9.892839	9.902419	10.097581	10.107161	10.204741	23
38	9.795417	9.892738	9.902679	10.097321	10.107262	10.204583	22
39	9.795575	9.892617	9.902938	10.097062	10.107363	10.204425	21
40	9.795733	9.892536	9.903197	10.096803	10.107464	10.204267	20
41	9.795891	9.892435	9.903455	10.096545	10.107565	10.204109	19
42	9.796049	9.892334	9.903714	10.096286	10.107666	10.203951	18
43	9.796206	9.892233	9.903973	10.096027	10.107767	10.203794	17
44	9.796364	9.892132	9.904231	10.095768	10.107868	10.203636	16
45	9.796521	9.892030	9.904491	10.095509	10.107970	10.203479	15
46	9.796679	9.891929	9.904750	10.095250	10.108071	10.203321	14
47	9.796835	9.891827	9.905008	10.094992	10.108173	10.203164	13
48	9.796993	9.891726	9.905267	10.094733	10.108274	10.203007	12
49	9.797150	9.891624	9.905525	10.094474	10.108376	10.202850	11
50	9.797307	9.891523	9.905784	10.093216	10.108477	10.202693	10
51	9.797464	9.891421	9.906043	10.093957	10.108579	10.202535	9
52	9.797621	9.891319	9.906302	10.093698	10.108681	10.202379	8
53	9.797777	9.891217	9.906560	10.093440	10.108783	10.202223	7
54	9.797934	9.891115	9.906819	10.093181	10.108885	10.202066	6
55	9.798091	9.891013	9.907077	10.092923	10.108987	10.201909	5
56	9.798247	9.890911	9.907336	10.092664	10.109089	10.201753	4
57	9.798403	9.890809	9.907594	10.092406	10.109191	10.201597	3
58	9.798560	9.890707	9.907852	10.092148	10.109293	10.201440	2
59	9.798716	9.890605	9.908111	10.091889	10.109395	10.201284	1
60	9.798872	9.890503	9.908369	10.091611	10.109497	10.201128	0
	Sine	Tangent		Secant		M.

51 Degrees.

39 Degrees.

Minutes	Sine		Tangent		Secant		
0	9.798872	9.890503	9.908369	10.091631	10.109497	10.201128	60
1	9.799028	9.890400	9.908627	10.091373	10.109600	10.200972	59
2	9.799184	9.890298	9.908886	10.091114	10.109702	10.200816	58
3	9.799339	9.890195	9.909144	10.090856	10.109805	10.200661	57
4	9.799495	9.890093	9.909402	10.090598	10.109907	10.200505	56
5	9.799651	9.889990	9.909660	10.090340	10.110010	10.200349	55
6	9.799806	9.889888	9.909918	10.090082	10.110112	10.200194	54
7	9.799962	9.889785	9.910177	10.089823	10.110215	10.200038	53
8	9.800117	9.889682	9.910435	10.089565	10.110318	10.199883	52
9	9.800272	9.889579	9.910693	10.089307	10.110421	10.199728	51
10	9.800427	9.889476	9.910951	10.089049	10.110524	10.199573	50
11	9.800582	9.889374	9.911209	10.088791	10.110626	10.199418	49
12	9.800737	9.889271	9.911467	10.088533	10.110729	10.199263	48
13	9.800892	9.889167	9.911724	10.088276	10.110833	10.199108	47
14	9.801047	9.889064	9.911982	10.088018	10.110936	10.198953	46
15	9.801201	9.888961	9.912240	10.087760	10.111039	10.198799	45
16	9.801356	9.888858	9.912498	10.087502	10.111142	10.198644	44
17	9.801511	9.888755	9.912756	10.087244	10.111245	10.198489	43
18	9.801665	9.888651	9.913014	10.086986	10.111349	10.198335	42
19	9.801819	9.888548	9.913271	10.086729	10.111452	10.198181	41
20	9.801973	9.888444	9.913529	10.086471	10.111556	10.198027	40
21	9.802128	9.888341	9.913787	10.086213	10.111659	10.197872	39
22	9.802282	9.888237	9.914044	10.085956	10.111763	10.197718	38
23	9.802435	9.888133	9.914302	10.085698	10.111867	10.197565	37
24	9.802589	9.888030	9.914560	10.085440	10.111970	10.197411	36
25	9.802743	9.887926	9.914817	10.085183	10.112074	10.197257	35
26	9.802897	9.887822	9.915075	10.084925	10.112178	10.197103	34
27	9.803050	9.887718	9.915332	10.084668	10.112282	10.196950	33
28	9.803204	9.887614	9.915590	10.084410	10.112386	10.196796	32
29	9.803357	9.887510	9.915847	10.084153	10.112490	10.196643	31
30	9.803510	9.887406	9.916104	10.083896	10.112594	10.196490	30
	Sine		Tangent		Secant		Minutes

50 Degrees.

39 Degrees.

Minutes	Sine	Tangent		Secant		Minutes
30	9.803510	9.887400	9.916104	10.083895	10.112594	10.196499	30
31	9.803664	9.887302	9.916361	10.083635	10.112696	10.196336	29
32	9.803817	9.887198	9.916619	10.083381	10.112802	10.196183	28
33	9.803970	9.887093	9.916876	10.083124	10.112907	10.196030	27
34	9.804121	9.886989	9.917134	10.082866	10.113011	10.195877	26
35	9.804276	9.886885	9.917391	10.082609	10.113115	10.195724	25
36	9.804428	9.886780	9.917648	10.082352	10.113220	10.195572	24
37	9.804581	9.886676	9.917905	10.082095	10.113324	10.195419	23
38	9.804734	9.886571	9.918163	10.081837	10.113429	10.195266	22
39	9.804886	9.886466	9.918420	10.081580	10.113534	10.195114	21
40	9.805038	9.886362	9.918677	10.081323	10.113638	10.194962	20
41	9.805191	9.886257	9.918934	10.081066	10.113743	10.194809	19
42	9.805343	9.886152	9.919191	10.080809	10.113848	10.194657	18
43	9.805495	9.886047	9.919448	10.080552	10.113953	10.194505	17
44	9.805647	9.885942	9.919705	10.080295	10.114058	10.194353	16
45	9.805799	9.885837	9.919962	10.080038	10.114163	10.194201	15
46	9.805951	9.885732	9.920219	10.079781	10.114268	10.194049	14
47	9.806103	9.885627	9.920475	10.079525	10.114373	10.193897	13
48	9.806254	9.885521	9.920733	10.079267	10.114479	10.193746	12
49	9.806406	9.885416	9.920990	10.079010	10.114584	10.193594	11
50	9.806557	9.885311	9.921247	10.078753	10.114690	10.193443	10
51	9.806709	9.885205	9.921503	10.078497	10.114795	10.193291	9
52	9.806860	9.885100	9.921760	10.078240	10.114900	10.193140	8
53	9.807011	9.884994	9.922017	10.077983	10.115006	10.192989	7
54	9.807163	9.884889	9.922274	10.077726	10.115111	10.192837	6
55	9.807314	9.884783	9.922530	10.077470	10.115217	10.192686	5
56	9.807465	9.884677	9.922787	10.077213	10.115323	10.192535	4
57	9.807616	9.884572	9.923044	10.076956	10.115428	10.192385	3
58	9.807766	9.884466	9.923300	10.076700	10.115534	10.192234	2
59	9.807917	9.884360	9.923557	10.076443	10.115640	10.192083	1
60	9.808067	9.884254	9.923814	10.076187	10.115746	10.191933	0

| | Sine | | Tangen | | Secant | Minutes |

50 Degrees.

40 Degrees.

Minutes	Sine		Tangent		Secant		Minutes
0	9.808068	9.884254	9.923813	10.076187	10.115746	10.191932	60
1	9.808218	9.884148	9.924070	10.075930	10.115852	10.191782	59
2	9.808308	9.884042	9.924327	10.075673	10.115958	10.191632	58
3	9.808519	9.883936	9.924583	10.075417	10.116064	10.191481	57
4	9.808669	9.883829	9.924840	10.075160	10.116171	10.191331	56
5	9.808819	9.883723	9.925096	10.074904	10.116277	10.191181	55
6	9.808969	9.883617	9.925352	10.074648	10.116383	10.191031	54
7	9.809119	9.883510	9.925609	10.074391	10.116490	10.190881	53
8	9.809269	9.883404	9.925865	10.074135	10.116596	10.190731	52
9	9.809419	9.883297	9.926121	10.073879	10.116703	10.190581	51
10	9.809569	9.883191	9.926378	10.073622	10.116809	10.190431	50
11	9.809718	9.883084	9.926634	10.073366	10.116916	10.190282	49
12	9.809868	9.882977	9.926890	10.073110	10.117023	10.190132	48
13	9.810017	9.882871	9.927147	10.072853	10.117129	10.189983	47
14	9.810167	9.882764	9.927403	10.072597	10.117236	10.189833	46
15	9.810316	9.882657	9.927659	10.072341	10.117343	10.189684	45
16	9.810465	9.882550	9.927915	10.072085	10.117450	10.189535	44
17	9.810614	9.882443	9.928171	10.071829	10.117557	10.189386	43
18	9.810763	9.882336	9.928427	10.071573	10.117664	10.189237	42
19	9.810912	9.882228	9.928683	10.071317	10.117772	10.189088	41
20	9.811061	9.882121	9.928894	10.071106	10.117879	10.188939	40
21	9.811210	9.882014	9.929196	10.070804	10.117986	10.188790	39
22	9.811358	9.881907	9.929452	10.070548	10.118093	10.188642	38
23	9.811507	9.881799	9.929708	10.070292	10.118201	10.188493	37
24	9.811655	9.881692	9.929964	10.070036	10.118308	10.188345	36
25	9.811804	9.881584	9.930219	10.069781	10.118416	10.188196	35
26	9.811952	9.881477	9.930475	10.069525	10.118523	10.188048	34
27	9.812100	9.881360	9.930731	10.069269	10.118631	10.187900	33
28	9.812248	9.881251	9.930987	10.069013	10.118739	10.187752	32
29	9.812397	9.881153	9.931243	10.068757	10.118847	10.187603	31
30	9.812544	9.881045	9.931499	10.068501	10.118955	10.187456	30
	Sine		Tangent		Secant		Minutes

49 Degrees.

40 Degrees.

Minutes	Sine	Tangent		Secant			
30	9.812544	9.881045	9.931499	10.068501	10.118955	10.187450	30
31	9.812693	9.880936	9.931755	10.068245	10.119062	10.187308	29
32	9.812840	9.880830	9.932010	10.067990	10.119170	10.187160	28
33	9.812988	7.880721	9.932256	10.067734	10.119279	10.187011	27
34	9.813136	9.880613	9.932522	10.067478	10.119387	10.186865	26
35	9.813283	9.880505	9.932778	10.067222	10.119495	10.186717	25
36	9.813430	9.880397	9.933033	10.066967	10.119603	10.186570	24
37	9.813578	9.880289	9.933289	10.066711	10.119711	10.186422	23
38	9.813725	9.880180	9.933545	10.066455	10.119820	10.186275	22
39	9.813872	9.880072	9.933800	10.066100	10.119928	10.186128	21
40	9.814019	9.879964	9.934056	10.065944	10.120037	10.185981	20
41	9.814166	9.879855	9.934311	10.065689	10.120145	10.185834	19
42	9.814313	9.879747	9.934567	10.065433	10.120254	10.185687	18
43	9.814460	9.879637	9.934823	10.065177	10.120363	10.185540	17
44	9.814607	9.879529	9.935078	10.064922	10.120471	10.185393	16
45	9.814753	9.879420	9.935333	10.064667	10.120580	10.185247	15
46	9.814900	9.879311	9.935589	10.064411	10.120689	10.185100	14
47	9.815046	9.879202	9.935844	10.064156	10.120798	10.184954	13
48	9.815193	9.879093	9.936100	10.063900	10.120907	10.184807	12
49	9.815339	9.878984	9.936355	10.063645	10.121016	10.184661	11
50	9.815485	9.878875	9.936610	10.063390	10.121125	10.184515	10
51	9.815631	9.878766	9.936866	10.063134	10.121234	10.184369	9
52	9.815778	9.878656	9.937121	10.062879	10.121344	10.184222	8
53	9.815923	9.878547	9.937377	10.062623	10.121453	10.184077	7
54	9.816069	9.878438	9.937621	10.062168	10.121562	10.183931	6
55	9.816215	9.878328	9.937887	10.062113	10.121672	10.183785	5
56	9.816361	9.878219	9.938142	10.061858	10.121781	10.183637	4
57	9.816507	9.878109	9.938397	10.061603	10.121891	10.183493	3
58	9.816652	9.877999	9.938653	10.061347	10.122001	10.183348	2
59	9.816797	9.877890	9.938908	10.061092	10.122110	10.183203	1
60	9.816943	9.877780	9.939161	10.060817	10.122220	10.183057	0
		Secant		Tangent		Secant	Minutes

40 Degrees.

A Table of Artificial Sines,

41 Degrees.

Minutes	Sine	Tangent		Secant		
0	9.816943	9.877780	9.939163	10.060837	10.122220	10.183057	60
1	9.817088	9.877670	9.939418	10.060582	10.122330	10.182912	59
2	9.817233	9.877560	9.939673	10.060327	10.122440	10.182767	58
3	9.817378	9.877450	9.939928	10.060072	10.122550	10.182621	57
4	9.817523	9.877340	9.940183	10.059817	10.122660	10.182477	56
5	9.817668	9.877230	9.940438	10.059562	10.122770	10.182332	55
6	9.817813	9.877120	9.940694	10.059306	10.122880	10.182187	54
7	9.817958	9.877010	9.940949	10.059051	10.122990	10.182042	53
8	9.818103	9.876899	9.941204	10.058796	10.123101	10.181897	52
9	9.818247	9.876789	9.941458	10.058542	10.123211	10.181753	51
10	9.818392	9.876678	9.941713	10.058287	10.123322	10.181608	50
11	9.818536	9.876568	9.941968	10.058032	10.123432	10.181464	49
12	9.818681	9.876457	9.942223	10.057777	10.123543	10.181319	48
13	9.818825	9.876347	9.942478	10.057522	10.123653	10.181175	47
14	9.818969	9.876236	9.942733	10.057267	10.123764	10.181031	46
15	9.819113	9.876125	9.942988	10.057012	10.123875	10.180887	45
16	9.819257	9.876014	9.943243	10.056757	10.123986	10.180743	44
17	9.819401	9.875904	9.943498	10.056502	10.124096	10.180599	43
18	9.819545	9.875793	9.943752	10.056248	10.124207	10.180455	42
19	9.819689	9.875682	9.944007	10.055993	10.124318	10.180311	41
20	9.819832	9.875571	9.944262	10.055738	10.124429	10.180168	40
21	9.819976	9.875459	9.944517	10.055483	10.124541	10.180024	39
22	9.820120	9.875348	9.944771	10.055229	10.124652	10.179880	38
23	9.820263	9.875237	9.945026	10.054974	10.124763	10.179737	37
24	9.820406	9.875126	9.945281	10.054719	10.124874	10.179594	36
25	9.820550	9.875014	9.945535	10.054465	10.124986	10.179450	35
26	9.820693	9.874903	9.945790	10.054210	10.125097	10.179307	34
27	9.820836	9.874791	9.946045	10.053955	10.125209	10.179164	33
28	9.820979	9.874679	9.946299	10.053701	10.125321	10.179021	32
29	9.821122	9.874568	9.946554	10.053446	10.125432	10.178878	31
30	9.821265	9.874456	9.946808	10.053192	10.125544	10.178735	30
		Secant		Tangent		Secant	Minutes

48 Degrees.

41 Degrees.

Minutes	Sine		Tangent		Secant		Minutes
30	9.821265	9.874456	9.946808	10.053192	10.125544	10.178735	30
31	9.821407	9.874344	9.947063	10.052937	10.125656	10.178593	29
32	9.821550	9.874232	9.947317	10.052683	10.125768	10.178450	28
33	9.821693	9.874120	9.947572	10.052428	10.125880	10.178307	27
34	9.821835	9.874008	9.947826	10.052174	10.125992	10.178165	26
35	9.821977	9.873896	9.948081	10.051919	10.126104	10.178023	25
36	9.822120	9.873784	9.948335	10.051665	10.126216	10.177880	24
37	9.822262	9.873672	9.948590	10.051410	10.126328	10.177738	23
38	9.822404	9.873560	9.948844	10.051155	10.126440	10.177595	22
39	9.822546	9.873448	9.949099	10.050901	10.126552	10.177454	21
40	9.822688	9.873335	9.949353	10.050647	10.126665	10.177312	20
41	9.822830	9.873223	9.949607	10.050393	10.126777	10.177170	19
42	9.822972	9.873110	9.949862	10.050138	10.126890	10.177028	18
43	9.823114	9.872998	9.950116	10.049884	10.127002	10.176886	17
44	9.823255	9.872885	9.950370	10.049630	10.127115	10.176745	16
45	9.823397	9.872772	9.950625	10.049375	10.127228	10.176603	15
46	9.823539	9.872659	9.950879	10.049121	10.127341	10.176461	14
47	9.823680	9.872547	9.951133	10.048867	10.127453	10.176320	13
48	9.823821	9.872434	9.951388	10.048612	10.127566	10.176179	12
49	9.823963	9.872321	9.951642	10.048358	10.127679	10.176037	11
50	9.824104	9.872208	9.951897	10.048104	10.127792	10.175899	10
51	9.824245	9.872094	9.952150	10.047850	10.127906	10.175755	9
52	9.824386	9.871981	9.952404	10.047596	10.128019	10.175614	8
53	9.824527	9.871868	9.952659	10.047341	10.128132	10.175473	7
54	9.824668	9.871755	9.952913	10.047087	10.128245	10.175332	6
55	9.824809	9.871641	9.953167	10.046833	10.128359	10.175191	5
56	9.824942	9.871528	9.953421	10.046579	10.128472	10.175051	4
57	9.825090	9.871414	9.953675	10.046325	10.128586	10.174910	3
58	9.825230	9.871301	9.953929	10.046071	10.128699	10.174770	2
59	9.825370	9.871187	9.954183	10.045817	10.128813	10.174630	1
60	9.825511	9.871073	9.954437	10.045563	10.128927	10.174489	0
	Sine		Tangent		Secant		Minutes

48 Degrees.

42 Degrees.

Minutes	Sine		Tangent		Secant		Minutes
0	9.825511	9.871073	9.954437	10.045563	10.128927	10.174489	60
1	9.825651	9.870960	9.954691	10.045309	10.129040	10.174349	59
2	9.825791	9.870846	9.954945	10.045055	10.129154	10.174209	58
3	9.825931	9.870732	9.955199	10.044801	10.129268	10.174069	57
4	9.826071	9.870618	9.955453	10.044547	10.129382	10.173929	56
5	9.826211	9.870504	9.955707	10.044293	10.129496	10.173789	55
6	9.826351	9.870390	9.955961	10.044039	10.129610	10.173649	54
7	9.826491	9.870276	9.956215	10.043785	10.129724	10.173509	53
8	9.826631	9.870161	9.956469	10.043531	10.129839	10.173369	52
9	9.826770	9.870047	9.956723	10.043277	10.129953	10.173230	51
10	9.826910	9.869933	9.956977	10.043023	10.130067	10.173090	50
11	9.827049	9.869818	9.957231	10.042769	10.130182	10.172951	49
12	9.827189	9.869704	9.957485	10.042515	10.130296	10.172811	48
13	9.827328	9.869589	9.957739	10.042261	10.130411	10.172672	47
14	9.827467	9.869474	9.957993	10.042007	10.130526	10.172533	46
15	9.827606	9.869360	9.958210	10.041754	10.130640	10.172394	45
16	9.827745	9.869245	9.958500	10.041500	10.130755	10.172255	44
17	9.827884	9.869130	9.958754	10.041246	10.130870	10.172110	43
18	9.828023	9.869015	9.959008	10.040992	10.130985	10.171977	42
19	9.828162	9.868900	9.959262	10.040738	10.131100	10.171838	41
20	9.828301	9.868785	9.959515	10.040485	10.131215	10.171699	40
21	9.828439	9.868670	9.959769	10.040231	10.131330	10.171561	39
22	9.828578	9.868555	9.960023	10.039977	10.131445	10.171422	38
23	9.828716	9.868440	9.960277	10.039723	10.131560	10.171284	37
24	9.828855	9.868324	9.960530	10.039470	10.131676	10.171145	36
25	9.828993	9.868209	9.960784	10.039216	10.131791	10.171007	35
26	9.829131	9.868093	9.961038	10.038962	10.131907	10.170869	34
27	9.829269	9.867978	9.961291	10.038709	10.132022	10.170731	33
28	9.829407	9.867862	9.961545	10.038455	10.132138	10.170593	32
29	9.829545	9.867747	9.961799	10.038201	10.132253	10.170455	31
30	9.829683	9.867631	9.962052	10.037948	10.132369	10.170317	30
	Sine		Tangent		Secant		Minutes

47 Degrees.

42 Degrees.

Minutes	Sine		Tangent		Secant			
30	9.829083	9.857631	9.962052	10.037918	10.138169	10.170317	30
31	9.829281	9.857515	9.962230	10.037694	10.132485	10.170179	29
32	9.829359	9.857379	9.962560	10.037440	10.132601	10.170041	28
33	9.830097	9.857203	9.952813	10.037187	10.132717	10.169903	27
34	9.832234	9.857167	9.603067	10.036013	10.132833	10.169765	26
35	9.830372	9.857051	9.963320	10.036680	10.132949	10.169628	25
36	9.830509	9.856935	9.963574	10.036426	10.133065	10.169491	24
37	9.830646	9.856819	9.963827	10.036173	10.133181	10.169354	23
38	9.830783	9.856703	9.964081	10.035919	10.133297	10.169216	22
39	9.830921	9.856586	9.964315	10.035665	10.133414	10.169079	21
40	9.831058	9.856470	9.964588	10.035412	10.133530	10.168942	20
41	9.831195	9.856353	9.964842	10.035158	10.133647	10.168805	19
42	9.831331	9.856237	9.965095	10.034905	10.133763	10.168668	18
43	9.831468	9.856120	9.965349	10.034651	10.133880	10.168531	17
44	9.831606	9.856004	9.965602	10.034398	10.133996	10.168394	16
45	9.831742	9.855887	9.965855	10.034145	10.134113	10.168258	15
46	9.831879	9.855770	9.966109	10.033891	10.134230	10.168121	14
47	9.832015	9.855653	9.966362	10.033638	10.134347	10.167985	13
48	9.832152	9.855536	9.966611	10.033384	10.134464	10.167848	12
49	9.832288	9.855419	9.966869	10.033131	10.134581	10.167712	11
50	9.832425	9.855302	9.967122	10.032878	10.134698	10.167575	10
51	9.832561	9.855185	9.967376	10.032624	10.134815	10.167439	9
52	9.832697	9.855068	9.967630	10.032370	10.134931	10.167303	8
53	9.832833	9.854950	9.967883	10.032117	10.135050	10.167167	7
54	9.832969	9.854833	9.968136	10.031864	10.135167	10.167031	6
55	9.833105	9.854716	9.968389	10.031611	10.135284	10.166895	5
56	9.833241	9.854598	9.958643	10.031357	10.135402	10.166759	4
57	9.833377	9.854481	9.968896	10.031104	10.135519	10.166623	3
58	9.833512	9.854363	9.969149	10.030851	10.135637	10.166488	2
59	9.833648	9.854245	9.969403	10.030597	10.135755	10.166352	1
60	9.833783	9.854127	9.969656	10.030344	10.135873	10.166217	0

	Sine		Tangent		Secant		Minutes

47 Degrees.

43 Degrees.

Minutes	Sine		Tangent		Secant		
0	9.833783	9.864127	9.969656	10.030344	10.135873	10.166217	60
1	9.833919	9.864010	9.969909	10.030091	10.135990	10.166081	59
2	9.834054	9.863892	9.970162	10.029838	10.136108	10.165946	58
3	9.834189	9.863774	9.970416	10.029584	10.136226	10.165811	57
4	9.834325	9.863656	9.970669	10.029331	10.136344	10.165675	56
5	9.834460	9.863538	9.970922	10.029078	10.136462	10.165540	55
6	9.834595	9.863419	9.971175	10.028825	10.136581	10.165405	54
7	9.834730	9.863301	9.971429	10.028571	10.136699	10.165270	53
8	9.834865	9.863183	9.971682	10.028318	10.136817	10.165135	52
9	9.834999	9.863064	9.971935	10.028065	10.136936	10.165001	51
10	9.835134	9.862946	9.972188	10.027812	10.137054	10.164866	50
11	9.835269	9.862827	9.972441	10.027559	10.137173	10.164731	49
12	9.835403	9.862709	9.972695	10.027305	10.137291	10.164597	48
13	9.835538	9.862590	9.972948	10.027052	10.137410	10.164462	47
14	9.835672	9.862471	9.973201	10.026799	10.137529	10.164328	46
15	9.835807	9.862353	9.973454	10.026546	10.137647	10.164193	45
16	9.835941	9.862234	9.973707	10.026293	10.137766	10.164059	44
17	9.836075	9.862115	9.973960	10.026040	10.137885	10.163925	43
18	9.836209	9.861996	9.974213	10.025787	10.138004	10.163791	42
19	9.836343	9.861877	9.974466	10.025534	10.138123	10.163657	41
20	9.836477	9.861758	9.974719	10.025281	10.138242	10.163523	40
21	9.836611	9.861638	9.974973	10.025027	10.138362	10.163389	39
22	9.836745	9.861519	9.975226	10.024774	10.138481	10.163255	38
23	9.836878	9.861400	9.975479	10.024521	10.138600	10.163122	37
24	9.837012	9.861280	9.975732	10.024268	10.138720	10.162988	36
25	9.837146	9.861161	9.975985	10.024015	10.138839	10.162854	35
26	9.837279	9.861041	9.976238	10.023762	10.138959	10.162721	34
27	9.837413	9.860921	9.976491	10.023509	10.139079	10.162587	33
28	9.837546	9.860802	9.976744	10.023256	10.139198	10.162454	32
29	9.837679	9.860682	9.976997	10.023003	10.139318	10.162321	31
30	9.837812	9.860562	9.977250	10.022750	10.139438	10.162188	30
	Sine		Tangent		Secant		Minutes

46 Degrees.

43 Degrees.

Minutes	Sine	Tangent		Secant		Minutes
30	9.837812	9.860502	9.977256	10.022750	10.139438	10.102188	30
31	9.837915	9.860442	9.977503	10.022497	10.139558	10.162055	29
32	9.838075	9.860322	9.977756	10.022244	10.139678	10.161922	28
33	9.838211	9.860202	9.978009	10.021991	10.139798	10.161789	27
34	9.838344	9.860082	9.978262	10.021738	10.139918	10.161656	26
35	9.838477	9.859962	9.978515	10.021485	10.140038	10.161523	25
36	9.838610	9.859842	9.978768	10.021232	10.140158	10.161390	24
37	9.838742	9.859721	9.979021	10.020979	10.140279	10.161258	23
38	9.838875	9.859601	9.979274	10.020726	10.140399	10.161125	22
39	9.839000	9.859480	9.979527	10.020473	10.140520	10.160993	21
40	9.839140	9.859360	9.979780	10.020220	10.140640	10.160860	20
41	9.839272	9.859239	9.980033	10.019967	10.140761	10.160728	19
42	9.839404	9.859119	9.980286	10.019714	10.140881	10.160596	18
43	9.839536	9.858998	9.980538	10.019462	10.141002	10.160464	17
44	9.839668	9.858877	9.980791	10.019209	10.141123	10.160332	16
45	9.839800	9.858756	9.981044	10.018956	10.141244	10.160200	15
46	9.839932	9.858635	9.981297	10.018703	10.141365	10.160068	14
47	9.840063	9.858514	9.981550	10.018450	10.141486	10.159936	13
48	9.840190	9.858393	9.981803	10.018197	10.141607	10.159804	12
49	9.840321	9.858272	9.982256	10.017944	10.141728	10.159672	11
50	9.840452	9.858150	9.982309	10.017691	10.141850	10.159541	10
51	9.840591	9.858029	9.982562	10.017438	10.141971	10.159409	9
52	9.840722	9.857908	9.982814	10.017186	10.142092	10.159278	8
53	9.840854	9.857766	9.983267	10.016933	10.142214	10.159146	7
54	9.840998	9.857665	9.983320	10.016680	10.142335	10.159015	6
55	9.841116	9.857543	9.983573	10.016427	10.142457	10.158884	5
56	9.841247	9.857421	9.983826	10.016174	10.142579	10.158753	4
57	9.841378	9.857300	9.984079	10.015921	10.142700	10.158622	3
58	9.841500	9.857178	9.984331	10.015669	10.142822	10.158491	2
59	9.841640	9.857056	9.984584	10.015416	10.142944	10.158370	1
60	9.841771	9.856934	9.984837	10.015163	10.143066	10.158229	0

| | Sine. | Tangent | | Secant | | Minutes |

44 Degrees.

Minutes	Sine		Tangent		Secant		Minutes
0	9.841771	9.850934	9.984837	10.015163	10.143000	10.158229	60
1	9.841902	9.856812	9.985090	10.014910	10.143188	10.158098	59
2	9.842033	9.856690	9.985343	10.014657	10.143310	10.157967	58
3	9.842163	9.856568	9.985596	10.014404	10.143432	10.157837	57
4	9.842294	9.856445	9.985848	10.014152	10.143555	10.157706	56
5	9.842424	9.856323	9.986101	10.013899	10.143677	10.157576	55
6	9.842555	9.856201	9.986354	10.013646	10.143799	10.157445	54
7	9.842685	9.856079	9.986607	10.013393	10.143922	10.157315	53
8	9.842815	9.855956	9.986860	10.013140	10.144044	10.157185	52
9	9.842946	9.855833	9.987112	10.012888	10.144167	10.157054	51
10	9.843076	9.855711	9.987365	10.012635	10.144289	10.156924	50
11	9.843206	9.855588	9.987618	10.012382	10.144412	10.156794	49
12	9.843336	9.855465	9.987871	10.012129	10.144535	10.156664	48
13	9.843465	9.855342	9.988123	10.011877	10.144658	10.156535	47
14	9.843595	9.855219	9.988376	10.011624	10.144781	10.156405	46
15	9.843725	9.855096	9.988629	10.011371	10.144904	10.156275	45
16	9.843855	9.854973	9.988882	10.011118	10.145027	10.156145	44
17	9.843984	9.854850	9.989134	10.010866	10.145150	10.156016	43
18	9.844114	9.854727	9.989387	10.010613	10.145273	10.155886	42
19	9.844243	9.854603	9.989540	10.010360	10.145397	10.155757	41
20	9.844371	9.854486	9.989893	10.010107	10.145520	10.155628	40
21	9.844502	9.854356	9.990145	10.009855	10.145644	10.155478	39
22	9.844631	9.854233	9.990398	10.009602	10.145767	10.155369	38
23	9.844760	9.854109	9.990651	10.009349	10.145891	10.155240	37
24	9.844889	9.853986	9.990903	10.009097	10.146014	10.155111	36
25	9.845018	9.853862	9.991156	10.008844	10.146138	10.154982	35
26	9.845147	9.853738	9.991409	10.008591	10.146262	10.154853	34
27	9.845276	9.853614	9.991662	10.008338	10.146386	10.154724	33
28	9.845404	9.853490	9.991914	10.008086	10.146510	10.154596	32
29	9.845533	9.853366	9.992167	10.007833	10.146634	10.154467	31
30	9.845662	9.853242	9.992420	10.007580	10.146758	10.154338	30

	Sine		Tangent		Secant		Minutes

45   Degrees.

44 Degrees.

Minutes	Sine		Tangent		Secant		
30	9.845662	9.853242	9.992423	10.007580	10.146758	10.154338	30
31	9.845790	9.853118	9.992672	10.007328	10.146882	10.154210	29
32	9.845919	9.852994	9.992925	10.007075	10.147006	10.154081	28
33	9.846047	9.852869	9.993178	10.006822	10.147131	10.153953	27
34	9.846175	9.852745	9.993430	10.006570	10.147255	10.153825	26
35	9.846304	9.852620	9.993683	10.006317	10.147380	10.153696	25
36	9.846432	9.852496	9.993936	10.006064	10.147504	10.153568	24
37	9.846560	9.852371	9.994189	10.005811	10.147629	10.153440	23
38	9.846688	9.852247	9.994441	10.005559	10.147753	10.153312	22
39	9.846816	9.852122	9.994694	10.005306	10.147878	10.153184	21
40	9.846944	9.851997	9.994947	10.005053	10.148003	10.153056	20
41	9.847071	9.851872	9.995199	10.004801	10.148128	10.152929	19
42	9.847199	9.851747	9.995452	10.004548	10.148253	10.152801	18
43	9.847327	9.851622	9.995705	10.004295	10.148378	10.152673	17
44	9.847454	9.851497	9.995957	10.004043	10.148503	10.152546	16
45	9.847582	9.851372	9.996210	10.003790	10.148628	10.152418	15
46	9.847709	9.851246	9.996463	10.003537	10.148754	10.152291	14
47	9.847837	9.851121	9.996715	10.003285	10.148879	10.152163	13
48	9.847964	9.850995	9.996968	10.003032	10.149004	10.152036	12
49	9.848091	9.850870	9.997221	10.002779	10.149130	10.151909	11
50	9.848218	9.850745	9.997473	10.002527	10.149255	10.151782	10
51	9.848345	9.850619	9.997726	10.002274	10.149381	10.151655	9
52	9.848472	9.850493	9.997979	10.002021	10.149507	10.151528	8
53	9.848599	9.850367	9.998231	10.001769	10.149633	10.151401	7
54	9.848726	9.850242	9.998484	10.001516	10.149758	10.151274	6
55	9.848852	9.850116	9.998737	10.001263	10.149884	10.151148	5
56	9.848979	9.849990	9.998989	10.001011	10.150010	10.151021	4
57	9.849101	9.849864	9.999242	10.000758	10.150136	10.150894	3
58	9.849232	9.849737	9.999495	10.000505	10.150263	10.150768	2
59	9.849359	9.849611	9.999747	10.000253	10.150389	10.150641	1
60	9.849485	9.849485	10.000000	10.000000	10.150515	10.150515	0

| | Sine | | Tangent | | Secant | | Minutes |

45 Degrees.

A TABLE of Angles, which every Rumb (or Point of the Compass) maketh with the Meridian.

North	South	Point	D.	M.	North	South
		¼	2	49		
		½	5	37		
		¾	8	26		
NE b E	S by E.	1	11	15	N b W	S by W
		1 ¼	14	4		
		1 ½	16	52		
		1 ¾	19	41		
N N E	S S E.	2	22	30	N N W	S S W
		2 ¼	25	19		
		2 ½	28	7		
		2 ¾	30	56		
NE b N	SE b S.	3	33	45	NW b N	SW b S
		3 ¼	36	34		
		3 ½	39	22		
		3 ¾	42	11		
N. East	S. East	4	45	00	N West	S. West
		4 ¼	47	49		
		4 ½	50	37		
		4 ¾	53	26		
NE b E	SE b E	5	56	15	NW b W	SW b W
		5 ¼	59	4		
		5 ½	61	52		
		5 ¾	64	41		
E N E.	E S E.	6	67	30	W N W	W S W
		6 ¼	70	19		
		6 ½	73	7		
		6 ¾	75	56		
E by N	E by S.	7	78	45	W b N	W by S
		7 ¼	81	34		
		7 ½	84	22		
		7 ¾	87	11		
East	East	8	90	00	West	West

F I N I S.

www.ingramcontent.com/pod-product-compliance
Lightning Source LLC
Chambersburg PA
CBHW020854210326
41598CB00018B/1664